Readings in Companion Animal

BEHAVIOR

Victoria L. Voith, DVM, PhD,

& Peter L. Borchelt, PhD, *Editors*

Published by Veterinary Learning Systems
Trenton, New Jersey

ISBN 1-884254-23-3

Published by Veterinary Learning Systems
Trenton, New Jersey

ISBN 1-884254-23-3

DEDICATION

Dr. Tuber (standing) with a student at The Ohio State University.

1942 –1995

David S. Tuber: Educator, humanitarian, and clinical animal behaviorist

If anyone deserves recognition as the Father of Clinical Animal Behavior, it is David Tuber. He was a quiet, humble person who would be astounded to know how far, and rapidly, his ideas have spread.

David had an excellent academic background. He received his PhD in psychology under Delos Wickens from The Ohio State University in 1979. For years before that, however, he contributed to a multitude of outstanding research studies on classical conditioning. In the late 1960s, he began applying principles of learning to the treatment of behavior problems in dogs, primarily those related to fears, phobias, and separation anxiety. His thorough understanding of learning principles allowed him to apply laboratory and theoretical knowledge to the "real world" with an uncanny degree of success.

After graduation, he worked with Gary Berntson on the learning capacities of premature and developmentally delayed human infants. David taught psychology at the Columbus and Mansfield campuses of OSU and at Otterbein College and Denison University. Because of his curiosity about canine behavior, he converted a warehouse into a canine research center, kennel, and a home for himself and any of his research dogs that needed a permanent place to live. He, and the many students he inspired, investigated attachment behaviors of dogs and conducted learning set studies. He had hundreds of dog toys and a K9 Wisconsin General Test Apparatus.

Sadly, most of his work has gone unpublished. A few years before his death, he submitted his work on learning sets of dogs to a prestigious journal. It met with mixed reviews, with one reviewer indicating that people weren't interested in that type of work anymore. The paper's rejection was unfortunate not only for David and others interested in canine behavior, but also for the Alzheimer research scientists who subsequently looked desperately for any work on learning sets in dogs to assess memory and learning abilities of geriatric canines.

David's first published paper on clinical animal behavior[1] almost met the same demise. The paper, describing the successful application of behavior modification techniques to treat fear-related behaviors in dogs, was submitted but he did not hear back from the publication's editors for about a year. Finally, a senior and well-respected member of OSU's Psychology Department called the editors and inquired about the paper's status. As the story goes, the editor had not sent it out for review because he thought it had been a hoax because it was such an unusual paper.

David's applied work included helping individual owners with their pets and, with his associate, Fran Linden, working in humane societies in Columbus and Dayton to prepare pets for successful adoption. He trained humane society staff to socialize pets and taught numerous students about animal behavior and treatment techniques using the applied and clinical cases he saw. Some people studied with him for extended periods, others for only a few weeks. No one who wanted to learn was turned away, regardless of how time consuming it was for him.

Although David lived and worked within a 200 mile radius, his idea of applying academically sound learning principles to treat behavior problems in companion animals has literally been accepted around the world. David was not married and had no genetic children, but he has a worldwide network of intellectual and academic siblings, children, and grandchildren. For example, let's trace the impact of David's teachings on the professional and academic lives of just two of the many people he influenced throughout his life: this book's editors. Keep in mind that the effect he had on both of our lives represents only two lines of influence that radiate from David. There are many, many others.

* * *

Dr. Borchelt:

Victoria worked with David in Columbus for several years in the early 1970s. Months would go by between clinical appointments because of public skepticism at that time regarding "animal psychologists." Papers Victoria presented at local veterinary meetings were looked upon initially with only moderate interest because veterinarians thought that behavior problems in animals were a reflection of behavior problems in the owner. The veterinarians who did not attend her lectures identified her as "The person who works with crazy people"; those who attended the talks realized that that was not at all what she, David, and David Hothersall (another colleague of theirs), were doing. Finally, however, referrals began to trickle in and techniques were applied and refined. Because the treatment procedures were based on proven principles, they usually worked right from the beginning—an interesting example of the advantage of education over experience.

In 1975, Victoria entered a graduate program at the University of California, Davis, and introduced and demonstrated behavior modification techniques to the students and faculty of the School of Veterinary Medicine. In 1977, she presented a paper on clinical animal behavior to the World Small Animal Congress in Amsterdam. The moderator's comments afterwards were, "We've been doing it ***backwards, haven't we?" Subsequently, she presented a paper in London, shortly after which Roger Mugford quit his job as a comparative psychologist with a pet food company and began seeing clinical behavior cases. A veterinarian from Norway with a longstanding interest in animal behavior, Torbjorn Owren, expanded his clinical services to include behavior consulting at his hospital in Oslo. Later, Victoria taught many students at the University of Pennsylvania, including three residents, Amy Marder, Barbara Chapman, and Karen Overall. Dr. Marder has a busy private practice limited to animal behavior in Boston, Dr. Chapman is practicing in Australia, and this year, Dr. Overall presented papers at the World Congress of Veterinary Medicine in Japan. These individuals are now teaching their own students and passing on to the next generation the behavior modification techniques inspired by David.

Dr. Voith:

Peter and a classmate of his at Michigan State University, Dan Tortora, first heard David and Dr. Hothersall discuss the idea of applied animal behavior at a Midwestern Psychological Association meeting in the early 1970s. The room was packed, the audience enthralled, and at least several people subsequently tried to start private practices but eventually gave up due to a lack of cases.

Peter and Dan took academic jobs in the New York City area. However, the Tuber, Hothersall, and Voith paper that was published in 1974 in *American Psychologist* inspired Dan to start a private practice shortly thereafter. In 1978, he and Peter started the Animal Behavior Clinic at The Animal Medical Center (AMC). From the first case, they followed the example of David's paper and looked at dog and cat behavior problems as animal behaviorists and scientists; the treatment techniques flowed from the behavioral analysis of the problem. Presentations and personal discussions at AMC and taking students on house call cases have carried the main message—behavior problems are amenable to behavioral analysis and treatment—widely throughout the veterinary profession and to the general public. Two other graduate school classmates of theirs have also entered the field of applied animal behavior: Dr. Mary Lee Nitschke, now in Oregon, and Dr. Henry Askew, now in Germany.

* * *

Much of the acceptance of clinical animal behavior in veterinary medicine, in psychology, and by the public is directly attributable to the insight and applications demonstrated by David Tuber.[2] He would be amazed how far his teachings have spread.

It is in recognition of his contributions and to encourage the continued development of the field of clinical animal behavior that a portion of the proceeds of this book will be donated to a fund established in David Tuber's name to be administered by the Animal Behavior Society.

Victoria L. Voith, DVM, PhD
Peter L. Borchelt, PhD

REFERENCES

1. Tuber DS, Hothersall D, Voith VL: Animal clinical psychology: A modest proposal. *Am Psychol*, 29:762–766, 1974.
2. Hothersall D: Eulogy: David S. Tuber. *Anim Behav Consultants Newsletter*. 1995. Vol. 12, No. 2, pp 1–2.

About the Editors

DR. VOITH obtained her DVM and an MA in psychology from The Ohio State University; she obtained her PhD in animal behavior and neuroanatomy from the University of California at Davis. Dr. Voith has been instrumental in developing diagnostic criteria and therapeutic approaches to behavior problems in companion animals. She has edited several books and written numerous articles in the fields of animal behavior and the human–animal bond. Dr. Voith developed and was Director of the Animal Behavior Clinic at the Veterinary Hospital of the University of Pennsylvania from 1979 to 1988. She was the recipient of the Leo Bustad Companion Animal Veterinarian of the Year Award, State of Pennsylvania, in 1986. She is a Certified Applied Animal Behaviorist by the Animal Behavior Society and is a charter diplomate of the American College of Veterinary Behaviorists. She currently is a practicing applied animal behaviorist providing services to industry, educational institutions, governmental agencies, and individual pet owners.

DR. BORCHELT received his PhD in comparative psychology and animal behavior from Michigan State University in 1973. He held a postdoctoral Fellowship in the Department of Animal Behavior at The American Museum of Natural History from 1972 to 1974, and taught as Assistant Professor in the Department of Psychology, Fordham University, from 1974 to 1978. Since then, he has worked as Director of the Animal Behavior Clinic at The Animal Medical Center in New York City and as President of Animal Behavior Consultants, Inc. in Brooklyn, New York. Dr. Borchelt is a Certified Applied Animal Behaviorist by the Animal Behavior Society, has published extensively, and is a frequent speaker.

INTRODUCTION

This book, a compilation of new and previously published papers, is designed to give a broad, scientifically-based overview of the growing field of applied animal behavior. New, previously unpublished material constitutes the majority of the content and has been built around the core series of popular *Compendium* behavior articles by the well-known writing team of Victoria L. Voith and Peter L. Borchelt. All ten of the original articles by Drs. Voith and Borchelt have been extensively updated for this volume.*

The readings in this book serve as a review of the relevant ethologic, psychologic, and veterinary literature. They also present behavior principles and terminology as they relate to a wide range of behavior problems in companion animals. What may make this book particularly valuable to the veterinary practitioner and to the applied animal behaviorist is that the basic principles discussed can be applied to all companion animals from dogs to cats, horses, birds, pigs, and other pets.

Of particular note are new articles on early experience and development, taste aversion conditioning, psychopharmacology and its applications in animals, and a full set of history taking forms that will be of immediate practical use to interested readers.

Some of the papers in this volume focus on one specific type of behavior problem while others present a more general overview of behavior. In either case, they are intended as a mix of practical and theoretical material serving as a starting point for the behavior student and practitioner. It is hoped that they will stimulate thought and discussion, encourage new ways of viewing behavior problems and new therapeutic approaches, and encourage further interest by veterinarians, behaviorists, trainers, breeders, and others in the art and science of applied animal behavior.

---

*Articles that were previously published in *Compendium* but extensively updated for this book are noted on the table of contents.

CONTENTS

ELEMENTS OF BEHAVIOR

HISTORY TAKING

LEARNING

Psychopharmacology and Endocrinology

Anxieties, Fears, and Stereotypies

Feeding and Elimination

Aggression

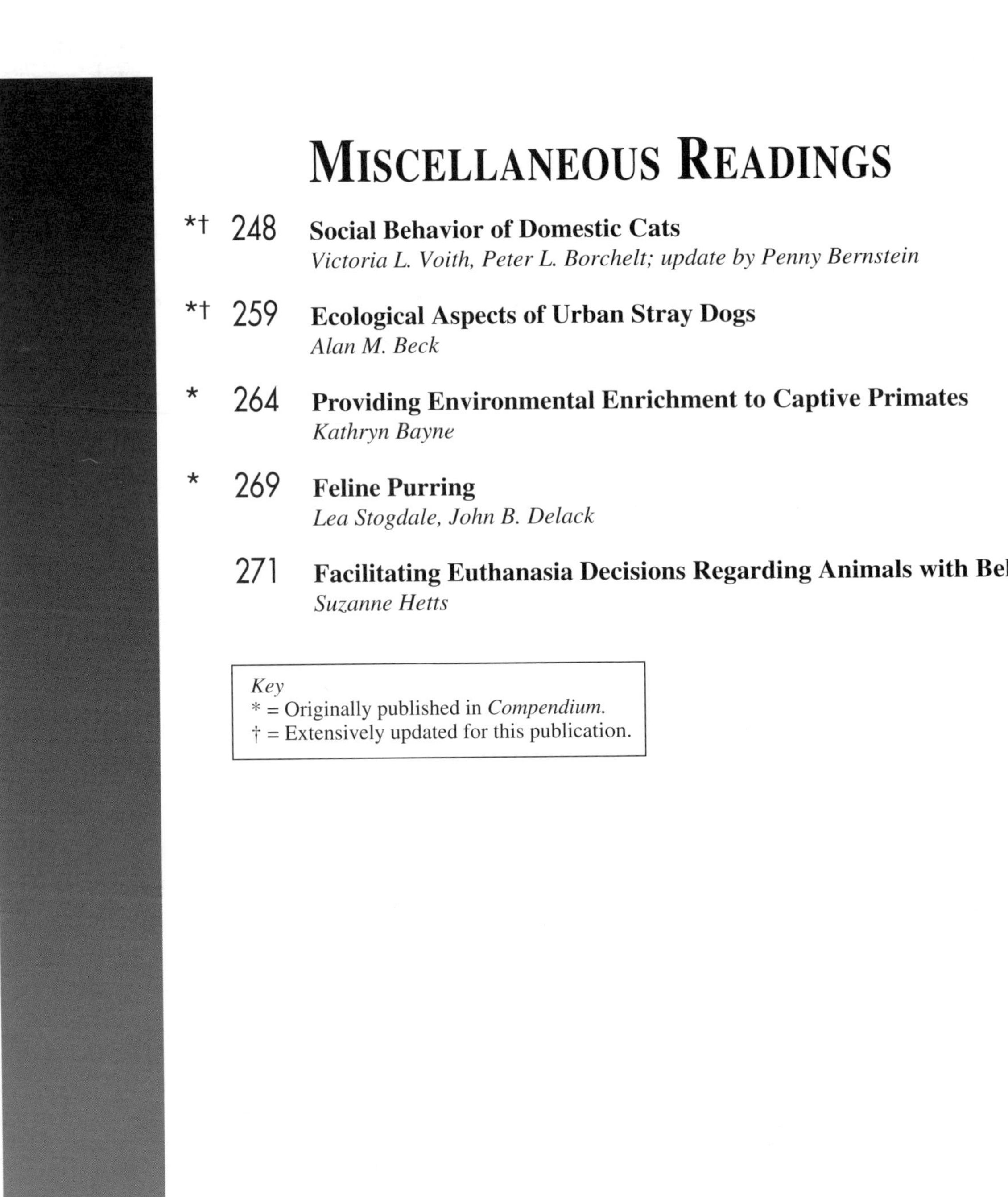

Miscellaneous Readings

Key
* = Originally published in *Compendium.*
† = Extensively updated for this publication.

ELEMENTS OF BEHAVIOR

History of Applied Animal Behavior

Department of Psychology
College of Social and Behavioral Sciences
The Ohio State University
David Hothersall, PhD

Throughout modern western science, animals have been widely studied in an attempt to understand basic psychological processes. In recent years, a dynamic field of applied animal behavior developed as humans sought to increase their understanding of the reasons animals behave as they do and how that behavior is adapted to achieve certain results. Several articles in this collection review recent developments pertaining to companion animal behavior. This article, which is a selective historical review of earlier work on applied animal behavior, provides a background for and gives perspective to the more recent studies.

WORKING ANIMALS

Fossils of humans and dogs have been found together at a number of sites in various parts of the world. Fossils of canine teeth and a jawbone estimated to be 14,000 years old were found with human remains in a cave in Palegawra, Iraq.[1] The dog's size suggests that it was used for hunting. The oldest-known fossil of a domesticated dog found in the New World is estimated to be 11,000 years old. The fossil was excavated at Jaguar Cave in Birch Creek Valley, Idaho, and probably is from the bands of Asian (Old World) hunters who were migrating to North America at the time.

Today, dogs continue to work with and for humans in a variety of settings. Shepherds, for example, use a repertoire of whistles, calls, and gestures to control their dogs and thus their flocks. The behavior of these dogs and their interaction with both the shepherd and sheep are impressive, as evidenced by the popularity of televised sheepdog trials and competitions in Great Britain.

Working dogs also function in noncivilian roles for security and for tracking.[2] For example, military dogs are trained to guard, patrol, and detect enemy presence[3]; in Vietnam, dogs trained to detect traps, mines, and tunnels were credited with saving many lives. Patrol dogs also are a fixture at major European airports. In addition, since 1970, U.S. Customs has used narcotic-detecting dogs. According to one FBI bulletin, 128 dogs participated in 400 seizures of illegal drugs during one year. The dogs searched 168,000 ships, aircraft, and vehicles; seven million units of mail; and seven million cargo shipments.[4] A U.S. Customs dog also detected 50 grams of heroin, two pounds of hashish, and 302 pounds of marijuana during one day at the

California–Mexico border. On the East Coast, another dog detected 851 pounds of hashish worth nearly $3.4 million in potential street sales. The training techniques used with these dogs are immediately recognizable to animal behaviorists: olfactory discrimination learning, immediate reinforcement of responses to a positive (drug) stimulus, and extensive use of conditioned reinforcers.

An example of working dogs in civilian roles is guide dogs for the blind. The trust and confidence shown by the handler are testimony to the effectiveness of training. The behavior of a guide dog, however, is dictated by more than a simple repertoire of responses to commands. Guide dogs must anticipate situations and respond appropriately. For example, a guide dog must be trained to avoid an obstacle under which it could freely pass but the blind handler cannot.

Although training procedures for guide dogs developed with little assistance from applied animal behaviorists, psychologists have made a contribution. In 1945, a committee chaired by C. J. Pfaffenberger evaluated the training of guide dogs and found that only 25% of the dogs successfully completed training. A majority of failures were attributed to failure of the dog to take responsible action or to respond independently, at times opposing the commands of the handler. Pfaffenberger and Scott[5] developed a number of primary socialization procedures and behavioral selection tests that improved the training success rate for guide dogs to 90%.

More recently, dogs have been trained to assist hearing-impaired people by alerting them to important auditory events—the ringing of a telephone or doorbell or the sound of a fire alarm. Other dogs have been trained as helpers for physically challenged people.

WILDLIFE

Applied animal behaviorists have made significant contributions to wildlife conservation and management. For example, trout raised in hatcheries and transferred to open streams were being caught by fishermen or natural predators within a few days after release. Bingham et al[6] investigated the reasons hatchery-raised trout were so vulnerable. The researchers found that, in the hatchery, food was being tossed into the fish tank. By accident, the trout were being classically conditioned to respond to the splash of the food (the conditioned stimulus, or CS) by swimming to the splash (conditioned response, or CR) because of its association with food (unconditioned stimulus, or US). In the hatchery, the conditioned approach response was adaptive; however, in an open stream, the splash often belonged to a fisherman's fly or a hunting otter or blue heron.

After they identified the classical conditioning, Bingham et al introduced a change in the hatchery feeding procedures. The splash of food dropped into the tank was paired with a mild electric shock delivered through the water. The trout learned a conditioned avoidance response by swimming away from the splash and eating the food later. The conditioned hatchery trout survived as well as native trout did in an open stream and at a much higher rate than did unconditioned hatchery fish.

Conditioning procedures also were the basis for control of coyote predation. In the western United States, significant economic losses have been linked to wild coyotes killing large numbers of sheep. On the other hand, coyotes are a beneficial contributor to the ecosystem, as they eat large numbers of mice, gophers, squirrels, and grasshoppers. Solutions preventing the loss of sheep while protecting the coyote from extinction were needed. One solution was derived from basic psychological research.

Garcia et al[7] found that rats would develop specific conditioned taste aversions when they drank sweet water and later were made sick. In contradiction to conventional classical conditioning principles, which stress a very short interval for optimum conditioning (0.5 seconds) between the conditioned stimulus and the unconditioned stimulus, this conditioned taste aversion was formed with a delay of one hour between drinking the sweet water (CS) and subsequent sickness (US). This experiment is now recognized as one of the most significant to be published in the field of animal learning in recent decades.[8,9]

In 1974, Gustavson and Garcia[10] developed a coyote predation control procedure based on conditioned taste aversions. Captive coyotes were fed minced lamb flesh, skin, and wool infused with lithium chloride, a powerful emetic that induces nausea and sickness. Subsequently, the coyotes refused to attack lambs, running away from them and actively avoiding any encounters. The coyotes did continue to attack rabbits, thereby indicating that a specific aversion and avoidance response had been conditioned. In a subsequent field test of the effectiveness of the procedure, Gustavson et al[11] reported a 60% reduction in the number of sheep killed by coyotes.

In 1977, Ellins et al[12] exposed wild coyotes to sheep carcasses laced with lithium chloride. Coyote predation was almost completely suppressed. When the lithium chloride bait was replaced with nontoxic sodium chloride, suppression of sheep attacks continued and persisted for the nine-week study period after termination of the lithium baiting. Because many mammals acquire food preferences from their parents,[13] avoidance of sheep might be expected to be transmitted from conditioned adult coyotes to nonconditioned offspring.

Another wildlife problem addressed by applied animal behaviorists resulted from the increasing number of visitors to national and state parks. One particular concern was encounters between resident animals, particularly grizzly bears, and human visitors. Although bear attacks are rare (one person per 1.5 million visitors[14]), they are widely publicized and have led to proposals to remove some wild animals from park settings.

Herrero, an animal behaviorist at the University of Calgary, analyzed the number of and circumstances surrounding all recorded grizzly bear attacks on humans in the national parks of the United States and Canada.[14] Herrero identified factors that contributed to the attacks. For example, at Yellowstone National Park, the policy was to feed bears in locations that afforded easy visibility by park visitors. In addition, bears were allowed to forage in garbage containers. Both policies increased the likelihood of bears coming into direct contact with humans, resulted in the bears associating humans with food, and decreased the flight distance of humans approached by bears. Herrero recommended that feedings at high-visibility locations be discontinued and that bear-proof garbage containers be installed in all campgrounds. Additional suggestions by Herrero need to be investigated experimentally but for obvious reasons have not been. One such recommendation centers around human flight behavior when confronted with a bear attack; instead of fleeing, adaptive behavior might include playing dead.

PERFORMING ANIMALS

Although the spectacle of performing animals often elicits negative reactions from psychologists and others, the techniques used to train the animals are worthy of study by applied animal behaviorists. Given its long history, circus training is of particular interest. Circus training techniques often are considered to have developed independently of psychology; in one Russian school, however, circus animal training is based on applied animal behavior. Early in the twentieth century, Vladimir Leonidovich Durov stressed that circus animals respond positively to kindness and rewards. When an animal responded correctly to a stimulus, the behavior was rewarded immediately—a familiar psychological technique known as shaping. Durov was influenced by Vladimir Bekhterev and by Ivan P. Pavlov, one of the founders of Russian research on conditioning.[15]

The most noted member of the Russian school is Gunther Gebel-Williams, who has established himself as a premier trainer and exhibitor of circus animals. Gebel-Williams exerts impressive control over the behavior of circus animals, particularly natural enemies (e.g., horses, tigers, and elephants) introduced simultaneously in the circus ring. Gebel-Williams establishes his territory and leadership role by always being in the circus ring before the animals enter.[16] In addition, Gebel-Williams follows behavior training techniques developed by the Russian school of circus animal trainers; these techniques focus on what is referred to as *gentle training*. Before Gebel-Williams came to the U.S. in 1969, circus training techniques had been based on punishment and animal submission.

The use of conditioning principles today also is apparent at marine parks. The behavior of dolphins, whales, penguins, and other animal performers is controlled by discriminative stimuli and the use of both primary and conditioned reinforcers.

THE BEHAVIOR AND MISBEHAVIOR OF ANIMALS

An explicit application of conditioning principles to behavioral control was made by two psychologists, Keller and Marion Breland, who were students of B. F. Skinner and had worked with him on an applied animal project called Project Pelican (to be described later). Impressed by the power of such Skinnerian techniques as shaping and schedules of reinforcement to mold and control animal behavior, the Brelands founded Animal Behavior Enterprises. Their aim was to use Skinnerian principles of operant conditioning in training animals for advertising and commercial display. Initially they were successful.[17] Examples included a hen trained to play a five-note tune on a piano, a second hen trained to dance, and a third one laying wooden eggs when the audience called out the number of eggs to be laid. Another act involved a pig entering a room, turning on the radio, eating breakfast at a table, picking up clothes and depositing them in a hamper, running a vacuum, and answering audience questions by flipping yes and no panels. All animals in these and other acts were trained by applying operant conditioning techniques. Breland and Breland concluded that this new field of applied animal behavior outstripped conventional animal training techniques. The couple also predicted that the new field would provide professional opportunities for psychologists and validate principles derived from laboratory studies of operant conditioning.

Ten years later the second Breland and Breland report was more guarded.[18] The couple reported continued success in developing a range of animal exhibits for commercials and television appearances. Thirty-eight species representing more than 6000 animals, including cockatoos, raccoons, porpoises, and whales, had been successfully trained. Breland and Breland, however, also had encountered situations in which animals could not be conditioned to perform desired behavior. Subsequent analysis showed that, in all such cases, the desired conditioned responses conflicted with instinctive animal behaviors. For example, one raccoon repeatedly washed a coin rather than deposit it in a box and a chicken scrabbled for food rather than standing still. The animals were responding to instincts that even powerful conditioning procedures could not overcome. This second report by Breland and Breland was an important early account of biological constraints on learning.

ANIMALS IN MILITARY AND INDUSTRIAL SETTINGS

Animals often have been used to assist humans during war.[3] In World War I, British-trained seagulls tracked German submarines in the English Channel. In World War II, the Russians trained dogs carrying satchels of explosives to run toward tanks. Also during World War II, Skinner devised Project Pelican,[19] which involved training pigeons to act as homing devices or missile guidance systems. Eventually, Skinner designed a multiple-bird unit that could determine the trajectory of a missile. Although the performance of the pigeons was impressive in laboratory simulators and nonlaboratory trials, U.S. military authorities never accepted the project. After the war, Skinner and his students designed Project Orcon and continued to investigate the possibility of organic control systems.

Herrnstein and Loveland trained pigeons to perform photographic reconnaissance.[20] Pigeons were operantly conditioned to inspect slides and respond only to those showing humans (i.e., conventional operant discrimination with a person as the discriminative stimulus). Pigeons responded with remarkable accuracy and had clearly developed a concept of person. Based on these findings, the U.S. Coast Guard began Project Sea Hunt during the 1970s. Pigeons were trained to recognize orange objects, such as life jackets, on the ocean surface. Placed in observation chambers beneath search-and-rescue helicopters, the trained pigeons proved to be better than human observers at detecting objects at first pass.[21]

Conditioning principles also have been applied to the use of animals in industrial settings. In 1955, Thom Verhave, a student of Skinner, was employed by a large pharmaceutical company that manufactured up to 20 million capsules a day, each of which had to be inspected visually for defects. At the time Verhave was employed by the company, 70 employees inspected the capsules as they moved at a fixed rate past an inspection point. All off-color, bumpy, dented, or otherwise defective capsules were to be rejected; however, a considerable number of defective capsules were missed by the inspectors.

Because pigeons have superior visual acuity to that of humans, can distinguish color, and had been extensively used in operant conditioning research, Verhave believed that pigeons might be conditioned to detect defective capsules. The training procedure was relatively simple.[22] Pigeons were trained to discriminate between perfect and defective capsules and to peck a key accordingly. When a defect was observed, the pigeon pecked the key indicating defect, the capsule was rejected, the pigeon could access food, and the next capsule was advanced. If the capsule was acceptable, the pigeon pecked a different key that advanced the next capsule for inspection but did not produce food. Each pigeon determined its own inspection rate. After one week of training, the pigeons had a 99% accuracy for selecting defective capsules. To make the system even more accurate, Verhave yoked two birds together. The success rate seldom deviated from 100%.[22]

A similar industrial application of operant conditioning was developed by Cumming, a psychologist at Columbia.[23] Cumming trained pigeons to inspect the paintwork on diodes manufactured by a leading electronic components company. Diodes with defective paint were to be rejected. The training procedures were similar to those used by Verhave. Ninety-eight percent of the diodes with imperfect paint were correctly rejected. The birds determined the inspection rates, with one bird inspecting up to 1000 diodes per hour. The pigeons were consistent and impervious to distraction. Both the Cumming and Verhave studies confirm that animals can respond in predictable and efficient ways.

ANIMALS IN SPACE

In November 1957, the Russians launched Sputnik II and its passenger, a small dog named Laika, which had been trained by classical conditioning (Pavlovian) techniques. During that same month, President Eisenhower ordered the formation of the Unusual Environments Section at Wright Patterson Air Force Base in Dayton, Ohio.[a] Headed by psychologist Frederick H. Rohles, Jr., the section was to explore the ability of humans to live and work in space.[24] It was proposed, however, that test animals be subjected to space environment before astronauts. Of particular importance was whether the animals could continue to perform learned responses during space flights. In 1958, Rohles conferred with two other psychologists, Charles Fester, who was a close associate of Skinner, and Donald Meyer, a specialist in primate learning and behavior. The three decided to develop a series of experiments that involved very simple behavior in

[a]Meyer DR: Personal communication, The Ohio State University, Columbus, OH, 1993.

mice and progressively introduced more complicated behavior, first in monkeys and ultimately in chimpanzees.[25] Operant conditioning procedures and reaction time measures were to be used in training animals for the experiments.

In December 1959 and January 1960, two trained rhesus monkeys, Sam and Miss Sam, made suborbital flights. The monkeys had been conditioned to avoid programmed electric shocks by pressing a bar (Sidman avoidance). Both monkeys successfully pressed the bar throughout the flight. The next space flights involved chimpanzees. The chimpanzee, Ham, made a suborbital flight on January 31, 1961.[26] He too had been trained to press a bar to avoid electric shocks and received only two shocks during his 18-minute space flight. On November 29, 1961, a second operantly conditioned chimpanzee, Enos, made two orbits around the earth. During the flights, he pressed the bar for both food and water on two schedules of reinforcement. Enos also responded to discriminative stimuli to avoid shock.

The space flights of these animals prepared the way for flights by Mercury astronauts and their successors. Reflecting on this area of applied animal behavior, Rohles wrote:

> The human flights were contingent upon the success of these first animal flights, but more important, they served as landmarks for comparative psychology. Reflecting back on the program, I can only say that Skinner was there. Every technique, schedule, and programming and recording device we used then and subsequently can be traced to him or his students.[24]

APPLIED ANIMAL BEHAVIOR IN ZOOS

Until the modern era of zoo management, zoo animals usually were housed in small cages and displayed with little regard for their behavior or needs. Visitors therefore had little opportunity to develop respect for zoo animals.[27] During the past 30 years, however, major changes have influenced the conditions under which zoo animals are housed. Applied animal behaviorists, including psychologists, zoologists, ethologists, and veterinarians, have played a major role in initiating the changes.

One environmental psychologist, Robert Sommer, has been a major influence on architectural redesigning of zoos. In his book,[28] Sommer described fundamental flaws in the design of zoo environments. He characterized zoos as examples of hard architecture, that is, sterile, inflexible, and dehumanizing. The modern trend has been to design zoos with soft architecture, that is, with open, natural spaces approximating those of the wild. Cages and concrete enclosures are replaced with large open areas shared by different species. Careful attention is paid to the behavior of the animals and to public education of each animal's environment.

Regardless of the improved architecture and environmental conditions, researchers have recognized that zoo animals need more than satisfaction of their primary needs. Basic research by experimental psychologists has shown that many animals are motivated to explore and manipulate their environments. In their classic experiments, Harlow and McClearn[29] found that rhesus monkeys would manipulate and solve mechanical puzzles independent of any primary reinforcer. Butler and Woolpy[30] found that monkeys were highly motivated to explore their environments and that the opportunity to observe was highly reinforcing. From a different perspective, Heini Hediger, an ethologist at the University of Zurich, arrived at a similar conclusion. As director of the Swiss zoo, Hediger concluded that in essence animals in captivity lack purpose and direction and require new challenges, similar to the challenges humans face with their occupations. Hediger believed that biologically suitable occupational training and therapy could furnish the challenge missing in the environment of captive animals.[31]

In 1972, Hal Markowitz, a psychologist at the Washington Park Zoo in Portland, Oregon, directed a program designed to teach the public more about animals and make zoo life more interesting and stimulating for the animals.[32]

First, zoo management designed an apparatus that would encourage activity from white-banded gibbons. In addition to regular feeding, the gibbons were reinforced for responding under stimulus control. Illuminated globes (stimulus lights) and two large levers were installed on both sides of the enclosure's rear wall. A food dispenser was next to the right lever. In the final form of the display, a zoo visitor drops a dime into a box outside the cage. The stimulus light on the left side of the cage flashes on. The gibbon presses the left lever; the right stimulus light comes on; and the gibbon swings over to the right side of the cage, presses that lever, and receives a piece of food. Because gibbons are agile, active animals, the entire sequence often took fewer than two seconds. The gibbons rarely fought over food, consumed only slightly more than they had before behavioral conditioning, and represented a popular exhibit for zoo visitors.[33] A similar exhibit with Diana monkeys proved equally successful.

Other exhibits designed by Markowitz and colleagues allowed zoo visitors to compete with mandrills, which are large terrestrial primates, in a reaction-time contest. The mandrills won 70% of the contests. Orangutans were trained to play tic-tac-toe with visitors and also proved to be excellent competitors. Similar innovative displays have been introduced at other zoos.

Applied animal behaviorists also have made important contributions in zoo management. One of the most impressive has been the work of psychologist Terry Maple at Zoo Atlanta. In 1984, Maple was named interim zoo director for Zoo Atlanta, which at that time was regarded as one of the 10 worst zoos in the United States.[34] Maple envisioned scientific zoos, whereby the zoo becomes a vehicle of science.[34] Eleven years later, Zoo Atlanta was transformed into a zoo with an international reputation for the display and conservation of animals. The zoo's gorilla exhibit, for example, consists of five contiguous habitats on 4½ acres of hillside woodlands.

Especially noteworthy is the work at Zoo Atlanta in rescuing and rehabilitating animals. A male gorilla, Willie B., had lived in isolation for 27 years. He was successfully resocialized, has lived among other gorillas for the past seven years, recently sired his first offspring, and has proved to be an exemplary parent. A lowland gorilla, Ivan, had lived most of his life in a solitary cage in a shopping mall. Ivan was moved to Zoo Atlanta, was slowly socialized, and will be introduced soon to the zoo's group of 19 gorillas.[34] Reflecting on his experience at Zoo Atlanta, Maple concluded that:

> Scientific zoos are no longer the exception; they are becoming the norm. At all levels, and around the world, science is alive and well at the zoo.[34]

PROFESSIONAL ORGANIZATIONS

Contributions to applied animal behavior have primarily been the result of the initiative, creativity, and dedication of specific individuals. Recently, the demand for applications of scientifically derived knowledge of animal behavior has led to the establishment of several professional organizations dedicated to applied issues.[35] In 1966, the Society for Veterinary Ethology was established in Europe. This multidisciplinary group was originally concerned with the behavior and welfare of large animals but now is increasingly attentive to companion animals.

In 1975, the American Veterinary Society of Animal Behavior was organized (originally named the American Society of Veterinary Ethology). Members are predominantly veterinarians interested in the diagnosis and treatment of behavior problems. A wide spectrum of interests—from laboratory to zoo animals, from farm to companion animals—is represented.

In 1993, the American Veterinary Medical Association (AVMA) recognized the American College of Veterinary Behaviorists as a specialty board. To date, there are 13 board-certified veterinary behaviorists. In 1990, the Animal Behavior Society, a society whose members are primarily zoologists, psychologists, and ethologists, established a method of certifying Applied Animal Behaviorists. In 1991, the first individuals were certified. Twenty-five Applied Animal Behaviorists have now been certified.[b]

CONCLUSION

This brief historical review has shown the work of applied animal behaviorists in a number of different areas. These individuals have made contributions to both the welfare of animals and the development of psychology, zoology, and veterinary medicine. Such work is often rewarding. In a behavioral collection dedicated to Dr. David Tuber, it seems appropriate to end with an informal account of one of his successes in the field of applied animal behavior.

As a result of their work with behavioral problems in companion animals,[36] animal behaviorists often receive interesting calls regarding animal behavior. One afternoon Dr. Tuber received a call from a lady who described what has been coined the Case of the Persistent Robin. A determined robin was flying repeatedly into the large front window of the lady's house. The robin appeared at the same time each day, and the caller was concerned that the bird might injure itself or damage the window. The lady had shown much ingenuity in trying to solve this problem: placing a mobile in the window, displaying a bright light, and encouraging her cat to lie on the window ledge—all to no avail. Dr. Tuber finally asked the caller to describe in detail the contents of the room. Nothing appeared significant except for a bulletin board on the wall opposite the window. On the bulletin board was a Valentine's Day card with a big red heart. Recalling David Lack's classic demonstration that intense fighting and attack are released in robins by presenting a bundle of red breast feathers,[37] Dr. Tuber asked that the red heart be removed. That was done. The robin made one last attack and never returned. Not all applied animal behavior problems are solved so easily; but in this case, the caller, Dr. Tuber, and one hopes the robin were all well pleased.

[b]Application materials and a pamphlet describing the objectives, requirements, and two levels of certification may be requested from Dr. Suzanne Hetts, Chair, Board of Professional Certification, 4994 South Independence Way, Littleton, Colorado 80123.

REFERENCES

1. Best friend: Science and the citizen. *Scientific American* 233(6):50,54, 1975.
2. Koehler WR: *The Koehler Method of Training Tracking Dogs.* New York, Howell Book House, 1984.
3. Lubow RE: *The War Animals.* New York, Doubleday, 1977.
4. FBI Law Enforcement Bulletin, January, 1976.
5. Pfaffenberger CJ, Scott JP: The relationship between delayed socialization and trainability in guide dogs. *J Genet Psych* 95:145–155, 1959.
6. Bingham WE, Adelman M, Maatsch JM: Predator control in trout. *Proceed Midwest Psycholog Assoc Annu Conven,* 1965.
7. Garcia J, Ervin FR, Koelling RA: Learning with prolonged delay of reinforcement. *Psych Sci* 5:121–122, 1968.
8. Seligman MEP: On the generality of laws of learning. *Psych Rev* 77:406–418, 1970.
9. Hinde RA, Stevenson-Hinde J (eds): *Constraints on Learning.* New York, Academic Press, 1973.
10. Gustavson CR, Garcia J: Pulling a gag on the wily coyote. *Psych Today* Aug:68, 1974.
11. Gustavson CR, Kelly DJ, Sweeney M, Garcia J: Pre-lithium aversion: Coyotes and wolves. *Behav Biol* 17:61–72, 1976.
12. Ellins SR, Catalano SM, Schechinger SA: Brief report: Conditioned taste aversions, a field application to coyote predation in sheep. *Behav Biol* 20:91–95, 1977.
13. Galef BG Jr, Clark M: Mother's milk and adult presence. Two factors determining initial dietary selection by weanling rats. *J Comp Physiol Psych* 78:220–225, 1972.
14. Herrero S: Human injury inflicted by grizzly bears. *Science* 170:593–598, 1970.
15. Horn JC: Gentle in the land of Pavlov. *Psych Today* Oct:33, 1983.
16. Hediger H: *The Psychology and Behavior of Animals in Zoos and Circuses.* New York, Dover Publications, 1968.
17. Breland K, Breland M: A field of applied animal psychology. *Am Psych* 6:202–204, 1951.
18. Breland K, Breland M: The misbehavior of organisms. *Am Psych* 16:681–684, 1961.
19. Skinner BF: Pigeons in a Pelican. *Am Psych* 15:28–37, 1960.
20. Herrnstein RJ, Loveland DH: Complex visual concept in the pigeon. *Science* 146:549–551, 1964.
21. Wasserman EA: The conceptual ability of pigeons. *Am Sci* 83:246–255, 1995.
22. Verhave T: The pigeon as a quality-control inspector, in Ulrich R, Stachnik T, Mabry J (eds): *Control of Human Behavior,* vol 2. Glenview, IL, Scott-Foresman & Co, 1966.
23. Cumming WW: A bird's eye glimpse of men and machines, in Ulrich R, Stachnik T, Mabry J (eds): *Control of Human Behavior,* vol 2. Glenview, IL, Scott-Foresman & Co, 1966.
24. Rohles FH Jr: Orbital bar pressing: A historical note on Skinner and chimpanzees in space. *Am Psych* 47:1531–1533, 1992.
25. Banghart FW: *Biological Payloads in Spaceflight.* ASTIA Document No. AD 204761, USAF Research and Development Command Report No. 58–58, Charlottesville, VA, 1958.
26. Eimerl S, DeVore I: *The Primates.* New York, Time-Life Books, 1963.
27. Ellenberger HF: Zoological garden and mental hospital, in Jeehn JD (ed): *Origins of Madness.* New York, Pergamon Press, 1979, pp 17–30.
28. Sommer R: *Tight Spaces: Hard Architecture and How to Humanize It.* Englewood Cliffs, NJ, Prentice-Hall, 1974.
29. Harlow HF, McClearn GE: Objective discrimination learned by monkeys on the basis of manipulation motives. *J Comp Physiol Psych* 47:73–76, 1954.
30. Butler RA, Woolpy JH: Visual attention in the rhesus monkey. *J Comp Physiol Psychy* 56:324–328, 1963.
31. Hediger H: *Wild Animals in Captivity.* New York, Dover Publications, 1964.
32. Markowitz H, Stevens VJ (eds): *Behavior of Captive Wild Animals.* Chicago, Nelson-Hall, 1978.
33. Markowitz H: *Behavioral Enrichment in the Zoo.* New York, Von Nostrand Rheinhold, 1982.
34. Maple TL: Psychology is alive and well at the zoo. *Psych Sci Agenda* 8:8–9, 1995.
35. Voith V: Applied animal behavior and the veterinary profession: A historical account. *Vet Clin North Am [Small Anim Pract]* 21(2):203–206, 1991.
36. Tuber DS, Hothersall D, Voith VL: Animal clinical psychology: A modest proposal. *Am Psych* 29:762–766, 1974.
37. Lack D: *The Life of the Robin.* London, Cambridge University Press, 1943.

Biologic Bases of Behavior of Domestic Dog Breeds

Hampshire College
Amherst, Massachusetts
Raymond Coppinger, PhD
Lorna Coppinger, AB, MS

Behavior is the functional component of evolutionary change. How well an animal runs is the selective force, not its legs. Paleontologists study the evolution of hard parts because those are what fossilize. Studying changes in femur lengths, however, leads to the misconception that it is legs that evolved, rather than running or jumping. For biologists, the evolution of dog behavior is found in the mechanisms of evolutionary change from the antecedent wolf behavior.

Studying the evolution of dog behavior implies some assumptions. First, one assumes that something called dog behavior can be delineated and that it varies somewhat among dogs. People often discuss dog behavior or dog training as if it were the same for all dogs. Second, the origin of some of these variations is assumed to be genetic. What evolves is the genotype, the specific arrangements of genes. The only material that can be passed from generation to generation is genes.

BREED-TYPICAL BEHAVIORS

Professional dog breeders and trainers believe that inherent differences in behavior exist among dog breeds. Breeders select animals that display the breed-typical behaviors, whereas trainers direct the expression of those innate behaviors. For example, trainers do not train border collies to eye sheep, or setters to point birds, or retrievers to retrieve. These behaviors emerge spontaneously during *ontogeny* (i.e., the course of the life of an individual). Training commences after the emergence of the innate behavior; the trainer teaches the dog when and where to display the behavior.

Breeds are often classified by the job they were selected to perform, such as hunting, herding, or guarding. They are further subclassified as coonhounds, pointers, or upland retrievers; heelers, headers, or huntaways; and people-guardians or flock-guardians. It is assumed that each breed has a unique set of behaviors that endow the members such that no other animal could be taught to perform the task as well. Not everybody agrees. Black and Green[1] hypothesized that social attachment and reduced interspecific aggression between dogs and sheep could be achieved using dogs of any lineage if they were raised properly. These conclusions were based on observations of Navajo dogs that were used to guard sheep but supposedly lacked long-term selection preparing them to perform this task.

UNNATURAL SELECTION

Selection in dog breeding differs from Darwinian or natural selection. Darwinian change results from differential survival and reproductive success among individuals in a geographically isolated population. In contrast, dog breeds are created by humans selecting dogs of a particular phenotype (i.e., with a particular set of observable characteristics and that perform a particular task) and separating them from the rest of the canine population for breeding purposes. Breeders identify a few single-gene characteristics, such as coat color or achondroplasia (short limbs), as a breed marker, which all animals of the breed must have—regardless of the adaptiveness of the characteristic.

Breeds are hybrid saltations, often perfected in a few generations. Darwinian selection takes longer, perhaps millions of years. Crossbreeding creates not averages but new phenotypes that can be maintained as new breeds.

DIFFERENT BREEDS—SAME GENES

Dogs of all breeds have the same karyotype (number and shape of chromosomes) and produce reproductively viable hybrid offspring. The differences among breeds are not *gene differences* (i.e., different arrangement, ordering, or number of genes). Phenotypic (including behavior) differences among dog breeds are produced by slight *allelic differences* in gene products in three categories: ontogenetic onset (timing of the activation of the gene), quantity of gene product, and ontogenetic offset (timing of deactivation of the gene).

No gene arrangement has yet been identified that would permit identification of any breed or any particular behavior. To our knowledge, attempts to detect "fingerprint" genes for pit bull terriers, or for aggressive behaviors in pit bull terriers, have so far failed. Mitochondrial DNA investigations give no clue to relatedness between breeds, nor do they group breeds in any of the well-known categories, such as sporting, working, or terrier. Rather, they reveal what has always been obvious but is usually ignored: dog breeds are local and temporal phenomena recently derived by crossing other breeds or local variants. Any handbook of dog breeds tells what breeds were used to create a new breed at the turn of the twentieth century or what breeds were used to improve an old breed.

In addition to chromosomal and mitochondrial similarities, most members of the genus *Canis* have conservative phenotypic similarities (e.g., palate to skull length[2]) and also fundamental similarities in social and reproductive behaviors.[3–5] This information suggests that despite the claim that *Canis familiaris* is one of the most varied species on earth, the evolutionary distance between dog breeds or within the genus is short—perhaps even nonexistent.

DIFFERENT BREEDS—DIFFERENT TEMPERAMENTS

Differences in behavior among dog breeds are often given as differences in temperament.[6–10] The combined results of these studies indicate some variation in emotionality, vocalization, activity, problem-solving, reaction to human handling, and trainability. Mahut,[11] Borchelt,[12] and Hart and Hart[13] reported breed differences in temperament among pet dogs raised in private homes. They assumed that these differences were largely innate because pet dogs of all breeds are raised similarly.

MOTOR PATTERN AND BREED BEHAVIOR

Motor pattern differences among dog breeds have also been reported.[14] Studies of motor patterns concentrate on the kind of behavior, its frequency, and the sequence in which the motor patterns appear.

The kind or quality of the behavior refers to the action the animal performs in reaction to a specific internal or external signal. Ideally, this would be done by sequencing anatomical events (e.g., muscle movement); but usually, the composite behavior is simply described: "eye" in herding dogs, "point" in pointers, heeling in heelers. Male dogs raise a hindlimb during urine marking. Because their relatives, male wolves, also display this behavior but members of other similarly shaped species (e.g., bears) do not, the behavior is assumed to be not only an inherited motor pattern but a taxonomic characteristic.

Two species, or two breeds, may differ not in the *kind* of motor patterns but in their frequency of expression. Wolves rarely bark, whereas barking is ubiquitous among dogs. Nevertheless, greyhounds in Argentinean villages rarely bark at strangers, whereas maremmas in Italian villages rarely refrain from barking at strangers. Behaviors also change frequency ontogenetically. Wolves bark more frequently as young adolescents and rarely as adults. Dog pups bark first at 7 days of age and frequently thereafter throughout ontogeny.

The behavior sequence in which motor patterns appear differs between species and between breeds. Dogs that work with sheep provide an excellent example.[14] Among border collies (a sheep-herding breed) a predatory eye–stalk motor pattern follows a

Figure 1—Border collie showing "eye" to sheep. This behavior, with its characteristic "stalk," causes sheep to bunch and move away from the dog.

predatory orientation 85% of the time, whereas an investigative or a social sequence commonly followed such an orientation among livestock-guarding dogs. Border collies displayed eye–stalk–chase behaviors both inter- and intraspecifically. Some authors believe that the eye–stalk–chase motor patterns are homologous to those of wolves. Most European sheep-guarding breeds (e.g., Great Pyrenees, maremma, and Anatolian shepherd) do not display these predatory sequences (Figure 1).

CRITICAL PERIODS OF BEHAVIOR

Environmental factors cause behavioral variation among dogs. Scott[15] and Fox[16,17] found that puppies undergo a *critical period* of socialization when they are approximately between 3 and 16 weeks of age. According to these authors, dogs are susceptible to permanent alterations of behavior as a result of environmental stimuli during this stage of development. Cairns and Johnson[18] and Fox[5] found that social conditioning with another species often results in unusually low levels of aggression toward the other species and development of a social attachment, even between species that might normally have a predator–prey relationship (e.g., dogs and sheep).

The critical period hypothesis is evidence for genetic predispositions for development of specific behavioral repertoires (see the article by Estep). For example, we raised livestock-guarding dogs with sheep and minimized pup–human contact during the critical period when social motor patterns emerge. Some of these pups became good flock guardians but were permanently shy around people. These dogs were difficult to manage, but they were less likely to harass strangers in or near the pasture. How much handling by humans dogs need in order not to become people-shy is also a genetic variable. We have had strains of dogs predisposed to this shyness; and within those strains, siblings raised in the same pen can be shy or not shy, with the same amount of human contact.

BIOLOGIC DETERMINANTS OF BEHAVIOR

To understand any behavior, whether motor patterns, critical periods, or temperament differences, one must discern the underlying biologic determinants. Differences in innate behavior between breeds can ultimately be traced to biologic differences. These may be gross differences in size or shape or minute differences in the chemical structure of a neurotransmitter or a hormone.

A breed is defined by a structural standard (e.g., a Siberian husky should not weigh more than 60 lbs [27 kg]). This standard was derived through a process of selecting breeding stock from among dogs that performed the task of pulling sleds. Thus, innate behavior implies the structural capacity to perform.

Behavior is shape and size moving through space and time. This definition may seem simplistic. The evolution of breeds, however, is the selection among allelic variations of genes; but genes simply produce enzymatic reactions for the production of proteins, which give the organism shape and size. The genetic basis of behavior is a gene-encoded shape that allows the animal to perform in a particular way. The shape allows, but also limits, the performance. In the following example, sled dogs provide a simple illustration of how size (and consequently shape) determines a particular kind of performance.

Racing sled dogs are the world's fastest quadrupeds for distances over 17 km. Dogs on championship teams weigh less than 25 kg. Sled dog drivers would like to use bigger dogs, because big dogs have longer gaits (more reach) and cover more ground with each stride, but bigger dogs do not seem to do well. Experiments by Phillips and coworkers[19] showed that bigger dogs suffered greater heat stresses because surface-to-volume ratio is geometric and not linear with size change. Big dogs have more heat-storing capacity because they have a larger volume compared to radiative surface area, which prevents the dissipation of heat during fast long-distance running.

Greyhounds (30 kg), which are among the fastest of dogs, are limited to sprints. Their racetracks are either ⅜ or 5/16 of a mile. Like human sprinters, they finish their races in a matter of seconds (31 seconds is a common finishing time); so heat load does not limit performance.

Sled dogs have *walking gaits* (pace, trot, or lope), in which at least one paw is always on the ground. Greyhounds have *running gaits*, a series of leaps into the air with all four feet off the ground. Running (leaping) is a fault in sled dogs. Dogs that run (called *floaters*) are at a severe disadvantage because the backstrap pulls them off balance when all four feet are off the ground (Figure 2).

Racing sled dogs have gait and size characteristics

Figure 2—A team of Alaskan huskies with two lead dogs. Note how similar their gait is.

that allow them to run long distances efficiently. These characteristics are innate because they are genetically (allelically) determined. The shape of their shoulders and pelvic girdles and the length of their back and their size are the products of their genes. All those differences in shape between the greyhound and the sled dog allow each of them to perform exceptionally well in their own environment but limit their performance in the other breed's environment.

One cannot teach the wrong breed to excel at the wrong task. A greyhound cannot pull a sled at racing speeds for long distances. Not only would the greyhound overheat, but it would be awkward and inefficient in harness because of its gait. It would be difficult (and probably uncomfortable for or even harmful to the dog) to train a greyhound to run with at least one foot on the ground. Similarly, no amount of training or conditioning would enable a basset hound to achieve the speeds of a greyhound. Structural variation is the biologic basis of behavioral variation.

LIMITS OF STRUCTURAL VARIATION

Breeds differ not only in gross conformation, size, muscle, bone, and hair, but also neurologically and hormonally. Arons and Shoemaker[20] searched for the structural differences (quantity and anatomical distribution of neurotransmitters) underlying differences in motor patterns among breeds. They found differences in motor pattern and motor sequence among sheep-herding dogs, sheep-guarding dogs, and Siberian huskies. They then attempted, with some success, to correlate these breed differences with neurotransmitter patterns in the brain.

Their data suggest that breed-typical motor patterns reflect differences in the distribution and quantity of neurochemicals. Their findings not only support the assumption of breed-typical anatomical differences but are parsimonious with the direction of those differences. Low levels of dopamine correlated with the more lethargic temperament of livestock-guarding dogs. This contrasted with higher dopamine levels in the hyperactive border collies and huskies. Additionally, a lack of this neurotransmitter in a particular region of the brain resulted in a lack of excitability in organs stimulated by that portion of the brain. The animal therefore cannot be expected to increase the rate of motor function through training or conditioning. Anatomy, whether structural or chemical, not only allows breeds to perform in a particular way but also limits behavior.

ONTOGENY OF BEHAVIOR

If a dog's anatomy changes over its lifetime, then the "innate" behavior must also change. Dog behavior, then, must also be defined ontogenetically.

As a dog changes anatomically (its size, shape, hormones, and neurons) from neonate to puppy to juvenile to adult, it goes through stages of behavioral changes. The synonyms *grow up* and *develop* misleadingly imply that ontogenetically a dog goes from a primitive to a more advanced state. In fact, neonates are behaviorally complex—just as or even more complex than adults. Many people think of ontogenetic changes as growth or maturation, but the changes might better be thought of as metamorphosis from one complex stage to another.

Each new structural organization of the animal produces a new set of innate behaviors. Thus, one must specify the dog's current ontogenetic stage when discussing its behaviors. Too often, people think of dog behavior as adult dog behavior, assuming that the puppy "develops" into the adult, that puppies are less than perfect, and that the purpose of life is to mature into an adult.

TWO COMPLEX INNATE BEHAVIORS

Alarm calls by neonates and puppy retrieval behaviors by dams are innate, reflexive, hard-wired, complex dog behaviors that are ontogenetically limited (i.e., limited to a particular time in the life cycle). Neonates of all breeds can give species-specific alarm calls from the moment they are born. One is a care-soliciting call, signaling "I am lost." This call is an innate behavior, a specific motor pattern. It is stereotyped and can be sonographically measured and described. It has a single "meaning" and is only given in one environmental context. Its onset is from the moment of birth. There is no learning period. We observed a puppy still attached to its placenta giving this call, having been abandoned by its mother in a pasture. She had departed to the barn to have the rest of the litter. Puppies continue to emit the call until they are retrieved or until they weaken and perhaps die. The call is unchanging from the first to the last vocalization.

Offset (cessation) of the "lost call" occurs before 5 weeks of age. Neither practice nor reward prolongs the behavior into adolescence or adulthood.

Puppy retrieval by mother dogs is similarly hardwired. In our experiments, the onset of puppy retrieval was after the last puppy was born. Onset varies in other species. Experiments with rats showed that they can exhibit pup retrieval several days before parturition. This motor pattern, displayed only by a dam (in dogs), results from the specific neonatal vocalization given by a pup. One of our border collies retrieved and returned to the nest a small tape recorder producing this vocalization. Other environmental signals also initiate the response, for dams sometimes retrieve dead puppies.

Offset of puppy retrieval averages 13 days after parturition. The dam no longer responds to these vocalizations from puppies after that day, even though puppies may emit the signal in contextually appropriate circumstances for several more weeks.

In both pup and dam, the motor patterns are intrinsic. There is no learning or development period. The behavior is "perfect" the first time it is displayed, the emergence is spontaneous, and it is contextually and motivationally relevant and specific. Neither behavior occurs in other behavioral contexts, such as play or courtship. Neither other puppies nor males respond to the retrieval call. Nor do adults, of either sex, emit them.

The fact that the mother dog retrieves a tape recorder during the critical period but not a calling pup after the critical period suggests that no cognitive process is occurring. The call is a reflex in neonates, and the retrieval of the calling pup is a reflex by the dam. Both motor patterns have a critical period of display. They are not elicited before or after the particular ontogenetic stage, although the retrieval behavior can be induced hormonally in experiments.

PHYLOGENY AND HOMOLOGIES

The care-soliciting calls of pups are not just species-specific but identical across the genus and similar across family and order. In other words, wolf cubs and dog puppies have the same set of care-soliciting calls, and mammals in general have similar care-soliciting calls. The similarity between them suggests that the ancestors of dogs and wolves had this same set of modulating tonal signals.

The care-soliciting calls of puppies and cubs are phylogenetically related and are said to be homologous. The genes that build the structures that allow this reflexive behavior to be displayed were inherited by dogs from their wolf ancestor and by wolves from their carnivore ancestor. That particular pattern of genes has probably been around for millions of years. Since the call induces the retrieval response, the two behaviors probably coevolved. Surely they are hardwired; if searched, canine anatomy would reveal neurologic and hormonal arrangements that predispose dogs to display these behaviors in the presence of an appropriate environmental stimulus.

Because these behaviors coevolved, one is not more primitive nor is the other more complex. Mammary glands evolved, and so did specialized neonatal mouths that suck on them. Care-soliciting behavior evolved with care-giving (maternal) behavior. The behavior of the neonate is just as old and just as complex as the coevolved behaviors of the adult. The main point is that the emission of the retrieval signal by the puppy is equally as complex, in evolutionary terms, as the retrieval response of the mother. In fact, mammalian neonatal behavior is a recent evolutionary advance in vertebrate evolution. The pup is not a primitive form that "develops" into a behaviorally advanced adult.

For this discussion, just one behavior each was picked from among many in the neonatal and adult repertoires. During parturition, females remove puppies from the placenta; and they do it perfectly (with rare exceptions) with their first pup. Pups released from the placenta find a teat and begin to nurse. Sucking is a complex motor pattern that adult dogs cannot perform.

HOMOLOGOUS BEHAVIORS AND MOTIVES

When describing a dog's behavior, an ethologist uses motor patterns to try to determine the animal's motivation. To say that a dog killed a sheep or bit someone does not help in understanding why the animal did it. Unless one knows why the dog killed the sheep, one cannot begin to correct the problem.

Dogs have many different innate bites; that is, they have several motor patterns that are described as bites. To understand what may have motivated a dog to bite, investigators need information on the quality of a bite: where it was directed and what other motor patterns were displayed in the sequence.

Canid behavioral repertoires include several predatory biting motor patterns. Wolves string these patterns together in feeding or foraging behavior. A typical stereotyped predatory sequence in a canid is: Orientation toward prey → eye → stalk → chase → grab bite → crush bite (kill) → dissecting bite → consuming bite.

A field biologist can often tell which predatory species killed an animal by observing where the grab bite, kill bite, or dissecting bite appears on the carcass. Typically, wolves and dogs direct the crush bite at a femoral artery—the victim then bleeds to death. Coyotes more frequently direct the kill bite toward the carotid arteries. Pumas suffocate their prey.

The dog family has other characteristic bite motor patterns besides those used in the predatory routine. Intraspecific (within species) dominance bites are directed to the top of the neck of an animal refusing to submit, to the loin in a fight, or to the abdomen of a submitting animal. When a farmer tells us that the dog he hoped would guard his sheep is biting or even killing them instead, we first try to determine the dog's motivation by noting the motor patterns. If the sheep are torn on top of their necks or their abdomens, we suspect that the dog is using intraspecific dominance bites interspecifically. Confidence is bolstered if further investigation reveals other playful motor patterns (e.g., either play bow or the dog's head and tail held up).

Directing these behaviors toward sheep may seem inappropriate until one realizes that the dog spent its critical period of socialization with sheep and "learned" to display normal intraspecific behaviors toward sheep. As dogs mature sexually, they sometimes show reproductive dominance routines toward sheep, including sexual mountings. Because these are not the motor patterns of a predatory routine, one can assume that the dog is not trying to kill the sheep but has successfully bonded with them and might still be trained to be a good sheep guardian. If the bite is directed at the hindlimb, however, then one suspects that the dog is displaying a grab or kill bite, thus indicating that the motivation is predatory. One could conclude that the dog is acting interspecifically with sheep and probably will never be reliable (with sheep) or retrainable, since the critical period for social development has passed.

Grab bites, crush bites, and dissect bites are specific motor patterns in dogs and are thought to be homologous with wolf predatory behaviors. Among dogs, these predatory behaviors differ with age and breed. Neonates do not display them, and the onsets vary between and within breeds. Pointers and retrievers have grab bite, but it is a fault for them to have crush bite ("hard mouth"). Grab bite is necessary for cattle-heeling breeds, and not having it is a fault. Grab bite is a fault in sheep-herding breeds. Livestock-guarding breeds must not have grab, crush, or dissecting bites. In fact, they should not even have chase as a motor pattern. Good livestock-guarding dogs will not chase a ball, a behavior one should try to elicit when judging a dog's fitness for protecting sheep. Good livestock-guarding dogs will not (because they do not have the dissect motor pattern) open a stillborn calf to feed. It is necessary to slice open dead calves for a pen full of livestock-guarding dogs to feed, but not for border collies.

CAN MOTOR PATTERNS BE MODIFIED?

As stated, livestock-guarding dogs develop their social behavior during the critical period from about 3 to 16 weeks of age. Evidently, the offset of this critical period is delayed in some animals. Emergence of predatory routines (if they appear) in livestock-guarding dogs occurs after six months and occasionally after a year. Seldom do any of the predatory motor patterns occur, although chase and grab bite are the most common. If, however, the dog has been properly socialized with sheep, the later-emerging motor patterns do not get directed toward sheep but may be directed to other species. This can lead to problems when, for example, dogs raised with sheep are transferred to goats or when they disrupt wildlife on the sheep range.

Border collies raised with sheep during the critical period of socialization can be ruined as herding dogs. Although such dogs might eye sheep, they seem unable to follow the eye with stalk. In a double-blind experiment in which the trainer did not know the background of the dogs, he reported that the collie raised with sheep "couldn't hold the eye."

An individual may be capable of a motor pattern that it never displays simply because it never gets the environmental signal that elicits it. Many puppies never use the lost call, for example, because they are never lost. Because of their early social experience, many livestock-guarding dogs never display grab bite.

Some motor patterns must be temporally reinforced to become functional. In many mammalian species, maternal behaviors must be reinforced within a few minutes of parturition or the dam will abandon the offspring. Similarly, sucking by neonates must be reinforced within a critical period or it is lost from the repertoire. Animals born with mouth infections or correctable abnormalities often lose the nursing motor pattern in the few days it takes to recover.

Other motor patterns are fixed in behavioral sequences and cannot be displayed until the preceding behavior is performed. Some species of cats cannot eat carrion because the motor patterns for consumption are connected sequentially to the whole predatory repertoire. In some situations, livestock-guarding dogs

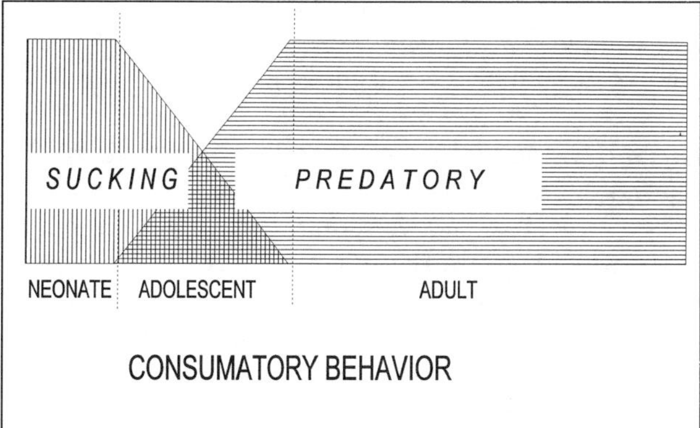

Figure 3—Schematic representation of consumatory behaviors in dogs. As the sucking motor patterns of the neonate wane, the predatory motor patterns of the adult wax. Similar representations could be made for all foraging, reproductive, and hazard-avoidance motor patterns. If all such sketches were then overlaid, the composite would be an ontogenetic ethogram—a depiction of an individual's behavior over the course of its life.

might be effective simply because they disrupt coyote predatory routines by placing the coyote in conflicting motivational states that do not allow it to complete the sequence. Border collies have strong connections between eye, stalk, and chase but weaker and modifiable connections with the grab bite and kill bite. Trainers can get a border collie to switch back and forth from chase to eye; but when it is allowed to proceed from chase to grab bite, it becomes more difficult to control the dog. Trying to get a wolf not to go from chase to grab bite is even more difficult.

NEW BEHAVIORS FROM OLD MOTOR PATTERNS

In the evolution of behavior, motor patterns can become detached from their ancestral sequence and reincorporated into a new sequence. Studies of dog breed differences give insights into how these evolutionary feats are accomplished. The reader must remember that breeds are allelically different and that allelic differences are differences in the onset, rate, and offset of gene products. Similarly, behavior should be described in terms of onset, frequency, and offset of motor patterns.

As mentioned, the care-soliciting behaviors occur for a few weeks after birth, whereas onsets of care-giving behaviors occur immediately after parturition and are displayed for the next few weeks. The frequency of display is a response to several internal and external variables.

Sucking (feeding behavior) is an innate behavior that appears shortly after birth (Figure 3). It is expressed for long periods throughout the day for the first few days, declines in frequency over the next few weeks, and disappears at about 8 weeks, never to reappear in the behavioral repertoire. Consumatory bites (feeding behavior) appear in the repertoire at 4 weeks of age and increase in frequency until an adult plateau is reached. They are maintained for the rest of ontogeny.

Figure 3 shows that sucking does not grow or mature into feeding. Sucking offsets at about 8 weeks. Feeding onsets at about 4 weeks. Hall and Williams[21] showed that sucking behavior has little to do with feeding in either anatomical or neurologic control. The display of sucking motor patterns is initiated in one section of the brain and that of consumatory bite motor patterns in another. They are under separate motor control. Each has separate motivations for performance. Satiation, for example, suppresses feeding but not sucking.

One way to change innate behaviors is to shift onset and offset times genetically. As one might expect, this is one of the differences between breeds. Border collies have early onsets of eye–stalk behaviors, whereas livestock-guarding dogs have late or no onset of these behaviors.

Another way to change behavior genetically is to disconnect motor patterns from other motor patterns. Some species of wild cats have grab bite, kill bite, dissect, and consume bites sequentially connected in the adult but not in the adolescent. These different adult motor patterns have separate onsets during adolescence and cannot be connected until they are all in place. The timing of the onset of these motor patterns may be due to developmental constraints or they may be the result of selection. These interconnected feeding motor patterns may be the most efficient way for the adult cat to feed; but at the same time, the connectedness prevents them from eating carrion.

If young cats that cannot yet kill their own food had feeding motor patterns sequentially connected, this sequencing would prevent them from eating dead prey provided by the parent. In this case, the adolescent may not be as efficient in killing prey but is more adaptable than the adult because feeding behavior is not rigidly innate.

THE ADAPTABLE ADOLESCENT

Neonatal and many adult species-typical behaviors have been discussed as innately rigid and reflexive. The adolescent, however, is quite a different story. Adolescence is so different that in many ways ethologists and psychologists have difficulty understanding exactly what effect it has on adult behavior. Nevertheless, this stage of ontogeny deserves attention because it is the most difficult stage of development to describe, it is the stage in which most mammals play and learn, and it is the stage in which dogs, as a species, are trapped.

The most simplistic definition of *adolescence* is the period of ontogeny between the neonate and the adult. It is the period when the organism is experiencing the offsets of neonatal motor patterns and concurrently the onsets of adult motor patterns. Juvenile mammalian behavior is a complex assemblage of waning neonatal and waxing adult motor patterns—the period of metamorphic remodeling.[22]

In the neonatal stage, behaviors are preprogrammed (hard-wired). Neonatal feeding behaviors are all accompanied by specialized anatomy that allows sucking to be performed. The neonate is not primitive or underdeveloped. It is a highly specialized organism in the sense that it is so perfectly adapted to its environment it does not have to learn anything. There is no selective advantage to learning, nor is there enough time or energy in the neonatal system to be able to experiment (learn) behaviorally.

The adult mammal is also well adapted to its environment. Feeding, reproduction, and hazard-avoidance behaviors have large innate components and often vary little or not at all within a species. In general, most adult mammals have difficulty rearranging sequences of motor patterns within a specific behavioral repertoire; it is therefore difficult to modify their behavior. They cannot afford the time or energy to experiment.

Adolescence is a period of metamorphosis—anatomical remodeling. The neonatal organism is taken apart and reconstructed into an adult. Behaviorally, the individual is remodeled from innate neonatal feeding and hazard-avoidance behaviors to the adult feeding, hazard-avoidance, and reproduction systems. Sucking feeding behaviors do not grow, or develop, into predatory feeding behaviors any more than the 18 feet of a caterpillar grow into the six legs of a butterfly. Instead, the animal is *de-differentiated*, to borrow the words of the embryologist. New organs are created de novo while old ones are discarded, just as the highly complex placenta and its associated behaviors are discarded at birth. Skulls do not grow from the neonatal skull (the sucking skull) into an adult predatory skull. The neonatal skull is resorbed while the adult skull is being laid down.[23]

Adolescents therefore have a changing access to neonatal and adult motor patterns. The sequencing of combinations of neonatal and adult motor patterns becomes the basis for play and learning. Sequences of motor patterns can be adjusted to the environment as the ever-changing adolescent uses them in contextually new and varied ways.

As pieces of neonatal motor patterns are offset, the remainder are displayed against the fragments of adult motor patterns. When pieces of neonatal motor patterns are displayed in sequence with adult fragments in nonfunctional behavior patterns, the behavior is called *play*. Putting fragmented neonatal motor patterns together with adult behaviors in functional ways is one definition of *learning*. The distinction between learning and play is simply whether the resulting motor sequence is functional and/or repeatable.

The two classes of vertebrates that play are mammals and birds, both of which have well-defined neonates. Play and learning are therefore considered artifacts of being able to use motor patterns that are not contextually appropriate to their phylogenetic origin. The hypothesis is that adolescents play and learn so well because their neonatal behaviors have become disconnected and their adult behaviors have not been sequentially aligned. Thus, they can use the dissociated motor pattern of two distinct ontogenetic stages in new ways.

THE PERMANENT ADOLESCENT

Dogs may be thought of as permanently adolescent. Genetically and evolutionarily they are "arrested" in the adolescent period. They have become reproductive as adolescents. Retaining juvenile characteristics into the adult period is called *neoteny* or *paedomorphism* by embryologists. Dechambre[24] suggested that canine breeds are differentially neotenic; whereas Fox,[17] Frank and Frank,[25] and we[26] suggested that not only are dogs behaviorally neotenic, but breeds of dogs are differentially neotenic behaviorally. The difference between the two types of sheepdogs might simply be that livestock-guarding dogs are arrested in a stage of development before the onset of adult predatory motor patterns (e.g., grab, kill, and dissecting bites), whereas border collies are in a stage that includes eye–stalk and chase but the grab and kill bites are weak and imperfect and not sequentially linked to the rest of the repertoire.[14]

It is common practice to try to understand dog behavior by searching for motor pattern homologues with wolves. But wolf ontogeny should be taken into account. People often say that dogs are territorial because wolves are and that wolves form packs with a leader, therefore dogs form packs and the human trainer has to become the pack leader. Wolf *cubs*, however, do not have reproductive territories or form packs and are not organized socially around a leader. If dogs are neotenic, then they should show behavior that is homologous to that of wolf *cubs*, not that of wolf *adults* (e.g., territory defense or pack formation). Nor should a person pretending to be the pack leader or dominant dog make any difference in their behavior. Wolf neonates and young adolescents do not hunt, nor do the neonates have any predatory motor pattern. Wolf adolescents have some of the predatory motor patterns, but they may not have them all, nor are the patterns connected sequentially, depending on the individual's age.

Exploration of the behavioral differences in breeds does support the hypothesis that dogs are arrested in an adolescent stage of wolf ontogeny. The difference between guarding and herding sheepdogs is illustrative. The guarding breeds never get to the ontogenetic stage when onsets of predatory behavior occur. The herding breeds get some of the predatory motor patterns, but not the final, wolf-like ones. Border collies and livestock guardians raised from puppyhood together in one large pen separated themselves into distinct "tribes" during play routines. The border collies played eye–stalk–chase games, whereas the guardians played at dominance–submission. The two types of sheepdogs segregated, like species, and did not play with each other. They are different temperamentally because of the kinds and frequencies of motor patterns they display.

The selection of type and intensity of play behaviors has differed among breeds. For example, the performance of sled dogs is based on selection for social play. What the driver thinks regarding his or her role as "pack leader" is irrelevant to putting together a team. The driver can train them and discipline them, but any attempt to become "dominant dog" is usually disruptive to the racing potential of the team. Many women with a gentle and supportive approach to training dogs excel with their racing teams. Many teams have female lead dogs. Many teams have two lead dogs, which if one were "dominant" over the other would not work. Many teams have several lead dogs that race back in the team and can be swapped for a tired leader. None of this would be possible with a team that behaved like adult wolves.

One should, however, treat the neoteny hypothesis with some caution. It would be silly to think that the ancestral wolf went through a maremma stage of behavior, then a border collie stage, then a husky stage, to become finally an adult wolf. It is just as silly to think of a maremma as representing a particular stage of wolf development, for the following reasons.

Puppies change (remodel) from neonates to adults. Heads grow (remodel) from little short-faced puppy heads to big long-faced adult heads. Change is allometric, not isometric. To arrest an animal at a particular growth stage is in effect to continue the allometric changes that are occurring at that stage. Since allometric changes are not prolonged in the ancestral stage, the resulting dogs are anatomically bizarre. Indeed, dogs are anatomically bizarre (e.g., head shapes or size variation).

If their anatomy is novel in terms of their ancestor, then their behaviors will also be novel and not directly equivalent to that of their ancestor. For this reason, one discussion of the evolution of working dogs was ready to drop the notion of behavioral neoteny, simply because although it might account for the heterochronic processes resulting in dogs, it was a poor predictor of either anatomical or behavioral result.[27] In the five years that have elapsed since the paper was written, optimism about the use of neoteny as a predictor of anatomy and behavior is returning.

CONCLUSION

The foregoing discussion of the selection process for behavioral differences among canine breeds has focused on breeds that were developed for work. This discussion underscores the biological bases of behavior, emphasizing details of physical structures that enable or limit behavior. We believe, however, that this information will also provide new ways of looking at the behavior of companion animals and offer alternative interpretations of problem behaviors. Such increased understanding of behavior should lead to improved techniques for the prevention, treatment, and management of canine behavior problems.

REFERENCES

1. Black HL, Green JS: Navajo use of mixed-breed dogs for management of predators. *J Range Management* 38:11–15, 1985.
2. Wayne RK: Cranial morphology of domestic and wild canids: The influence of development on morphological change. *Evolution* 40:243–261, 1986.
3. Bekoff M (ed): Social play in mammals. *Am Zoologist* 14(1): 266–427, 1974.
4. Scott JP: Evolution and domestication of the dog, in Dobzhansky T, Hecht MK, Steere WC (eds): *Evolutionary Biology*, vol 2. New York, Appleton-Century-Crofts, 1968, pp 243–275.
5. Fox MW: A comparative study of the development of facial expressions in Canids, wolf, coyote and foxes. *Behaviour* 36:49–73, 1970.
6. Fuller JL: Hereditary differences in trainability of purebred dogs. *J Gen Psychol* 87:229–238, 1955.
7. Freedman DG: Constitutional and environmental interactions in rearing of four breeds of dogs. *Science* 127:585–586, 1958.
8. Elliot O, Scott JP: The development of emotional distress reactions to separation, in puppies. *J Gen Psychol* 99:3–22, 1961.
9. Elliot O, Scott JP: The analysis of breed differences in maze performance in dogs. *Anim Behav* 13:5–18, 1965.
10. Scott JP, Fuller JL: *Genetics and the Social Behavior of the Dog.* Chicago, University of Chicago Press, 1965.
11. Mahut E: Breed differences in the dog's emotional behaviour. *Can J Psychol* 12:35–44, 1985.
12. Borchelt P: Aggressive behavior of dogs kept as companion animals: Classification, influence of sex, reproductive status, and breed. *Appl Anim Ethol* 10:45–61, 1983.
13. Hart BL, Hart LA: *Canine and Feline Behavioral Therapy.* Philadelphia, Lea & Febiger, 1985.
14. Coppinger RP, Glendinning JI, Torop E, et al: Degree of behavioral neoteny differentiates canid polymorphs. *Ethology* 75:89–108, 1987.
15. Scott JP: The domestic dog: A case of multiple identities, in

Roy MA (ed): *Species Identity and Attachment: A Phylogenetic Evaluation.* New York, Garland STPM Press, 1980, pp 129-143.
16. Fox MW: *Integrative Development of Brain and Behavior in the Dog.* Chicago, University of Chicago Press, 1971.
17. Fox MW: *The Dog: Its Domestication and Behavior.* New York, Garland STPM Press, 1978.
18. Cairns RB, Johnson DL: The development of interspecies social attachments. *Psychonometric Sci* 2:337–338, 1965.
19. Phillips CJ, Coppinger RP, Schimel DS: Hypothermia in running sled dogs. *J Appl Physiol* 51:135–142, 1981.
20. Arons CD, Shoemaker WJ: The distribution of catecholamines and ß-endorphin in the brain of three behaviorally distinct breeds of dogs and their F_1 hybrids. *Brain Res* 594:31–39, 1992.
21. Hall WG, Williams CL: Suckling isn't feeding, or is it? A search for developmental continuities. *Advances in the Study of Behavior* 13:219–254, 1983.
22. Coppinger RP, Smith CK: A model for understanding the evolution of mammalian behavior, in Genoways H (ed): *Current Mammalogy,* vol 2. New York, Plenum, 1990, pp 335–374.
23. Enlow DH: *Handbook of Facial Growth.* Philadelphia, WB Saunders Co, 1975.
24. Dechambre E: La théorie de foetalization et la formation des races de chiens et de porc. *Mammalia* 13:129–137, 1949.
25. Frank H, Frank MG: On the effects of domestication on canine social development and behavior. *Appl Anim Ethol* 8:507–525, 1982.
26. Coppinger R, Coppinger L: Livestock-guarding dogs that wear sheep's clothing. *Smithsonian* April:64–73, 1982.
27. Coppinger R, Schneider R: The evolution of working dogs, in Serpell JS (ed): *The Domestic Dog: Its Evolution, Behaviour and Interactions with People.* New York, Cambridge University Press, 1995.

The Ontogeny of Behavior

Animal Behavior Associates, Inc.
Littleton, Colorado
Daniel Q. Estep, PhD

Ontogeny, or development, is concerned with the changes that occur during the lifetime of an organism from conception to death. The study of behavioral ontogeny primarily focuses on changes from birth to sexual maturity of the organism; however, behavior begins before birth, and important behavioral changes occur throughout the lifetime of an organism.

The study of behavioral ontogeny has predominantly been a study of individual differences and the sources of those differences. Why, for example, would one puppy be fearful of humans while another would be very friendly? It is generally believed that there are only two major sources of differences: genetic and environmental. The differences between the fearful and nonfearful puppies may be attributable to genetics, environments, or an interaction of these factors.

Historically, genetic and environmental differences also have been the two major subdivisions of research on behavioral development. Behavioral geneticists study the degree to which genes influence behavior as well as the manner in which genes exert their influences. Other developmental psychobiologists examine the ways in which the environment and, most specifically, experience influence behavior.

This article reviews some basic concepts and general findings from these two areas of research and then shows how they apply to companion dogs and cats. Such knowledge can be useful in helping professionals and pet owners gain an understanding of the normal developmental processes and changes that occur in companion animals. Basic normative development data also are necessary to judge when development has become abnormal. An understanding of behavioral developmental processes may help to prevent some behavioral problems from occurring, such as housesoiling in dogs as a result of inadequate housetraining. Such understanding also may help in the treatment of other behavioral problems, including an animal's fear of humans.

NORMATIVE CHANGES IN THE BEHAVIOR OF DOGS AND CATS

Figures 1 and 2 present information on the early behavioral development of dogs and cats. The time frame considered is from birth to onset of adulthood, which has been the most intensively studied developmental stage for most mammals. The figures group behavioral changes into senso-

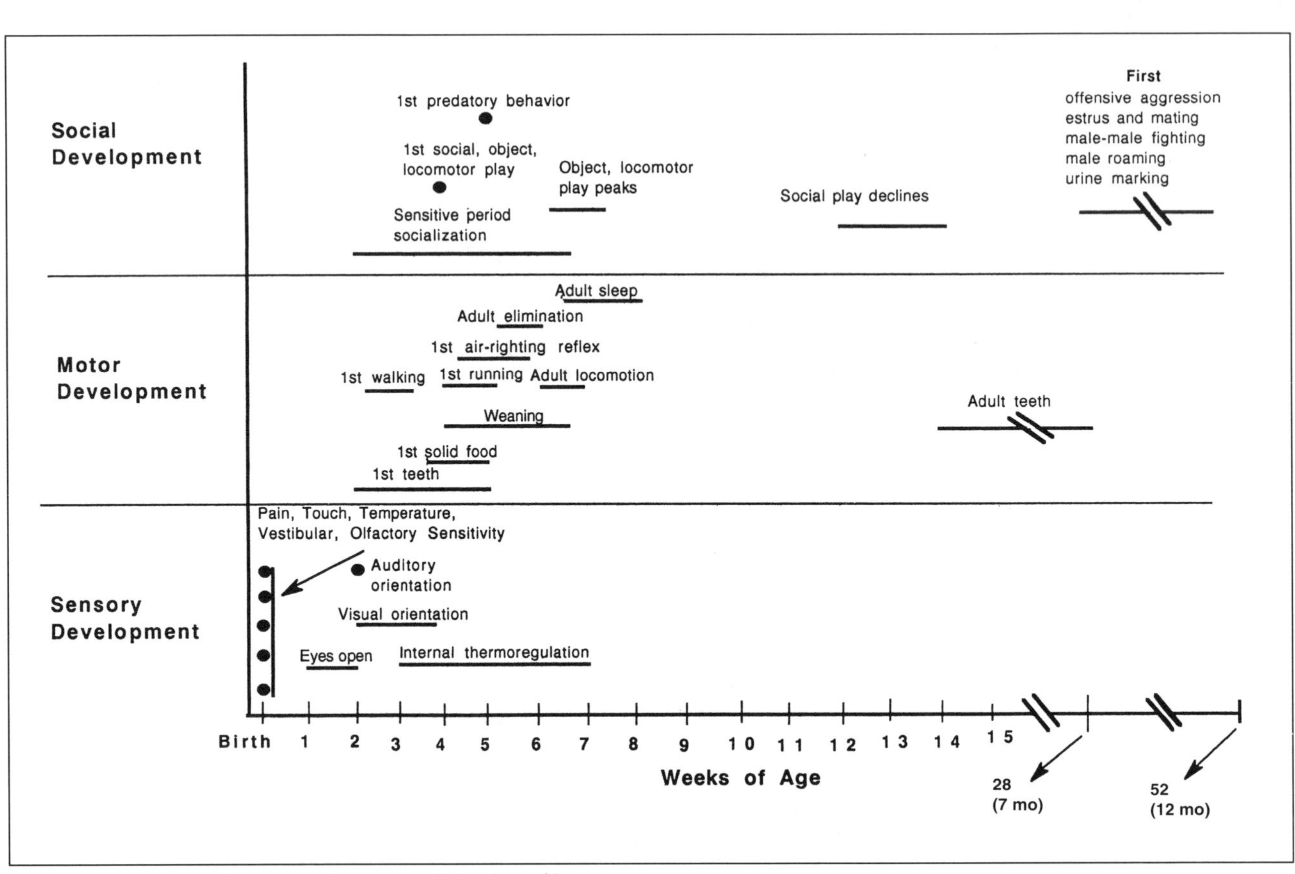

Figure 1—Early behavioral development of the dog.[1,4,5]

ry, motor, and social categories. Not all behavioral changes are presented—only the ones most related to social development. Summaries of early developmental changes can be found in the literature[1–5]; these reviews also provide the basis for the following discussion.

Although the figures start at birth, the choice of this point was arbitrary. Behavior does not begin at birth for mammals; it begins before that point. There is evidence that some species, such as rats, can perceive such stimuli as olfactory cues,[6] engage in complex coordinated movements,[7] and even learn.[8] In cats, tactile sensitivity is known to be present by day 24 of gestation and the vestibular righting reflex is present by day 54.[9,10] The nature of other prenatal behavioral capacities in dogs and cats is unknown at present.

Behavioral change does not stop with the achievement of sexual maturity. Animals continue to change as a result of experiences, other environmental influences, and activation and deactivation of specific genes. Behavioral changes in older animals have received very little systematic study. As companion animals live longer, however, owner complaints about behavioral changes and behavioral problems are becoming more frequent. Studies of the normal behavioral changes that occur in older dogs and cats are needed.

Early Developmental Changes in Dogs

The period from birth to sexual maturity in dogs can be divided into four distinct phases or periods[11]: neonatal, transitional, socialization, and juvenile (Figure 1).

The *neonatal period* is organized around maternal nourishment and care. The puppy is born with a relatively undeveloped brain, few sensory capacities, and poor motor abilities. Tactile, vestibular, pain, taste, and smell senses are available to the newborn dog; hearing and vision, however, are not.[4] Some basic reflexes are present at birth, and puppies 3 to 10 days of age are capable of learning simple associations between an aversive stimulus and suckling.[12] Locomotion is restricted to crawling, and vocalizations are limited to distress calls. Young dogs are completely dependent on the mother for survival. They cannot internally thermoregulate, urinate, or defecate without external stimulation. No new behaviors arise during this phase, but there is rapid growth in size and strength.

The *transitional period* is believed to be a period of reorganization of behavioral capacities. The period

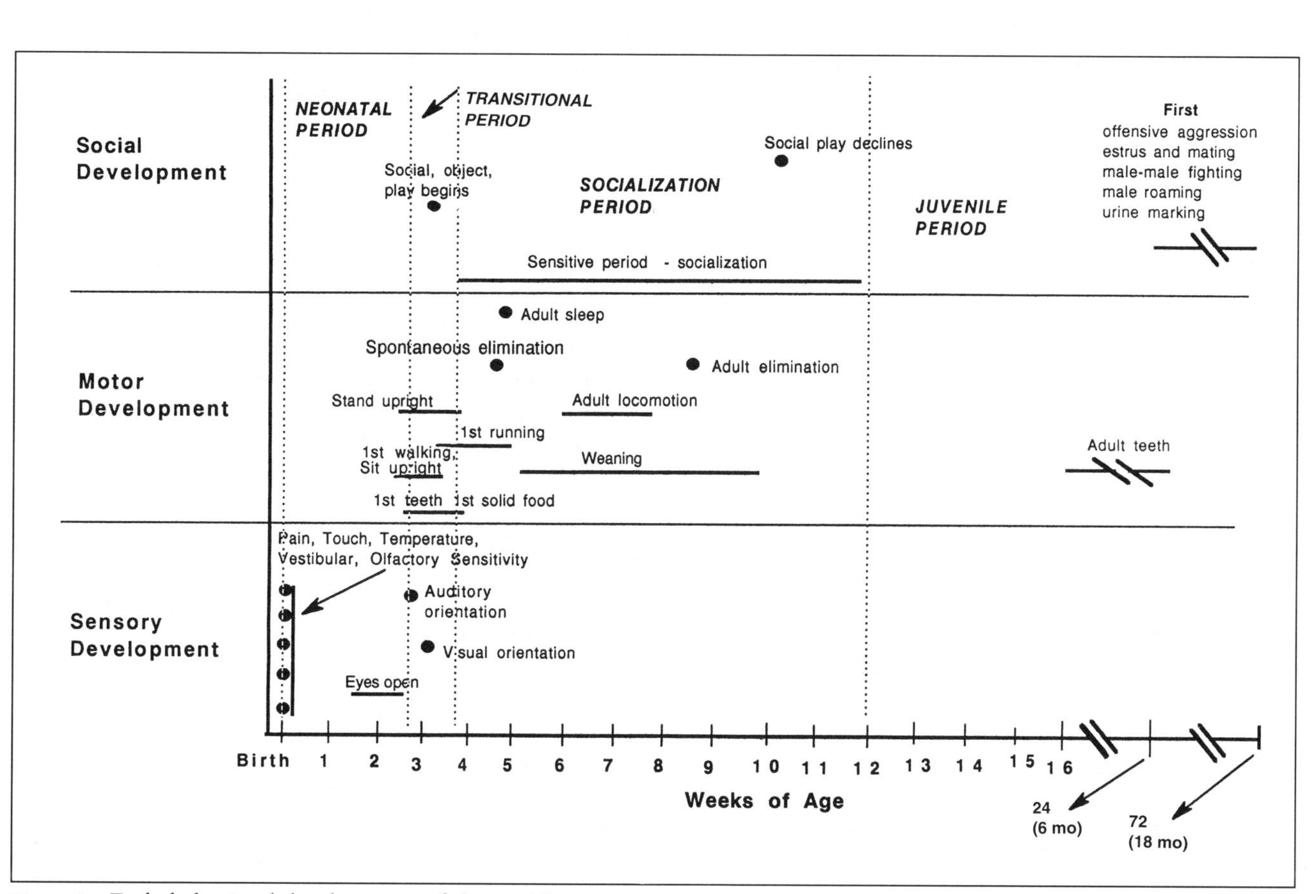

Figure 2—Early behavioral development of the cat.[1–3]

starts with opening of the eyes on or about day 13 and ends with the startle response to sounds on or about day 20. Visual and auditory capacities become available, as do many new behaviors. The first teeth begin to come in, puppies begin to lap and chew food, and they no longer require licking from the mother to urinate and defecate. Puppies begin to stand and walk, wag their tails, and show the first signs of agonistic behavior, even though social behavior tends to remain very simple.

The *socialization period* starts with the startle response to sounds and ends with the first long excursion from the den or nest box at about 12 weeks of age. This is a period of rapid behavioral change, particularly in social behavior. All the sensory organs are functional by the start of this period, and puppies begin to run as well as show other highly coordinated movements. Weaning begins around 5 weeks of age and is usually complete by 7 to 10 weeks. By 8 to 9 weeks of age, puppies are attracted by the odors of urine and feces to specific areas for elimination. These facts have practical importance for housebreaking dogs[13] (see Voith and Borchelt's article on elimination behavior and related problems in dogs).

Investigation of novel objects and explorations away from the nest area begin to occur. Learning capacities become better developed, although they are limited by incompletely developed motor abilities. Some complex visual and auditory discriminations are possible by the end of this period.

Important changes in social behavior occur during this period. Social play begins, as does social following and other aspects of pack behavior. Between 3 and 12 weeks of age, puppies learn with whom they should affiliate and form strong attachments to others; social experiences during this time can influence later adult behavior.[11]

The *juvenile period* extends from the end of primary socialization to sexual maturity. A few new behaviors develop during this period, but it is mainly a period of refining existing capacities and skills. The permanent teeth begin to come in by 16 weeks of age and are usually all in by 5 to 6 months of age. Motor skills become better developed and more adultlike. Learning abilities are fully adult, but complex tasks may not be learned because of short attention spans and remaining deficits in motor skills. At sexual maturity, female estrous cycles and sexual behavior as well as male sexual behavior and the androgen-facilitated behaviors of urine marking, aggression, roaming, and mounting begin to occur.[14] Other adult behaviors, such as territorial, protective, and dominance aggression, occur after sexual maturity in both males and females.

Early Developmental Changes in Cats

Unlike dogs, early development in cats has not been divided into developmental phases. In general, kitten development proceeds in a similar fashion to puppy development. A summary of published information concerning the development of cats is illustrated in Figure 2.

BEHAVIORAL GENETICS

Behavioral genetics is a relatively young discipline, with the first textbook on the topic appearing in 1960.[15] The discipline has subsequently grown dramatically in both quantity of research activity and diversity of research questions asked. It is impossible to review the breadth and depth of the field in this article. Interested readers are referred to general texts.[16,17] This section focuses on several questions that have direct implications for professionals and owners.

Do Genes Influence Behavior?

The earliest question that behavioral geneticists addressed was whether genes had any effect on behavior. Although this question may seem inappropriate today, in the early part of the twentieth century, many behaviorists were embroiled in the nature versus nurture debate. In one sense, the debate has been a pseudoquestion. There would be no behavior and indeed no organism without both genes and environment. All behavior depends on physical structures and physiologic processes for its occurrence. Because these structures and processes depend on genes, it can be said that all behavior is influenced by genetics, even if indirectly.

Because such logical arguments have not been enough to convince some skeptics, more direct empirical evidence that genes influence behavior has been sought. Such evidence has come from a variety of experimental techniques.

One of the earliest analyses in behavioral genetics was a selection study for maze-learning ability among domesticated Norway rats.[18] Tryon, a psychologist, was interested in the genetics of intelligence. He tested laboratory rats in a multiple-unit T-maze, counting the number of errors that the rats made in finding their way to a food source over several sessions or trials. Tryon then took the ones that made very few errors (the maze-bright rats) and mated them. The rats who made many errors (maze-dull) were also mated. The intermediate rats were discarded from further study. The offspring of both groups or lines were then tested in the maze when they reached adulthood. The brightest of the maze-bright offspring were again bred to each other, as were the dullest of the maze-dull offspring. Thus, two distinct lines were created. The offspring of every generation for 18 generations were tested, with the brightest of the maze-bright and the dullest of the maze-dull selected for breeding. By the seventh generation of selective breeding, Tryon was able to show statistically significant differences between the lines in their maze-learning performance. Because Tryon housed and tested both lines under identical environmental conditions throughout the study, the logical conclusion is that the behavioral differences between the lines are attributable to genetic differences.

A subsequent study by Searle[19] showed that Tryon had not selected simply for learning ability or intelligence. Searle compared maze-bright and maze-dull rats on a variety of behavioral tasks. Maze-bright rats did not always outperform maze-dull rats. In fact, maze-dull rats were found to be equal to or superior to maze-bright rats on three different learning tasks: a water escape task, visual discrimination task, and complex elevated maze. Searle concluded that motivation (maze-bright rats were more food motivated) and emotion (maze-dull rats were more timid of mechanical apparatus) better explained the differences in performance between the two selected lines than did differences in intelligence.

Selection studies have shown that genetics can influence a variety of behavioral patterns in a number of animal species, including aggression in chickens,[20] courtship and mating in fruitflies,[21] and emotion in rats.[22] The great diversity of species and behavioral patterns that have successfully undergone selection provide ample evidence that genes can influence not only highly stereotyped patterns, such as mating behavior, but even more variable patterns, such as ability to perform in a maze.

Because of the long generational time of companion dogs and cats and the expense of breeding and housing these animals over several generations, few selection studies have been done. One notable selection study performed on companion animals was the work of Murphree and colleagues on the fearful behavior of pointer dogs.[23,24] Starting with several breeding pairs, the intention was to develop two strains of dogs: one that was *nervous* or *unstable* and one that was *normal* or *stable*. Within a few generations, these researchers were able to produce dogs that differed on a number of behavioral and physiologic measures. Among the behavioral differences were that nervous dogs were consistently less active in exploring a novel room, tended to freeze in response to loud noises, and avoided both threatening and friendly people.[25] Over the years, as the investigators have sought the causes and correlates of the behavioral differences, a number of physiologic differences were found, including differences in heart rate[26] and neurochemistry.[27]

Some of the differences between nervous and normal pointers were discernible by eight weeks of age, while others emerged more slowly as the dogs reached six months of age. Once the differences emerged, they were highly stable. Cross-fostering of nervous dogs onto normal mothers and vice versa had no effect on behavior, demonstrating that differences in postnatal behavior and environment were minimal.[28]

Attempts to treat or rehabilitate the nervous dogs met with only limited success. Benzodiazepines, such as chlordiazepoxide, could greatly diminish fearful behavior but only when the dogs were under the influence of the drugs. Intensive positive handling of the nervous dogs, in-home rearing, and systematic desensitization produced some improvement but did not produce normal behavior.[28–30]

Another useful technique in demonstrating genetic influences on behavior is to compare behavioral differences between and within groups of animals that are known to differ genetically. The most common procedure is to compare inbred strains of a given species. An inbred strain is a population of animals that, through breeding of close relatives over many generations, has become genetically identical or nearly so. Such strains exist for laboratory mice, rats, guinea pigs, and fruitflies among others. When different strains are housed and tested under identical conditions and behavioral differences are greater between strains than within the strains, it can be assumed that genetic differences underlie the behavioral differences. Such studies have shown genetic influences on a variety of behavioral patterns in several mammalian species, including the mating behavior of guinea pigs,[31] aggressive behavior in mice,[32] and learning in rats and mice.[33]

Strain comparisons need not be performed on genetically pure inbred lines to yield important information about genetic influences on behavior. Non-inbred lines or breeds can be compared as well. Such comparisons are valid but more risky because the genetic differences between lines may be no greater than differences within lines. An excellent example of breed comparisons is the work of Scott and Fuller on the behavior of dogs.[11]

Scott and Fuller compared five breeds of dogs on a variety of behavioral tasks. The breeds were basenjis, beagles, American cocker spaniels, Shetland sheepdogs (shelties), and wirehaired fox terriers. The animals were reared, housed, and tested under identical conditions up to one year of age. Breed differences were found for a variety of behavioral tasks, including emotional reactivity, trainability, and several different kinds of problem-solving tasks.

Reactivity was measured when the dogs were restrained in a harness that allowed measurement of several physiologic parameters. The researchers found breed differences on 31 of 42 behavioral and physiologic measures, with terriers, beagles, and basenjis showing consistently higher reactivity than did shelties and cocker spaniels. Puppies were trained to remain quiet on a scale for weighing, to walk on a leash, and to stay on a table and then jump down when called. Again breed differences were found, with cocker spaniels being the easiest to train on all three tasks while basenjis and beagles were consistently harder to train. The terriers and shelties did better on some tasks than on others.

Problem-solving included performance in a maze, discrimination learning, delayed response tests, a tracking test, and spatial orientation. Again breed differences were found on many test measures, but no one breed consistently performed better (or worse) than the others on all the tests.

The results of this work illustrate that genetics can influence a variety of different kinds of behavior in dogs. These results do not indicate how important genetic influences might be for the behavior of dogs or how genes might interact with the environment in the production of behavior.

Genetic influences can sometimes be deduced by examining the occurrences of behavior between relatives. Two examples illustrate this point. Thorne[34] was able to trace the occurrence of fearfulness of humans in a group of 178 dogs that had been bred and studied for their approach and withdrawal behavior. Eighty-two of the 178 dogs were fearful, and 52% of these were traced back to a single basset hound that was a known fear-biter. Examination of the genealogies convinced Thorne that genetics influenced the responses of the fearful dogs, although such environmental effects as maternal influences could not be ruled out.

The second example concerns a paternal effect on the friendliness of domestic cats toward humans. Cats in two different colonies were rated on their willingness to approach and make contact with humans.[35] Examination of breeding records showed that friendly males were more likely to have offspring that were friendly. What led these researchers to suspect genetic influences was that the offspring never had contact with their fathers and there was no relationship between the mother's friendliness and offspring friendliness.

In a second study, Reisner et al[36] examined the role of paternity and early weaning (with or without early handling by humans) on fear-induced aggression in 13 litters of kittens. The results suggested that the paternal genetic effects had more influence on the kittens' subsequent friendliness to humans and tolerance of restraint than did early weaning, handling by humans, or isolation effects.

The mechanism by which genes influence friendliness is unknown, but it could be indirect. For example, Robinson[3] discusses a hypothetical case where feline fathers do not directly provide genes that influence friendliness but rather influence the coat colors of offspring, with the friendlier fathers providing more attractive coat colors to their offspring. A more attractive coat color leads human caretakers to handle the more attractive kittens more often. Because more frequent handling as kittens can lead to more friendly subsequent behavior (as discussed later), the kittens that received more attractive coats from their fathers would be more friendly to humans. This example points out that genetic effects may be quite indirect and raises the general question of how genes exert their influences.

How Do Genes Influence Behavior?

There is a fundamental misunderstanding among most people (including some behaviorists) about how genes influence behavior. The misunderstanding is that genes determine behavior in a very direct fashion. This is implied when people say there is a "gene for fearfulness" or "pit bull terriers are instinctively aggressive." Bluntly, genes do not directly determine any behavioral trait, nor do they exclusively determine behavior to the exclusion of environmental influences.

Early ethologists, such as Lorenz,[37] believed that highly stereotyped behavioral patterns called *instincts* or *innate behaviors* (e.g., some courtship displays in birds) were determined by one or a few genes with very little environmental input. This is not the case. Genes produce enzymes that regulate metabolic processes and structural proteins that are used to build cellular structures.[17] The enzymes and structural proteins then create anatomic structures and physiologic processes that ultimately produce behavior. There are many steps along this path from gene to behavior. At each stage, such environmental influences as temperature changes, fluctuations in nutrients, or particular experiences can alter the products produced and ultimately affect the final behavioral outcome. McFarland[38] has indicated that it is better to think about development as a process in which genetic and environmental influences are inextricably bound together and each developmental event sets the stage for but does not dictate the next stage. This view of development is called *epigenesis.*

According to this model, at any given stage of development the phenotype (P; whatever is being measured) is a product of genetic influences (G) and environmental influences (E) acting on the present phenotype. Starting with a fertilized zygote as P_1, the next stage in development would be $P_1 + G_1 + E_1 = P_2$. If P_2 is a developing embryo, the environmental influences (E_1) will mainly be biochemical factors, nutrients, temperature fluctuations, and the like. The next stage (P_3) will then be a product of new genetic influences (G_2) and environmental influences (E_2) acting on the embryonic phenotype (P_2), or $P_2 + G_2 + E_2 = P_3$.

Note that neither genetic influences nor environmental influences are static during development. Genes are activated and deactivated at different times during development,[17] and different environmental influences may come or go or become more salient as the organism develops. For example, soon after the eyes open in the neonate, visual stimuli from the environment can influence the organism. As the nervous system develops, a variety of experiences, including learning, that could not influence the organism before begin to influence it now.

If genes do not directly determine behavior, how do they exert their influences? Physiologic and biochemical studies in a variety of species have attempted to trace the pathways between genes and behavior.[17,39] One example concerns the nervous pointer dogs described previously. Studies of their physiology and biochemistry have been undertaken and have suggested problems with opioid receptors and dopaminergic functioning in the brain.[40,41] This research area is still in its infancy, but a better understanding of gene influences on behavior will undoubtedly be revealed by such studies.

One way to conceptualize how genes influence behavior is to think of genes setting the limits on how much of a given behavior occurs or that genes set the relative thresholds for the occurrence of a behavior. For example, the limits of the scent-tracking ability of a bloodhound are set by genotype. The bloodhound can do no better (or worse) than the physical limitations to the sensory and nervous systems that direct the behavior. Within those limits, however, how good a tracker the bloodhound is depends on the interaction of genotype with environmental influences, such as early neonatal environment, nutrition, training, and experiences.

For some behavioral patterns, the limits may be very narrow, producing very stereotyped behavior without much variation across a wide range of environments. In other cases, the limits may be very broad, allowing considerable behavioral variability. The former could be viewed as *innate* or *instinctive* behaviors, while the latter could be viewed as more flexible learned acts. It should be clear from this discussion that there is no dichotomy between learned and innate behaviors; both are influenced by genes and environment. Rather, they are on a continuum, varying only in the breadth of the genetic limits.

The genotype may help to set thresholds for behavior as well. For example, all dogs seem to be capable of biting in response to a painful stimulus. Yet there are enormous differences between dogs in how much pain they can tolerate before biting. These differences in thresholds to bite may be attributable in part to genetic differences.

What Are the Relative Influences of Genes and Environment on Behavior?

While it may be clear that both genes and environment are important for the expression of behavior, it still may be reasonable to ask whether there is any way to determine the *relative* influences of genotype and environment. Within a single individual, the question of which is more important, genes or environment, is nonsensical, as both are required for production of the organism and behavior. It would be similar to asking which is more important to the creation of a water molecule, the hydrogen or the oxygen?

If, however, it is the *differences* between organisms that are the focus of inquiry, then some meaningful questions can be asked. Because it is assumed that the phenotype (P) is a result of the action of genes and environment, it follows that differences in phenotypes (phenotypic variance or V_P) are attributable to genetic differences between the organisms (genetic variance or V_G), the differences in environments that the organisms experience (environmental variance or V_E), and the nonadditive effects of genes and environment (gene–environment interaction variance or $V_{G \times E}$). Stated more formally, the model is $V_P = V_G + V_E + V_{G \times E}$. If it can be assumed that the gene–environment interaction is negligible (this is discussed later), it is possible to estimate the proportion of phenotypic differences that are attributable to genetic differences. This is known as heritability in the broad sense or the coefficient of genetic determination and is given by the formula:

$$H^2 = \frac{V_G}{V_G + V_E}$$

A related value is known as heritability in the narrow sense (h^2) and is defined as the proportion of phenotypic variance that is attributable to additive genetic variance (V_A) or:

$$h^2 = \frac{V_A}{V_G + V_E}$$

Genes within the organism can interact with each other in different ways to produce genetic differences, and additive genetic variance is when each gene that influences a given trait acts in a simple incremental way to increase or decrease the value of the trait. Additive genetic variance is that proportion of genetic variance that responds most quickly to artificial or natural selection; so knowledge about narrow-sense heritability has been important in breeding programs to determine how quickly traits might respond to genetic selection.

Heritability scores are specific to the particular populations and environmental conditions under which they were collected. The scores cannot be generalized to other groups of animals or to other environmental situations. Nevertheless, heritabilities have been calculated for a wide variety of behavioral traits in both humans and animals. Heritability scores take values from 0 to 1.00. The higher the value, the higher the heritability and the greater the proportion of phenotypic variance attributable to overall genotypic variance (H^2) or additive genetic variance (h^2). For example, broad-sense heritabilities (H^2) have been found to vary from 0.09 for avoidance learning in rats to 0.94 for preferences for saccharine-flavored water in mice.[17] Narrow-sense heritabilities (h^2) have been found to vary between 0.05 for general activity in mice to 0.95 for emotional reactivity in rats to a novel environment.[17]

Heritabilities have been calculated for several different traits in dogs. Reuterwall and Ryman[42] calculated heritabilities for dogs in the Swedish Army Dog Training Center. These researchers looked at behavioral traits, such as affability, deposition to self-defense, play fighting, courage, startle and fear responses, and adaptability to new situations. The narrow-sense heritabilities were low for all the traits. Goddard and Beilharz[43,44] calculated narrow-sense heritabilities for nervousness in prospective guide dogs for the blind in Australia. Their heritabilities were high, varying from $h^2 = 0.58$ for Labrador retrievers to $h^2 = 0.47$ for purebreds and crossbreeds between German shepherd, Australian kelpie, boxer, and Labrador retriever.

Mackenzie and colleagues[45] calculated narrow-sense heritabilities for German shepherds bred for U.S. Army service. The researchers calculated heritabilities for temperament (a composite of ability to chase and attack a decoy and tendency to use olfactory abilities) of $h^2 = 0.51$ and hip dysplasia of $h^2 = 0.26$. Estimates of the genetic correlations between the two traits were $r = -0.33$, suggesting that dogs with good temperaments were unlikely to have hip dysplasia.

Scott and Fuller[11] calculated narrow-sense heritabilities for the dogs in their study for a variety of behavioral and morphologic traits. The heritabilities varied between 0.00 and 0.66. For example, they found h^2 = 0.44 for a composite score on leash training, h^2 = 0.20 – 0.30 for running speed in a T-maze (a problem-solving task), and h^2 = 0.00 for performance on an obedience test.

There is great danger in overinterpreting heritability scores. Some people have believed that heritabilities were a way to resolve the nature-nurture controversy. It has been assumed, for example, that a high heritability (greater than 0.50) meant that the trait was strongly controlled by genes (an innate or instinctive behavior) and that environmental influences, particularly learning, had little influence. This is not necessarily true. As pointed out earlier, heritabilities are specific to the animals and environments studied.[19] Different groups of animals or different environments may produce different heritabilities. According to Scott and Fuller,[11] researchers should be cautious in applying quantitative figures for heritability outside the specific context in which they were obtained. It is questionable whether, once some genetic control of a character has been demonstrated, precise estimates of heritability have any value. Selection programs for military or police dogs or for assistance animals, such as guide dogs for the blind, are the types of programs in which heritabilities may be useful for predicting success or evaluating progress. More recently, the limited value of heritabilities outside these narrow purposes has been echoed by others.[46–48]

Are Gene–Environment Interactions of Significance to Behavior?

In the quantitative genetic model discussed previously ($V_P = V_G + V_E + V_{G \times E}$), allowance was made for sources of phenotypic variance resulting from nonadditive interactions of genes and environment ($V_{G \times E}$). It is often assumed that gene–environment interactions are relatively uncommon and therefore unimportant. In addition, also discussed was that to calculate heritabilities, gene–environment interactions must be near zero. A finding of a significant gene–environment interaction makes interpretation of heritabilities impossible.

Significant gene–environment interactions have been found for a variety of behavioral traits and species.[17,39] For example, Cooper and Zubeck[49] studied the maze-learning performance of two strains of rats reared in three different environments. The strains were the offspring of maze-bright and maze-dull rats selectively bred for maze performance at McGill University (not the same rats or type of maze used by Tryon). In this study, different groups of maze-bright and maze-dull rats were reared in the standard laboratory environment in which they were selected, in an enriched environment, or in a restricted environment. After these different early experiences, all of the rats were tested in the maze for which they were selected. As expected, bright rats reared in the standard environment performed better on the maze than did dull rats reared in the standard environment. Maze-bright and maze-dull rats reared in the enriched environment, however, did equally well, with the dull rats performing like bright rats. Maze-bright and maze-dull rats reared in the restricted environment did equally poorly, with the bright rats performing as poorly as the dull rats. There was a gene–environment interaction here because maze-bright rats did not always outperform their maze-dull counterparts. Maze-bright rats sometimes behaved like maze-dull rats and vice versa, depending on their early experiences.

Another study demonstrating gene–environment interactions was performed by Freedman[50] on the basenjis, shelties, fox terriers, and beagles bred by Scott and Fuller at the Jackson Laboratory. Freedman reared pups from all four breeds in one of two rearing and training conditions. *Indulgent rearing* consisted of encouraging pups to engage in all the activities they initiated, such as pawing and climbing on the experimenter. *Disciplinary rearing* consisted of discouraging free play and teaching pups basic commands, such as sit, stay, come, and walk on a leash. At eight weeks of age, all the pups were tested in their response to punishment. A bowl of food was presented to each pup individually, and the pup was punished for approaching the food. The pup was then left alone with the food and observed. All the basenjis ate the food immediately after the experimenter left the room, while the shelties tended to leave the food alone. In both the beagles and fox terriers, the indulgently reared animals took longer to return to the food after punishment than did the disciplinary reared animals. Thus, different kinds of early experience and training affected only two of the breeds, while the other two were unaffected.

Findings of gene–environment interactions have implications for behavior genetics as well as early experience research. First, it means that uncovering breed, strain, or other genetic differences in behavior may depend on the particular environment in which the animals are reared and tested. Second, the finding of various early experience effects on animals may depend on the genotypes examined. Clearly, an understanding of developmental processes depends on knowledge of both genetic and environmental factors.

Practical Implications of Behavior Genetic Research

Breeders especially need to understand that genetics can influence all aspects of behavior from fearfulness to trainability. Most differences in behavioral traits are influenced by many genes, not just one or a few genes, which means that inheritance will likely be complex.[17] This, in turn, means that genetically influenced behavior problems or undesirable traits may not show up quickly. Thus, breeders should follow the offspring (and succeeding generations) of their breeding stock very closely to identify problems as well as desirable traits that come from particular breedings.

Behavioral traits should be considered heavily in breeding decisions. It is a disservice to the public to breed animals that look good but behave poorly. Breeders who are specifically breeding animals for companion purposes have an ethical obligation to breed and place animals with stable temperaments and with low risks of aggression, fears, and other behavior problems.

Breeders should be cautious in interpreting heritability scores, recognizing they cannot be generalized to other populations or other environments. For example, a narrow-sense heritability for temperament in U.S. Army German shepherds of $h^2 = 0.51$ allows no predictions about the temperament of any other dogs at any other facility.[45]

Prospective dog and cat owners frequently ask about breed differences in behavioral and physical traits. Many publications give breed profiles with the intent of helping people select a pet.[51–54] Such profiles should not be relied on to tell how a particular pet will behave because breed profiles are often based on human opinion with all its biases rather than on quantitative behavioral measurements of the animals (see Goodloe's article on description and measurement of temperament in companion dogs) and because there is considerable behavioral variation within breeds.[11] Some of the differences between dogs of the same breed are due to environmental influences, but some are due to genetic differences. Scott and Fuller[11] found great within-breed genetic variation in their study of the behavioral genetics of dogs. Furthermore, the existence of gene–environment interactions suggests that a particular breed may not always act the same way in all environments. Great care must be taken in using or recommending breed profiles as a means of picking a companion animal.

Basic introductory information on behavioral genetics is available in the literature.[38,55]

ENVIRONMENTAL INFLUENCES ON ONTOGENY

The environment can exert influences on behavioral development in at least two ways: physical means (such as alterations in temperature, moisture, or sunlight) and chemical means (such as changes in pH or the presence or absence of certain chemicals). One subset of such chemical influences is nutrition—the provision of essential nutrients to the organism.

Physical and chemical stimuli that activate the sensory organs and ultimately produce changes in the nervous system are called *experiences.* While all of the different physical and chemical influences are important to behavioral development, only those labeled experiences are discussed in this article.

Are Specific Experiences More Influential at Certain Times Rather Than Others?

A stage of development during which a given event can more strongly influence characteristics than at other stages is called a *sensitive period.*[56] This concept also is known as critical period, optimal period, vulnerable point, susceptible period, and crucial stage.

The idea of sensitive periods first gained prominence in animal behavior in the study of imprinting, a process by which social preferences are influenced by experience.[56] Lorenz[57] popularized imprinting through his description of young precocial birds developing social attachments to their mother and siblings. In chickens, ducks, and geese, there is a brief period of a few hours to a few days after hatching when they approach, follow, and stay near any conspicuous object or stimulus in the environment. After this exposure, the young birds show a preference for this stimulus over other stimuli to which they may be exposed. These preferences are long lasting and can extend into adulthood. In natural situations, it is the mother and siblings to which the young birds are exposed after hatching; and it is in this way that they learn with whom to affiliate. Recent research has shown that there are no finite limits to imprinting but it is much easier at certain times rather than others.

The concept of sensitive periods also has been applied to the formation of social attachments in dogs and cats, development of song in passerine birds,[58] development of visual perception in mammals,[59] and development of sexually dimorphic physiology and behavior in birds and mammals.[60] Other sensitive periods may yet be identified. Bateson[56] points out that mechanisms underlying different sensitive periods are probably not the same and that caution should be used in generalizing from the phenomena of imprinting.

Are There Sensitive Periods for Socialization in Companion Animals?

In companion animals, the focus of the concept of sensitive periods has been on the development of social attachments.[11,61] The term *socialization* has been used to describe the process of forming primary social attachments to mother and littermates. The term, however, is also broadly used to describe the formation of attachments to other dogs, other species (including humans), and environmental elements[62] as well as the process of exposing animals to novel people, environments, or other stimuli. Given these differences in use, readers should be cautious in interpreting results, conclusions, or applications from socialization research. Different authors may be defining their terms differently.

For both dogs and cats, reports of experiments regarding the formation of interspecific and intraspecific social attachments can be found in the literature.[11,61–85] Evidence that formation of social attachments can occur outside a sensitive period comes from the work of Woolpy and Ginsburg[71] on wolves. These researchers found that wild wolves captured as adults can form positive affiliative relationships if much time and great effort are devoted to developing them. The process took several months, and great care had to be taken to avoid eliciting from the wolves fearful or aggressive responses that would interfere with the socialization process. Anecdotes abound about adult wild animals and unsocialized domestic animals forming attachments to humans after a traumatic illness or during close confinement, as in a zoo setting.[72,73]

Does Early Environmental Enrichment or Restriction Influence the Behavior of Companion Animals?

Most research on enrichment and restriction has been conducted on laboratory rodents living in standard laboratory cages. The kinds of stimuli manipulated include social isolation, handling by experimenters, enhancement or deprivation of specific sensory experiences (e.g., rearing in the dark), and more general environmental enrichment or restriction.[86]

The effects of general environmental enrichment have been reported to produce enhancements in behavior, while environmental restriction has been reported to produce deficits in behavior.[39] For example, restriction-reared Scottish terriers showed more stereotypies[87] and were inferior in problem-solving ability[88] compared with nonrestricted dogs.

Handling of young animals has been proposed to produce a variety of beneficial effects, including acceleration of development, increase in resistance to certain diseases, reduction in emotionality, and increase in activity in a novel environment.[55] Not all studies find consistency in these effects on laboratory rodents.[86] There have been few studies of the effects of handling on dogs and cats, but these studies do report positive effects. Fox and Stelzner[77] reported that handled puppies were more active, showed more exploratory behavior, and were more sociable to humans and more dominant toward other dogs in play. Meier[89] found that daily handling of young Siamese kittens accelerated eye opening, development of the breed-typical coloration, growth, and emergence from the nest. Meier and Stuart[90] reported that cats handled and raised in a stimulating environment performed better in a visual discrimination task.

Unfortunately, research on environmental enrichment and restriction suffers from contradictory evidence and has often been confounded with manipulations of the social environment, making it difficult to sort out the causes of these effects.

Do Certain Behavioral Patterns Change Over Time Without Specific Experiences?

Certain behavioral patterns change through *maturation*, defined as an irreversible developmental process not dependent on specific experiences.[58] This does not mean that the environment plays no role but only that it produces relatively small changes.

A classic example of maturational change is the development of flying ability in pigeons. Grohmann[91] found that young birds, reared in tubes so small that the birds could not open their wings, were able to fly when released despite the lack of any experience or practice with flying. Flying skills did improve after the birds were released, demonstrating that experience can modify behaviors once they occur.

The most studied examples of maturation in companion animals are those dimorphic patterns that are correlated with gonadal hormones. Courtship and copulatory behavior in males and females, male patterns of urine marking, and intermale aggression are correlated with adult blood levels of gonadal hormones and first appear at puberty.[55] Note that social isolation can alter male copulatory behavior in dogs.[74] It is likely that the declines in social and object play observed in both cats and dogs as they age also are attributable to maturational changes.

The development of some problem behaviors in dogs and cats also may be related to maturational changes. Territorial protective and dominance aggression, territorial barking, urine marking in males and females, and roaming and escape motivated by male sexual desire do not appear until puberty or thereafter.[14,92]

Practical Applications of Environmental Influences

Breeders in particular should be aware of the concept of sensitive periods for dogs and cats. Exposing young animals to humans during this time can achieve the greatest results in terms of producing animals that can form strong attachments to people and are friendly toward and unafraid of unfamiliar people encountered later in life.

Because it is known that social isolation and rearing animals in restricted environments can affect a number of behavioral patterns other than those related to attachment, breeders and owners should strive to provide social partners and some physical stimulation to young companion animals. For example, keeping a dog by itself in a barren crate for 20 hours a day probably will not produce a quality companion.

One misunderstanding of Scott and Fuller's work is the interpretation that one or two 20-minute exposures per week can produce adequate socialization in dogs. Fuller[93] found that semi-isolated puppies could be socialized to the level of nonisolated laboratory-reared dogs with such exposures. Scott and Fuller[11] point out, however, that the laboratory-reared dogs were behaviorally deficient compared with home-reared dogs. Laboratory-reared dogs had relationships with humans that were considered to be shallow, showed more fear of humans, showed less play, and did more poorly on behavioral tasks where less fear and a wider base of experience would have helped. The standard for socialization of pets should be that of home-reared dogs and not laboratory-reared animals.

If owners and breeders want a companion animal to be friendly with other members of the same species and with other species (dogs with cats, cats with birds, and so on), the animal should be given that exposure during the sensitive period. Because many households contain several species of animals, it would behoove breeders to socialize their animals to other common companion animals. It also appears that the more handling and positive experiences the animal has with many different people, the greater the effects on attachment and friendly behavior toward humans.

Similarly, if owners and breeders want their animals to be relaxed and unafraid in specific situations (e.g., riding in an automobile, swimming, being bathed and groomed, visiting strangers' houses), the animals should be exposed to those experiences from an early age. As with the other aspects of socialization, the experiences should always be positive from the animal's perspective. Food treats and/or play may encourage such positive experiences. Shy or fearful animals should never be forced to experience something in a fearful emotional state.

Puppy classes, puppy parties, and kitten carrier classes are useful ways to help socialize dogs and cats and to expose them to novel environments and new experiences in a nonthreatening way. Puppy classes are often part lecture, part socialization of puppies, and part obedience training. There is a common belief that puppies younger than six months of age cannot learn or that at least it is better not to expose them to formal learning experiences. The research on learning in dogs suggests this is not true. Puppies as young as 12 weeks of age can learn basic commands, such as sit and come.

Puppy parties can be a one-time socialization and education experience for dogs and owners in which basic developmental processes are explained and puppies can interact in a controlled setting with unfamiliar people, puppies, and other stimuli. Obedience training usually is not emphasized. Kitten carrier classes are analogous to puppy parties, where young cats can be exposed to other cats, people, and other stimuli. One goal of the classes usually is to show owners how to acclimate the cats to traveling in the carriers to such places as the veterinary clinic. Veterinary clinics may want to sponsor puppy classes, puppy parties, and kitten carrier classes or refer clients to programs run by competent animal behaviorists or animal trainers.

The risk of disease in puppy and kitten classes is small if all the animals have been started on their vaccinations. In contrast, the advantages of early socialization, early education of new owners, and development of consistent behaviors on the part of both owners and animals are very great and may outweigh risks of disease. More details about how veterinarians can provide practical information on the development of behavior and prevention of behavioral problems can be found in the literature.[13,94]

A final bit of practical information concerns the usefulness of puppy tests for predicting future behavior. A variety of tests have been developed over the years. Some have made far-reaching claims about their usefulness in helping owners and breeders predict the temperament of the animals, their suitability for particular tasks (such as hunting or herding), and the likelihood of future behavioral problems. All of these tests have suffered from a lack of scientific rigor. Measures of reliability and validity have rarely been performed. One study that examined the validity of puppy tests[95] found that test results at six to eight weeks of age could not predict adult activity levels or fearful, aggressive, or dominance behavior. Great caution should be exercised in using such tests as predictive instruments until more reliable and valid tests are developed.

REFERENCES

1. Houpt KA: *Domestic Animal Behavior*, ed 2. Ames, Iowa State University Press, 1991.
2. Martin P, Bateson P: Behavioral development in the cat, in Turner DC, Bateson P (eds): *The Domestic Cat.* New York, Cambridge University Press, 1988, pp 9–22.
3. Robinson I: Behavioral development of the cat, in Thorne C

(ed): *The Waltham Book of Dog and Cat Behaviour.* New York, Pergamon Press, 1992, pp 53–64.

4. Nott HMR: Behavioral development of the dog, in Thorne C (ed): *The Waltham Book of Dog and Cat Behaviour.* New York, Pergamon Press, 1992, pp 65–78.
5. Fox MW: *Integrative Development of Brain and Behavior in the Dog.* Chicago, University of Chicago Press, 1971.
6. Pedersen PE, Stewart WB, Greer CA, et al: Evidence for olfactory function in utero. *Science* 221:478–480,1983.
7. Smotherman WP, Robinson SR: Environmental determinants of behaviour in the rat fetus. *Anim Behav* 34:1859–1873, 1986.
8. Stickrod G, Kimble DP, Smotherman WP: In utero taste/odor aversion conditioning in the rat. *Physiol Behav* 28: 5–7, 1982.
9. Corionos JD: Development of behavior in the fetal cat. *Genet Psychol Monog* 14:283–383, 1933.
10. Windle WF, Fish MW: The development of the vestibular righting reflex in the cat. *J Comp Neurol* 54:85–96, 1932.
11. Scott JP, Fuller JL: *Dog Behavior, The Genetic Basis.* Chicago, University of Chicago Press, 1965.
12. Stanley WC, Cornwell AC, Poggiani C, et al: Conditioning in the neonatal puppy. *J Comp Physiol Psychol* 56:211–214, 1963.
13. Hetts S, Estep DQ: Behavior management. Preventing elimination and destructive behavior problems. *Vet Forum* Nov: 60–61, 1994.
14. Borchelt PL: Development of behaviour of the dog during maturity, in Anderson RS (ed): *Nutrition and Behaviour in Dogs and Cats.* New York, Pergamon Press, 1989, pp 189–205.
15. Fuller JL, Thompson WR: *Behavior Genetics.* New York, John Wiley & Sons, 1960.
16. Ehrman L, Parsons PA: *The Genetics of Behavior.* Sunderland, Sinauer Associates, 1976.
17. Fuller JL, Thompson WR: *Foundations of Behavior Genetics.* St. Louis, CV Mosby Co 1978.
18. Tryon RC: Genetic differences in maze-learning ability in rats. *39th Yearbook of the National Society for the Study of Education.* Bloomington, Public School Publishing Co, 1940, pp 111–119.
19. Searle LV: The organization of hereditary maze-brightness and maze-dullness. *Genet Psychol Monog* 34:279–325, 1949.
20. Guhl AM, Craig JV, Mueller CD: Selective breeding for aggressiveness in chickens. *Poultry Sci* 39:970–980, 1960.
21. Manning A: The effects of artificial selection for mating speed in *Drosophila melanogaster. Anim Behav* 9:82–93, 1961.
22. Hall CS: Emotional behavior in the rat. I. Defecation and urination as measures of individual differences in emotionality. *J Comp Psychol* 18:385–403, 1934.
23. Dykman RA, Murphree OD, Reese WG: Familial anthropophobia in pointer dogs? *Arch Gen Psychiat* 36:988–993, 1979.
24. Lucas LA, DeLuca DC, Newton JEO, et al: Animal models for human psychopathology: The nervous pointer dog, in Gershon ES, Mytthgase S, Breakefield XO, et al (eds): *Genetic Research Strategies for Psychobiology and Psychiatry.* Pacific Grove, CA, Boxwood Press, 1981, pp 241–252.
25. Murphree OD, Dykman RA, Peters JE: Genetically-determined abnormal behavior in dogs: Results of behavioral tests. *Cond Reflex* 2:199–205, 1967.
26. Newton JEO, Murphree OD, Dykman RA: Sporadic transient atrioventricular block and slow heart rate in nervous pointer dogs. A genetic study. *Cond Reflex* 5:75–89, 1970.
27. Angel C, DeLuca DC, Murphree OD: Probenecid-induced accumulation of cyclic nucleotides, 5-hydroxyindoleacetic acid and homovanillic acid in cisternal spinal fluid of genetically nervous dogs. *Biol Psychiat* 11:753–763, 1976.
28. Murphree OD, Angel C, DeLuca DC: Limits of therapeutic change: Specificity of behavior modification in genetically nervous dogs. *Biol Psychiat* 9:99–101, 1974.
29. McBryde WC, Murphree OD: The rehabilitation of genetically nervous dogs. *Pav J Biol Sci* 9:76–84, 1974.
30. Murphree OD, Newton JEO: Crossbreeding and special handling of genetically nervous dogs. *Cond Reflex* 6:129–136, 1971.
31. Valenstein ES, Riss W, Young WC: Sex drive in genetically heterogeneous and highly inbred strains of male guinea pigs. *J Comp Physiol Psychol* 47:162–165, 1954.
32. Southwick CH, Clark LH: Interstrain differences in aggressive behavior and exploratory activity in inbred mice. *Comm Behav Biol [A]* 1: 49–59, 1968.
33. Wahlsten D: Genetic experiments with animal learning: A critical review. *Behav Biol* 7:143–182, 1972.
34. Thorne FC: The inheritance of shyness in dogs. *J Genet Psychol* 65:275–279, 1944.
35. Turner DC, Feaver J, Mendl M, et al: Variation in domestic cat behaviour towards humans: A paternal effect. *Anim Behav* 34:1890–1891, 1986.
36. Reisner IR, Houpt KA, Erb HN, Quimby FW: Friendliness to humans and defensive aggression in cats: The influence of handling and paternity. *Physiol Behav* 55(6):119–124, 1994.
37. Lorenz KZ: The establishment of the instinct concept, in Martin RD (trans): *Studies in Animal and Human Behaviour,* vol 1. Cambridge, Harvard University Press, 1970, pp 259–315.
38. McFarland D: *Animal Behavior.* Menlo Park, Benjamin/Cummings Publishing Co, 1985.
39. Dewsbury DA: *Comparative Animal Behavior.* New York, McGraw-Hill, 1978.
40. Angel C, DeLuca DC, Newton JEO, et al: Assessment of pointer dog behavior. Drug effects and neurochemical correlates. *Pav J Biol Sci* 17:84–88, 1982.
41. Angel C, McMillan DE, Newton JEO, et al: Differential sensitivity to morphine in nervous and normal pointer dogs. *Eur J Pharmacol* 91:485–491, 1983.
42. Reuterwall C, Ryman N: An estimate of the magnitude of additive genetic variation of some mental characteristics in Alsatian dogs. *Hereditas* 73:277–283, 1973.
43. Goddard ME, Beilharz RG: Genetics of traits which determine the suitability of dogs as guide dogs for the blind. *Appl Anim Ethol* 9:299–315, 1983.
44. Goddard ME, Beilharz RG: A multivariate analysis of the genetics of fearfulness in potential guide dogs. *Behav Genet* 15:69–89, 1985.
45. Mackenzie SA, Oltenacu EAB, Leighton E: Heritability estimates for temperament scores in German shepherd dogs and its genetic correlation with hip dysplasia. *Behav Genet* 15: 475–482, 1985.
46. Wahlsten D: Insensitivity of the analysis of variance to heredity–environment interaction. *Behav Brain Sci* 13:109–161, 1990.
47. Hirsch J: A nemesis for heritability estimation. *Behav Brain Sci* 13:137–138, 1990.
48. Chiszar DA, Gollin ES: Additivity, interaction and developmental good sense. *Behav Brain Sci* 13:124–125, 1990.
49. Cooper RM, Zubeck JP: Effects of enriched and restricted early environments on the learning ability of bright and dull rats. *Can J Psychol* 12:159–164, 1958.
50. Freedman DG: Constitutional and environmental interactions in the rearing of four breeds of dogs. *Science* 127:585–586, 1958.

51. Tortora DF: *The Right Dog for You.* New York, Simon & Schuster, 1980.
52. Hart BL, Hart LA: *The Perfect Puppy.* New York, WH Freeman Co, 1988.
53. AVMA: *The Veterinarian's Way of Selecting a Proper Pet.* Schaumsburg, IL, American Veterinary Medical Association.
54. AVMA: *Your Role in Pet Selection.* Schaumsburg, IL, American Veterinary Medical Association.
55. Hart BL: *The Behavior of Domestic Animals.* New York, WH Freeman Co, 1985.
56. Bateson P: How do sensitive periods arise and what are they for? *Anim Behav* 27:470–486, 1979.
57. Lorenz KZ: The companion in the birds' world. *Auk* 54: 245–273, 1937.
58. McFarland D (ed): *The Oxford Companion to Animal Behaviour.* New York, Oxford University Press, 1987.
59. Hubel DH, Wiesel TN: The period of susceptibility to the physiological effects of unilateral eye closure in kittens. *J Physiol* 206:419–436, 1970.
60. Nelson RJ: *An Introduction to Behavioral Endocrinology.* Sunderland, Sinauer Associates, 1995.
61. Karsh EB, Turner DC: The human-cat relationship, in Turner DC, Bateson P (eds): *The Domestic Cat: The Biology of Its Behaviour.* New York, Cambridge University Press, 1988, pp 159–177.
62. Scott JP: The process of primary socialization in the dog, in Newton G, Levine S (eds): *Early Experiences and Behavior.* Springfield, Charles C Thomas, 1968, pp 412–439.
63. Freedman DG, King JA, Elliot O: Critical periods in the social development of dogs. *Science* 133:1016–1017, 1961.
64. Fuller JL: Experiential deprivation and later behavior. *Science* 158:1645–1652, 1967.
65. Stanley WC, Elliot O: Differential human handling as reinforcing events and as treatments influencing later social behavior in basenji puppies. *Psychol Rep* 10:775–788, 1962.
66. Fisher AE: The effects of differential early treatment on the social and exploratory behavior of puppies. Unpublished Doctoral Dissertation, Pennsylvania State University, 1955.
67. Fox MW, Stelzner D: The effects of early experience on the development of inter- and intraspecies social relationships in the dog. *Anim Behav* 15:377–386, 1967.
68. Scott JP: The domestic dog: A case of multiple identities, in Roy MA (ed): *Species Identity and Attachment.* New York, Garland STPM Press, 1980, pp 129–143.
69. Fox MW: Behavioral effects of rearing dogs with cats during the "critical period of socialization." *Behaviour* 35:271–280, 1969.
70. Cairns RB, Werboff J: Behavior development in the dog: An interspecific analysis. *Science* 158:1070–1072, 1967.
71. Woolpy JH, Ginsburg BE: Wolf socialization: A study of temperament in a wild social species. *Amer Zool* 7:357–363.
72. Bateson P: The interpretation of sensitive periods, in Oliverio A, Zappella M (eds): *The Behavior of Human Infants.* New York, Plenum Press, 1983.
73. Hediger H: *The Psychology and Behavior of Animals in Zoos and Circuses.* New York, Dover Press, 1968.
74. Beach FA: Coital behavior in dogs. III. Effects of early isolation on mating in males. *Behaviour* 30:218–238, 1968.
75. Pfaffenberger CJ, Scott JP: The relationship between delayed socialization and trainability in guide dogs. *J Genet Psychol* 95:145–155, 1959.
76. Thompson WR, Melzack R: Early environment. *Sci Am* 194:38–42, 1956.
77. Fox MW, Stelzner D: Behavioral effects of differential early experience in the dog. *Anim Behav* 14:273–281, 1966.
78. Thompson WR, Heron W: The effects of restricting early experience on the problem-solving capacity of dogs. *Can J Psychol* 8:17–31, 1954.
79. Melzack R, Scott TH: The effects of early experience on the response to pain. *J Comp Physiol Psychol* 50:155–161, 1957.
80. Karsh EB: Factors influencing the socialization of cats to people, in Anderson RK, Hart BL, Hart LA (eds): *The Pet Connection: Its Influence on Our Health and Quality of Life.* Minneapolis, University of Minnesota Press, 1983, pp 207–215.
81. Rosenblatt JS, Turkewitz G, Schneirla TC: Early socialization in the domestic cat as based on feeding and other relationships between female and young, in Foss BM (ed): *Determinants of Infant Behaviour.* London, Methuen, 1961.
82. Kuo ZY: The genesis of the cat's response towards the rat. *J Comp Psychol* 15:1–35, 1930.
83. Kuo ZY: Studies on the basic factors in animal fighting. VII. Interspecies coexistence in mammals. *J Genet Psychol* 97: 211–225, 1960.
84. Seitz PFD: Infantile experience and adult behavior in animal subjects. III. Age of separation from the mother and adult behavior in the cat. *Psychosomat Med* 21:353–378, 1959.
85. Konrad KW, Bagshaw M: Effects of novel stimuli on cats reared in a restricted environment. *J Comp Physiol Psychol* 70:157–164, 1970.
86. Henderson ND: Effects of early experience upon the behavior of animals: The second twenty-five years of research, in Simmel EC (ed): *Early Experiences and Early Behavior.* New York, Academic Press, 1980, pp 39–77.
87. Thompson WR, Melzack R, Scott TN: "Whirling behavior" in dogs as related to early experience. *Science* 123:939, 1956.
88. Clarke RS, Heron W, Fetherstonhaugh ML, et al: Individual differences in dogs: Preliminary report on the effects of early experience. *Can J Psychol* 5:150–156, 1951.
89. Meier GW: Infantile handling and development in Siamese kittens. *J Comp Physiol Psychol* 54:284–286, 1961.
90. Meier GW, Stuart JL: Effects of handling on the physical and behavioral development of Siamese kittens. *Psychol Rep* 5:497–501, 1959.
91. Grohmann J: Modification oder Funktionsreifung? Ein Beitrag zur Klarung der wechselseitigen Beziehungen:zwischen Instinkthandlung und Erfahrung. *Z Tierpsychol* 25:132–144, 1939.
92. Borchelt PL, Voith VL: Classification of animal behavior problems. *Vet Clin North Am [Small Anim Pract]* 12:571–585, 1982.
93. Fuller JL: Effects of experiential deprivation upon behaviour in animals, in *Proceedings of the 3rd World Congress of Psychiatry.* Montreal, University of Toronto/McGill University Press, 1961, pp 223–227.
94. Hetts S, Estep DQ: Preventing behavior problems. A developmental approach. *Vet Forum* Sept:96–97, 1994.
95. Young MS: What do "puppy tests" test? Paper read at the Annual Meeting of the Animal Behavior Society, Raleigh NC, 1985.

Issues in Description and Measurement of Temperament in Companion Dogs

Animal Behavior Consultants, Inc.
New York, New York
Linda P. Goodloe, PhD

People love to describe their dogs. I recently completed a survey in which over 2000 dog owners completed an eight page questionnaire that included rating each dog on 126 specific behaviors. This opportunity was not sufficient for many subjects to believe that they had adequately characterized their dogs. Along with scribbled notes in the margins and in some cases additional pages, many owners attached a photo of their dog, thus enlivening for us the tedious task of wading through raw data. Their response, however, illustrates the difficulty of describing the "personality" of a dog.

This article is concerned with how to best describe a dog in terms of behavior. Behavior characteristics of an individual are considered expressions of "temperament" or "personality." In a review of animal temperament studies (at that time studies of rats and mice only), Calvin Hall noted: "... temperament is the raw stuff of individuality. Temperament accounts for the original uniqueness of each organism."[1] Although he then tries to distinguish "basic" temperament from "...the conditioning effects of the physical environment and the social culture ...," this article will not be concerned with this distinction. Temperament/personality will be considered to be the result of the interaction between the "raw material" and all combined internal and external environmental influences at a particular point in time.

How can the temperament of a particular dog best be described? Some guidance can be found in studies of human personality. This past century has seen an explosion in the exploration of personality, aided in large part by statistical techniques that provide insight on relationships among behaviors. So many instruments have been developed to measure aspects of human personality that the list is encyclopedic. No respectable researcher would use one of these tests without first reviewing evaluations of the development of the instrument and statistical data relevant to the reliability and validity of the test. Standards for acceptability have become increasingly rigorous, and the burden is on the test developer to provide evidence that the test does what it claims to do (e.g., predict aptitude for a career in science, provide an index for depression, measure spatial ability, or rate an individual's degree of "optimism"). For describing and measuring companion dog behavior, much can be learned from methods in studies of human personality.

DESCRIPTION OF TEMPERAMENT

Labels abound for elements of temperament: "sociability," "playfulness," "excitability." We use such labels when discussing dogs we know, assuming the listener has a general sense of what we mean. If we plan to evaluate or measure a dog's temperament, however, such labels are less acceptable. There are many different behaviors that could be considered "playful." Some behaviors might be labeled "playful" by one observer and "destructive" by another. Any such ratings, even when examples or brief definitions are provided, are highly subjective. For measurement, more objectivity is required.

For any dog there is a nearly infinite set of specific behaviors that can be described. Are they all of equal importance or can they be organized or grouped in some meaningful way? "This dog is aggressive" is not a particularly helpful statement, at least to a behaviorist, although it may be sufficient for a potential owner. It is well known that many different kinds of aggression have been described.[2] "This dog is aggressive to strangers" is a bit more specific, but many additional qualifications could provide even greater specificity, such as the sex, race, or size (adult or child) of the stranger and whether the encounter occurs at home or away from home and in the presence or absence of the owner. Aggression also could be more objectively defined by physical descriptions (e.g., lips retracted to expose canines, vocalization ["growl," "snarl"] or dominance posture [ears erect, tail up, posture rigid], etc.).

A level of description must be found that fits a particular purpose and tested to see if it produces meaningful results. One way of simplifying description and also of gaining insight into the structure of temperament or the way behaviors are organized is to try to identify traits, each of which may account for a number of related behaviors. The goal is to provide an empirical rather than a popular or intuitive basis for elements of personality.

FACTOR ANALYSIS

The primary method used in studies of human personality to identify traits underlying specific behaviors is factor analysis, developed by Thurstone[3] and further elaborated and used in the study of human personality by psychologists Hans Eysenck,[4] Raymond B. Cattell,[5] and others. Factor analysis provides a way to identify clusters of behaviors that are related to each other but not to different behavior clusters. Each cluster of related behaviors is considered to indicate a higher level factor or element of temperament, the meaning of which is inferred through examination of the behaviors in the cluster.

A factor is, thus, one level of organization higher than the specific behaviors the factor subsumes. Eysenck identified the familiar human personality dimensions "introversion-extroversion" and "neuroticism" using this method.[6] An individual scoring high on the factor "introversion" will express this factor in numerous specific behaviors, such as being less talkative, more secretive, somewhat reclusive, and more cautious than others. A single factor can account for many different behaviors.

Analysis starts with a set of behaviors. Each behavior can be expected to vary among individuals. In human personality studies, subjects are rated (or rate themselves) on each behavior (e.g., an attitude, an emotion, the likelihood of expressing a particular behavior, a preference). In animal studies, the subject is rated by an observer. Individual scores on each behavior provide the material for a factor analysis. Next, the investigator constructs a correlation matrix in which each behavior rating is correlated with every other behavior rating. From this matrix, clusters of related behaviors are identified. Each cluster represents a "factor." A small number of factors can explain relationships among a larger set of behaviors. Interpretation of a factor involves examining those behaviors strongly related to that factor as well as noting those behaviors that are *not* related. The strength of the relationship between each behavior and the factor is provided by the "factor loading," which is similar to the correlation of the behavior with that factor. We never actually see "factors." We infer their existence from relationships among specific behaviors.

Unfortunately, factors do not emerge with useful labels. We have only the list of behaviors related to a factor to guide our interpretation of what that factor might represent. In a recent factor analysis of companion dog behaviors that we conducted, high positive loadings were obtained on one factor for "Tries to initiate play by bringing 'toys' to owner," "Chases after thrown objects," and "Has favorite toys or objects and keeps them a long time." This factor is clearly related to play, and we gave it the fairly neutral label "Play 1" to differentiate it from two other qualitatively different factors also related to play.[7] Not all factors are so easy to interpret. Often, examination of the array of behaviors loading on a particular factor does not suggest any unitary construct or label. A factor label represents a hypothesis that can and often should be subjected to further testing. Thus, while factor analysis is an effective tool in determining relationships among behaviors, interpretation of identified factors is seldom simple or straightforward. Factors are dependent on the behavior variables used, decisions concerning the setting of certain parameters for a specific analysis (i.e., the kind of rotation employed, the number of factors to extract), and characteristics of the sample tested.

PREVIOUS FACTOR ANALYSES OF TEMPERAMENT IN DOGS

For many years the colony of dogs at the Roscoe B. Jackson Memorial Laboratory in Bar Harbor, Maine, made famous by Scott and Fuller,[8] provided ideal subjects for carefully controlled factor analytic studies. Using several distinct breeds raised under uniform environmental conditions, researchers used factor analysis to identify genetic differences in traits.

Influenced by early studies by Thurstone and others using rat subjects, Anastasi et al[9] tested dogs for 17 behaviors related to learning (e.g., number of errors in a T-maze, latency to solve discrimination problems, performance on a leash). Examples of factors identified in this study include "persistence of positional habits" and "activity and impulsiveness."

Royce[10] studied "emotionality" in dogs using 32 observed behaviors. Observations were made during specific laboratory procedures (i.e., "vocalization" and "timidity-confidence" measured during regular weighings) and staged tests (e.g., "dominance" measured by placing two pups in a room with a bone) and on physiologic variables (e.g., sinus arrhythmia, systolic blood pressure, temperature). Of 11 described factors, five were uninterpretable. Factor labels included Activity Level, Audiogenic Reactivity, Aggressiveness, and Heart Reactivity to Social Stimulation. These factors are difficult to interpret in terms of companion dog behavior since they are tightly bound to the operational definitions of behavior in each test. For example, items loading high on the Aggressiveness factor primarily describe a high activity level.

Cattell conducted a study that is of particular interest because it went beyond the identification of factors to the use of "scores" on those factors to classify individual dogs according to breed (a factor score for an individual dog on a single factor is the weighted sum of ratings the dog received for all behaviors loading on the factor). The study consisted of two parts: (1) identification of factors[11] and (2) use of a computer program, Taxonome, developed by Cattell to assign individual dogs to taxonomic categories based on factor scores.[12] The factor analysis used 42 physiologic and behavior variables and identified 15 behavioral factors, several of which resembled factors described by Royce[10] and Anastasi et al.[9]

Although the classification was not perfect, an impressive proportion of dogs were grouped appropriately using the Taxonome program. Cattell had clearly identified some basic temperament domains in these laboratory dogs, and the hypothesis of a genetic basis for differences in these behaviors was supported by the success of his classification program. Basenjis, with their African lineage, might be suspected to be the most distinct and were the most easily classified of the breeds, which included beagles, cocker spaniels, Shetland sheepdogs, and fox terriers.

These are all excellent studies for the purpose of identifying genetic differences in behavior utilizing precise measurements in a tightly controlled laboratory setting. For those looking for insight into the temperament structure of companion dogs, however, these studies provide little assistance. We do not know the relationship between behaviors measured in the laboratory and the more familiar behaviors of companion dogs. "Aggressiveness," as defined in the Royce study, is certainly a legitimate factor as defined operationally and statistically, but this may or may not relate to behaviors that most people would identify as aggressive in a companion dog.

Cattell commented in both of these papers on the desirability of adding more naturalistic behavior measures to the analyses. Unfortunately, this did not happen. The years of research at the Jackson Laboratory under Scott and Fuller surely constitute a "golden age" in the study of dogs. The likelihood of establishing such a research endeavor today under current funding constraints seems rather dim, so it is unlikely that the studies needed to relate these previous findings to companion dog behavior will ever be done. The methods used to study temperament in the Jackson Laboratory dogs can be applied in other contexts, however, including the home environment.

APPLICATION OF FACTOR ANALYSIS TO COMPANION DOGS

A review of the existing studies raises a fundamental question regarding the application of factor analysis to the study of companion dogs. Given the dependence of identified factors on the behaviors chosen for analysis, how can we determine which are the "real" factors?

On one level, all identified factors are real in that they represent correlations among the behaviors. However, not all factors are equally "meaningful" or helpful in understanding and describing behavior. For example, in our factor analysis of companion dog behavior, we used descriptions with what we considered an intermediate amount of specificity, including some contexts but omitting others, aiming for behavior descriptions simple enough to be understood by the average dog owner. Using aggressive behavior as an example, a minimal amount of specificity would be "... is aggressive." We chose item descriptors to allow for inclusion of contexts that we believed were relevant to the described behavior. "Growls and/or bares teeth when away from home and an unfamiliar person approaches" is one of the items we used (response categories were "never," "rarely," "sometimes," "often," and "always"). A description could also have included whether the owner was present or absent, details of posture, presence or absence of food, whether the dog was on or off leash, etc. Each added detail, however, would greatly increase the number of items needed to provide a balanced representation of this particular set of behaviors.

From our final total of 126 behavior items, our factor analysis produced four factors relating to aggression from the total of 22 factors identified[7]: (1) aggression toward a family member in response to some action by that person, (2) aggression in response to approach by a stranger, (3) aggression initiated by the companion dog against another dog, and (4) biting behavior across a variety of contexts.

A factor analysis of all 22 factor scores—a "factor analysis of the factors" (a second-order factor analysis)—produced a smaller number of factors including a single aggression factor that contained all four factors mentioned above. It is important to consider the level of organization represented by each factor. The level of description determines the specificity of extracted factors. A separate factor analysis of just the items loading on aggression factor 2, aggression toward strangers, suggested several lower-order factors. Possibly, several different kinds of aggression toward strangers (e.g., "fear-related aggression," "territorial aggression," or "protective aggression") are collapsed into this single factor just as the four aggression factors collapsed into a single factor in the second-order analysis. The number of items was insufficient to adequately represent this new domain of "aggression toward strangers," but the exercise was informative in suggesting that aggression factor 2 may itself have several components.

A new study, with more finely elaborated items to represent the domain of "aggression toward strangers," would be needed to separate the varieties of aggression toward strangers. Increased sophistication of descriptive items might then require that trained observers rather than owners rate dogs. The effects of a chosen level of description include the number of behavior items needed, the complexity of the rating system, and, finally, the identity of the factor that is extracted.

Aggression factor 1, aggression toward a family member, provides a somewhat different example. The items loading on this factor mirror the clinical symptom list for the construct of "dominance aggression."[2,13–15] All of these items specifically indicate aggression directed toward a family member in such contexts as being restrained, being subjected to verbal discipline, or having something taken from it. Factor analysis of just the items loading on this factor produced a single factor, indicating a unitary construct. Here the analysis supported a clinical hypothesis and has helped differentiate behaviors associated with this construct from those that are not. For example, some writers have suggested that certain kinds of rough play, such as "tug of war" (especially when the dog is allowed to win!), or allowing the dog to sleep with the owner can contribute to dominance aggression. Our analysis included items describing these behaviors, and we found no relationship between this factor and the items describing "tug of war," other kinds of rough play, or sleeping with or near the owner.

Our temperament study provides a contrast with the Jackson Laboratory studies. Lack of controlled history and environment for the broad assortment of companion dogs we recruited, our use of owner's rating of behavior rather than scores on carefully controlled manipulations, and our use of descriptions of naturally occurring behaviors that required the owners to recall how their dog had behaved in assorted contexts all contribute to measurement error. We sacrificed precision of behavior descriptions and environmental control for a combination of comprehensibility and generalizability. Assuming that measurement error would occur randomly, we obtained a very large sample to attempt to absorb this error. We now have 22 factors or temperament domains that can be comprehended by average dog owners and can provide profiles of individual dogs or groups of dogs, based on their scores on these factors. The value of these scores as descriptors or predictors of future performance, however, is yet to be determined.

Factors can guide understanding of the organization of temperament. We believe this methodology provides a path to an empirically based classification system for companion dog behaviors, but it is a first step rather than a final one.

TESTING SPECIFIC BEHAVIORS

A number of diverse groups have been grappling with assessment of temperament in individual dogs. Shelters want to assure that dogs offered for adoption have a maximum chance of remaining in a particular home. Organizations involved with animal-assisted therapy want to assure that their dogs are able to behave appropriately in the settings they will encounter. Puppy testing for temperament traits or for specific aptitudes, such as suitability for guide dog training, has been occurring for many years. While reasons for testing vary, the problems of test development are similar. Factor analysis can give insight into the structure of temperament and, thus, suggest measurement domains, but there is also a need to evaluate dogs for more specific aptitudes and skills. These may be as basic as general socializing skills, such as the ability to behave in an acceptable manner around humans and other dogs, or as complex as determining aptitude for performance as a guide dog.

Using available instruments, an extremely detailed profile of a human can be created for a set of personality factors. However, none of these may be particularly informative if we want to know whether that individual possesses enough relevant skills and knowledge to be able to drive a car. Similarly, with dogs, factor scores on a number of personality domains may not provide relevant information if we want to know whether a dog can behave appropriately when visiting a nursing home. For some evaluations, more specific task-relevant behaviors need to be assessed.

In 1989 the American Kennel Club (AKC), in response to perceived anticanine sentiment, approved a "canine good citizenship (CGC) test" to demonstrate that dogs can be trained to behave appropriately in and out of the home.[16,17] The test contains a series of staged encounters with other humans and dogs under specific conditions and includes demonstration that the dog responds to the basic commands "sit," "down," and "stay." This test has been widely used and has been adapted by other organizations (e.g., the Delta Society's Pet Partners Program[18]) for assessment purposes. A certificate is awarded if the dog successfully passes each requirement.

Particularly important to the development of such a test is establishing the reliability of the test. "Reliability" is a technical term referring to consistency of measurement. There are many sources for variation in behavior resulting from environmental conditions rather than the temperament of the dog. Experience of the observer, the identity of the handler, location of the test (enclosed versus open space), interpretations of scoring criteria, and whether or not the dog is hungry are only a sample of the many influences that could lead to measurement error. A test must be developed and administered in ways that minimize this kind of error. If the test is intended to sample a relatively stable component of temperament, it should be expected that the same dog tested on more than one occasion should show similar scores. Also, and particularly important in assessment of canine temperament where scores depend on the judgments of observers, different observers should produce similar scores when simultaneously judging the same dog. These are issues of reliability.

Both the AKC and the Delta Society are presently concerned with making their tests more reliable. This can be accomplished by standardizing tests by providing: detailed instructions on test administration, detailed criteria for making judgments, and training for test administrators. The Delta Society currently has such a training program under their Pet Partners Program. Individuals who complete training are certified as "animal evaluators" to administer their screening tests. The American Kennel Club is reevaluating the CGC out of concern that criteria may vary in different locations and that some evaluators are providing shortened or otherwise modified versions of the test. Data from multiple evaluations are under review, and training of evaluators is being considered to tighten standards.*

Providing conditions that maximize reliability, however, is not sufficient to indicate that a test has acceptable reliability. Reliability must be demonstrated using appropriate statistical measures. This means each dog must have more than one evaluator, who should *independently* score the dog on each procedure; after many such measurements, the correlations between observers should be reviewed for each test item.

A particular test can have very high reliability, observers can agree on all the ratings, and the subject can obtain similar scores over repeated tests; however, if the test does not measure what it is claimed or intended to measure, then it has no "validity." Establishing the validity of a test also requires evidence. The kind of proof required depends on the kind of test created.

Evaluations of social behavior, such as the CGC test, are similar to the test for a motor vehicle license. Passing the test indicates that the subject has demonstrated at least a minimum level of ability and knowledge of relevant tasks. In addition to reliability, a basic requirement for such a test is that it has "content validity"; the items must adequately and accurately represent the performance domain (the desired behaviors/abilities/knowledge). The usual process for establishing content validity is to assemble an independent panel of experts to evaluate a test to ensure that all appropriate domains are adequately covered.[19]

If certification affects the status of the dog outside the test situation, occasional recertification procedures may be appropriate. Maturational and environ-

*Burch M (AKC field representative in charge of the Canine Good Citizenship Test.) Personal Communication.

mental changes can influence behavior in both humans and dogs. People must periodically demonstrate an acceptable level of visual acuity to be allowed to continue to drive a car legally. Dogs certified for use in animal-assisted therapy gain privileged access to vulnerable individuals and, unlike other companion animals, also are allowed access to public facilities and transportation. It is similarly appropriate to reevaluate their basic social skills at periodic intervals.

TESTS TO PREDICT FUTURE BEHAVIOR

Problems occur when tests are used to draw inferences about behavior in situations apart from the test environment. A person taking a test for a motor vehicle license may demonstrate considerable driving skill and correctly respond to all traffic control signals such as stop signs and speed limits. The resulting test score does not, however, predict driving behavior under different conditions, such as the absence of an authority figure, heavy traffic, ice, or a late appointment. To be able to make such inferences from a particular test, the test developer must *prove* that test scores actually have the hypothesized relationship with these external events. This kind of validity—"predictive validity"— must be evaluated by measuring behaviors that the test claims to predict and determining the correlations (hopefully high ones) of these behaviors with test scores.

An effective procedure for screening shelter dogs for problem behaviors prior to adoption might increase the likelihood of their being retained in adoptive homes. In a recent shelter study, 81 dogs were evaluated on 21 separate measures for four problem-related characteristics plus some additional problem behaviors.[20] Unfortunately, the criteria for scoring problem behaviors for prediction decisions and many other details of the methodology were not described. Outcome variables, such as "...(potential) problem behavior" and "real problem behavior," are not clearly described. The authors claim a 75% success rate in predicting "(potential) problem behaviors." This kind of study (i.e., connecting test results with external behavior measures) is very much needed, but this particular study cannot be adequately evaluated because of the missing information.

Puppy testing is a popular and widespread practice in which future behavior is inferred from responses to test items. William Campbell[21] described five procedures "designed to reveal critical behavioral tendencies of puppies at the time of selection." He was careful to state that a prediction of adult behavior cannot be made, although he indicates "behavioral *tendencies* are predictable." The Puppy Aptitude Test (PAT) devised by Wendy Volhard[22,23] and widely distributed is similar to Campbell's test but has some additional procedures to evaluate sensory capacities and conformation; it is intended to select dogs for both obedience and companionship. Both tests involve judgments of behaviors in terms of dominance, submission, and independence.

There have been no published accounts concerning reliability of these and other similar tests, and only recently has there been an effort to test the validity of some items. Using Campbell's procedures, Beaudet et al[24] evaluated puppies at the recommended seven weeks[8,24] and again at 16 weeks. Each puppy received a total score for "social tendencies" from each evaluation. They found no significant correlations between the "social tendency" scores at the two periods. The report by Beaudet et al[24] also cites Young, who reported that scores for puppies tested at seven weeks did not predict social behavior at later ages.

A better example of the development of puppy selection tests can be found in Clarence Pfaffenberger's account[25] of the development of selection tests for Guide Dogs for the Blind in San Rafael, California, in the late 1940s and early 1950s. The breeding, raising, and training of Guide Dogs requires a tremendous investment of time and effort. Pfaffenberger and others worked to develop an aptitude test for the puppies they raised to increase the rate of successful completion of training.

Their test had a specific purpose, namely, to select those puppies most likely to pass all future tests in Guide Dog training. Puppies were tested once a week from their eighth through twelfth week. Test items, created to simulate experiences that would be encountered as Guide Dogs, were continually refined. Working from a list provided by Guide Dog instructors of the reasons why dogs failed in training, items were designed to try to detect these faults. A chart was created to show correlations between puppy test scores and scores obtained later during Guide Dog training, thus demonstrating the ability (or lack of ability) of each procedure to predict future performance. Their success rate (the proportion of dogs successfully completing training) went from 8% in 1946 to over 90% by the time the article was published (1963).

During the development of the puppy tests, the suggestion was made that a "fetch" training test be included in the selection battery. Pfaffenberger recounts that the test was added but without any expectations of particular significance. Several years later, when the test data were finally analyzed, the fetch test was shown to be one of the two best predictors of successful completion of training! This illustrates the importance of constructing initial tests without preconceptions concerning the outcome. We cannot rely on intuition and experience alone to guide the creation of a test. Test development involves providing a variety of items or procedures and using feedback from outcome data to guide pruning and refining those procedures. Items on a widely used objective diagnostic test for psychopathology in humans, the Minnesota Multiphasic Personality Inventory (MMPI), were chosen because of their ability to discriminate between certain patient groups regardless of content. It is not surprising to find that paranoid individuals tend to agree with the statement "Someone has been trying to poison me." It is less clear, however, why the same individuals are also more likely to *disagree* with the statements "I think most people would lie to get ahead" and "Most people are honest chiefly through fear of being caught." We can produce numerous theories of why puppies who excel at fetching in early tests are more likely to complete Guide Dog training or why paranoid individuals have this particular view of honesty, but for test purposes it does not matter. Both have an established ability to provide information about probabilities of specific behaviors beyond the test environment.

TEMPERAMENT DIFFERENCES AMONG BREEDS

Many books have been written describing in detail the differences in temperament among breeds, yet for companion dogs there is little or no empirically based knowledge. Stereotypes and prejudices abound, and there is considerable economic incentive to provide favorable profiles for each existing breed. Scott and Fuller[8] used five different breeds in their investigation of the genetic basis of dog behavior and described many specific behavior differences in the laboratory setting. Their sample was developed, however, from a small founder group for each breed. As with the factor analysis studies described above, it is not possible to generalize their findings to the larger and more diverse population of companion dogs of the same breeds.

Hart and Miller[26] describe behavior profiles of breeds using judgments of individuals with extensive experience with dogs (i.e., veterinarians in small animal practices, obedience trial judges, dog show judges, and handlers of dogs at shows). Each rater was required to provide a rank ordering of seven different breeds on 13 "types of behavior" (e.g., "playfulness," "general activity," "housebreaking ease"), and the data were analyzed to produce breed profiles for each of these 13 behavior types. The "behavior types" were defined by the interviewers providing hypothetical examples of characteristic behaviors. Ranking of breeds on each behavior type was determined by the subjective judgments of the chosen experts. The profiles obtained, therefore, describe *beliefs* about breed behavior types rather than the distribution of the behavior types themselves. Information about the distribution of these beliefs is certainly of interest in itself, but it should not be confused with objective data that document expression of particular traits.

With the exception of behaviors that are dependent on a particular size and structure of dog, no knowledge is available for *any* breed concerning the variability of *any* single behavior within that breed to compare with the variability of the same behavior across other breeds. We have all seen exceptions to current beliefs about breed-specific behaviors, such as an occasional marauding golden retriever or a wimpish rottweiler, but we have no information about the extent of such variability. A study of breed behavior differences in companion dogs using a large and geographically diverse sample of each breed is needed to fill this huge vacuum of knowledge.

CONCLUSION

Methodologies exist to develop and refine temperament description and evaluation of companion dogs. In a field characterized to a considerable extent by hype and dogma, we need to establish higher standards. We need to insist that theories of temperament and tests to evaluate puppies and dogs be accompanied by appropriate data and statistical evidence to support specific claims.

REFERENCES

1. Hall C: Temperament: A survey of animal studies. *Psych Bull* 38(10):909, 1941.
2. Boas PL: Aggressive behavior of dogs kept as companion animals: Classification and influence of sex, reproductive status and breed. *Appl Anim Ethol* 10:45–61, 1983.
3. Thurstone LL, 1947: The factor problem. Reprinted in Jackson DN, Messick SR (eds): *Problems in Human Assessment.* New York, McGraw-Hill, 1967, pp 279–287.
4. Eysenck HJ, 1953: The logical basis of factor analysis. Reprinted in Jackson DN, Messick SR (eds): *Problems in Human Assessment.* New York, McGraw-Hill, 1967, pp 288–299.
5. Cattell RB, 1952: The three basic factor analytic research designs—Their interrelations and derivatives. Reprinted in Jackson DN, Messick SR (eds): *Problems in Human Assessment.* New York, McGraw-Hill, 1967, pp 300–304.
6. Eysenck HJ as cited in Gleitman H: *Psychology.* New York, WW Norton & Co, 1981.
7. Goodloe L, Borchelt P: Identification of temperament factors in companion dogs. Manuscript in preparation, 1995.
8. Scott JP, Fuller JL: *Dog Behavior: The Genetic Basis.* Chicago, University of Chicago Press, 1965.
9. Anastasi A, Fuller JL, Scott JP, Schmitt JR: A factor analysis of the performance of dogs on certain learning tests. *Zoologica* 40(3):33–46, 1955.
10. Royce JR: A factorial study of emotionality in the dog. *Psychol Monogr Gen Appl* 69:22 (Whole No. 407), 1955.
11. Cattell RB, Korth B: The isolation of temperament dimensions in dogs. *Behav Biol* 9:15–30, 1973.
12. Cattell RB, Bolz CR, Korth B: Behavioral types in purebred dogs objectively determined by Taxonome. *Behav Genet* 3(3):205–216, 1973.

13. Crowell-Davis SL: Identifying and correcting human-directed dominance aggression in dogs. *Vet Med,* pp 990–998, Oct 1991.
14. Line SL, Voith VL: Dominance aggression of dogs towards people: Behavior profile and response to treatment. *Appl Anim Behav Sci* 16:77–83, 1986.
15. Voith VL, Borchelt PL: Diagnosis and treatment of dominance aggression in dogs. *Vet Clin North Am Small Anim Pract* 12(4):655–663, 1982.
16. Dearinger J: Canine Good Citizen. The American Kennel Club. Press release, 1990.
17. Lane M: Getting started: Canine citizenship test. *Pure-Bred Dogs/American Kennel Gazette,* pp 92–96, June 1988.
18. Fredrickson MA: Temperament testing procedures for animals involved in nursing home, school and hospital visiting programs through Delta Society Pet Partners (abstract). *Applied Anim Behav Sci* 37:83, 1993.
19. Crocker L, Algina J: *Introduction to Classical and Modern Test Theory.* New York, Holt, Rinehart and Winston, 1986.
20. van der Borg JAM, Netto WJ, Planta DJU: Behavioral testing of dogs in animal shelters to predict problem behaviour. *Appl Anim Behav Sci* 32:237–251, 1991.
21. Campbell WE: A behavior test for puppy selection. *Mod Vet Pract* 53(13):29–33, 1972.
22. Bartlett M: Puppy aptitude testing: A new look. *Pure-Bred Dogs/American Kennel Gazette* 102:30–34, 1985.
23. Fischer GT, Volhard W: Puppy personality profile. *Pure-Bred Dogs/American Kennel Gazette* 102:36–42, 1985.
24. Beaudet R, Chalifoux A, Dallaire A: Predictive value of activity level and behavioral evaluation on future dominance in puppies. *Appl Anim Behav Sci* 40:273–284, 1994.
25. Pfaffenberger CJ: *The New Knowledge of Dog Behavior.* New York, Howell Book House, 1963.
26. Hart BL, Miller MF: Behavioral profiles of dog breeds. *JAVMA* 106:1175–1180, 1985.

History Taking

History Taking and Interviewing

Victoria L. Voith, DVM, PhD
Department of Clinical Studies
School of Veterinary Medicine
University of Pennsylvania
Philadelphia, Pennsylvania

Peter L. Borchelt, PhD
Animal Behavior Consultants, Inc.
Forest Hills, New York
The Animal Medical Center
New York, New York

History taking is of utmost importance to clinicians in making a diagnosis and developing a treatment plan for an animal's behavior problem. Establishing the history enables a clinician to determine why an animal is engaging in a problem behavior and to isolate the factors and circumstances that influence the behavior. For example, the client's statement "my cat doesn't use the litter box" does not provide sufficient information for diagnosis and treatment. What is required is information or observations about the cat's present elimination behavior patterns (in and out of the litter box); the chronologic development of the elimination behavior problem; and other pertinent factors, such as the cat's interactions with other cats and how the litter box is maintained.

The diagnostician is faced with the task of either observing the behavioral sequences over extended periods and in all of the contexts in which they occur or obtaining accurate descriptions from the owners. In the authors' experience most owners are excellent observers of their animals' behaviors, although the owners' explanations of the behavior are colored by misconceptions, anthropomorphism, and inappropriate terminology.

Location of the Interview

Home visits have the advantage of allowing a clinician to see the environment in which the behavior problems usually occur. This insight can be helpful in developing and demonstrating a treatment plan, particularly regarding elimination behavior problems or exit-related aggressive behaviors. Even with home visits, however, good history taking is imperative. Frequently, circumstances are such that the pet is unlikely to replicate the problem behavior in the presence of the clinician; or it can be undesirable to have the pet replicate the problem. It is the authors' opinion that adequate history taking, detailed and careful explanations, and demonstrations of a behavioral technique in an appropriate office setting can produce excellent therapeutic results. One strategy may be to conduct initial visits in an office setting where laboratory and medical diagnostic resources are present. If an office setting is inappropriate for the particular case, a subsequent home visit can be scheduled. For example, some neophobic dogs are so frightened in strange surroundings that they cannot relax sufficiently to demonstrate desensitization and counterconditioning techniques.

A typical veterinary examination room is not an ideal

office setting for history taking and observation of an animal. Usually clinical examination rooms are too small, and most pets are too frightened in them to relax. Optimum conditions include a large room with comfortable chairs for all participants and a decor that is not visually the same as that of a veterinary treatment or examination area.

The pet should be allowed to roam freely in the room during the interview. After a short period, the animal usually habituates to its surroundings and often begins interacting with the owner, strangers, and environment in a manner somewhat similar to its typical behavior. It may, for example, place its paws on the owner's lap, bark at people or dogs that pass by, or try to play with the clinician. Actions such as these should not—at least initially—be disturbed or interrupted but should be observed, which allows the clinician to identify stimuli that initiate and terminate the behavior, learn how the pet actually behaves, and how the owner responds to the behavior. For instance, when the dog barks at a passerby, does the owner call to it and pet it? If relevant, the clinician may point out what he or she has observed and discuss it with the client.

Setting the Tone for the Interview

The atmosphere of the interview should be congenial, relaxed, and nonjudgmental. Most owners are somewhat embarrassed about behavior consultations regarding their pets, are under the misconception that *they* are going to be "psychoanalyzed," and feel guilty because they think they may have caused the behavior problems. Books written for the general public about dogs are replete with the statement or implication that there are "no bad dogs—only bad owners." Despite a pet's behavior problem, an owner still loves the animal and is distraught at the possibility that, if a treatment cannot be found, the animal may have to be euthanatized.

Data collection in a series of 100 clients presenting a dog or cat for a behavior problem[1] indicated that the prominent reason most owners kept problem pets was that they were emotionally attached to them, or for humanitarian reasons. Because they believed no one else would want an animal with a behavior problem, they felt that either they had to keep problem pets or the animals would have to be euthanatized. There were no statistical differences between dog owners and cat owners as to why they had kept problem pets. Most owners also said that when friends learned that they were seeking clinical help for their pets because of a behavior problem, the friends laughed and sometimes made statements to effect that the the owners were the ones who needed to be analyzed and not the pets and/or that such an approach was a waste of money. Most owners must overcome some social ridicule when they seek professional help for a pet's behavior problem. In the years since these data were collected, a trend has developed toward more acceptance of the concept of seeking help from a professional animal behaviorist or veterinarian for an animal behavior problem. More often owners now say that their friends are curious and would like to know what transpires, especially if the friends have pets that also have behavior problems.

History Taking

Although a standard history-taking form is helpful to assure that the clinician does not forget to ask specific questions and is necessary to retrieve data for analysis, the interview itself should not be rigidly structured. Unexpected information that should be pursued may be uncovered. It may be prudent to interrupt the interview to draw attention to and discuss a behavior that was just demonstrated by the pet or owner, and a checklist approach may inhibit important voluntary information the owner might give. Regardless of the precise style of the history-taking sessions, specific information must be collected and recorded in an organized manner.

After a brief signalment is obtained, the first step is to ask the owners what they consider the problem to be, then let them describe it. Any attempt to structure or interfere with the initial complaint and description usually meets with resistance. The owners must tell you what they consider the problem to be and what they think about it. This usually takes only 5 to 10 minutes, after which the clinician can structure the interview. The initial time is not wasted because it allows the clinician to begin formulating some ideas as well as observe the pet and its owner; it also enables all participants to start relaxing.

The overall objectives of the history taking are to identify the events (stimuli and circumstances) that elicit the behavior problem, what the immediate consequences of the behavior are, and the development of the behavior and to uncover any related behavior problems that exist. Even if the owner rarely sees the actual behavior of the animal (for example, as is often true with an elimination behavior problem), it is still possible to determine when, where, and what occurs. This information should lead the clinician to deduce a likely motivational or functional classification of the problem. A frequency and/or intensity baseline of the occurrence of the behavior problem also should be determined so that progress can be measured at subsequent visits or evaluation rechecks.

After the owners' initial description of the problem, it is helpful to ask when the last incident occurred and then to have them describe the incident in detail—the location where it occurred, who was present, and the sequence of events that led up to the behavior as well as what happened immediately after. The last incident is likely to be remembered and described accurately. At first, owners may describe behaviors in subjective terms, such as the dog got "angry" or the cat looked "mad." When this occurs, ask the owners to tell you what the animal actually looked like or actually did that causes them to describe the animal as "angry" or "mad." Gradually, an owner can be shaped to describe objectively the animal's postures and behaviors and the prevailing circumstances. The owner usually has to be taught to describe in detail the sequence of events that precedes the behavior of the animal. For example, an owner might state, "I wasn't doing anything. All of a sudden the dog attacked me for no reason." The owner should then be asked where he or she was in the room, what position he or she was in (e.g., sitting, standing, lying down), where the dog was in the room, exactly what the owner was doing before the attack, what the dog was doing, and so on. Subsequently, the owner might describe the same incident: "It was about 8 PM. I was the only person in the room and was sitting on the floor watching TV and petting the dog when he suddenly bit me." Petting now has been identified as the stimulus immediately preceding the attack. Then the owner is asked to describe the next to last incident in detail, the third to last, and so forth and any other significant occurrences or incidents that are different from those already described.

The chronologic development of the problem should be established and pertinent information collected about incidents at various stages of the problem. By this time a tentative diagnosis has probably been made.

Several ways of cross-checking information and obtaining additional essential information are available. One is to obtain a detailed description of the pet's daily activities and interactions with people. The owner could start at the point when the dog wakes up in the morning. As the day is described, the clinician can take opportunities to substantiate the tentative diagnosis. If, for example, the dog sleeps on the owner's bed and dominance aggression is a tentative diagnosis, the clinician should ask what the dog's reaction is if the owner happens to touch it when it is sleeping or resting, e.g., does the dog mutter, growl, snap, or bite? If the dog gets on the bed before the owner does, the owner should be asked whether the dog mutters or growls when the owner enters the room and approaches the bed. If a urination or defecation behavior is the problem, the owners can be asked if they find any urine or feces in the house in the morning on arising. The dog's daily schedule provides such information as the amount and frequency of time the dog is let outdoors, its feeding and exercise schedule, and with whom it interacts and how often. Information of this type not only helps confirm a tentative diagnosis but also gives the clinician ideas about how to incorporate a treatment plan into the owner's daily schedule.

Another cross-checking system is to discuss behaviors other than the presenting problem behavior. For instance, the description of the play behavior of a dominant aggressive dog may reveal circumstances during which the dog is likely to "stand over" the owner; or the description of the greeting behavior of a dog with separation anxiety may reveal prolonged attention-seeking behaviors.

At some point early in the interview, information should be obtained on the human and animal constituents of the household, behavioral histories of related animals, and general background information about the pet. The latter two informational areas usually do not affect a treatment plan but may influence the prognosis or help predict the possibility of genetic influences on the behavior problem.

Although the interview should be flexible, the clinician should follow a format and remain in control. For instance, it is important always to get an answer to a question. Owners may begin to answer the question but get sidetracked on other issues or may think they are answering the question but actually are talking about an irrelevant matter. It may be necessary to interrupt their narration and refocus on the original question. It is unnecessary to let owners talk about anything and everything they feel inclined to discuss. Although some of the avenues may be worth exploring, much of the rambling is inapplicable and simply a waste of time. The direction of the behavioral interview and the ability to diagnose behavior problems depends on considerable knowledge of animal behavior and, in particular, behavior patterns of the species involved. A diagnostic and classification scheme is also necessary.[2] A practitioner cannot conduct a good interview if he or she does not thoroughly understand behavior.

Before the behavioral interview is concluded, the owner should always be asked if anything has not been mentioned that he or she thinks is important. By this time, the owner has begun to understand how to describe behaviors and has learned what information is important. Importance may now be ascribed to events that previously were considered insignificant, thus the owner should have the opportunity to present this information.

Finally, the clinician should share with the owner his or her views regarding the diagnosis, prognosis, and treatment plan. If the clinician is not yet sure of the diagnosis or is unable to develop a logical treatment plan spontaneously, he or she should tell the owner to schedule a second appointment. This will allow the clinician time to formulate a diagnosis and to provide the owner with a detailed outline of the treatment procedure. A short time allotment between the initial interview and complete discussion of the problem with the owner is a good approach when a clinician is beginning to see animal behavior cases. Several years are sometimes required to develop the ability to make a diagnosis and develop a specific treatment plan simultaneously.

At the end of the interview, if the clinician believes it important, a physical examination and/or laboratory tests can be conducted. To perform those procedures at the beginning of the interview may frighten the animal or arouse it to an extent that the subsequent interview and observation period are difficult or uninformative.

The amount of client contact time to allow for a history-taking session depends on the type of problem, the clinician's experience, and the individual personality of the client. An experienced clinician can usually evaluate and explain the preliminary treatment procedures for a feline elimination behavior problem in about one hour and a canine separation anxiety problem in about one and one-half hours. Aggressive behavior problems, however, may require more than two hours. As a rule, a behavioral interview should be allotted a two-hour slot. If the case does not take that long, the extra time can be used to organize records or prepare for the next case. For a case that requires longer than two hours, the clinician can have the owner come back to resume the history-taking and evaluation session.

It is optimal, but not always necessary, that all persons involved with the pet be present during the interview. With the entire family present, the clinician can obtain different observations of the animal's behavior, observe how each member interacts with the animal, and discuss his or her observations directly with all members of the family. In this way, nothing is lost through interpretation or relaying of information. It is also helpful if the clinician determines how strongly the owners feel about the pet and how serious they consider the problem. Owners' expectations and impressions vary considerably. Some owners are willing to assume the risk of treating a dog even if the prognosis is poor. Other owners may be amenable to euthanatizing an aggressive animal that has a poor prognosis. When some owners go to a behaviorist, they may not even want to keep the animal but have been discouraged from giving it away or euthanatizing it because of peer or social pressures. All of these factors should be considered before detailed treatment procedures are discussed.

Saving an animal's life through treatment of a behavior problem can be as satisfying as saving an animal's life via routine medical and surgical procedures. Treating behavior problems is rewarding in other ways also. It allows the clinician to use knowledge in the areas of physiology, pathology, anatomy, psychology, and animal behavior. Solving a behavior problem requires the ability to integrate information from many disciplines simultaneously. In many ways it is detective work. Working with animal behavior problems also increases basic knowledge of canine and feline behavior as well as pet-owner interactions. The field of clinical animal behavior not only provides a needed service but also is an area of research.

REFERENCES

1. Voith VL: Profile of 100 animal behavior cases. *Mod Vet Pract* 62(6): 483-484, 1981.
2. Borchelt PL, Voith VL: Classification of animal behavior problems, in Voith VL, Borchelt PL (eds): *The Veterinary Clinics of North America Small Animal Practice* (12:4). Philadelphia, WB Saunders Co, 1982, pp 571-585.

UPDATE

Comparison of Household Versus Office Practices

Debra F. Horwitz, DVM
Veterinary Behavior Consultations
Bridgeton, Missouri

In the diagnosis of a medical condition, a thorough history is important for evaluating signs, making a diagnosis, and establishing a treatment plan. In behavior cases, however, the complaint by the owner is merely a symptom.

In describing problem behavior, pet owners often relate the end result of what the pet did. Statements like "he bit me," "urine is everywhere," and "my house is destroyed" are merely describing symptoms. They are parallel to owners reporting that their dog is coughing. To diagnose and treat the cough, more information is necessary and extensive questioning is needed. Because of the depth of information needed in understanding and addressing behavior problems, adequate time should be allowed for history taking, diagnosis, and treatment recommendation. This usually requires 1 to 2 hours of a clinician's time. A clinician must be prepared to listen carefully and nonjudgmentally to the problem history. Prepared history forms aid in obtaining consistent and thorough histories. History taking can be augmented by video or audio tapes of the problem behaviors.

Voith and Borchelt discuss the need to establish baseline parameters in behavior consultations. Information should be gathered as to household routine, family members, and other pets in the home. Early influences can be important diagnostic clues and should be explored. Previous training attempts and other homes or owners can have an impact on companion animal behavior.

Time must be set aside to discuss the problem behavior. Owners should be asked to describe the most recent occurrence of the behavior. Encouraging owners to describe the event as if they were drawing a picture allows visualization of the location, the people present, and the actions and reactions of the participants. Owners may need to be prompted to include all pertinent information. Each specific action, the behavior of all participants, and the location of people present help contribute to the understanding of the event. By proceeding backward in time through previous episodes of the behavior, a tentative diagnosis may be possible.

Asking about different situations in the home can aid in establishing a diagnosis. Areas of interest include how the animal responds when family members handle its food, how it reacts if disturbed while sleeping or resting, how it responds to discipline, and whether it follows commands. Each opportunity for the owner to add information allows the clinician to gather salient points on which to base a diagnosis. Discussion of other behaviors, such as how the animal behaves with visitors or unfamiliar people or when left alone, may reveal additional problem areas that need exploration and treatment. Often the information that an owner gives will prompt additional questions that aid in understanding the problem.

Questioning owners about their feelings about the pet and the problem behavior can be very useful. Understanding the pet-owner interaction aids the clinician in creating a treatment plan that will fit the owners temperament, ability, and life-style. These considerations can make the difference between a treatment plan that is followed and one that is discarded. Each treatment plan should be tailored to the client's individual needs.

The location of the behavior consultation may or may not have an impact on history taking. The alternatives available to veterinarians are home visits, office visits, and telephone consultations. Each type offers advantages and disadvantages.

Home visits allow the clinician to visualize the daily environment of the pet. Food, water, and elimination sites are available for examination. Home visits can be scheduled when most family members are likely to be present. This allows observation of family interactions with the pet. People and animals are often more relaxed in their home environment. Influences outside the home that may have an impact on the problem behavior can be seen and evaluated. These include other animals nearby, the traffic on the street, and other noises and activities in the area. Often, dogs do not exhibit territorial behavior away from home; therefore, home visits may afford an opportunity to observe this behavior. Home visits require little investment in equipment and have low start-up costs.

Home visits can have drawbacks, however. A home visit can be very time consuming for the veterinarian. First, time that could otherwise be spent on other tasks is needed for travel. Second, owners are often distracted in the home environment. Telephone calls, televisions, young children, visitors, and any number of household disruptions and tasks can interrupt the consultation. Some owners attempt to turn the visit into a social occasion. Others have trouble focusing on the problem at hand.

The statement is commonly made that home visits allow you to see the pet in its own environment and therefore see a more representative sampling of the animal's behavior, including problem behavior. My experience in 12 years of home visits is that, as a visitor, you disrupt the home routine. In many cases, my visit was enough of a distraction to cause the animal to behave differently than it would have with family members alone. In most behavior cases, I did not see the animal engage in the problem behavior while I was at the home. The primary areas where home visits were helpful were for the observation of territorial behavior and for cases involving house soiling.

For the past year in my behavior practice, I have limited home visits and have been available primarily for office-based consultations. These also have their own benefits and pitfalls.

In office consultations, owners seem more focused. Perhaps this is the result of making the effort needed to schedule and attend the consultation. Certainly, there are less distractions for the owner in the office setting. I now mail a history form to clients prior to the visit, another change that may have affected client behavior. Clients fill out the form at home and bring it with them to their scheduled appointment. This seems to help them direct and focus their thoughts on the areas of discussion. In filling out the form, clients need to answer the questions with a minimum of anthropomorphizing. When owners focus on what the animal did, rather than on what they think the animal felt, the information they provide is more accurate. In the office, we go over and elaborate on their answers.

One concern mentioned about office calls is the possibility that the pet will behave in an unnatural manner. My observations have not supported this concern. In my home visits, my arrival in the home always resulted in behavior changes in the pet. These included excitation, aggression at the door, demanding behavior toward the owner, hiding, agitation, or fear. Usually after I had been in the home for a period of time, usually 10 to 60 minutes, the animal would calm down. Some animals became reactive again at the slightest noise. During office consultations over the past year, I have noticed the same trend. Initially, animals are agitated. Once off the leash or out of carriers, they may investigate the room, take a treat, interact with the owner, hide, or lie down and rest. Many animals react to the slightest noise; others do not. It takes 10 to 60 minutes for animals to relax in the office. In general, I have found that the pet/owner interaction is evident. This interaction, the owner response to pet behavior, and the pet response to owner actions do not seem to be greatly altered in the office setting.

Office settings do seem to have some effect on canine aggressive behavior. Dogs that are territorial at home may not exhibit this behavior or may show a diminished version in the office. Excessively fearful dogs may be more so in the veterinary setting. Dominant aggressive dogs seem to act the same or they may behave better in the office setting. Overall, the changes do not appear to be problematic nor do they affect diagnosis and treatment.

Most cats adjust to the office when there for long periods. On occasion, cats seems very frightened but these may be cats that hide in their own home when unfamiliar people are present.

For house-soiling cats and dogs, because visualization of the environment is not possible, maps of the household with illustrations of soiled areas are requested from the owner. These renditions are usually adequate.

One distinct advantage that I have found in office-based visits is realized in the setting up of training scenarios to teach owners counterconditioning and desensitization techniques. The presence of office staff and animals allows the creation of training situations to illustrate what owners need to do at home. Both desirable and undesirable responses can be elicited so that owners learn what to reward and what *not* to reward.

Office-based visits also can create opportunities to demonstrate behavior products available at the clinic. There are many different products on the market that can be adjuncts to therapy. Transporting these products on home visits in quantities sufficient to disburse to the client can be cumbersome. Office visits also allow for laboratory testing, which may be necessary to rule out metabolic disease or establish baseline values prior to drug therapy.

The time involved in office visits is the same as home visits, 1 to 2 hours; however, the clinician gains time by not having to travel to the home. Follow-up in my practice is accomplished by telephone at 7- to 10-day intervals. In cases with multiple problems, the most urgent of these are addressed initially, and additional office visits are scheduled as needed. If necessary, a home visit can be scheduled.

Telephone consultations also may have a place in a behavior practice. House-soiling cases in both dogs and cats can often be treated by telephone consultation but such discussions still take 1 to 2 hours. Prior to telephone consultation, a history form can be mailed or faxed to the client, who can then return it with a diagram of house-soiling locations. Aggression cases can be difficult to diagnose and treat over the telephone, although some clinicians may choose to do so. Liability is always a concern in aggressive animals; it may be increased in telephone situations or if a valid client-patient relationship is deemed not to exist. An additional method of telephone consultation involves conferring with a specialist. The referring veterinarian receives a history form and meets with the client to fill it out. The veterinarian then consults over the telephone with the behavior specialist, who aids in establishing a diagnosis and treatment plan, which the referring veterinarian oversees.

In my practice, all clients sign a release form prior to treatment. The release form covers potential surgery, drug therapy, and behavioral treatment. Owners leave with written instructions, appropriate handouts detailing treatment, and any behavioral products that may be appropriate. Instructions are written on carbonless forms; the original form is given to the owner, and the duplicate is placed in the patient file. In aggression cases, consultations are audiotaped, and the tape is filed with the patient's records.

Interview Forms

Victoria L. Voith, DVM, PhD

These questionnaire and interview forms represent 20 years of development. They were initially designed and implemented in 1980 based on more than 10 years of experience in learning what information is required in order to solve most behavior problems and how to organize questions in order to obtain that data. Minor adjustments were made to the forms periodically between 1980 and 1988 to accommodate specific research interests and to enhance follow up. The questionnaires also are formatted to facilitate data tabulation for publication and research purposes. Asterisked items on the forms are research questions that periodically are replaced with other research questions. Some questions currently on the forms are research queries that were modified and permanently incorporated. For example, "Had the owner considered euthanasia of the pet prior to coming to the behavior clinic?" was initially a research question that ultimately proved to be a valuable introduction to subsequent questions about dog behaviors that had not surfaced during other parts of the interview. The question also uncovered previously unvoiced owner concerns about keeping a particular pet.

The forms primarily are designed to collect information that assists in arriving at differential diagnoses, advising the owner regarding management, and developing a treatment plan. As lengthy as the forms undoubtedly appear to someone unfamiliar with behavioral history taking, additional information is still needed to confirm a diagnosis and to develop a definite management and treatment plan for the animal. For example, if the initial description of the problem is suggestive of dominance aggression, a clinician will want to pursue questions similar to those on the *Dominance Aggression Profile* form. Answers to questions on this profile, as well as the cascade of questions and answers that follow, leads to either supporting or dismissing a dominance aggression diagnosis. The details derived from specific profiles assist in the development of treatment plans. If the interview and admittance forms suggest separation anxiety or a fear response, detailed profiles of the dog's behaviors and the pertinent eliciting stimuli should be obtained for these clinical entities.

Behavior Fact Sheets A through F are for the owner to complete prior to being interviewed. Recording signalment and historical information in advance saves precious time during the interview and assessment session. They also, somewhat, assist the owner in organizing his or her thoughts regarding the behavior problem and facilitates the interview.

Behavior Fact Sheets I through VII are completed during the interview. The recorder is the person filling out the form. Some of the critical questions on the admittance forms are repeated during the interview. This is to confirm that these data are correct and complete and to serve as a springboard to related questions. A clinician should not restrict questions and observations to those on these forms. Every case is different and the unique variables of each case need to be identified. Skipping questions can result in failure to uncover behaviors that are not directly explored based on the primary complaint. Such behaviors may be important to address.

Animal Behavior Clinic Admittance Questionnaire (All Animals at Time of First Visit)

Victoria L. Voith, DVM, PhD, 1980

Please fill out the following questionnaire for each animal you are bringing to the behavior clinic. The information will be kept confidential. Thank you.

Date ____________________

Please fill in the following information about your pet.

A. Name ______________________________

B. Breed ______________________________

C. Age______________________________

D. Sex ______________________________

E. Spayed or castrated? ____________________

F. Color ______________________________

G. Weight ______________________________

How old was this pet when you acquired him/her?

How long have you owned this pet?

Who is accompanying the pet to the behavior clinic today?

Has your pet had, or does it now have, any medical problems that you are aware of? Please describe.

******

BEHAVIOR FACT SHEET A

THE FAMILY: Please describe all the people living in the household now:

Sex	*Age*	*Relationship (husband, wife, son etc.)*	*Occupation*

Has the household changed since the pet was acquired?
yes ____ no ____
If so, how has household changed?

How many times have you moved with this pet since acquiring it? _________
List the previous owners of this pet.

Animals: List all of the animals in the household in the order that they were acquired:

Name	*Breed*	*Sex*	*Age Acquired*	*Age Now*

******

BEHAVIOR FACT SHEET B

From which of the following sources did you acquire this pet?

01. SPCA or similar agency ______
02. Pet shop ______
03. Professional breeder/kennel ______
04. Advertisement of nonprofessional (home environment) ______
05. Your veterinarian ______
06. Friend, neighbor, or relative ______
07. Found somewhere, such as in the park or street ______
08. From a litter of a pet already present in your home ______
09. Other (please describe briefly below) ______________
____

Please indicate which of the following best describes the area in which you live.
1. Urban (city) ____ 2. Suburban ____ 3. Rural ____

Which of the following best describes your home?
1. Efficiency/studio apartment
2. 1–2 bedroom apartment
3. 3 or more bedroom apartment
4. Attached house - twin - duplex
5. Attached house - 3 or more units
6. Single home
7. Trailer
8. Other (please describe)

*Why did you select this particular pet out of the litter or from the animals available?

*Do you know how many males and females were in this pet's litter? yes ____ no ____

*Why did you choose this particular breed?

*Has anyone in the family had this breed before?
yes ____ no ____

Have you owned pets before? yes ____ no ____

Have you owned dogs before? yes ____ no ____

As an adult? ______ As a child? ______

Have you owned cats before? yes ____ no ____

As an adult? ______ As a child? ______

******

BEHAVIOR FACT SHEET C

DAILY ROUTINE

Feeding:

canned ____ moist ____ dry ____ human food ____

supplements? ______________________________

feeding times? ______________________________

who feeds? ______________________________

appetite: normal _____ excessive _____ poor _____

water intake: normal ____ excessive ____ poor ____

Where does pet sleep? ______________________________

Where is pet during the day? ______________________________

hours indoors __________ hours outdoors __________

Exercise: (list each time of day exercised and duration of exercise)

walks ______________________________

loose in yard______________________________

tied outside______________________________

runs loose______________________________

plays games indoors______________________________

outdoors ______________________________

DEPARTURES:

Please describe, in detail, how you prepare to leave the house when the pet will be left alone. Do you ignore your pet, do you seek it out and say goodbye, do you make a fuss over it, etc.?

What does the dog do as you prepare to leave?

******

BEHAVIOR FACT SHEET D (Dogs)

TRAINING

Has the dog been to obedience school? yes ____ no ____

group lessons ______ private ______

how many lessons? ______

Age when dog started obedience school ______________

Why did you take the dog to obedience school? _________

____

Who took the dog to obedience school?_______________

How did the dog do in obedience school? _______________

Does the dog have any obedience titles? _______________

Did you consider attending obedience school worthwhile/helpful, and in what way?_______________

____

or not helpful and why?______________________________

____

Have you taken other dogs to obedience school?
yes ____ no ____
How many and what kind? ______________________________

Have you ever taken the dog to a trainer?
yes ____ no ____
Why? ______________________________

Has a trainer ever worked with the dog at your home?
yes ____ no ____

What commands does the dog know and how well?
sit ___ stay ___ lie down ___ come ___ other ___

Does the dog know any tricks? yes ____ no ____
What?

******

BEHAVIOR FACT SHEET E (Cats)

How many litter boxes do you have?_______________

Describe the litter boxes. ______________________________

What kind of litter material do you put in the box(es)?

____

Where are the litter box(es)? ______________________________

Describe, in detail, how your cat uses the litter box.

For example, does she/he scratch in the litter before

eliminating? Cover up feces? Scratch outside box?

Does your cat "scratch" on furniture? When and what?

Are the front feet declawed?

yes ____ no ____ age declawed ______

Are the back feet declawed?

yes ____ no ____ age declawed ______

Does your cat eat plants? When and what kind?

BEHAVIOR FACT SHEET F

What is the problem(s) that brings you to the behavior clinic and how much of a problem do you consider the behavior to be?

Problems	*Very Serious*	*Serious*	*Not Serious*

Comments:

Why have you kept the pet despite its behavior problems?

Have you considered euthanasia (putting your pet to sleep)?
yes ____ no ____

*Did you tell anyone that you were taking your pet to see an animal behaviorist/psychologist? yes ____ no ____

*Who did you tell and what were their reactions?

******

Animal Behavior Clinic Interview Forms

Victoria L. Voith, DVM, PhD, 1980

BEHAVIOR FACT SHEET I

Date______________________________
Recorder ____________________________
Clinician ____________________________
Species _____________________________
Breed_______________________________
Age of pet now ________ Age pet acquired________
Weight ______________ Color ______________
Neutered: Yes ________ No ________________
Sex: Male _____ Female _____
Age neutered __________ Date neutered __________
Reason for neutering _____
Any behavioral changes after neutering? ____________
Does the pet have any pre-existing, or current medical problems? ______________________________
____
Persons accompanying pet ______________________
Other people in the room ________Men _______ Women
Location of interview ______________________

THE PROBLEM
Chief Complaint(s):

Precipitating reason for visit:

BEHAVIOR FACT SHEET II

Date______________________________
Recorder ____________________________

THE PROBLEM: Record *detailed* description of events and how long ago each occurred.

Last incident:

2nd to last incident:

3rd to last incident:

Chronological development of the problem; other significant incidents:

Duration of problem: ____ days ____ months ____ years

Age of animal when it first began showing signs of the problem ______

******

BEHAVIOR FACT SHEET III

Date______________________________
Recorder ____________________________

THE PROBLEM: Frequency of occurrence of the undesirable behavior(s):
problem: ____________ daily ______ weekly ______
monthly ____________ % ______________
problem: ____________ daily ______ weekly ______
monthly ____________ % ______________
problem: ____________ daily ______ weekly ______
monthly ____________ % ______________

CORRECTIONS AND/OR MEDICAL THERAPY TO DATE AND OUTCOME:

******

BEHAVIOR FACT SHEET IV

Date______________________________
Recorder ____________________________

THE PET:
Age obtained__________________________
Source______________________________
Behavior patterns of mother and father: Ask specifically if parents engaged in behaviors similar to those of the presented animal.

Does owner know if any littermates are engaging in same behaviors?

List other pets in the household *now* ______________

Other pets that were in the household at the time this one was acquired ______________

Describe interactions between pets in the household:

How does pet react to strangers?

How does pet behave in veterinary offices and while being examined?

Has pet ever been in a boarding kennel?
yes ____ no ____

How did pet behave at boarding kennel?

Has pet even been to a groomer? yes ____ no ____
How did pet behave at the groomer's? (Use back if necessary).

******

BEHAVIOR FACT SHEET V

Date______________
Recorder ______________

PET'S DAILY ACTIVITIES
Ask client to describe, in detail, 24 hours of a typical day in the pet's life starting with where the pet is when it wakes up in the morning.

**Don't* ask the following, but if at any time during the interview either of the following occurs check *yes* and record the comment. If none of these topics is mentioned, check *no* at the end of the session. Record *who* makes the comments as well as the comment.

*a. Did owner express guilt that he/she caused the problem? yes ____ no ____

*b. Did client refer to pet as a person during interview? yes ____ no ____ (quote exactly)

******

BEHAVIOR FACT SHEET VI

Date______________
Recorder ______________

BEHAVIORAL OBSERVATIONS
The Pet:
Describe general activity level of pet and interesting behaviors engaged in over time as the interview progresses.

How did the animal interact with the owner?

How did the animal interact with students/clinician during physical and interview?

How did the animal react to strangers entering room, noises in the room, being left alone in the room, other dogs/pets in the room, etc.

The People:
Describe the general activity level and behaviors engaged in over time as the interview progresses.

How did owners interact with the pet?

How did owners interact with each other?

How did owners interact with students/clinician?

******

BEHAVIOR FACT SHEET VII

Date______________

Recorder ______________

RECOMMENDATIONS
Treatment:
Diagnosis/summary of problems:

*WERE LABORATORY SAMPLES SUBMITTED? WHAT AND WHERE? ______________

____

Outline of procedures to be practiced at subsequent office visits.

Prognosis: ______________

Was any advice given to the owner regarding getting rid of the dog?

Was euthanasia discussed/recommended? Under what circumstances?

DRUG THERAPY PRESCRIBED:

What handouts were given to the owner?

Were owners instructed to call back?
yes ____ no ____ when ____

Were owners instructed to make a follow-up appointment?
yes ____ no ____ when ____

******

PROGRESS CHECK SHEET

Date______________
Owner______________
Recorder ______________

Ask how things are going. Exactly what has the client done daily with pet—program or drug therapy? If incidents have happened, describe *exactly* what happened (as done in original history).

Dominance Aggression Profile Form

Victoria L. Voith, DVM, PhD, and Amy R. Marder, VMD, 1984

List family members or others who spend at least 4 hours a day with the dog. For each person, (1) determine if he or she ever engages in the following activities and if so (2) indicate if dog has ever threatened (i.e., lift lip, snarl, growl, snap, or bite) in the following circumstances: + = yes, engages in activity; – = activity does not occur; ✓ = if threats have ever occurred; 0 = never occurred; N/A = not applicable (e.g., if owner never uses leash or adds food to dish); DK = don't know.

Family Member	Try to take meal away	Add food to meal	Try to take away bone or steal food item	Try to take away coveted object (list what)	Walk by eating dog	Pet dog	Suddenly reach for dog	Hug dog	Groom dog	Dry dog with towel	Trim nails or try to	Disturb dog while sleeping	Disturb dog while resting (touched, pushed, etc.)	Walk by resting dog	Stare at dog or give dog an order	Try to make lie down (pull, push)	Pull back by neck	Put on leash	Catch dog	Scold dog	Punish or threaten dog	Other circumstances

LEARNING

Treating Canine Behavior Problems: Behavior Modification, Obedience, and Agility Training

Cambridge, Massachusetts
Amy Marder, VMD

Toronto, Ontario, Canada
Pamela J. Reid, PhD

A treatment program to correct canine behavior problems may include components derived from three main areas: environmental manipulation, physiologic intervention, and behavior modification.[1] In general, environmental manipulation is easily accomplished, even by the most inexperienced owners. Physiologic intervention through surgical procedures (e.g., castration) or through drug therapy also is easy to implement. Neither environmental manipulation nor physiologic intervention requires much time, motivation, or understanding from the owner. In many cases, however, behavior modification is necessary to realize long-term change in a dog's behavior. In contrast to environmental manipulation and physiologic intervention, a behavior modification program does require considerable time, motivation, and understanding on the part of the owner.

BEHAVIOR MODIFICATION

Behavior modification involves the implementation of learning principles to change a behavior (see the article on fears and phobias by Voith and Borchelt elsewhere in this book). Most behavior modification programs rely heavily on positive reinforcement and on techniques involving operant and classical conditioning. Both avoidance and actually changing the animal's response to stimuli may be involved. For example, in the treatment of a dominant-aggressive dog, an owner might be advised to first avoid provoking aggressive behavior. Avoidance both prevents the owner from being injured and averts aggressive encounters that might result in the dog asserting a dominant position over the owner.[2] Physical punishment is usually avoided in these cases, as it might cause a dog to become more aggressive and potentially dangerous. The behavior modification program for a dominant-aggressive dog also might include conditioning the dog to assume positive reactions and nonaggressive behavior with specific situations that could otherwise elicit aggressive responses. For example, if the dog displays aggressive behavior during grooming, the owner first can teach the dog to sit or stand and stay for delicious food treats. Grooming implements then can be introduced gradually (desensitization). The dog receives treats for tolerating first the sight and then the use of the implements (counterconditioning). A soft brush might be used initially. After the dog tolerates being groomed with a soft brush, firmer brushes can be introduced. The dog should receive a treat during each stroke of grooming and as a reward after the stroke.

Although behavior modification may be the most important part of a treatment program, it is the most difficult for owners to implement. Most programs are time-consuming and require an owner to understand some simple learning principles. Many owners abandon the behavior modification program early, while others fail to initiate any program. Wanting an easy quick fix, owners may change the environment, use a drug, or avoid confronting problem behaviors but actually do little to change the animal's responses to stimuli that elicit the problem. By doing so, an owner may fail to understand the cause of the behavior problem and the reason the problem is continuing or becoming worse. For example, a dog that exhibits territorial aggression when people enter the house may be handled in two ways: avoiding the problem or modifying the behavior. An owner may avoid the problem by confining the dog to the backyard every time a person visits. Because this method does little to actually change the dog's response at the door, however, if the dog is subsequently allowed in the house when people enter, the aggressive behavior is likely to recur. It may even be more severe, as territorial aggression often becomes worse with time. Only by initiating a program that modifies the dog's behavior toward people who enter the house results in a long-term change.

LEARNING PRINCIPLES AND BEHAVIOR MODIFICATION

Both professionals and owners must understand some elementary learning principles to either design or implement an effective behavior modification program. The following is a brief introduction to some frequently used behavior modification techniques. For more detailed information, readers can refer to recent textbooks on learning theory.[3,4]

Counterconditioning, *systematic desensitization*, *flooding*, and *habituation* are defined in the article on Fears and Phobias in Companion Animals. *Positive reinforcement* refers to application of a stimulus or event (called a positive reinforcer) that follows an animal's response and increases the probability that the response will recur. *Negative reinforcement* refers to removal of an unpleasant stimulus (called a negative reinforcer) following an animal's response and also increases the likelihood that the response will recur. Reinforcers can be used to modify behaviors that already exist in an animal's repertoire or to establish new ones.

Acquisition of a new behavior is facilitated by a continuous reinforcement (CRF) schedule in which each and every response is rewarded. Acquisition also is promoted when the time interval between the response and reward is short (about 0.5 seconds for most behaviors). On the other hand, maintaining a behavior is best achieved when an intermittent reinforcement (IR) schedule is used; the response is reinforced only some of the time. Intermittent reinforcement schedules are either ratio or interval. A ratio schedule means that reinforcement follows a certain number of responses. An interval schedule means that reinforcement follows the first response that occurs after a specific time interval has elapsed since the animal was previously reinforced.

Schedules are also fixed or variable. For example a fixed-ratio (FR) schedule provides reinforcement after a specified number of responses has occurred (i.e., an FR 5 schedule means reinforcement occurs after every fifth response). A fixed-interval (FI) schedule provides reinforcement for the first response that occurs following a specified time interval (i.e., an FI 5 schedule means that reinforcement occurs after a five-second time interval has elapsed since the previous reinforcement). A variable-ratio (VR) schedule provides reinforcement after a variable number of responses (i.e., a VR 5 schedule means that, on the average, every fifth response is reinforced but the actual number of responses before reinforcement may vary). A variable-interval (VI) schedule provides reinforcement for the first response that occurs following a variable time interval (i.e., a VI 5 schedule means that reinforcement occurs after the first response following a variable time interval but that, on the average, the time interval is five seconds).

In general, all types of intermittent reinforcement schedules produce stronger responses than continuous reinforcement schedules do. Likewise, variable schedules (both VR and VI) produce much higher rates of response than the fixed schedules (FR and FI) do. Behavioral histories usually reveal that dogs with problem behaviors are actually on intermittent reinforcement schedules. For example, a dog that barks to be fed at the table is likely operating on a mixed variable-interval and variable-ratio schedule. The dog barks a variable number of times and for variable lengths of time before the owner periodically gives in and feeds it. Fortunately, intermittent reinforcement schedules also can be effectively incorporated into behavior modification programs to change problem behaviors. Using food rewards, a continuous reinforcement schedule can be implemented to teach a dog new, appropriate behavior (such as sitting when people enter the home). Once the behavior has been established, the dog can be switched to an intermittent reinforcement schedule of rewards that maintain the good behavior.

Extinction occurs when reinforcement for behavior is withdrawn and, as a consequence, the animal stops the behavior. Many owners are quick to use punishment to eliminate undesirable behavior (see the article on punishment by Borchelt and Voith and the article on taste aversion conditioning by Gustavson elsewhere in this book); however, serious drawbacks may occur with punishment. Extinction can be used as an alternative. Admittedly, some inherent difficulties do exist with extinction; but generally if the owner is warned about what to expect, the technique is well accepted. For instance, typically when all reinforcement for a behavior is withdrawn, the frequency of the behavior increases dramatically (called an *extinction burst*). The burst may be accompanied by temporary appearance of displacement activities and other behaviors. If the reinforcement continues to be withheld, however, the unwanted behavior begins to decrease in frequency and eventually disappears. After a behavior has been eliminated through extinction, it may reappear if the animal is not exposed periodically to the stimulus. This is called *spontaneous recovery.*

Shaping is a procedure commonly used to establish new behaviors by reinforcing successively closer approximations to the desired response. With each successful increment, previous levels of the behavior are no longer reinforced.

Prompting and *fading* techniques also are useful for teaching new behaviors. A prompt is a stimulus that manipulates the animal into performing the desired response so it can then be reinforced. The strength of the prompt is gradually reduced in intensity (faded) until the animal performs the behavior without being prompted. For example, a dog can be taught not to take objects by using the prompt of a sudden movement toward the dog's face with the hand holding the object. The command "leave it" is given simultaneously. The dog is rewarded with a treat when it moves away from the training object. The threatening movement to the face is faded over time, and the dog is taught to stay away from the training object (or "leave it") and subsequently to stay away from other objects.

Secondary reinforcers (second order or conditioned reinforcers) are stimuli that have been associated with primary, or biologic, reinforcers and have taken on reinforcing properties of their own as a result of this association. The most common secondary reinforcer is verbal praise ("good dog!") because it often is paired with food treats and petting. If it is impossible to deliver a primary reinforcer immediately after the animal's response, a secondary reinforcer can be used to reinforce the correct response and signal the forthcoming delivery of a primary reinforcer. Periodic pairing of secondary and primary reinforcers is necessary to maintain the reinforcement properties of the secondary reinforcer.

Chaining is a procedure that relies on conditioned reinforcers to maintain a series of responses. *Backward chaining* involves conditioning the final response in a series first, conditioning the response that immediately precedes it next, and so on. This process is continued for the entire response series. The final response is always reinforced with a highly motivating primary reinforcer. Backward chaining can be used to teach a dog to sit and stay when people enter the house. First the dog is rewarded for sitting while being greeted by a visitor and then the approach of the person is introduced, followed by the person walking through the door. At each stage, only the final response of sitting while being petted is reinforced with a powerful primary reinforcer. *Forward chaining* involves working in the reverse order; the first response is initially established, then the second response, and so on. The last response in the chain is reinforced with the primary reinforcer.

A behavior modification program is easier, more enjoyable, and more likely to be implemented if the owner understands learning principles and if the program design allows the owner to be frequently rewarded with small successes. The owner is also likely to view the dog more favorably after working with it in pleasurable ways and seeing frequent, even if small, positive changes. As a result, the owner may be less likely to abandon the program and the animal.

Dr. Victoria Voith has developed a program called The Do-It-Yourself Sit-Stay Program (see boxed insert for a shortened version). This autotutorial program teaches an owner how to use positive reinforcement and shaping techniques. The program is easy and fosters immediate success for both owner and dog. For completing each step, the dog is rewarded with a delicious tidbit of food that keeps the pet interested, motivated, and happy. The program is repetitious to enable both owner and dog to proceed with few, if any, mistakes.

As owners progress through the program, they feel immediately successful and are usually eager to go further. After they have finished the program, owners may incorporate desensitization and countercondi-

DO-IT-YOURSELF SIT-STAY*

by Amy Marder, VMD

This program is designed to teach your dog to sit and stay and to make learning easy for both owners and animals. No punishment is used at the beginning—only verbal praise and food.

First a few basics are necessary. The best way to teach a new behavior is by continuous reinforcement. Thus, every repetition should initially be reinforced with food. Because you cannot always have food available, however, the food is paired with verbal praise. This pairing of food with your voice makes the latter a more powerful reinforcer. Eventually intermittent reinforcement is introduced, and the food treat is not offered every time. This approach makes the sit-stay behavior more permanent.

You should select a food treat your dog loves, preferably something that can be broken up into little pieces and is low in calories. Dog beef jerky treats, boned chicken, and cheese are reasonable choices. My dog works well for little pieces of rice cakes.

The reward must be given immediately. Timing is crucial. You should reinforce only desired behavior. If reinforcement is delayed, the dog may bark, at which point the food rewards the bark rather than the stay.

You must proceed exactly in the order of the program. The steps are designed to make learning very easy. Dogs learn by being successful. If you proceed too rapidly, your dog will make mistakes and both of you will become frustrated. If the dog makes a mistake (e.g., breaks the stay), you should sternly say "no" or "uh-uh," should not give the food reward, and should reposition the pet in the sit mode and repeat the task.

Sometimes your dog may become bored or difficult and refuse to work for you regardless of what you do. On these days, do not push yourself or your pet. Instead, end the training session on a good note with a few very easy short stays (e.g., count 5) and try again tomorrow.

You should end a session by giving your dog a release word. Most people just say "ok," but each owner can coin his or her own phrase. A food reward should not be given when the dog is released, as doing so trains the pet to stand up.

Daily practice is important. It is best if every person in the household practices with the dog. Each session need only last 10 minutes.

You should teach the sit-stay command first in the house and then outside in a fenced yard. If the yard is not fenced, the dog can be tied to a long lead for outside practice.

When rewarding the dog, only verbal praise and food should be offered. It may be hard for you to resist petting the animal, but petting can be very distracting and should be avoided for now.

TEACHING THE SIT (using prompting and fading techniques and positive primary and secondary reinforcement)

You should place the food in one closed fist and pass your hand over the end of the dog's nose, over its head, and toward its hindend as you say the word "sit." Usually when the dog's head goes up, the rear end goes down. When the dog's rear hits the ground, praise the dog immediately and give the food reward. Repeat this exercise 20 times and then omit passing your hand over the dog's head. Instead, now use the sit hand signal (an open hand held horizontally and raised); repeat this exercise 20 times, giving a food reward each time. After your dog understands the sit command, you can start teaching the pet to stay.

TEACHING THE SIT-STAY (using prompting)

You should begin with your dog in the sit position. Hide a handful of food tidbits in one hand (do not dangle the food in front of the dog; keep it hidden). With your other hand, give the stay signal (an open hand with the palm toward the dog) while saying "stay." Immediately praise the dog as soon as it completes the task successfully (e.g., after count one or after taking one step); you can say "good dog," "good boy," good girl," or whatever words of praise you prefer and then with the hand you used for signaling offer the pet a food reward. You should keep your dog in the sit position.

The Sit-Stay Exercise

Owner Task	*Dog Completed*	*No. of Times Performed*
Count one	________	________
Count two	________	________
Count three	________	________
Count five	________	________
Count seven	________	________

(continued next page)

DO-IT-YOURSELF SIT-STAY *(continued)*

Owner Task	*Dog Completed*	*No. of Times Performed*
Count ten	______	______
Count fifteen	______	______
Count twenty	______	______

Now you are ready to do some dancing!

Owner Task	*Dog Completed*	*No. of Times Performed*
One step left and return	______	______
One step right and return	______	______
One step back and return	______	______
Two steps left and return	______	______
Two steps right and return	______	______
Two steps back and return	______	______
Three steps left and return	______	______
Three steps right and return	______	______
Three steps back and return	______	______
Count ten	______	______
Two steps right, count ten	______	______
Two steps left, count ten	______	______
Two steps back, count ten	______	______
Five steps right and return	______	______
Five steps left and return	______	______
Five steps back and return	______	______
One step back, about face, return	______	______
Two steps back, about face, return	______	______
Three steps back, about face, return	______	______
Count twenty	______	______
Ten steps right and return	______	______
Ten steps left and return	______	______
Ten steps back and return	______	______
Count ten	______	______

Congratulations! Now you know how smart your dog is!

*Modified and shortened version of Do-It-Yourself Sit-Stay Program by Victoria Voith, DVM, PhD.

tioning procedures. The stimuli that provoke problem behaviors are gradually introduced; an example would be the doorbell ringing, the door being opened, a person entering the house, and the person approaching the dog. In each case, the dog continues to sit or down-stay and receives food treats. While the dog is in a nonaggressive mood, it is exposed to stimuli that had previously elicited an aggressive response. The same program can be adapted using stimuli that evoke fear or anxiety. A continuous reinforcement schedule is applied first, followed by intermittent reinforcement introduced for behavior maintenance.

OBEDIENCE CLASSES

Before Clinical Animal Behavior became a recognized specialty, obedience classes were routinely recommended for every dog with a behavior problem. Obedience classes, especially puppy classes, probably do help prevent development of some behavior problems and also help owners get some control over their dogs. Once a serious behavior problem has developed, however, routine obedience classes may contribute little to change problem behaviors.[5] Most classes use combinations of positive reinforcement, negative reinforcement, and punishment techniques to teach commands. Many instructors, however, continue to rely extensively on punishment and have minimal knowledge of the proper application of positive reinforcement or learning principles. Thus, the classes may become unenjoyable for both owners and their pets.[6]

The most common commands taught in an obedience class are sit, down, stay, heel, and come. The commands are rarely applied, however, in situations where a behavior problem occurs. For example, commanding a dog to down-stay for a long period may prevent a dog from being destructive when an owner is present but will do very little to stop destruction that occurs when an owner is absent (e.g., separation anxiety). Similarly, obedience training is often recommended to prevent dogs from engaging in behaviors without addressing the underlying motivation for the behaviors. For instance, teaching a dog to stay might prevent a fearful dog from biting visitors as they enter the house (just as a gate would) but does nothing to reduce the dog's fear. In contrast, a good behavior modification program not only teaches an owner how to control a dog through commands but uses counterconditioning to change the dog's response to specific stimuli related to the problem.

AGILITY TRAINING

Agility training is a new sport in which dogs are taught by their owners to navigate obstacles. It teaches an owner how to apply learning principles effectively and may maintain an owner's interest in continuing the behavior modification program. Typically, positive reinforcement in the form of food and play is used in the training process. Because the dogs must be taught to negotiate the obstacles gradually, owners learn how to implement and then understand the learning principles involved in most behavior modification programs.

Shaping, chaining, prompting, and fading are commonly used in agility training. For example, teaching the seesaw, one of the most difficult obstacles, requires the implementation of several principles. As most dogs are afraid to walk on narrow, unsteady surfaces, this behavior modification must be gradually shaped. The dog is first taught to walk on a narrow plank by using positive reinforcement. Next, the dog is introduced to a narrow inclined plank, again giving positive reinforcement. After the dog successfully walks the inclined plank, the dog is taught to walk up the plank until it tips and then to tip the plank unassisted. Finally, by using the forward chaining technique, the dog is rewarded for walking up the seesaw, tipping it, and then walking off the seesaw after it touches the ground.

Agility training also involves teaching a dog to run through a tunnel, burrow through a fabric tunnel, walk on an elevated narrow plank, climb and descend a six-foot A-frame, jump over and through a variety of obstacles, weave through poles in slalom fashion, and do a short down-stay on a table. After the dog learns to maneuver each obstacle, both owner and dog walk or run through a course that comprises numerous obstacles. At this point, backward chaining can be used. The dog first should be rewarded for completing one obstacle, then for completing a series of two obstacles, for completing a series of three, and so on until the dog can successfully complete approximately 20 obstacles between reinforcements. Most dogs and owners inherently enjoy running the course. The dog not only receives deliberate positive reinforcement from the owner but is being self-reinforced.

Agility training also is a useful method of counterconditioning a variety of problems. The method can be helpful with dogs that are aggressive to other dogs, dogs that are afraid of people, and dominant-aggressive dogs that resist being handled and manipulated. Agility training also has an additional benefit, as it teaches dogs to pay attention to their owners. Many owners have difficulty with their dogs paying attention during obedience classes or when implementing a behavior modification program. During agility training, however, the dogs receive reinforcement only for following owner directions on the course; thus, learning to pay attention becomes an easier task for the animals.

SUMMARY

Several owner-clients have attended obedience classes, implemented a behavior modification program, and enrolled in agility training. In one case, the owner of a dominant-aggressive wirehaired fox terrier that had bitten its owner twice, was aggressive to dogs, and was afraid of people had implemented all three. The owner remarked that the obedience classes and behavior modification program helped resolve the dominance aggression, while aggression to other dogs and fear aggression to people were helped more by agility training classes. The owner felt that the agility tasks also had changed his relationship with his dog. Although the dog continued to growl at the owner on occasion, the owner no longer viewed the pet as a major behavior problem but instead as a companion he could have fun with and enjoy.

REFERENCES

1. Voith VL: Clinical animal behavior. *Calif Vet* June:21–25, 1979.
2. Voith VL: Aggressive behavior and dominance. *Canine Pract* 4:8–15, 1977.
3. Domjan M: *Domjan and Burkhard's The Principles of Learning and Behavior*, ed 3. Pacific Grove, CA, Brooks/Cole Publishing Co, 1993.
4. Schwartz B, Robbins SJ: *Psychology of Learning and Behavior*, ed 4. New York, WW Norton & Co, 1995.
5. Voith VL, Wright JC, Danneman PJ: Is there a relationship between canine behavior problems and spoiling activities, anthropomorphism, and obedience training? *Applied Anim Behav Sci* 34:263–272, 1992.
6. Myles S: Advances in companion animal behavior. Trainers and chokers: How dog trainers affect behavior problems in dogs. *Vet Clin North Am [Small Anim Pract]* 21:239–246, 1991.

Learning

Animal Behavior Services
Toronto, Ontario, Canada
Pamela J. Reid, PhD

Animal Behavior Consultants, Inc.
Brooklyn, New York
Peter L. Borchelt, PhD

Learning can be defined as an enduring change in behavior as a result of prior experience. This process enables animals to adjust to an ever-changing environment.[1] Analyzing the effects of learning in animals is challenging because learning must be inferred from the animal's behavior. For any particular behavior, there may be many ways by which it came to be established and many ways by which it is now maintained. These different ways are by no means incompatible; in any given situation, learning occurs via several processes simultaneously.

The modification of behavior problems typically involves the intelligent manipulation of events, or stimuli, in the animal's world so that sequences of problematic behaviors are unlikely to recur. A thorough understanding of the principles of learning is necessary for accurately analyzing situations and developing effective strategies for addressing behavior problems. This article reviews learning theory for those interested in developing their skills as agents of behavioral change.

HABITUATION AND SENSITIZATION

Animals are constantly learning about the stimuli in their environment. This is learning in its simplest form. Animals learn to attend to stimuli that are important to them and to ignore stimuli that are not. Typically, animals become alert to and orient themselves toward anything novel in the environment; but with repeated exposure to the stimulus, even in the absence of any consequences, they either begin to ignore the stimulus or attend to it more. A decrease in responsiveness produced by repeated stimulation is *habituation*, whereas an increase in responsiveness is *sensitization*. Habituation and sensitization are fundamental forms of behavior change that occur in response to repeated presentations of most stimuli and reflect how animals sort out what to ignore and what to attend to in the world.

Habituation and sensitization apparently occur simultaneously in any given situation; but typically, one process will override the other, thus determining whether responsiveness will decrease or increase. The behavioral changes that follow repeated presentations of a stimulus always reflect the combined actions of habituation and sensitization. If a stimulus arouses the animal, repeated presentations elicit progressively stronger responses; whereas if a stimulus fails to arouse the animal, repeated presentations elicit progressively weaker responses.

There are two types of habituation: one that takes place within a few presentations of a stimulus (over a few minutes) and one that takes much longer. For instance, an animal is startled by a loud sound but becomes habituated after hearing it a few times. Hours or days later, however, the sound

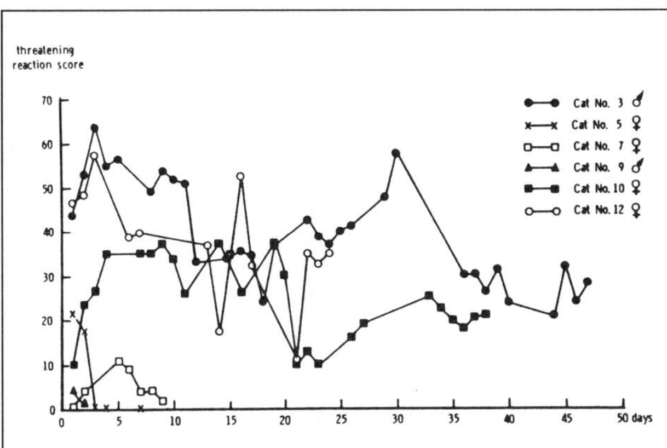

Figure 1—Habituation curves of the threatening display of six adult cats individually confronted by a dog. Each point represents the number of times each component of the response occurred during one session. The highest possible score is 110 (all 11 components in each of 10 confrontations). (Courtesy of *Acta Neurobiologiae Experimentalis.*)

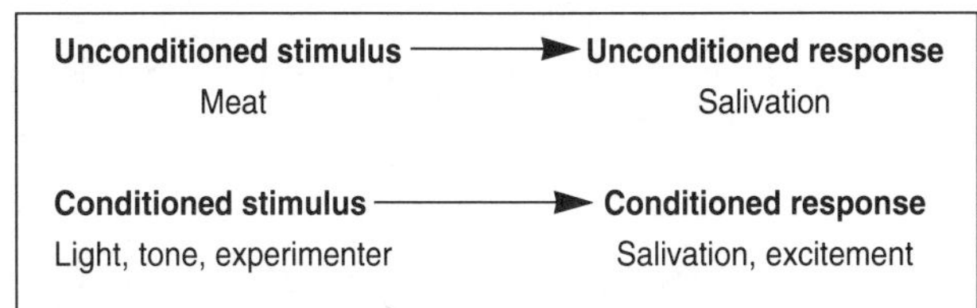

Figure 2—Classically conditioning a dog to salivate to neutral stimuli.

will once again elicit the startle response. The return of responsiveness is called *spontaneous recovery.* Only after many sessions does long-term habituation with no spontaneous recovery occur.

Zbrozyna illustrated the interaction of these two processes in an experiment in which adult cats were repeatedly exposed to a dog.[2] Six adult cats, housed individually, were each exposed to a docile dog that was brought to within a few inches of the cat's home cage for about 1 minute 10 times per day at about 1-minute intervals. In such confrontations, cats typically react with threat displays that consist of the following, in order of increasing intensity:

1. Pupillary dilation
2. Pricking up of the ears
3. Raising of the tail
4. Piloerection
5. Arching of the back
6. Crouching
7. Growling
8. Hissing
9. Spitting
10. Pawing or swatting with extended claws
11. Springing forward.

The occurrence of each of these 11 responses was added to yield a "threatening reaction score" for each confrontation session. Figure 1 depicts the scores for each cat over time and contains a wealth of information about individual differences, the effect of initial level of response, and the interaction of habituation and sensitization.

Cats 5 and 9 showed initially low-level responses and became habituated quickly. Cat 9 showed a lower initial level than did Cat 5 and became habituated quicker. Cat 7 did not overtly respond at all to the first confrontation, became sensitized during the next two sessions, and then became habituated over five more sessions. Cat 10 showed a low initial level of threat response but then became steadily sensitized over the next three sessions to a moderate level, which was maintained for nine sessions before decreasing somewhat. Cats 3 and 12 started off with high levels of threat response that then sensitized for a few sessions before beginning to decline, but only over many sessions and with great variation. Changing the situation in some way, such as making a sudden noise or introducing the dog in a different room or cage, would have led to a major increase in threat responses. This is called dishabituation.

Sensitization effects are less stimulus-specific in that any change in stimulation could lead to sensitization. If the stimulus change led to a high level of arousal, the animal could become responsive to a wider range of stimuli or to less-intense stimuli. A good example is a cat that has been aroused by the sudden appearance of a dog and then suddenly becomes aggressive toward its owner.

Habituation and sensitization are both influenced by the intensity and frequency of the eliciting stimulus. More-intense stimuli are more likely to activate sensitization processes, whereas weaker stimulation usually leads to habituation. Rapid repetitions of a stimulus are more likely to enhance whatever process is predominant. Despite these generalities, it is difficult to predict whether a specific response to a particular stimulus will escalate or diminish.

Species, breed, and individual differences certainly influence habituation and sensitization, and various behaviors may also be enhanced or diminished in different ways.[3] In the Zbrozyna study, the previous experience of the adult cats was unknown. Differences in previous experiences with dogs or basic temperament differences may have influenced habituation. If the experimental conditions had been different—for example, a different dog, a different duration of confrontation, a different interval between confrontations within a session (interstimulus interval), or a different

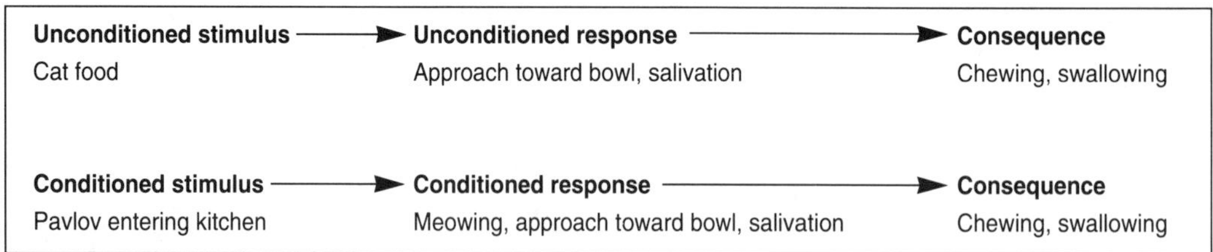

Figure 3—Pavlov's cat salivates when Pavlov enters the kitchen because the cat has learned to associate Pavlov's presence in the kitchen with feeding (classical conditioning). Through instrumental or operant conditioning, the cat has learned to meow when Pavlov enters the kitchen—such meowing often elicits cat-feeding.

number of confrontations per session—then different levels of initial response and degree of habituation and sensitization might be expected.

When one is using a habituation technique for an animal behavior problem, it generally helps to be familiar with the particular animal and the relevant stimuli (eliciting and concurrent) and then test several variables (e.g., stimulus intensity, interstimulus interval, and number of stimuli per session) before deciding on a treatment program.

DESENSITIZATION

Animals with extreme reactions (usually fear) to specific stimuli can be desensitized so that the reaction diminishes. A desensitization procedure serves to encourage habituation over sensitization. It involves repeatedly presenting the stimulus at such a low level that the animal's state of arousal is kept low and habituation predominates. Gradually, the intensity of the stimulus is increased while the animal's state of arousal is maintained at a low level and the response continues to decline. Eventually, the animal can tolerate the stimulus at the original intensity without displaying the problematic behavioral or emotional responses.

Attempts to desensitize an animal demand a delicate balancing of habituation and sensitization. Too low a stimulus level or too gradual an increase in stimulation wastes time. Too high a stimulus level or too rapid an increase in stimulation causes sensitization. For instance, desensitizing Cat 3 from the Zbrozyna study would require that the dog (or perhaps even a smaller or quieter dog) be initially presented at a greater distance from the cage and for a shorter length of time or more exposures per day. The goal would be to ensure that some small degree of habituation would occur rapidly in the first session. At the beginning of the second session, spontaneous recovery would be expected; but the animal's response would show habituation more quickly. Only after little or no spontaneous recovery occurs should the stimulus intensity be increased slightly.

CONDITIONING

Two forms of conditioning are distinct in theory but in reality are completely interdependent. *Classical conditioning* involves learning about relations between two or more stimuli in the environment. Events usually occur somewhat predictably or consistently, and animals can learn this through classical conditioning. Figure 2 illustrates the standard paradigm, which is derived from Pavlov's prototypical study of dogs that learned to associate neutral stimuli with food. This situation contains the basic elements of classical conditioning: a meaningless stimulus that initially elicits no response signals the occurrence of some meaningful stimulus that does elicit a response.

The animal learns that the first stimulus predicts the second and begins to show the same or a similar response to both stimuli. Everyday examples abound: the dog that dances with anticipation when the owner picks up the leash or that runs to the door when it hears the family car pull into the driveway; the cat that runs to the food dish and meows when someone enters the kitchen or that runs under the bed when the doorbell rings; the owner who jumps up to walk the dog when it gets restless and scratches at the door. In each case, there is a predictable relation between two events and the animal responds to the first in anticipation of the second. The animal learns a *stimulus–stimulus* association.

Notice that these examples differ somewhat from Pavlov's study. Pavlov studied the conditioning of "simple reflexes," such as salivation; whereas everyday examples involve more complicated behaviors: locomotion, exploration, greeting, feeding, and exhibiting fear. Although all of the examples involve learning about stimulus relations, the sequences of behavior also have consequences or results. The dog runs to the door and is greeted by the family coming home; the owner walks the dog and avoids a house-soiling accident in the living room.

Learning about a situation in which events directly *result* from an individual's own behavior is called *instrumental* or *operant conditioning*. Suppose Pavlov had a cat that ran to the food dish and meowed when Pavlov walked into his kitchen. The cat has learned the association between Pavlov's presence in the kitchen and feeding; but the cat has also learned that meowing often leads to feeding (Figure 3).

Instrumental learning involves learning *stimulus–response* and *response–consequence* relations. In other words, in instrumental conditioning, the animal learns a classical association between (1) a stimulus and (2) a relation between a response and a consequence, as though the response–consequence relation were a separate stimulus (*stimulus–[response-consequence]*). In any situation, learning involves some stimulus–stimulus relationships (classical) and some stimulus–response–reward relationships (instrumental). Thus, it is pointless to ask whether classical or instrumental conditioning is more important because both play a role.

Breland and Breland[4] provided probably the best evidence of classical and instrumental conditioning both contributing to behavior when they demonstrated that the two conditioning mechanisms are sometimes at odds with each other. They tried to use positive reinforcement to train various animals to perform behaviors; but occasionally, their efforts resulted in dramatic behavior patterns that were inconsistent with what they were trying to elicit.

In one case, food reinforcement was used to train pigs to pick up large wooden coins and place them in a "piggy bank." The pigs quickly learned the response (a response–reward association) but then began to have difficulty dropping the coins into the bank. They would repeatedly drop the coin, root it, pick it up and toss it in the air, and so on. This behavior became more prevalent as time went on, even though the behavior delayed the food reward. The instrumental response of dropping the coin was disrupted because the coins became associated with food reinforcement and thus elicited food-related behaviors.

The classically conditioned response of rooting was incompatible with the instrumental response, so the stimulus–stimulus association won out over the stimulus–(response–consequence) association. The Brelands termed this phenomenon *instinctive drift*. In the complex world of pet behavior, *stimulus–stimulus*, *stimulus–response*, and *response–consequence* associations are conditioned in any situation, and each affects how easily learning takes place and how easily behavior can be modified.

In instrumental conditioning, there are four types of response–consequence relations or contingencies, two of which *increase* the likelihood that the response will be repeated (reinforcement) and two of which *decrease* the likelihood that the response will be repeated (punishment) (Figure 4). *Positive reinforcement* occurs when a behavior produces a pleasant consequence. Typical positive reinforcers are food, water, petting, praise, and play. *Positive punishment* occurs when a behavior produces an aversive consequence. Typical punishments include a startling sound (e.g., "No!"), a swat, a squirt of water, a jerk on the leash, or any other stimulus that the animal finds aversive.

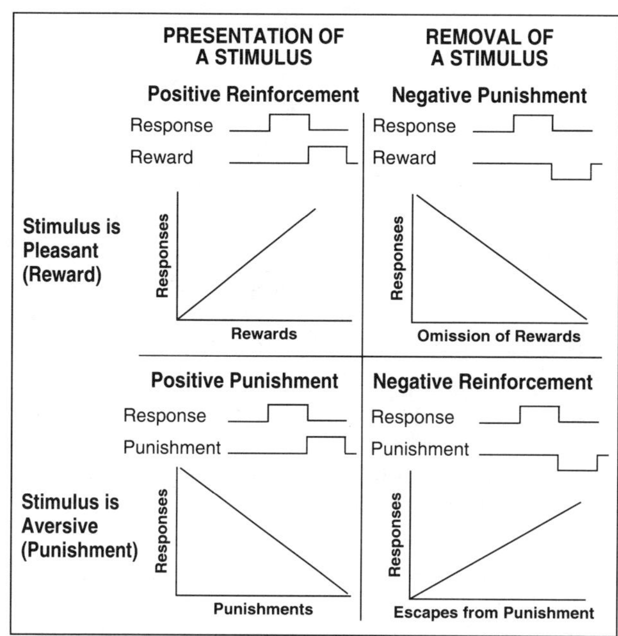

Figure 4—The four types of response–consequence relations.

Negative reinforcement occurs when a behavior results in the prevention or termination of an unpleasant stimulus that was imminent or already present. A typical example is when an animal learns to escape and even avoid an aversive stimulus. For example, after a few squirts of water for jumping up on the couch, a cat might first learn to run away when the squirt gun is picked up and then learn to avoid the couch when the owner is present.

Finally, *negative punishment* occurs when a behavior results in the prevention or termination of a pleasant stimulus that was imminent or already present. For example, a puppy that playfully nips the owner's hand might be negatively punished by cessation of all hand and toy movements or even brief removal of the toy. Numerous short play sessions that cease whenever nipping occurs might reduce future nipping. Further discussion and examples of reinforcement and punishment are provided elsewhere in this book in the article on punishment and in the article by Marder and Reid.

STIMULUS CONTROL OF BEHAVIOR

Conditioned behavior comes to be controlled by the particular stimuli that are present when a response is reinforced or when an unconditioned stimulus is presented. In the language of instrumental conditioning, the stimuli that control responding are called *discriminative stimuli*; they signal that responding will result in reinforcement. In the language of classical conditioning, the stimuli that control responding are called *conditioned stimuli.*

Stimulus control, in essence, means that the animal responds differentially in the presence of different stimuli. Stimulus control has two component processes: *discrimination* and *generalization.* Discrimination between stimuli enables the behavior to come under the control of the appropriate stimulus. A well-trained dog, for example, has learned to discriminate between the word "fetch" and all the other words its trainer uses for commands. Conversely, stimulus generalization exists whenever an animal fails to respond differently to various stimuli. A novice dog will retrieve a tossed object, regardless of the command given by the trainer.

Programs for changing problem behavior must consider the process of generalization because what is learned during treatment should generalize to everyday situations. To facilitate generalization, the treatment situation should be as similar as possible to the natural environment. Generalization is also increased if the treatment procedures are conducted in varied settings. As the behavior change is conditioned to one setting, the animal should be switched to a new setting for further conditioning. Using a large variety of stimuli is also useful. For instance, when attempting to eliminate a cat's fearful response to unfamiliar people, the cat should be exposed to a sufficient number of different types of people to encourage generalization. Generalization is also facilitated if new responses are conditioned to stimuli that are common across different situations. The fearful cat, for example, could be conditioned to approach an outstretched hand, which is a stimulus that any person can replicate.

Often, discrimination between stimuli rather than generalization is desired. One stimulus (the *stimulus*$^+$) signals reinforcement; another stimulus or stimuli (the *stimulus*$^-$) signals nonreinforcement. With sufficient exposure, the animal will come to respond whenever the *stimulus*$^+$ is present and withhold responding whenever the *stimulus*$^-$ is present. Likewise, a *conditioned stimulus*$^+$ can be used to signal the unconditioned stimulus and a *conditioned stimulus*$^-$ can be used to signal no unconditioned stimulus. Dog trainers call discrimination training "proofing." The trainer might set the dog up for a retrieve, toss the object, and then say some irrelevant word to teach the dog to attend specifically to the word and not to other stimuli that are present. Discrimination learning is a process that influences all behavior problems, for example, how the animal learns which surfaces and locations to use for elimination or what events predict its being left alone.

Stimuli in the natural environment tend to be complex. Accordingly, responding tends to be controlled by specific features or elements of such *compound* stimuli. The elements of a visual stimulus, for instance, include various aspects of the stimulus' appearance, such as shape, color, brightness, and so on. Although all components of a stimulus are present at the time of the reinforced response, and all may gain some control over the behavior, all elements do not gain *equal* control. Some elements exert a much stronger influence on the response than others. Whenever a compound stimulus is used, the presence of an element that is easy to condition may impair learning about other elements. This is called *overshadowing.* For instance, a dog that is trained to sit using both a verbal command and a tactile prompt (lifting the leash) is often controlled more by the tactile than the auditory element. If the dog is later tested with the verbal command only, the response is not as likely to occur as when the physical prompt is also incorporated. This is why dogs often fail to sit unless the owner lifts up the leash. During conditioning, the auditory element was overshadowed by the more salient tactile element.

Variables that Influence Conditioning

Timing

The relative timing of stimuli can have profound effects on the rate and amount of learning. In classical conditioning, the time between the occurrence of the conditioned stimulus and the occurrence of the unconditioned stimulus is called the *interstimulus interval.* For most situations, the best procedure is to present the conditioned stimulus a bit before the unconditioned stimulus but with some overlap. This is *short-delay* presentation. *Simultaneous* presentation, in which the conditioned and unconditioned stimuli occur at the same time, is less likely to lead to the animal learning anything about the conditioned stimulus. *Backward* presentation, in which the conditioned stimulus is presented shortly after the unconditioned stimulus, is almost never effective.

Teaching a dog to eliminate on command is a practical application of classical conditioning in which timing the conditioned stimulus in relation to the unconditioned stimuli is critical. The unconditioned stimuli consist of internal and external events that may be difficult for the owner to pinpoint. The conditioned stimulus is a spoken signal, such as "Do your business" or "Hurry up." If the conditioned stimulus is not delivered until the dog is already in the act of eliminating, the presentation is backward and the dog will not learn the association between the signal and the feelings that precede elimination.

If, however, the conditioned stimulus is delivered during the preceding stage, when the dog is sniffing and circling, and is continued until the dog begins eliminating, the conditioned stimulus will have been presented prior to and during the onset of elimination (the unconditioned stimulus). This would be termed *long-delay* presentation. Of course, adding praise ("Good dog!"), a treat, or a playful petting immediately after the elimination adds an instrumental component to the learning and makes it more likely that the dog will learn what to do.

The importance of timing the reinforcer in instrumental conditioning is equally critical for effective learning. In most situations, learning is best when there is little or no delay between the response and the consequence.

Predictability

The predictability of events is extremely important in establishing classical and instrumental conditioning. In classical conditioning, maximum conditioning results when the conditioned stimulus always predicts the unconditioned stimulus. If picking up the leash (the conditioned stimulus) is always followed by a walk (the unconditioned stimulus), the dog will display great excitement (the conditioned response) whenever the leash is picked up. As the correlation between conditioned stimulus and unconditioned stimulus declines, so does the strength of the association.

Pairing the conditioned stimulus with the unconditioned stimulus but also presenting the conditioned stimulus by itself, without the unconditioned stimulus, also leads to a weaker association between the conditioned and unconditioned stimuli because the conditioned stimulus is not a reliable predictor of the unconditioned stimulus. For instance, if picking up the leash is always followed by a walk but the dog is taken for a walk just as frequently without the leash, the dog may show some excitement when the leash is picked up but its excitement will be less intense than if the dog were always taken for a walk on the leash. If the conditioned stimulus and the unconditioned stimulus are sometimes paired but just as often presented independently, there is virtually no learning. If the dog often goes for a walk on the leash but just as often goes without the leash, and if the leash is also frequently picked up to get at the umbrella hanging underneath, the dog will likely not respond at all when the leash is picked up.

If the conditioned stimulus and the unconditioned stimulus are presented but the conditioned stimulus is *never* followed by the unconditioned stimulus, *conditioned inhibition* results. In other words, the animal learns the negative correlation between the two stimuli. Imagine that the dog always goes for a walk without the leash; furthermore, the dog never goes out at the same time as the puppy in the household, which is always taken out on the leash. The dog quickly learns that when the leash is picked up, there is no chance for a walk at that time.

Not only must a stimulus predict something about the occurrence of another stimulus, but the information must also be valuable or relevant to the animal. This is best illustrated through a phenomenon called *blocking*. Blocking occurs when a new stimulus is paired with an already established conditioned stimulus. Because the animal has already learned that the conditioned stimulus signals the occurrence of the unconditioned stimulus, adding a second stimulus provides no additional information for the animal. Consequently, the new stimulus is "blocked" from learning; the animal does not pay attention to the second stimulus.

Blocking is similar to, yet distinct from, overshadowing. Both blocking and overshadowing are very important concepts because in the real world of pet behavior and behavior problems, the stimuli are complex and multiple stimuli are often associated with specific responses. It is impossible to predict which stimuli will come to control the behavior. During learning, some stimuli are overshadowed and some are blocked; even though these stimuli are obvious to an owner or trainer, they play no role in learning and have no information value to the animal.

For instance, people generally try to teach animals by using words as signals, but the animal will likely have a long history of paying attention to nonverbal signals from the owner or trainer. Thus, the verbal signals may be blocked.

Novelty

The novelty of the stimuli has a major impact on learning because the animal must first notice the stimuli. Animals are always more alert to novel stimuli. If a highly familiar conditioned stimulus is used, the rate of learning is dramatically slower than if a novel stimulus is used. If an animal has already been exposed to a stimulus numerous times before this stimulus is paired with an unconditioned stimulus or with a response–reinforcer relation, it is much more difficult for the animal to learn the association than if the stimulus had never been encountered before. This phenomenon is called *conditioned stimulus–preexposure effect* or *latent inhibition* in classical conditioning or *learned irrelevance* in instrumental conditioning.

Everyone knows how quickly a cat can learn the association between the sound of a can opener and dinnertime. If, however, a cat has lived in a home where the can opener was used frequently but never for cat food (because the cat had always been fed kibble), the cat will take noticeably longer to learn that the can opener noise signals dinner (after the owners start providing canned cat food) than if the cat had never heard a can opener before.

Intensity

An association between two stimuli or between a stimulus and a response–reinforcer relation is learned more rapidly and more extensively when the stimuli are more intense. An intense stimulus is simply easier for the animal to notice; also, the animal is more likely to notice relations among intense events. In addition, responses to intense stimuli are less likely to diminish through habituation.

Relevance

The relevance of stimuli to each other affects learning because certain types of stimuli are more easily associated than others. It is as though some stimuli are more relevant to, or belong better with, some other stimuli. This is termed *belongingness.*[5]

For instance, it is easy to teach an animal to escape an aversive stimulus by running but exceedingly difficult to each the animal to avoid the stimulus by manipulating an object. Running is a component of normal predator-avoidance behavior; manipulating an object is not. It is as if different reinforcers belong to different behavior systems; if the instrumental response does not belong in the behavior system of the reinforcer, it is difficult to condition the response.

Controllability

Animals that are rewarded noncontingently (irrespective of their behavior) often exhibit responses called *superstitious behavior. Adventitious reinforcement* causes the accidental strengthening of a response that coincided with reinforcement. For example, pets occasionally acquire seemingly bizarre behaviors because they are inadvertently rewarded with attention from owners.

Whereas noncontingent positive reinforcement often leads to superstitious conditioning, noncontingent and therefore uncontrollable aversive stimulation leads to a phenomenon called *learned helplessness.* The animal apparently learns that the aversive stimulation cannot be escaped; the result is a general suppression of behavior. Once an animal has been exposed to uncontrollable aversive stimulation, it is exceedingly difficult for the animal to learn, during subsequent conditioning, that it can escape or avoid the aversive event.

EXTINCTION

Eliminating a learned association is accomplished through *extinction.* For classical conditioning, extinction involves repeatedly presenting the conditioned stimulus by itself, without the unconditioned stimulus and in the absence of any kind of signal relation with the unconditioned stimulus. This process results in a gradual decline of the conditioned response. For instrumental conditioning, extinction is the nonreinforcement of a response that was previously reinforced. Withdrawing reinforcement gradually reduces the likelihood of the instrumental response as the animal learns that its behavior no longer leads to reinforcement. During extinction, the animal actually learns the new *lack* of relation between the conditioned stimulus and the unconditioned stimulus or between the stimulus and the response–reinforcer relation. (See the article by Marder and Reid for more information.)

AVERSIVE CONTROL OF BEHAVIOR

Instrumental conditioning procedures that involve aversive stimulation take the form of either positive punishment (presenting an unpleasant stimulus contingent upon a response) or negative reinforcement (removing an unpleasant stimulus or the threat of an unpleasant stimulus contingent upon a response). Punishment serves to decrease the likelihood that the response will occur, whereas negative reinforcement serves to increase the likelihood that the response will occur. For a comprehensive discussion of punishment, refer to the article on punishment by Voith and Borchelt.

Escape/Avoidance Conditioning

Negative reinforcement involves a negative contingency between a response and an aversive event and is called *escape/avoidance conditioning.* There are two types of escape/avoidance procedures. In a *signaled* procedure, a warning signal is followed by an aversive stimulus. At any point during the aversive event, the animal may perform the instrumental response to terminate the aversive stimulus. If the response is performed during the warning signal, before the aversive event, the stimulus is avoided altogether. During early stages of learning, the animal escapes the stimulus. After a few repetitions, however, the animal learns to avoid the aversive stimulus.

For example, training a dog not to pull on the leash can be accomplished through signaled escape/avoidance conditioning. The owner gives the dog a warning that it is about to reach the end of the leash. If the dog continues to accelerate, the owner jerks on the leash in such a way that it is aversive to the dog. The dog can escape the aversive stimulation by decelerating. If, however, the dog decelerates in response to the early warning, it can avoid the aversive stimulation from the leash altogether. Eventually, if the stimulation is sufficiently averse, the dog will avoid pulling on the leash by slowing down in response to the warning signal.

In an *unsignaled* procedure, there is no discrete warning signal and the animal must perform the instrumental response periodically in order to avoid the aversive stimulus. Contextual cues become the warning signals. More often, training a dog not to pull on the leash is established through an unsignaled procedure. The dog comes into contact with the aversive stimulation from the owner jerking on the leash whenever it moves out of a defined range around the owner. In order to avoid the aversive stimulation, the dog periodically slows down to ensure that it does not stray out of the acceptable range. (Note that a dog can be effectively trained not to pull on a leash through the use of a halter, which does not appear to be aversive.)

Escape/avoidance responses underlie many pet behavior problems. The pet that flees in response to a real or imagined threat is performing an escape/avoidance response. The dog that reacts aggressively toward a feared stimulus is performing an escape response. Escape/avoidance behavior is extremely reliable. Once an animal has learned an avoidance response that successfully prevents an aversive event, the behavior is remarkably persistent. Even if the contingency between the response and the aversive stimulus is removed, the animal will continue to avoid.

The traditional extinction procedure of simply removing the contingency between the response and its consequence is ineffective for eliminating avoidance responses, partly because the animal rarely, if ever, encounters the aversive event and possibly because the animal is still rewarded by the fear reduction produced by performing the response. In other words, the agoraphobic dog that runs home from the park in a panic continues to do so, even though no threat exists, because it has left the park and cannot possibly know that there is nothing frightening there. Furthermore, running away allows the dog to feel less fearful because it has avoided the anticipated threat; thus, the behavior of running away is reinforced.

Flooding

Flooding or *response prevention* is an effective procedure for extinguishing an avoidance response.[6] The animal is exposed to the avoidance situation but is prevented from performing the avoidance response. Thus, the animal is exposed to, or "flooded" with, the warning stimuli without being permitted to escape and the aversive stimulus does not occur.

A pet that experiences a particularly aversive event in a specific location (for instance, a dog that was severely scolded for a transgression in the basement) may come to avoid that location or may avoid the responsible person when in that location. Ignoring the behavior and hoping that the animal will eventually forget the experience is unlikely to be effective because the dog's avoidance behavior is reinforced by the reduction in fear associated with avoiding the area. In contrast, forcing the animal to come into contact with the location or to come into contact with the person in the location, while preventing the animal's escape, can produce rapid extinction of the avoidance response. Flooding usually produces an initial increase in fear, but the fear declines if the level is not so high that the animal cannot learn that the aversive event is no longer forthcoming.

Species-Specific Defense Reactions

Escape/avoidance conditioning can be useful for establishing appropriate behavior in pets. Most dogs can easily be conditioned to stay within the boundaries of "electronic fencing" through the use of escape/avoidance conditioning, with the result that responding is highly reliable and persistent. The exact nature of the avoidance response, however, is highly specific to the species.

Aversive stimuli elicit strong unconditioned responses in animals, and these responses are assumed to have evolved because they enable the animal to defend itself. These responses are called *species-specific defense reactions*.[7] Some species are more likely to flee from an aversive stimulus, other species are more likely to freeze and become immobile, and other species to become aggressive and attack the stimulus. Within a species, however, individuals may vary somewhat in these responses.

The response may also depend upon the situation. If an escape route is available, the animal may flee; whereas if there is no escape route, the animal may freeze or fight. It may be difficult to establish a behavior through avoidance conditioning if that behavior is inconsistent with the animal's species-specific defense reactions.

Training a dog to move from one location to another or to come when called can easily be accomplished through avoidance conditioning because running away from an aversive stimulus is a species-specific defense reaction. Training some dogs to lie down through escape and avoidance conditioning may also be easy because assuming a prone position is part of the dog's natural submissive response and is therefore a species-specific defense reaction. For some dogs, however, the use of an aversive stimulus may cause an escalation of the submissive response—the dog may roll over and expose its belly.

COUNTERCONDITIONING

In counterconditioning, an animal is conditioned to make a response that is behaviorally, emotionally, or physiologically incompatible with an undesired response.[8] A dog displaying fear of thunderstorms can, for instance, be counterconditioned to associate the sound of thunder with the delivery of food. Anticipation of food elicits responses that are incompatible with fear. In classical conditioning terms, a conditioned stimulus is conditioned from an undesirable unconditioned stimulus–unconditioned response to a new, incompatible unconditioned stimulus–unconditioned response (see Figure 5).

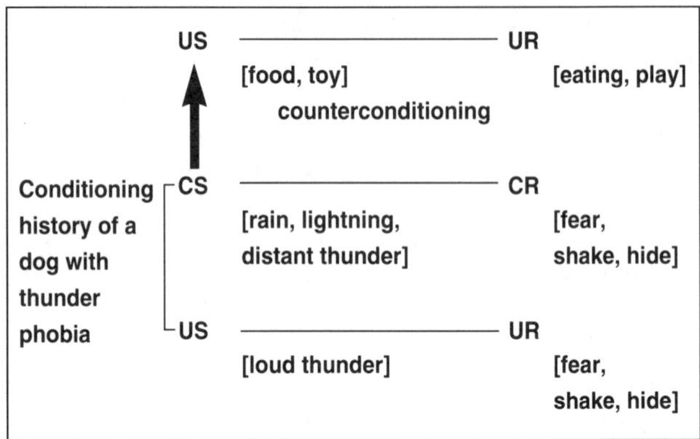

Figure 5—Counterconditioning for a dog with a thunderstorm phobia. *US* = unconditioned stimulus; *UR* = unconditioned response; *CS* = conditioned stimulus; *CR* = conditioned response.

Counterconditioning is often confused with countercommanding. Countercommanding is instructing the dog to do an obedience command (e.g., sit-stay) that is incompatible with an undesired behavior (e.g., jumping up or lunging). Countercommanding works for a well-trained dog *and* a relatively unmotivated behavior *and* in a context of control but will probably be insufficient to prevent a motivated behavior or a behavior out of the context of obedience (e.g., not in training class, off leash, or not in the presence of the trainer).

Counterconditioning requires that the conditioning process occur in the *context* in which the undesirable behavior occurs. It entails a process of shaping and conditioning (see the article by Marder and Reid) that results in the animal learning not to do the undesirable behavior and instead do a different behavior when in the very stimulus conditions that previously made the undesirable behavior highly likely.

For counterconditioning to be effective, the precise eliciting and contextual stimuli that control the behavior must be discovered and the most relevant stimuli must be manipulable. For thunderstorm phobias (see the article on fears and phobias), the sound of thunder can be easily simulated and controlled with high-quality recordings, but the placement of the speakers, bass-to-treble ratio, ambient light or time of day, presence of lightning or rain, or speed of onset of the simulated storm may all be relevant. Counterconditioning is important for treating most problem behaviors, and the relevant stimuli for its use are discussed in other articles in this volume.

CONDITIONED EMOTIONAL REACTIONS

Emotional responses, such as fear, relief, and excitement, are often components of the unconditioned response and, following conditioning, these emotions also become a part of the conditioned response. For instance, a dog that fears thunder (the unconditioned stimulus) may become fearful and agitated with the drop in air pressure (the conditioned stimulus) that precedes a thunderstorm. Emotional responses become conditioned in exactly the same manner as do overt behavioral responses. Dogs, probably because their emotional states are so recognizable to us, appear extremely susceptible to associating emotions with various stimuli. Hence the dog owner that spells out the word *cookie* or *walk* in conversation so that the dog does not become excited and active in anticipation.

Many pet behavior problems are manifestations of conditioned emotional responses. Counterconditioning a new, incompatible emotion to stimuli which elicit the problem behavior is an effective form of intervention. The dog that becomes fearful in the presence of children can be counterconditioned to elicit a different emotion by repeatedly pairing the presence of children with, for instance, a phrase that already triggers a different emotion. If the phrase "Where's your ball?" reliably elicits an excited, happy response by the dog, this can be incorporated into a program of pairing children with the phrase. Although care must be taken to ensure that the phrase is frequently followed by the desirable activity in order that the phrase doesn't lose the original association, with sufficient repetitions of children and the phrase paired together, the dog will come to behave with excitement whenever it sees children.

SOCIAL LEARNING

Social learning involves one animal learning from another. Certain theoretical accounts place social learning within the frameworks of classical and instrumental conditioning while other accounts argue that social learning is a specialized ability. Whether social learning is qualitatively different from other forms of learning will not be addressed here. Social learning is, however, relevant to a discussion of learning principles and how learning shapes behavior because pets often display behavior patterns that are influenced by the presence of other animals.

Learning from others occurs via several mechanisms. In a recent review paper, Whiten and Ham[9] identified several basic processes whereby one animal apparently copies the behavior of another. *Stimulus* or *local enhancement* refers to the tendency for one animal to approach or attend to another animal's behavior. Animal 2 may end up copying the behavior of Animal 1 because Animal 2 approaches the same location or becomes attentive to the same stimuli as Animal 1 and therefore comes to display the same behavior.

For instance, a dog that had never dug in the yard before may be sparked into action if a visiting dog begins digging. The resident dog is attracted to the area and is exposed to stimuli that initiate digging because of the behavior of the visiting dog. Local enhancement is likely responsible for the claim that owners occasionally make regarding the older pet "house-training" the younger one. The younger pet simply follows the older one around and is exposed to stimuli that trigger elimination, and a pattern for elimination is established.

Social facilitation or *contagion* causes one animal to display the same behavior as another because Animal 1 has affected the motivational state of Animal 2. For instance, a dog that has eaten until it is satiated is likely to begin eating again if it is exposed to another dog eating. A dog might try a novel food that it had previously rejected if it sees another dog eating it.

COGNITIVE PROCESSES

Traditionally, the study of learning has been for the purpose of better understanding behavior; but more recently, attention has been directed toward better understanding how cognitive processes (thoughts) influence behavior. Behavior that is influenced by cognition is distinct from conditioned behavior because the behavior cannot be explained on the basis of external stimuli.

Technically, cognition is the use of a neural representation of some past experience as a basis for behavior. In other words, the behavior appears to be caused by some thought or memory that has occurred, independent of the stimuli that are present at the time. Although it appears that behavior can, on occasion, be guided by internal representations of events and relations rather than by concrete external stimuli, our current understanding of animal cognition is too limited to be applied.

REFERENCES

1. Domjan M (ed): *Domjan and Burkhard's The Principles of Learning and Behavior.* Pacific Grove, CA, Brooks/Cole Publishing, 1993.
2. Zbrozyna AW: Habituation of the threatening response in cats and kittens. *Acta Neurobiol Exp* 43:183–192, 1983.
3. Leibrecht BC, Askew HR: Habituation from a comparative perspective, in Denny MR (ed): *Comparative Psychology: An Evolutionary Analysis of Animal Behavior.* New York, John Wiley and Sons, 1980, pp 208–229.
4. Breland K, Breland M: The misbehavior of organisms. *Am Psychol* 61:681–684, 1961.
5. Seligman MEP, Hager JL: *Biological Boundaries of Learning.* Appleton-Century-Crofts, New York, 1972.
6. Baum WM: Extinction of avoidance responses through response prevention (flooding). *Psychol Bull* 74:276–284, 1970.
7. Bolles RC: Species-specific defense reactions and avoidance learning. *Psychol Rev* 77:32–48, 1970.
8. Wolpe J, Lazarus AA: *The Practice of Behavior Therapy.* New York, Pergamon, 1969.
9. Whiten A, Ham R: On the nature and evolution of imitation in the animal kingdom: Reappraisal of a century of research. *Adv Stud Behav* 21:239–283, 1992.

Punishment

Peter L. Borchelt, PhD
Animal Behavior Consultants, Inc.
Forest Hills, New York
The Animal Medical Center
New York, New York

Victoria L. Voith, DVM, PhD
Department of Clinical Studies
School of Veterinary Medicine
University of Pennsylvania
Philadelphia, Pennsylvania

The concept of punishment probably causes more confusion than any other in the field of applied animal behavior. The commonsense, everyday use of the term *punishment* and the way the term is described and used in the popular dog-training literature indicate widespread misunderstanding about its definition as well as the requirements for effective use of punishment and the potential side effects of its misuse. This article defines punishment, presents and discusses the multitude of variables that influence its effectiveness, and critiques examples of common punishment techniques. For many pet behavior problems a punishment technique is neither appropriate nor likely to be effective, although for some problems it can be a humane and effective procedure *if certain requirements are met.*

Definition of Punishment

Punishment is formally defined as the *application* of an aversive stimulus, contingent on a behavior, which *lowers* the probability of that behavior in the future. This definition has two main points. The first is that punishment involves the application of a stimulus that is *aversive*, for instance, a loud yell, a swat, or an electric shock. Punishers are usually stimuli that are obviously aversive—the animal will withdraw from, avoid, or even become frightened of the stimuli. Sometimes, however, a stimulus can be a punisher even if it is not obviously aversive but only startles the animal or stops or inhibits a behavior. The second point in the definition is more critical and requires that the effect of the application(s) of the punisher be a *reduction in the future probability of the response* that is punished. For example, if yelling at a dog when it runs to the door does not decrease the likelihood of its running to the door, yelling was, by definition, not a punisher in the situation. The process that leads to a change in behavior is referred to as *punishment*, and the stimulus that is used in a punishment technique is called a *punisher* or *punishing stimulus.*

Variables That Influence the Effectiveness of Punishment

Many variables have been well researched in laboratory studies of punishment and have been shown to influence significantly the effectiveness of a punisher. These variables also apply to the use of punishment in practical situations.

Contingency

A punishing stimulus must be delivered contingent on the behavior to be punished. This means that the punishing stimulus must be delivered when the problem behavior(s) occur(s). The animal must be "caught in the act." If the punisher is delivered at any other time, the result is punishment of a different (nonproblem) behavior. A good way to remember this is to keep in mind that the *animal* is not punished, but a *behavior* (specific response) is punished.

Immediacy

Some discussions in the popular literature state that the punisher should be delivered quickly after an undesirable behavior occurs but either state or imply that punishment will be effective up to a minute or more after the behavior has occurred. Laboratory research, however, clearly indicates that punishment is most effective when delivered within one second (or less) after the occurrence of the behavior. If there is a delay of even a few seconds the animal is engaging in another, different behavior from the one that was intended to be punished. Thus, the wrong behavior is punished.

In people, however, a delayed punishment *can* be effective. People are able to associate the onset of punishment with a specific behavior that has occurred in the past and do not necessarily associate the punishment with their present behavior. People can understand what they are presently being punished for because of their ability to comprehend language. Dogs cannot make the connection between punishment now and a specific behavior in the past, even if we tell them what we are doing and why. Although a dog may know what several words mean, it does not comprehend the use of past, present, and future tenses.

This does not mean that dogs cannot *remember* things that have occurred in the past. All owners daily observe that their pets remember something learned in the past, for example, commands such as "sit," something that frightened them, or people who were particularly friendly to them. **The ability to *remember* something that was learned in the past is, however, not the same as the ability to *learn about* something that happened in the past.** For instance, if a 10-year-old child is *told* that he is being punished for having broken a lamp while his parents were gone, he is able to associate the punishment with the behavior of having broken the lamp. On the other hand, a dog punished for having broken a lamp while the owner was gone only learns to associate the punishment with the present situation, e.g., what the dog is doing at the time the punishment is initiated, the approach of the owner, or perhaps even with the sight of the broken lamp. Even if the dog is "told" why it is being punished or shown the lamp as it is punished, it will not associate the punishment with the *act* of having broken the lamp.

Many owners are convinced that their dog "knows" when it has done something wrong while they are gone. For example, they say the dog "looks guilty" when they return home only if it has destroyed something. The guilty look described is one of ears down, head lowered, tail tucked, and the dog either slinking away or approaching the owner in a very subdued and submissive manner. These postures are not really "guilty looks" but are species-typical submissive signals. A dog can readily learn that it will be punished when the owner returns, if destroyed items are present (e.g., a broken lamp). The dog does not, however, remember that it has done something wrong earlier in the day. The dog only learns that when two conditions occur simultaneously—(1) destroyed items are present and (2) the owner returns—something unpleasant happens. Therefore, only when destroyed items are present does the dog "look guilty."

The use of punishment after the fact is more like an act of revenge than effective punishment. This really is not punishment, but is more like an attempt to exact retribution and is ineffective in animals. Owners may "punish" a dog for eliminating in the house while they were gone by yelling at it, hitting it, shoving the dog's nose in the mess, or placing the dog in the bathroom for half an hour. The fact that the owner may "feel good" about engaging in this type of behavior to deal with a problem is not sufficient justification for doing it because it is likely to have no effect on the probability of the animal's inappropriate behavior in the future.

Intensity

Almost all studies have found a greater effect with higher intensities of a punishing stimulus than with lower intensities; but extremely high levels of punishment may evoke fear, submission, or aggression from animals. To be practical, the intensity of a punisher has to be high enough to be effective yet low enough so that it is safe for both the animal and the person administering the punishment.

The appropriate intensity of the punishing stimulus is most accurately and easily judged by the animal's response to it. If the animal ignores the stimulus and no behavior change occurs, or if the animal slightly startles and then continues the behavior, the stimulus level is too low. If the animal cowers, runs away and avoids or stays away for a long time, or shows any other signs of fear, the level may be too high. Whether a fear response is too high depends on the response that is being punished. If the goal is for the animal to stay away from an area or stop approaching an object, then a strong withdrawal response and low to moderate level of fear may be the appropriate response to the punisher. In this situation, the punisher should be effective in one or two presentations. If the response to be punished is simply jumping up on strangers, though, such a strong withdrawal response may be excessive because it is likely to lead to fear of strangers. In this situation, the appropriate response level to a punisher would be a startle or withdrawal response of very short duration.

It requires considerable judgment to use a high enough intensity punisher to achieve an immediate effect but not so high as to induce unwanted side effects. The idea is to use the lowest intensity punisher that will achieve the necessary results.

Manner of Introduction

The manner in which punishment is delivered is critical. Sudden introduction of a punisher produces a much larger reduction in the punished behavior than if the punishment intensity is increased gradually. For instance, one study reported that cats would continue to respond for food even under high levels of punishment if the punishing stimulus (electric shock) was introduced at a low intensity and the intensity gradually increased over a period of time.[1] In pigeons, the key-pecking behavior to obtain food was likely to be suppressed completely and irreversibly very quickly if key pecks were punished initially with high-intensity shock, defined as 80 volts or more.[2] If, however, the initial introduction of punishment involved lower intensities (60 volts or less), key-pecking behavior usually was not suppressed even when the punishment was later increased to intensities as high as 130 volts.[3] Thus, to be effective, punishment has to be introduced initially at a high enough level to produce behavioral suppression, rather than introduced at a low level and then gradually increased. Gradual escalation of punishment requires a much higher final punishment level than is necessary if the punishment is initially introduced at an appropriately high level.

Schedules

Punishment techniques used according to a variety of punishment schedules have been studied. In some schedules every response is punished, and in others a certain frequency or percentage of responses is punished. The general finding is that the most suppression occurs when the behavior is punished every time it occurs. The higher the probability of punishment, the greater its effectiveness. In the real world this means that punishing a dog's barking response 1 time of every 10 times it barks is less effective than punishing it for 5 of 10 times it barks. This, in turn, is less effective than punishing it for 9 or 10 of 10 times it barks.

Motivation

The higher an animal's motivation to engage in a specific behavior, the lower the likelihood that punishment will be effective in suppressing it. Therefore, before attempting to use a punishment technique, it is important to understand why an animal is engaging in a problem behavior and to lower the motivation. For example, before a punishment technique for urine marking is tried, an attempt should first be made to lower the motivation for urine marking by castration.

Obviously, any behavior that is rewarded is most effectively treated by removing the source of positive reinforcement before or concurrently with initiating a punishment technique. For instance, if a dog that jumps up at the dining room table to get food is punished by a startling noise and is also prevented from obtaining any food, it will stop jumping up sooner than a dog that jumps up and is punished but still gets the food.

In treating many behavior problems, particularly phobias and separation anxiety, a punishment technique is unlikely to be effective because aversive stimuli can actually increase the animal's motivation for the problem behavior by increasing its level of fear or anxiety.

Repeated Use

Repeated use of the same low-intensity punisher within a short time span can lead to habituation to that punisher—particularly if the animal is highly motivated to engage in the behavior. For instance, owners or trainers may start off with low-intensity punishment (e.g., a leash correction or "no") that is effective in getting the dog's attention and briefly suppressing a behavior. The dog may respond to that stimulus for a while but eventually habituate to it, making the punisher no longer effective. This phenomenon of habituation to low-level punishers that are repeatedly presented is probably the basis for the mistaken idea that a dog has a short attention span. Because of the tendency to habituate to a continuing stimulus, a dog's "attention span" to low levels of the punisher is, indeed, relatively short. This means in practical terms that a low-intensity punisher should be used only for a short period. Practical ways to avoid habituation to a low-intensity punisher are to practice for only short periods of time, to alternate types of low-intensity punishers, and (most effective) to substitute a reward-based technique to maintain the animal's attention and motivation to perform appropriate behaviors.

Alternative Responses

Punishment is always more effective in stopping a particular behavior if an animal can perform a competing behavior for which it is rewarded. Punishment alone simply gives the message "do not do this behavior" but does not signal the animal what to do. A much better effect is obtained if the message is "*do not* do this behavior, but *do* that behavior." For example, teaching a dog to sit for praise, food, or petting when strangers enter a home is likely to be more effective in preventing the dog from jumping up than simply trying to punish the jumping behavior. Using both punishment and reward techniques jointly might work even better.

Effectiveness of Punishment

In view of the preceding factors, it should be obvious that the effective administration of punishment is not a simple matter. It is necessary to consider *all* of the previously mentioned variables simultaneously for the punishment technique to have a high probability of working. Suboptimum conditions for any of the previous variables can make a punishment ineffective even if the parameters of all the other variables are optimum. Probably the most common reasons punishment does not work are failure to reduce the animal's motivation to engage in a particular behavior and delay of the punishment (which results in the failure to punish the actual misbehavior).

Many "punishment techniques" used by the average person and explained in some popular dog literature are often ineffective. This is not an indictment of punishment as a procedure for behavior change but rather a statement that punishment is used incorrectly. A large number of behavior modification "tips" for reducing problem behaviors (e.g., use of mousetraps, balloons, or water pistols or tying chewed objects in the dog's mouth) have naively been offered as solutions because they work in some cases. Authors who suggest these methods fail, however, to specify the circumstances in which they work or the variables that influence their effectiveness. Punishment techniques *can* be extremely effective as shown by laboratory studies (where control of variables is easy) and clinical experience with humans and animals (limited to those instances where optimization of relevant variables is possible). In fact, implemented properly, punishment can work faster and have a more enduring effect than many other procedures that are used to reduce behaviors, e.g., physically restraining the dog from engaging in the behavior, satiation (allowing the dog to engage in the behavior until it is satisfied), extinction (withdrawal of rewards that maintain the behavior), or changing something about the situation in which the behavior occurs.[4]

Effective use of punishment also necessitates understanding the characteristics of the punisher as well as the effects the punisher is going to have on an individual. It is desirable to be able to characterize the relevant aspects of the punisher. A stimulus like a loud noise can be characterized by its intensity, measured in decibels. A stimulus like an electric shock can be measured in terms of amperage and voltage. A stimulus like a hit or a swat administered by a person is hard to characterize; this type of physical contact varies and, at best, is limited to a subjective evaluation as to whether it is mild or severe. Ideally, the intensity of the punishing stimulus should be able to be varied over a wide range of values to allow the appropriate degree of intensity that will achieve the desired result. For example, you cannot alter the intensity of a mousetrap, balloon, or water pistol. It may be that the intensity of these punishers is just right for some dogs for some problems; but if it is not, nothing can be done to raise or lower it to the appropriate level for an individual animal.

It is important to know that an effective punisher reduces the probability, frequency, or magnitude of a behavior very quickly. Laboratory studies generally find complete suppression of a response within two or three presentations of a punishment. Application of effective punishers in the real world of animal behavior problems should also yield behavior changes relatively quickly—usually within 2 to 10 applications of the punisher. If a change does not occur quickly, the punishment is not optimum relative to one or more variables; and the parameters should be adjusted or another behavior modification technique used. When punishment works, it works quickly. When it does not work quickly, it should be discontinued or changed in some way so that it does work quickly.

It is difficult to make blanket statements about the use of punishment for a particular behavior problem because of the individual characteristics of each situation and each dog, including why the dog is motivated to engage in a problem behavior. The decision to use a punishment technique is based on the nature of the problem, prevailing circumstances, and the individual nature of a dog.

Generalization and Discrimination

When an animal is confronted with an intense aversive stimulus, it is motivated not only to withdraw and escape but also to learn to anticipate the occurrence of and avoid similar circumstances in the future. For example, a dog frightened by a sudden blast from a loudspeaker in a show ring may either freeze or try to run away from it. The next time it nears the same place, the dog may try to avoid the area even if no sounds are coming from the speaker. The dog's response may generalize. The dog may avoid other loudspeakers in the show ring or loudspeakers in other show rings. Alternately, the dog may discriminate and avoid only the specific loudspeaker that frightened it. It is not always easy to predict when an animal is going to *generalize* or *discriminate* or even what the animal is going to *associate* with the particular aversive stimulus.

Some of the factors that play a part in these processes are the intensity of the stimulus, motivation of the animal, what it is attending to (paying attention to) at the time the aversive event occurs, and how familiar the animal is with various aspects of the event and circumstances (novel stimuli are likely to be noticed and discriminated). For instance, a dog that has been frightened by the explosion of a firecracker when it pushes off the lid of a garbage can[a] may avoid all objects that look like garbage cans (generalization). Generalization is likely to occur if the dog is extremely frightened of firecrackers

[a]"Pull-type" firecrackers are detonated by pulling on a string rather than lighting a fuse. The string can be attached so that the firecracker explodes when the lid of the garbage can is moved. The authors do not necessarily recommend use of firecrackers as punishers.

and is not highly motivated to get into garbage cans. On the other hand, the dog may only avoid garbage cans of a specific shape or smell or those in a particular location (discrimination). Discrimination is likely to occur if the dog is not extremely frightened of firecrackers and is highly motivated for food or if garbage cans containing firecrackers can be easily detected.

To further complicate the concept, if the dog is relatively unafraid of firecrackers and its level of motivation for food is relatively high, and if the owner only puts firecrackers in garbage cans that contain the best or most food scraps, then the dog might actually learn to seek out garbage cans with firecrackers in them. The dog may have learned to investigate garbage cans containing a punishing stimulus because firecrackers have been associated with a discriminable and predictable source of reinforcement (food). Thus, low-level punishers and high levels of motivation can lead to a supposedly "punishing" stimulus becoming a discriminative stimulus or signal for the presence of a reinforcement.

Conditioned Punishment

If motivation is low and/or the punishment intensity is relatively high, other stimuli associated with the punishment can lead to the animal anticipating punishment if they are present. Stimuli *associated* with punishers can become *conditioned punishers.* Conditioned punishers can be used when it is difficult or impossible to deliver an effective primary punisher. For instance, it is ideal if the owner's voice, which is almost always readily available to be quickly delivered, can function as an effective punisher. The owner's voice may be of such a quality, however, that the dog does not find it sufficiently aversive and thus does not listen when the owner says "no." In this circumstance, the association of the owner's "no" with a stimulus (e.g., air horn) that *is* aversive to the dog may result in the dog paying attention and appropriately responding to the owner's voice. If the owner's voice is to be used as a conditioned punisher, it is necessary that the dog be able to distinguish that tone of voice or word from words that the owner uses in other, nonpunishment situations. Otherwise, the dog will either habituate to the conditioned punisher or become frightened any time the owner speaks.

Punishment and Dogs

The intricate social nature of dogs and other canids (wolves, foxes) is based on a general dominance/subordinance system in which specific postures and vocalizations are used to signal relative ranking in the social hierarchy and function to modulate much of the social behavior. Failure to behave appropriately to these signals may lead to aversive consequences, such as a bite from another animal.

The domestic dog is thus likely preadapted to be responsive to subtle human vocal and gestural signals. In addition, through domestication, dogs have been selected for thousands of generations for sensitivity to subtle human social signs. Most dogs are described as "wanting to please" and being "aware of the owners' wishes or moods" (at least some of the time). Many human social signals affect the behavior of companion dogs and can serve as inhibitors or suppressors (e.g., punishers) of behavior, even in the absence of apparent aversive properties.

Subtle voice cues, or even facial and postural expressions for some sensitive dogs, can serve to stop a behavior in which a dog is engaging or reduce the future probability of its occurrence. These signals *function* as punishers, even though they are not (or may not be) aversive or interpreted by the owner as "punishing." Most dogs, as sensitive social animals, are disposed to some degree to use the owner's actions (voice, gestures) as cues for modulating their own behavior. A dog, particularly when engaging in a behavior that is not highly motivated, generally takes a subordinate role and is motivated to avoid threats, punishers, or social aggression from the owner. Of course, if the dog is engaging in a behavior that is highly motivated, human social signals may not be effective or even perceived. In such situations, the owner must use a signal that is effective or reduce or change the animal's motivation for engaging in the problem behavior.

The fact that most companion dogs are socially responsive to subtle inhibitory/punishing stimuli, such as voice ("no," "don't do that," "hey") and subtle gestures (pointing, a harsh look) accompanied by human speech makes it easy for the average person to believe, mistakenly, that the dog *actually understands* what is said (or meant). Thus, when a highly motivated problem behavior occurs, the owner is often uncertain or confused as to why the dog does not understand and behave "correctly" to the same words and gestures that "worked" before (when the dog was *not* highly motivated to engage in the problem behavior). The degree to which the owner believes the dog understands or responds to the linguistic/symbolic aspects of human language hampers the owner's effectiveness in using social signals. Owners with these beliefs often rely on *telling* a dog what not to do ("shame on you, get off," "how many times have I told you not to do that?") rather than using the social signaling aspects of their voices or gestures to which a dog is likely to respond.

Moreover, the effectiveness of the "punishing" verbalizations may be further diminished by their repeated use in other contexts, such as social play and affectionate behavior. If an owner's verbalizations during play and affection (as is usually the case) are low, soft, repeated, and prolonged, a punishing verbalization should be loud, short, and abrupt so that it is easily discriminable. For instance, "hey," "hup," or other guttural explosive sounds seem to work well.

Punishment and Cats

The basic principles regarding the effective use of

punishment are the same regardless of species; but the type of punisher, the potential side effects, and the type of behavior problem may dictate different approaches among species. The social system of cats differs from that of dogs; therefore, the same social signals that people find effective with dogs may not be effective with cats. For example, cats do not appear to be attentive to human facial expressions; cats pay attention to different types of sounds than dogs do; and cats do not have the same dominance/subordinance system that dogs have.[5-8] Canids usually live in social groups and have developed a communication system, which shares some features with that used between people. The ancestors of domestic cats usually lived solitary adult lives. Feral domestic cats continue to follow this social pattern. The only social units that exist are temporary courtship and mating units, and the mother-offspring association. The latter lasts for a considerable period (six months to one year) depending on the resources available. By contrast, cats in households, farm cats provided with food and shelter, and laboratory colonies of cats can continue to live together harmoniously throughout their lives, develop some form of a dominance hierarchy, and engage in such mutually reciprocal behaviors as playing, grooming, and sleeping together. Domestic cats also play with and sleep with people as well as engage in behaviors that reflect attachment, such as crying when people leave and greeting/following behaviors when people return.

Cats can assume a solitary life-style or a communal one depending on the environmental circumstances. The solitary life-style exists in most natural conditions, and the communal style only exists when an abundance of resources is available through human intervention. By capitalizing on infantile behaviors that kittens exhibit toward the mother or to littermates, domestication has probably led to selection for the ability to assume the communal system. Even though cats appear to have a flexible social system, they still do not have the intricate cooperative or communication system that has evolved through natural selection among animals that live in groups.

People are accustomed to having dogs pay attention to facial expressions, body gestures, and tones of voice. When the same human signals are used to communicate with a cat, the cat does not respond the same way. Staring at and scolding with a deep voice usually will elicit a response from even a naive dog (one that has never been punished before) because these signals resemble intraspecific signals used among dogs. Such signals do not evoke the same responses from a naive cat. A cat is unlikely to respond to such signals unless the actions of the person have been previously paired with an aversive stimulus, such as chasing, hitting or swatting. The social signals people use with dogs can be powerful punishers, but the same signals directed toward a cat are usually meaningless. This difference often leads people to believe that cats cannot be taught anything, that cats are aloof, or that cats are stupid. These statements are untrue. Rather, it is a matter of using different signals to get the cat to attend and respond.

Many canine behaviors for which a person uses a punisher involve social interactions (jumping up or barking) or learned behaviors. These types of behaviors are likely to respond to punishment delivered by a person. Most behavior problems of cats are not these types. Such problems as spraying, not using a litter box, eating grass, and destructive scratching are highly motivated, instinctual behaviors that are unlikely to be affected directly by aversive stimuli. In such cases it is more effective to try to lower the animal's motivation to engage in the behavior (e.g., by castration or by reducing the number of cats in the household) or to provide the animal with acceptable alternative outlets for the behavior (e.g., a scratching post of more desirable material, different litter material, or its own grass plots).

Cats can, however, learn to avoid *locations* when aversive stimuli occur. For instance, an aversive stimulus could be used to teach a cat to avoid locations where spraying occurs, but it is unlikely to prevent spraying when the cat is in that location.

Stimuli that cats find aversive are usually those such as loud noises and sudden or looming motions that elicit a fear response. A spray of water may be frightening or only irritating, depending on the cat and on how the water is administered. A cat highly motivated to engage in a particular behavior may endure or ignore being sprayed with water. A cat also may adopt an escape strategy rather than an avoidance strategy; that is, it will engage in the problem behavior until sprayed and then escape the spray but will not learn to avoid the situation. In other cases, a cat may discriminate circumstances and not engage in the behavior only when there is a high likelihood of being sprayed, for instance, when a person is in the room or nearby.

Owners who use a frightening punisher on a cat should be warned not to touch or reach for the cat until it assumes a calm appearance. Cats often react in a defensive-aggressive manner when approached while frightened, and the owner could be bitten or scratched. A frightened dog often assumes a submissive posture when approached, while a cat usually becomes defensively aggressive.

Negative Reinforcement

The concept of negative reinforcement, and its distinction from punishment, is widely misunderstood. Negative reinforcement is formally defined as the *withdrawal* of an aversive stimulus, contingent on a behavior, which *increases* the probability of that behavior in the future. Even though both punishment and negative reinforcement techniques employ the use of aversive stimuli, the result of negative reinforcement is the *opposite* of that of punishment. Negative reinforcement *increases* behavior; punishment *decreases* behavior. An easy way to remember this distinction is to note that reinforcement, whether positive *or* negative, tends to *increase* behavior.

Negative reinforcement is the basis for escape-and-avoidance learning. For instance, a dog that is frightened by a loudspeaker at a show ring may pull the owner or run away until it reaches a quiet area. The behavior of running away is reinforced because the aversive stimulus of the loudspeaker is reduced in intensity. If the dog is repeatedly brought to the show ring and is repeatedly frightened by the loudspeaker, it will likely learn after only a few exposures the most effective way to escape and will do so more and more quickly. Eventually, the dog also will learn to *avoid* entering show rings. It may balk, resist being led to the ring, or engage in any other behavior that works to prevent it from being brought there.

A common example of escape-and-avoidance learning is a dog's fear of traffic or noises in urban areas. In one case seen by the authors, as a large adult dog was walked in the city, it was extremely frightened by the explosive noise associated with the blowout of a bus tire. The dog broke loose from its leash and ran four blocks to its home. One successful escape from the aversive, frightening noise then resulted in avoidance of the area where the blowout occurred. Furthermore, the fear and escape responses generalized and the dog also would try to run home during its walks whenever it heard firecrackers explode or car doors slamming.

Behaviors that are the result of escape-and-avoidance learning do not extinguish readily. Once an animal has learned that an escape response removes it from an aversive stimulus (e.g., exploding bus tires), subsequent avoidance responses prevent it from ever learning that the aversive stimulus is no longer there. In the case just described, successful escape and then avoidance of the area prevented the dog from learning that exploding bus tires were no longer in that location.

Negative reinforcement and escape-and-avoidance learning also can be used to teach animals certain behaviors. An aversive stimulus is used to motivate the animal to engage in the desired behavior, rather than used to *stop* a behavior (punishment). To be effective, certain requirements must be met. First, it is necessary that the aversive stimulus be of proper intensity. It must motivate escape behavior(s) but not elicit aggression, intense fear, or any other behavior that interferes or competes with the desired response. Second, it is necessary that the aversive stimulus be controllable so that it can be administered only at the proper time. Third, in some circumstances it may be necessary that the person administering the aversive stimulus not be associated with its delivery or else the desired behavior will occur only when that person is present.

Commonly used training techniques involving negative reinforcement are the release of an ear pinch or decreased tension on a leash when a correct behavior occurs. If the pinch or tension is just the right intensity to be mildly aversive and does not elicit responses that interfere with the desired response, the release of the pinch or tension could act as a negative reinforcer. For most behavior problems, however, these are not practical solutions because it is difficult or impossible to time their occurrence correctly. In many cases, they are not sufficiently aversive to motivate the correct response and an increase in their intensity can easily elicit other responses, such as fear or aggression toward the person administering the stimulus. Moreover, the presence of the owner or trainer becomes an important component of the learning context and the desired response may not occur in the absence of that person.

An interesting and potentially useful tool for the effective application of negative reinforcement, escape learning, and avoidance learning is the electric training collar.[9] This equipment delivers electrical stimulation controlled from a distance, can be adjusted to deliver the stimulation at appropriate levels for individual dogs, and enables the trainer to time the electrical stimulation precisely. Because electrical stimulation can be highly aversive, considerable care, skill, and knowledge about principles of learning are required for its effective use. This electric equipment is used widely for training hunting dogs, but it is important to note that few individuals have used it to treat behavior problems and the efficacy of its use for this purpose has not been independently established.

Aversion Conditioning

Aversion conditioning is yet another procedure involving aversive stimuli; an aversive stimulus is associated with the stimuli that elicit the problem behavior. The goal is to reduce or prevent a problem behavior by making aversive the stimuli that elicit the problem behavior, rather than (as in punishment) applying the aversive stimulus when the undesired behavior is occurring. Aversion conditioning (aversion therapy) has been used with mixed success in human behavior therapy, primarily for the treatment of obsessions, compulsions, fetishes, habits of attraction to inappropriate objects, and addictions.[10] It is important to note, however, that it is *not* the behavior treatment of first choice, because many human behavior problems have an emotional base (often anxiety) that must also be addressed.

A number of techniques described in the popular dog-training literature attempt to employ aversion conditioning. For instance, tying a dead chicken around the neck of a dog and leaving it there for a day or so is supposed to reduce or eliminate chasing and killing chickens. It has been suggested that a dog that chews objects can be treated if the object is put in the dog's mouth, which is then tied shut for several hours or more. It is also reported that a dog can be conditioned not to destroy a couch if an aversive odor is introduced there after the dog is forcibly exposed to the odor source by having its nose rubbed in it or the substance placed in its mouth.

If aversion conditioning techniques work, they are likely to result in a dog avoiding only the specific object that was made aversive. This may suffice if the problem is merely an attraction to a particular object. If the dog's misbehavior is the result of a general emotional state, such as separation anxiety, exuberant play, or fear, the dog is likely to direct the behavior to another object or manifest the problem in another form. A problem behavior resulting from an animal's emotional state can be treated only by addressing the emotional base of the problem and lowering the motivation for that behavior. Aversion conditioning techniques are not the first choice of treatment for most pet behavior problems.

Even if used appropriately, the effectiveness of aversion conditioning techniques and the parameters that influence their effectiveness are not well understood for animal behavior problems. Thus, when an aversion conditioning technique fails to work, it is unclear how to modify it in order to increase its efficacy.

Critique of Some Common Punishers

A variety of punishers are described and recommended in the popular dog-training literature and are widely used in obedience training and problem solving. These punishers are subject to the basic rules and parameters described in this article, although the popular literature does not provide this information. While these techniques work for some behavior problems, their universal application for all behavior problems is inappropriate and even has the potential for making a problem worse.

Leash Correction

For leash correction, a dog's leash is given a jerk that tightens a choke collar to signal the dog when an error or incorrect behavior has occurred. Only occasionally in discussions of the leash correction technique is it noted that the release of the tension may be used to signal a *correct* behavior. It should be obvious that the release of the tension can function as a signal for correct behavior only if the correct behavior is occurring as the tension is released. If the jerk is of appropriate intensity, the leash correction can be an effective method in many cases for teaching behaviors related to locomotion (heel) or change in posture (sit, down) as long as the dog is not highly motivated to engage in other behaviors. If the dog is already highly trained, leash corrections can be used to stop a behavior problem involving locomotion, e.g., pulling, lunging, and some other behaviors that occur while the dog is on the leash.

Leash correction is not, however, an effective method for treating many behavior problems, such as those related to anxiety and fears, housebreaking and urine marking, excessive grooming, and general overactivity. Even with problems for which leash correction could be appropriate, the technique may fail to work for several reasons. If the dog is highly motivated to engage in a behavior (e.g., lunging and growling at another dog), the jerk on the leash may be ignored even by a well-trained dog. A jerk on the leash may not be aversive enough to inhibit the problem behavior as it occurs and is highly unlikely to prevent future occurrences of that behavior. Some trainers recommend increasing the severity of the leash correction to the point of "hanging" the dog by lifting if off its feet and holding it in the air until it ceases struggling. This procedure may cause tissue damage, as well as lead to fearful or intense aggressive behavior.

Inherent in a leash correction is the presence of the owner and the owner's behaviors, such as approaching the dog and reaching for and grabbing the leash. The dog may be able to discriminate the situations in which the leash correction is likely from those in which leash correction is not.

Leash corrections also require a degree of physical strength and coordination. Further, many pet owners are reluctant to use techniques that employ force but will use reward-based behavior modification techniques to treat their pets' behavior problems.

Shake Can

A tin can containing coins or pebbles, when shaken, makes a startling, rattling noise that some dogs find aversive, at least initially. A startling noise could inhibit some behavior problems, such as barking at a window. Several drawbacks of the shake can are that it is inconvenient to carry and is unlikely to be available when needed. When carried, the can may make low-level sounds to which the dog may habituate. In addition, some dogs do not find the sound of the shake can sufficiently aversive. Better alternatives to the shake can include small, shrill, compressed-air whistles or air horns, which are louder and can be easily carried, silently, in the owner's pocket. It is important to note that startling, loud noises can be used as effective punishers only for some behavior problems.

Time Out

One of the most common "punishment" techniques used by dog owners is to isolate the dog from social contact by putting it in another room or outside for extended periods of time. This technique is sometimes referred to as *time out*, which technically is shorthand for a laboratory procedure called *time out from positive reinforcement*.

Time out from positive reinforcement has been found to be an effective punisher in the laboratory under some conditions and to be relatively ineffective under other conditions.[4] Consider as an example laboratory operant conditioning wherein an animal has learned to engage in a certain behavior (e.g, pressing down on a lever) for a food reward. The animal presses the lever at a steady rate to obtain food. If, after every tenth press of the lever, the lights in the cage go out and pressing the lever no longer delivers food for a one-minute period, it would be expected that the rate of lever pressing would decrease if this time out from food reward functioned as a punisher. In fact, no such decrease is likely to occur.

If the situation is changed so that *two* levers are available that deliver food when pressed and pressing lever A results in occasional time out (lights out, lever inoperative) but pressing lever B consistently leads to food reward, then a dramatic decrease in the pressing of lever A will occur. In this situation, time out functions as a punisher that stops the pressing of lever A. The result of research on animals in the laboratory clearly indicates that time out can function as a punisher only if the animal has available an *alternative response* that is unpunished and that produces a positive reinforcement.

The common technique of isolating a dog from social contact is ineffective as "time out" in that it provides no alternative response to be reinforced. Social isolation actually takes the dog out of the situation in which reinforcement would be available for an alternative response to the problem behavior.

Time out is commonly used by pet owners because it seems to parallel a technique that is used for *human* behavior modification. Probably the most common form of punishment used for children in a home is withdrawal of the opportunity to engage in rewarding activities, such as watching television or playing with a favorite toy. This procedure, however, is *not* the same as the time-out procedure used for animals in the laboratory, because with people the withdrawal of the reinforcing activity usually does not immediately follow the misbehavior. This technique works with people because it involves the cognitive, linguistic element of *explaining* why the reinforcement is unavailable.

Time out may even escalate an animal behavior problem. If a dog is isolated for a behavior involving exuberant play or greeting, social isolation deprives the dog of social contact and is likely to increase the problem behavior when the dog is released.

Summary

Punishment is not as simple a procedure as it may appear to be; however, punishment is often the first, and many times the only, technique on which the average person relies to deal with pet behavior problems. Sometimes the owner mistakenly attributes the resolution of a problem to punishment, but in reality success was achieved because of other techniques unknowingly employed. In other situations, punishment does work because it is appropriately applied to a particular problem. In most cases, punishment fails to work; and, in some cases, it makes a problem worse.

REFERENCES

1. Masserman JH: *Principles of Dynamic Psychiatry*. Philadelphia, WB Saunders Co, 1946.
2. Azrin NH: A technique for delivering shock to pigeons. *J Exp Anal Behav* 2:161-163, 1959.
3. Azrin NH: Effects of punishment intensity during variable interval reinforcement. *J Exp Anal Behav* 3:123-142, 1960.
4. Azrin NH, Holtz WC: Punishment, in Honig WK (ed): *Operant Behavior: Areas of Research and Application*. New York, Appleton-Century-Crofts, 1966, pp 380-447.
5. Leyhausen P: *Cat Behavior*. New York, Garland STPM Press, 1979.
6. Laundre J: The daytime behavior of domestic cats in a free-roaming population. *Anim Behav* 25:990-998, 1977.
7. Baron A, Stewart CN, Warren JM: Patterns of social interaction in cats (*Felis domestica*). *Behavior* 11(1):56-66, 1957.
8. Liberg O: *Predation and Social Behavior in a Population of Domestic Cats. An Evolutionary Perspective.* University of Lund, Sweden, 1981.
9. Tortora DF: *Understanding Electronic Dog Training*. Tucson, AZ, Tri-Tronics, Inc., 1981.
10. Wolpe J: *The Practice of Behavior Therapy*, ed 2. New York, Pergamon Press, 1973.

UPDATE

In reference to the *Immediacy* section, in laboratory testing situations where only one or a few responses are available to the animal, punishment applied immediately after behavior can be quite effective. In the "real world" of dogs and cats where animals engage in continuous sequences of behavior, punishers are more effective when applied as soon as possible after the onset of the behavior or even during the "intention" movements that often are the initial components of a behavior sequence.

Descriptions of commercially available products that may have application as punishers are provided in "Products for preventing or controlling undesirable behavior" by Gary Landsberg, *Veterinary Medicine*, Vol 89, 1994, pp 970-983 and in *Practitioner's Guide to Pet Behavior Problems*, 1995, American Animal Hospital Association, Denver, Colorado, by Wayne Hunthausen and Gary Landsberg.

Modifying Unruly Breeding Behavior in Stallions*

University of Pennsylvania
Sue M. McDonnell, PhD
Regina M. Turner, VMD

Centre Equine Practice
Pleasant Gap, Pennsylvania
Nancy Kate Diehl, MS, VMD

KEY FACTS

- **Most stallions can be trained and handled to behave in an orderly, safe manner for in-hand breeding.**
- **Unruly stallions are those that are overly aggressive in the breeding shed or that exhibit specific behaviors that make in-hand breeding inefficient or unsafe.**
- **Simple behavior modification can be effective in retraining unruly stallions; pharmacologic aids are rarely required.**
- **Truly savage stallions should be distinguished from simply unruly stallions; attempting to retrain savage stallions is not recommended.**
- **Successful behavior modification involves eliminating undesirable habits without suppressing normal sexual behavior.**

Although most stallions are readily trained and handled in a manner that allows safe and efficient collection of semen or in-hand breeding of mares, more than 10% of stallions presented to our sexual behavior clinic exhibit what we refer to as unruly behavior.[1] This is characterized by overly aggressive breeding, usually coupled with habits that make the breeding process dangerous for the mare, stallion, and personnel. Specific unruly behaviors include rearing; kicking or striking the mare or handlers; resisting examination of the penis or testicles by kicking or moving away when touched; forcefully approaching the mare or dummy mount before being signaled to do so or despite signals to wait; thrusting the pelvis excessively, kicking, striking, or biting when the penis is washed; and thrusting with excessive force once mounted, which can lead to injury of the mare or stallion and malpositioning on the dummy mount.

In our experience, unruly breeding behavior is more common among stallions with inherently high energy and libido; however, bad breeding habits can develop in a stallion with a relatively calm temperament. The manner in which the stallion has been restrained and handled for breeding apparently often exacerbates or provokes unruly behavior in the breeding shed. Related handler factors include giving unclear signals to the stallion, applying discipline based on unreasonable expectations for breeding-shed demeanor (e.g., jerking on the shank when the stallion vocalizes or prances in place), providing inconsistent breeding-shed protocol, failing to reinforce desirable behavior, and applying inappropriate discipline (ill-timed, too harsh, prolonged, too little, or unsuited to the type of disobedience).

Unlike simply unruly or misbehaved stallions, savage (so-called killer) stallions viciously attack handlers or other horses.[2] For savage stallions, we advise euthanasia rather than ordinary handling or retraining. If owners insist on breeding savage stallions, we recommend management in bull-stud conditions, in which people are at minimal risk of direct interaction with the

*This study was a Dorothy Russell Havemeyer Foundation project conducted at the Georgia and Philip Hofmann Research Center for Animal Reproduction at the New Bolton Center.

stallion. In our experience, most savage stallions are described as well-mannered in breeding situations, with generally calm, compliant, and gentle temperaments. Vicious attacks are sporadic, sudden, and usually unprovoked and are not necessarily associated with breeding.

Relatively simple behavior modification procedures are typically effective in retraining stallions presented to our clinic with unruly behavior problems. Usually, less than one week of brief (5- to 15-minute) daily sessions are required, at the home farm or at our specialized facility. Even in the most fractious stallions, the training time for horse and handler often totals less than one hour. This article describes general principles and procedures for modifying unruly behavior and presents two cases of unruly stallions retrained at our facility.

BEHAVIOR MODIFICATION

The successful modification of unruly behavior of breeding stallions is challenging because undesirable behaviors must be eliminated without suppressing normal sexual behavior. Such behavior to be allowed in the breeding shed includes vocalization, prancing gait, eagerness to approach a mare, stomping or pawing, olfactory investigation and/or gentle nipping, flehman response, erection, willingness to mount, insertion of the penis, thrusting, and organized dismount. It is not always intuitive or easy for horse handlers to recognize and tolerate the normally vigorous sexual behavior of stallions, particularly the high-pitched vocalization or prancing gait when approaching a mare. Rushing up to a mare, wheeling and kicking, rearing, striking, and biting are elements of equine sexual interaction that are generally not safe for in-hand breeding and that can be selectively eliminated.

Stallions can be readily trained to proceed at the handler's pace and direction as well as to tolerate procedures and associated delays during the precopulatory sequence. As a minimum, we expect most stallions to be trainable to tease in a safe and controlled manner, to tolerate washing of the penis or examination of the testicles, to approach a mare and mount at a controlled pace, and (if required) to serve an artificial vagina and use a dummy mount. We also expect training and routine handling to be done in a calm and systematic manner, without rough handling and commotion.

For initial evaluation of the behavior of an unruly stallion, we proceed with the specific routine at our clinic for natural breeding or semen collection. The mare is confined in stocks located in one corner of a 40- × 40-foot breeding shed. The stallion is brought into the shed and teased near the mare until erection occurs. He is then backed away from the mare, and the penis is washed with warm water. For natural breeding, the mare is removed from the stocks and positioned in the center of the room. For semen collection, the stimulus mare is positioned next to the dummy mount if the stallion will not mount the dummy. The stallion is again led forward to the mare (natural breeding) or to the dummy mare (collection of semen).

During the initial session, specific desirable and undesirable behaviors are recorded. Based on the observations, and with reasonable expectations, a strategy for training is designed. The behaviors to be encouraged or maintained and the behaviors to be eliminated are listed. The overall goal of the course of behavior modification is to train the stallion to be safely manageable in the breeding situation to which it will be discharged. We thus consider what facilities and equipment will be available, what type of breeding will be done, and whether the farm personnel are experienced in handling breeding stallions.

Depending on what type of training the handlers need to manage a particular stallion, we may simply discuss our procedures and recommendations, bring the handlers in for training at our facility, or work with them at their farm. We recommend that a particular time be set aside for each training session, with minimal distractions and pressures for breeding or obtaining semen for breeding on a particular time schedule.

In our experience, stallions learn the desired responses to a handler's cues most efficiently when a routine is established and handling is consistent. Multiple, brief sessions that end with the successful completion of specific short-term goals (as opposed to attempting to perfect the entire breeding sequence performance) may avoid frustrating the handlers and stallion. The behavior modification techniques we use are routinely applied in training horses in general. These techniques are directed toward altering specific unruly responses without discouraging the expression of acceptable sexual behavior. The stallion learns to associate the judicious use of punishment (simple verbal reprimands, pressure on the lead shank chain, and occasional slaps to the shoulder or belly) with specific undesirable behaviors.

Negative reinforcement is the application of an aversive stimulus until a desired response occurs. For a stallion that will not stand still for washing, for example, the verbal command "stand" is immediately followed by steady pressure on the lead shank (over the tongue) until the stallion stands still; pressure is then released. Simple hand or postural signals are consistently paired with punishment or reinforcement so as to become conditioned stimuli. During all phases of teasing, washing, standing, and approaching, controlled behavior is positively reinforced by the handler's relaxed posture and verbal and tactile praise. As with all training, the timing of application of punishment or positive reinforcement is important. The more closely the handler's response follows a particular behavior, the more readily the horse will learn.

The physical facilities and staff used to train a difficult stallion are important considerations. We prefer an area with plenty of space that is free of obstacles, extraneous personnel, and animals. Although some handlers feel more secure and are effective with stallions confined to small spaces or with numerous handlers and assistants, we recommend a large breeding enclosure and a minimum of personnel.

During initial training of a stallion, and especially during retraining of an unruly stallion, the handler must have the skill and facilities necessary to prevent the stallion from forcefully gaining access to the mare and copulating. Ineffective handling that leads to such a situation inadvertently rewards the stallion for undesirable behavior and thus counteracts the training effort. At our clinic, the stallion handler initially works with the horse away from the breeding shed to familiarize the horse with the chain shank and to teach appropriate responses to simple, calm voice commands (e.g., "stand", "back", and "walk") and associated postural signals. This can usually be accomplished in a brief period just before entering the breeding shed; in particularly difficult cases, we devote at least one training session to establishing basic communication and discipline.

Particularly if the stallion has been charging ahead, it is essential to teach it to walk close beside the handler, not advancing the shoulder ahead of the handler (like training a dog to heel). At the beginning of the breeding situation, we routinely test the horse's willingness to obey commands or signals when it is still at a distance from the mare; a disobedient horse thus can be safely diverted and removed from the breeding area for further training. When a stallion exhibits an undesirable behavior, he is immediately returned to the last successfully completed step in the prebreeding procedure. If necessary, we move the stallion to a corner to stand quietly and await the handler's next cue, or we remove him from the breeding shed for a brief period. This technique is similar to the so-called time-out procedure used with children.[3] Access to a mare and eventual copulation apparently are the ultimate positive reinforcements for tolerating handling and prebreeding procedures.

Our laboratory experiments and clinical examinations often require numerous, varied procedures to be conducted in the breeding shed before the stallion is permitted to mount. With few exceptions, stallions readily learn these new routines and comply with the handler's cues. It is thus evident that stallions can learn to tolerate a complex prebreeding sequence if they are eventually permitted to copulate. We proceed with the expectation that each step in the prebreeding process will be accomplished. In some instances, we may elect to bypass one step during the initial sessions (e.g., washing of the penis) if it is causing undue delay.[4]

To learn efficiently the handler's cues and associate the handler's responses with its own behavior, the stallion must focus on a human handler despite the intrinsic excitement and distraction of the situation. Intense sexual interest and excitement, as is typically observed in unruly breeding stallions, can complicate the training process. Conditions that temper sexual arousal can help in maintaining the stallion's focus on handling. A novel environment usually partially suppresses sexual arousal in stallions.[5] Suboptimum sexual stimuli (e.g., an ovariectomized teaser mare, removal of the teaser mare, or use of a dummy mount) may contribute to an unruly stallion's early and rapid progress by reducing the level of sexual arousal and thus permitting the stallion to focus on the training.

Frequent breeding or semen collection (as often as two or three times daily) is typically helpful in retraining an unruly stallion. Frequent ejaculation and the associated simple physical fatigue apparently temper libido and increase manageability.[6] Keeping the attention of an especially difficult stallion is usually less difficult for an experienced handler than for a novice, at least during initial retraining.

Pharmacologic aids are sometimes considered in calming unruly stallions, particularly if an experienced handler will not be doing the initial retraining. The use of progestins has been suggested for handling aggressive or unruly stallions.[7] There is apparently minimal advantage associated with the use of an α-adrenergic agonist (detomidine hydrochloride) in training unruly stallions.[8] In light of our success with behavior modification alone as well as the physiologic and behavioral disadvantages associated with progestins and other sedative agents, we avoid pharmacologic aids at our facility.

We use and recommend simple methods of restraint and physical disciplinary aids. An integral aid to behavior modification and routine handling of stallions is the chain lead shank (Figure 1). The chain is run through the mouth and over the tongue or, for more effective control, through the mouth and back across the nose. This affords excellent directional control of the head. The handler usually requires training to regulate the tension on the chain bit effectively. A lightweight, light-colored plastic bat, used as

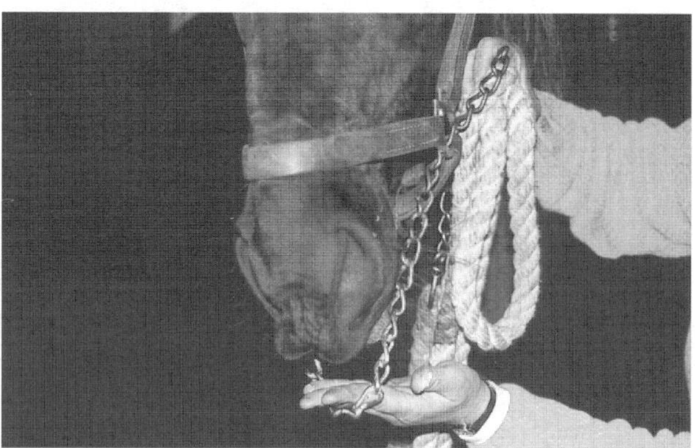

Figure 1A

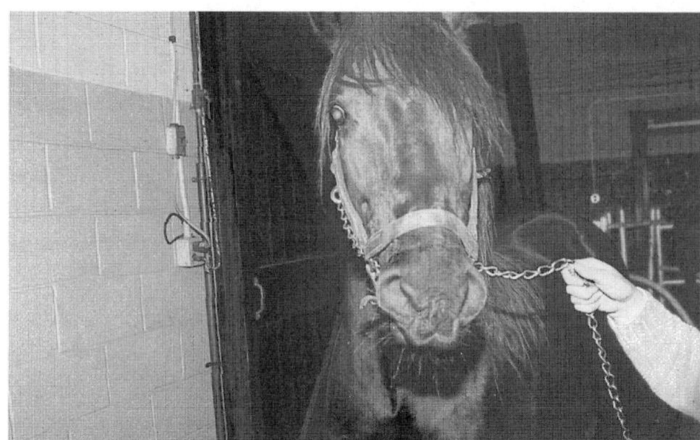

Figure 1B

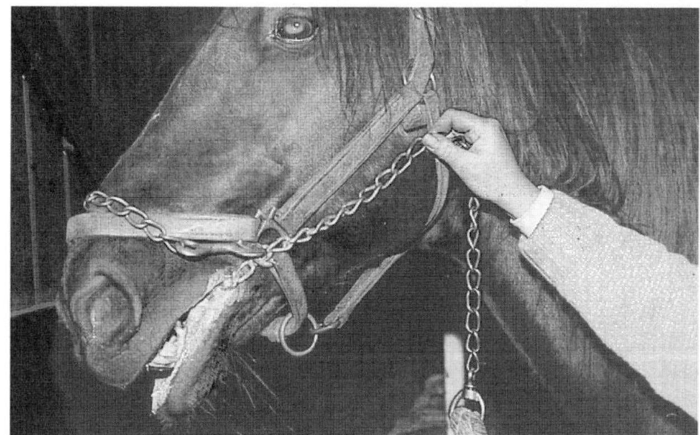

Figure 1C

Figure 1—(**A** and **B**) The chain lead shank is passed through the mouth, over the tongue, and through the lower halter ring and is clipped to the upper halter ring on the far side. (**C**) Alternatively, the chain lead shank is passed through the mouth, over the tongue, through the lower ring on the far side, and back over the nose to the lower halter ring on the near side.

an extension of the handler's or assistant's arm, may be waved in front of the stallion to partially divert its attention from the mare or dummy mount (Figure 2). The bat is infrequently used to deliver a relatively painless but effectively loud slap to the shoulder, chest, or belly just behind the forelegs.[9] Hitting the head typically elicits rearing and may lead to head-shyness.

We videotape the initial and subsequent training sessions, whether at our facility or at a farm. The videotapes often help in devising a behavior modification strategy, in reviewing the stallion's progress, or in showing owners and handlers who do not attend the retraining how we accomplished and maintained control of the stallion.

CASE 1

A 10-year-old Thoroughbred stallion was donated to the School of Veterinary Medicine of the University of Pennsylvania because he was not controllable for in-hand natural breeding. The specific complaint was that the stallion rushed and leaped at the mare and could not be satisfactorily restrained by the handler wielding a crop or using a chain through the horse's mouth. No difficulties in general handling of the stallion were reported, although it was noted to be prone to lose weight, particularly during the breeding season.

At the initial training session, four people were present, including an experienced stallion handler. During subsequent sessions, the horse was handled by a person in training to handle stallions for breeding but with no specific stallion experience. A seasoned stallion handler was present to give advice.

During washing of the penis, the stallion was kept in the corner of the shed opposite the mare, facing away from her. If the stallion kicked or began pelvic thrusting, he was punished by a single word reprimand paired with a sharp tug on the lead shank. The person washing the stallion's penis used cool water, changed tension on the penis, or deflected the penis downward to discourage thrusting.

After the penis was washed, the stallion was encouraged to stand quietly in the corner facing the center of the shed until signaled to approach the dummy mount. This was done by standing relaxedly if the horse stood quietly; when the horse did not stand quietly, gentle pressure was applied on the lead shank and left shoulder while the word "wait" was calmly repeated. After the horse stood quietly for at least a few seconds, the handler signaled the horse to approach the dummy by saying "OK" and simultaneously dropping the hand from the shoulder of the horse and by taking a step toward the dummy mount. If the stallion pulled or charged ahead of the handler, he was punished with a single, short, downward jerk on the lead shank along with a single word command ("no" or "wait"). The horse was then

Figure 2—A plastic bat used as a training aid to hold the attention of the stallion or to deliver punishment for specific undesirable behaviors. Protective head gear (not shown) is used at some facilities and recommended to clients, particularly for working with unruly breeding stallions.

Figure 3—Method of stabilizing the artificial vagina against the dummy to discourage the stallion from traveling forward on the side of the dummy mount during thrusting.

backed or walked to the place where he was last reasonably responsive to the handler. When the stallion reared, tension on the lead shank was decreased; when the forefeet returned to the ground, the horse was immediately returned to the last successfully executed position in the procedure.

Steady downward pressure on the lead shank was used if the stallion rushed toward the dummy mount. The pressure was accompanied by a single-word command ("easy" or "slow"). When the horse proceeded at the handler's pace, the pressure was released; it was resumed if he rushed forward again. Pressure on the lead shank simultaneously punished rushing and negatively reinforced a controlled approach. The handler walked facing the stallion at an oblique angle on the stallion's left side, with the left hand on the lead and the right hand at the stallion's shoulder. An assistant, positioned just to the front and on the left side of the stallion, waved a yellow plastic bat slowly in front of him at eye-level during the approach to the dummy mount to partially divert the stallion's attention.

A desired modification of the stallion's breeding behavior in response to these techniques was achieved during the first session. The procedure was repeated once daily for five days, and semen was collected each day. The total time for each session of teasing, washing, and semen collection averaged four to five minutes. After these initial sessions, only two people were needed to collect semen from the stallion efficiently and safely.

Although the stallion had been used for natural breeding, he readily accepted the artificial vagina as well as the dummy mount on the first attempt. To train the stallion to mount the dummy, an ovariectomized stimulus mare was positioned next to the dummy. After the first five collection sessions, the stimulus mare was no longer used. On every occasion, the stallion readily mounted the dummy. His positioning on the dummy mount was rarely ideal; his hindlegs progressed under the dummy during thrusting, or he tended to move up the side toward the front of the dummy during his extremely vigorous thrusting. This forward movement was modified by a more controlled approach and mounting squarely from the rear and was further reduced by stabilizing the artificial vagina against the side of the dummy mount (Figure 3).

During the next two months, semen was collected at one- to four-day intervals and the stallion became as safe and controllable during breeding as the other stallions used in our research. The horse's demeanor in the breeding shed became less volatile after several sessions, and he was less likely to rear. He eventually did not need to be positioned at a distance from the stimulus mare during washing of the penis, although he continued some nondirected kicking with his hindlegs and moderate pelvic thrusting. The plastic bat was no longer used. The stallion occasionally adopted a sedate attitude and required more than one approach to the dummy before mounting. After two months of regular semen collection, 36 days elapsed before the next collection session. Nevertheless, there was no recurrence of undesirable behaviors. At this time, the stallion was reintroduced to a live mount mare for collection of semen; again, the original unruly behavior did not recur.

CASE 2

A 10-year-old Thoroughbred stallion that had been used for natural breeding was presented to our facility because he was reportedly dangerous and difficult to handle during breeding and exhibited self-mutilative flank-biting behavior. The specific behaviors that made the horse difficult to handle were not detailed

TABLE I
Behavior Modification Progress in Case 2

				Undesirable Behaviors			
Session	*Total Time (minutes)*	*Stimulus Mare Present*	*Wash Attempts (per time)*	*Kicking*	*Rearing*	*Striking*	*Corrections Required*[a]
1	20	Yes	15	>40	>20	>15	>75
2	14	Yes/no	6	3	23	4	~30
3	2.5	No	1/45 seconds	0	6	0	~5
4	8[b]	No/yes	1/100 seconds	0	0	0	0
5	7.3[b]	No	1/60 seconds	0	4	0	0

[a]Verbal corrections; stallion redirected with lead shank.
[b]Stallion reluctant to mount and thrust on dummy mare; abrasions on medial carpi.

further. To limit injury due to self-mutilation, the horse wore a metal muzzle that prevented self-biting but allowed eating and drinking.

The handling team for the initial and subsequent sessions consisted of an experienced stallion handler, a semen collection technician, and an assistant for washing the penis. The results of the first five behavior modification sessions are summarized in Table I. Specific undesirable behaviors were frequent rearing, wheeling (rapidly swinging the hindquarters around), intolerance of handling and washing of the penis, striking, kicking with the hindlimbs, and attempts to bite the handler.

After approximately 10 aborted attempts to touch the stallion's penis, goals for the initial session were modified from washing the penis to having the stallion stand quietly and permit the technician to simply touch, hold, and wrap a warm, wet towel around the penis. The handler helped introduce the stallion to this process by rubbing the horse's shoulder and belly with the right hand while holding the lead shank in the left hand. This allowed the stallion to habituate to tactile stimulation along the belly before the wash technician approached. When the stallion kicked, he was punished with a single, sharp, downward jerk on the lead shank; a single, firm verbal command ("no"); and/or a sharp, loud slap to the belly. Rearing was handled as in Case 1. When the horse wheeled, the handler countered the circular movement by turning the stallion's head and neck to oppose the hindquarters and by applying firm pressure to the shoulder.

After successful handling of the stallion's penis, the behavior modification session continued with the goal of controlled semen collection via a dummy mare. As in Case 1, the immediate goal was to eliminate the undesirable rearing, striking, kicking, and biting. The stallion was encouraged to stand quietly until signaled by the handler to approach the dummy mount at a controlled pace (as described in Case 1). With every inappropriate behavior, the stallion was returned to the last controlled location in the procedure; progress toward breeding was permitted only when the stallion tolerated the preceding step in the procedure. At the end of the first session, which was rather prolonged, semen was collected from the stallion. The degree of control during the approach to the dummy was considered to be unsatisfactory.

During the second session, the stallion was even more sexually aroused and rearing was more frequent (Table I). The stimulus mare was thus removed from the breeding shed (the stallion remained sexually excited but at a more manageable level), and the chain was repositioned through the mouth and back over the nose to improve control of the head. After these changes were implemented, the stallion reared only once. The penis was washed on the second attempt. The stallion waited quietly in the corner as directed by the handler; it was then permitted to approach and mount the dummy for collection of semen.

For the third and subsequent sessions, no stimulus mare was used. The chain was placed only through the stallion's mouth. The horse's breeding behavior was considered to be acceptable and controllable, and the procedure was completed within a reasonable period.

By the fourth session, the stallion exhibited none of the undesirable behaviors. It had developed superficial skin abrasions over the medial side of each carpus (Figure 4), incurred during vigorous gripping and thrusting on the dummy mare. The stallion appeared to be reluctant to grasp the dummy with his forelegs, squealing and backing off when the lesions contacted the dummy. This apparent discomfort reduced his vigor and willingness to grasp the dummy mare appropriately with his forelegs. While mounted, he

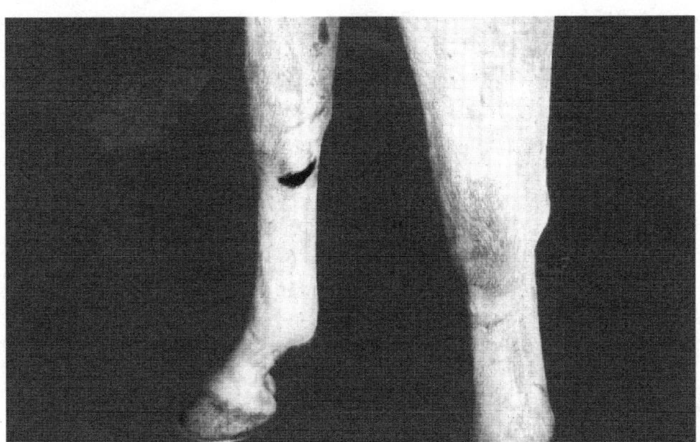

Figure 4—Rub sore of the medial carpus, incurred during aggressive mounting of a dummy mare.

tended to put his forefeet on top of the dummy and stand with his chest high above the dummy. (We have observed similar postures in other stallions with carpal rub sores.) The horse mounted four times before the artificial vagina was safely applied and ejaculation occurred. For subsequent sessions, bandages were applied to both carpi; the stallion's mounted position improved.

Before presentation to our clinic, this stallion had been bred by natural cover only. Nevertheless, he readily accepted the artificial vagina and dummy mount on the first attempt. Because of his aggressive thrusting, he tended to travel up the near side of the dummy. The handler discouraged this shifting of position by attempting to hold the artificial vagina firmly at the rear of the dummy mare with the left hand while simultaneously applying pressure to the stallion's left flank with the right hand.

The stallion was discharged to the owner and manager before reintroduction to a live mount. Although significant progress had been made, further behavior modification should have been completed before discharge (including gradual reintroduction of the mare to the breeding shed). For the first month at the farm, the stallion's behavior was acceptable, even for natural cover. Subsequently, the behavior regressed to an unsafe level, which the referring veterinarian attributed to inconsistent and rough handling at the farm.

CONCLUSION

These two cases exemplify specific concepts that we stress when advising a client on training an unruly breeding stallion.[9] The stallion handler should establish basic control using the chain lead shank before entering the breeding shed. The breeding situation must be arranged so that the stallion is unable to breed unless the prebreeding procedures have been performed to the satisfaction of the handlers. This usually requires a breeding area large enough to allow the handler to direct the stallion away from the mare if he attempts to charge forward.

For each stallion, the specific responses to be prevented or eliminated should be identified and a strategy devised for modifying them. Each person involved in the breeding situation should be apprised of the training goals and strategies and should know what behaviors are desirable or at least acceptable. Normal expression of sexual arousal should be permitted. The use of suboptimum stimuli and gradual reintroduction of standard stimuli can be helpful. Our experience suggests that a person who is good with horses can generally learn to handle aggressive stallions; nevertheless, an experienced stallion handler may be required for the initial training session. Additional information concerning the training of stallions is available in the literature.[4]

ACKNOWLEDGMENTS

The authors thank Malgorzata Pozor, DVM, for her assistance in stallion handling and semen collection in Case 1 as well as Jim Morris and George Dreisbach for their assistance in stallion handling in Case 2.

About the Authors

Drs. McDonnell and Turner are affiliated with the Section of Reproductive Studies, School of Veterinary Medicine, University of Pennsylvania, New Bolton Center, Kennett Square, Pennsylvania. Dr. Diehl is affiliated with the Centre Equine Practice in Pleasant Gap, Pennsylvania.

REFERENCES

1. McDonnell SM: Normal and abnormal sexual behavior. *Vet Clin North Am Equine Pract* 8(1):71–89, 1992.
2. McDonnell SM: Reproductive behavior in the stallion. *Vet Clin North Am Equine Pract* 2(3):535–555, 1986.
3. Blackham GJ, Silberman A: *Modification of Child and Adolescent Behavior*, ed 2. Belmont, CA, Wadsworth Publishing Co, 1975, pp 75–77.
4. Conboy HS: Training the novice stallion for artificial breeding. *Vet Clin North Am Equine Pract* 8(1):101–109, 1992.
5. McDonnell SM, Kenney RM, Meckley PE, Garcia MC: Novel environment suppression of stallion sexual behavior and effects of diazepam. *Physiol Behav* 37(3):503–505, 1986.
6. Wierzbowski S: Badania nad zachowaniem plciowym ogierow i buhajow. *Wydawnic Wlas Instyt Zootech* 154:1–60, 1962.
7. Roberts SF, Beaver BV: The use of progestins for aggressive and for hypersexual horses, in Robinson NE (ed): *Current Therapy in Equine Medicine*, ed 2. Philadelphia, WB Saunders Co, 1987, pp 129–131.
8. Weiher JH, McDonnell SM: Unpublished data, University of Pennsylvania, New Bolton Center, Kennett Square, PA, 1993.
9. McDonnell SM: Coping with the unruly breeding stallion. *Mod Horse Breed* 5(5):10–11, 1988.

UPDATE

Editors' Comments: Even though this paper deals with unruly breeding behavior of stallions (an atypical companion animal problem), it is worth reading in detail by anyone interested in behavior modification techniques for any species. The examples illustrate several important points. The authors are thoroughly knowledgeable about the range of normal desirable behaviors of their subjects. They know exactly which behaviors *should be* suppressed and which *should not*. They carefully observed the sequence of behaviors exhibited by the unruly stallions and identified stimuli associated with the undesirable behaviors. Identification and subsequent control of relevant environment stimuli allowed the handlers to manipulate the level of the stallions' libido, thus enabling handlers to suppress certain behaviors through desensitization and to negatively reinforce or punish others. The aversive stimuli were judiciously selected, and appropriate behaviors were concurrently shaped and positively reinforced.

The behavior modification techniques were initially implemented in situations where the stallions' motivation was lower and the techniques were therefore more likely to succeed. Furthermore, the handlers often worked on only one component of the sequence at a time. The handlers did not attempt a complete sequence until each component was under control and it was predicted that the stallion would be able to complete the entire sequence of required behaviors. Attempting to complete an entire breeding sequence each time the stallion was handled would not only have failed but would have made matters worse. The stallions would have repeated the undesirable behaviors, thus further strengthening them.

In the case of unruly stallions, the handlers are faced with the delicate situation of having to eliminate or, trickier yet, modify only specific behaviors in a sequence without interfering with other components of the sequence and without introducing new behavior problems. Negative reinforcement techniques and punishment used incorrectly can escalate defensive aggression or induce fear and suppress breeding behaviors.

Notice how rapidly and successfully the cases in this article and the one described in the update section of the article on fears and phobias were treated when the behavior modification techniques were carefully implemented. The handlers did not proceed too quickly in the early stages, nor were the animals permitted to engage in the undesirable behaviors between the treatment sessions.

Taste Aversion Conditioning Versus Conditioning Using Aversive Peripheral Stimuli

Bio-Behavioral Technology Inc.
Tempe, Arizona
Carl R. Gustavson, BS, MS, PhD

Taste aversion conditioning in animals describes the process of learning to avoid consuming flavors that, when consumed, are followed by illness. When an animal is again exposed to the particular flavor that was previously paired with illness, the animal will respond to the flavor as unpalatable and refuse to eat it.

Taste aversion conditioning was discovered in the early 1950s by Dr. John Garcia who was studying whether animals could learn to avoid the toxic effects of gamma radiation. More than 3000 articles on taste aversion conditioning have appeared in the scientific literature since 1955. At first glance, taste aversion conditioning seems to be a relatively simple process; over time, however, it has become clear that an understanding of this conditioning procedure is not as simple as initially perceived. Although much remains to be learned about taste aversion conditioning, our current knowledge has many useful applications.

Conditioning using aversive peripheral stimuli also has been well researched and involves the presentation of an aversive stimulus (e.g., shock, a loud sound, or a swat) contingent upon a specific behavior. The use of aversive peripheral stimuli involves punishment, negative reinforcement, and aversion conditioning (see the article on punishment by Voith and Borchelt elsewhere in this book). These techniques have been widely applied to ingestive and noningestive behaviors with varying degrees of success.

This article describes the underlying processes and potential applications of taste aversion conditioning for managing behavior problems in companion animals. In addition, taste aversion conditioning is compared with the use of aversive peripheral stimuli for behavior control.

SKIN VERSUS GUT DEFENSE SYSTEMS

Taste aversion conditioning and conditioning using aversive peripheral stimuli differ because two systems have evolved in vertebrates to defend against various types of environmental assaults: the skin defense system and the gut defense system.[1] These two systems are anatomically and functionally divided, and the underlying processes for producing changes in behavior within each of these systems are remarkably different.

From a functional perspective, the skin defense system protects animals from mechanical, thermal, and chemical peripheral assaults by enabling the animal to use specific responses to react quickly to a wide range of circumstances. When a peripheral environmental assault occurs, the skin defense system relies on quick response timing, appropriate response form, and attention to stimulus context and circumstances for a successful defense. These features allow the animal to learn the cues that predict an aversive event.

By contrast, the gut defense system protects animals from unwholesome and/or toxic items that are ingested. The tongue is the first line of gut defense to detect an error in eating. When the tongue is stimulated by foods containing sugars, salts, and proteins, the taste buds that sense sweet, salty, and sour signal chewing and swallowing (unless these flavors are too strong). If the taste buds that sense bitter are stimulated, gagging, spitting, or wiping of the mouth is signaled and swallowing is prevented. Most bitter-tasting substances are poisonous, and the refusal to swallow them prevents tragic consequences. Some poisons, however, are not bitter, but nerve and vascular connections among the tongue, stomach, and brain have not left animals helpless in avoiding these toxic substances. After food has passed from the mouth into the gut, running, jumping, and attacking will not help the animal avoid negative toxic consequences. Vomiting, nausea, and rapid learning to avoid ingesting these flavors, however, provide protection. From an adaptive perspective, stimulus context should have minimum impact on gut defense, and quick response timing and appropriate response form are relevant only to preswallowing behaviors.

A major problem in the use of aversive peripheral stimulation for controlling ingestive behaviors in applied settings is that users of this conditioning procedure fail to understand and appreciate that the associative aspects of skin and gut defense systems differ dramatically. When learning occurs in the skin defense system, the associations represent "if–then" relationships. For if–then relationships to develop, stimuli must be associated in spatial proximity (inches and feet, not yards or miles) and temporal proximity (milliseconds or seconds, not minutes or hours). Most if–then associations require frequent and consistent repetitions of temporal and spatial pairings. After conditioning, the animal treats if–then associations in spatial and temporal terms, but the desirability of the goal of the behavior remains unaltered. For example, food paired with the presentation of pain is still food and remains palatable and nutritious. A chewed couch paired with a painful stimulus remains soft, warm, and comfortable. A shoe paired with sharp vocal reprimands remains chewable. In a more subtle fashion, an effective repellent placed on an item subverts the desirability of the treated item; however, the item without the repellent remains desirable.

Any stimulus in the environment that consistently occurs in the absence of the if–then relationship will be used by the animal to adjust its behavior. Dogs can be readily trained not to reach onto a kitchen counter to obtain food that has been left there. This is accomplished by use of loud vocal reprimands and swats (if the owner remains in the area). On the other hand, temptation often overwhelms the dog when the owner leaves the room. Commercial shock pads placed on the counter to train the animal not to jump up for food can remain effective even when the current is turned off; if the pad is removed, however, a motivated dog will jump up to get the food.

High motivation is likely to subvert if–then aversive associations. In the early morning just after breakfast, foot shock or anticipation of the shock (anticipatory or conditioned fear) may be sufficient to prevent a cat from stealing meat from the kitchen counter. By the middle of the afternoon, hunger may increase the incentive to test the if–then association. On a warm, sunny afternoon, electric shock from electronic fencing may be sufficient to keep a virile male dog in the yard, but with the appearance of a female in estrus, a brief pain in the neck may seem little penalty to pay for access to a receptive sexual partner. After consummation, however, the shock or anticipation of shock may be sufficient to prevent the dog from returning to the yard.

A mildly repellent (aversive) stimulus may terminate an ongoing series of activities and establish an if–then association. When the incentive value of an item exceeds the repellent qualities of the stimulus, however, an animal can begin to use the repellent itself as a cue to the presence or desirability of the item. Salient visual repellents can be used to signal the presence of an otherwise cryptic or ignored item. Animals can follow the concentration gradient of olfactory repellents directly to an otherwise hidden item. Gustatory repellents can become desirable in and of themselves if consumption of the mild repellent is followed by nutritive feedback.[1]

Historically, it has been common to discuss both skin defense and gut defense associative processes within the procedural framework of Pavlovian, or

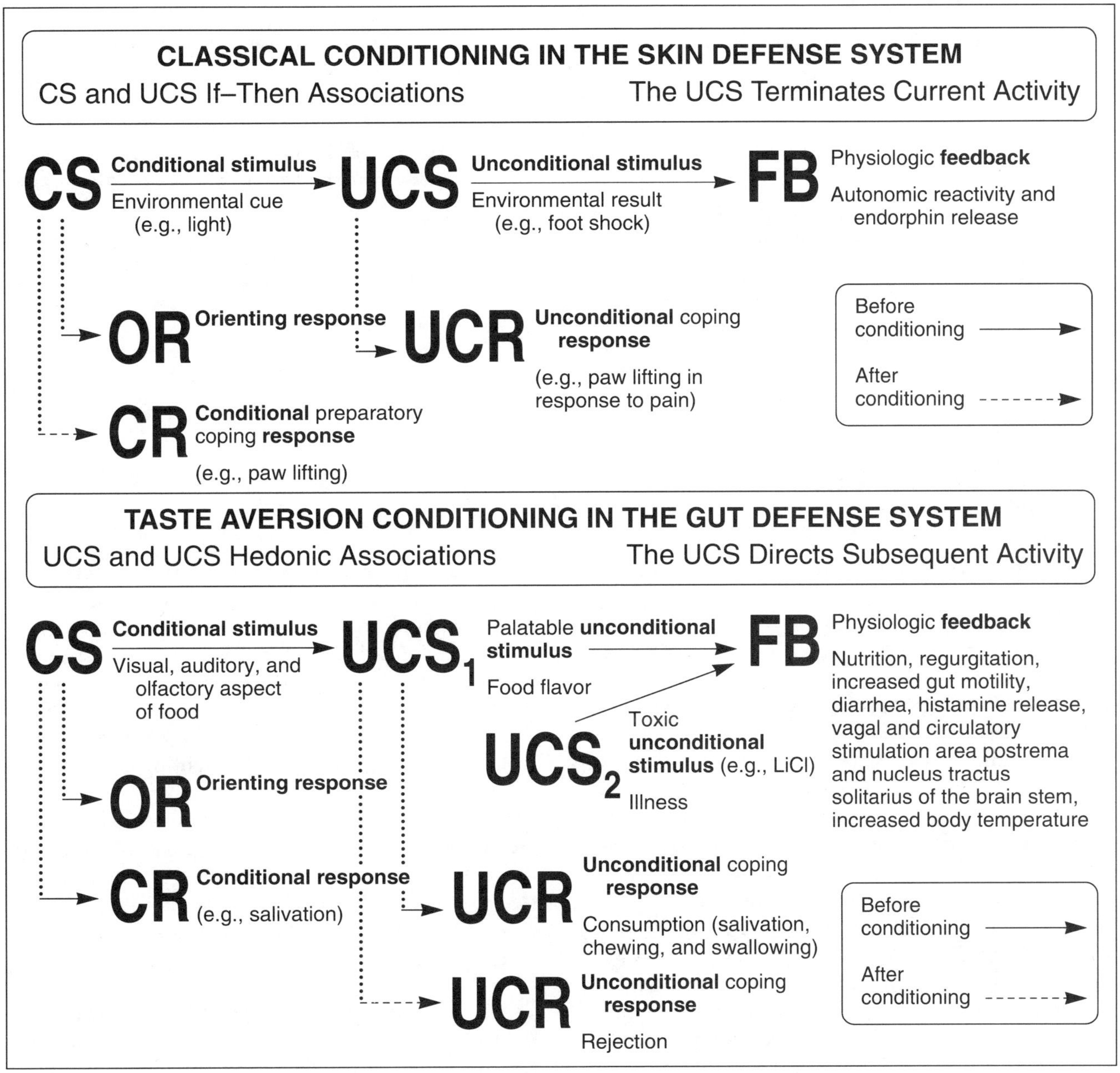

Figure 1—A comparison of the associative processes in the skin and gut defense systems.

classical, conditioning. Garcia[1] argued that skin defense associative processes are best discussed as associations between the conditional stimulus and the unconditional stimulus in traditional Pavlovian processes. Gut defense associations are better understood and represented as associations between one unconditional stimulus (e.g., taste) and a second unconditional stimulus (e.g., a toxin) in which the physiologic changes produced by the second unconditional stimulus feed back on the first unconditional stimulus to alter the hedonic value of the first unconditional stimulus and change the taste from eliciting consumption to eliciting rejection. Figure 1 compares the associative processes in the skin and gut defense systems.

A LABORATORY MODEL

A direct comparison of aversive peripheral stimulus conditioning and taste aversion conditioning for ingestive behavior is provided in experiments by Gustavson and Gustavson.[2] Five procedures were tested on different groups of rats to determine which procedure(s) prevented rats from eating a highly preferred food, Oreo® cookies. In one part of the experiment, the procedures were applied while the rats were approaching the cookies. In another part, the proce-

TABLE I

Effects of Different Aversive Stimuli on Cookie Eating in Rats as a Function of When the Stimuli Were Previously Presented and Where the Rats Were Later Tested

Training Phase	Testing Phase						
		Subsequent Eating Behavior in Different Test Environments as a Function of Previous Aversive Stimulus Experience					
Time at Which Aversive Stimulus was Presented	***Environment in Which Rat was Later Presented with Cookie***	*No Stimulus*	*Shock*	*Ammonia*	*Mustard*	*Quinine*	*Lithium Chloride*
While the rat approached the cookie	Training cage	+	h+	+	+	+	+
	Home cage	+	+	+	+	+	+
	Novel cage	+	+	+	+	+	+
While the rat ate the cookie	Training cage	+	h+	+	+	+	–
	Home cage	+	+	+	++	++	–
	Novel cage	+	+	h+	+	+	–

+ = ate cookie, – = did not eat cookie, ++ = ate more cookie, h = hesitated before eating cookie.

dures were applied while the rats were eating the cookies.

Rats were trained in a distinct box (with a brass rod floor) and room. An Oreo® cookie was placed at the far end of the box. One aversive stimulus was brief foot shock. A second aversive stimulus was a noxious odor (ammonia). For the approaching group, the ammonia was on a swab brought close to the rat's nose through the floor bars when the rat reached the cookie. For the eating group, the ammonia was on the cookie. A third aversive stimulus was a "hot" flavor (hot mustard). For the approaching group, the hot mustard was on a swab that was placed briefly in the rat's mouth when the rat reached the cookie. For the eating group, the mustard was on the cookie. A fourth aversive stimulus was a bitter flavor (quinine), which was presented in a manner similar to that used for the hot mustard. The fifth aversive stimulus was an injection of lithium chloride, which produced illness. For the rats in this group, after either approaching or eating the cookie, the rats were picked up and immediately given an intraperitoneal injection of lithium chloride.

Rats in all groups were immediately returned to their home cages. Two control groups of rats were simply picked up and returned to their home cages after approaching the cookie (first control group) or after eating the cookie (second control group).

After training, all rats were given the opportunity to approach and eat a cookie in three different environments: the training cage, the home cage, and a novel cage and room. Table I summarizes the results of these experiments.

In general, none of the aversive stimuli (peripheral or lithium chloride injection) presented as the rats approached the cookie had any affect on subsequent cookie-eating behavior in any of the test environments. Rats that received a shock stimulus for approaching the cookie, however, did hesitate before eating in the training cage. None of the peripheral aversive stimuli presented as the rats ate the cookie reduced subsequent cookie-eating behavior in any of the test environments; however, shock caused the rats to hesitate before eating in the training cage, ammonia caused the rats to hesitate before eating in the novel cage, and mustard and quinine caused an increase in eating in the home cage. Taste aversion conditioning (the rats given injections of lithium chloride after eating the cookies), however, caused the rats to totally suppress eating in all three test environments.

THE AVERSIVE STIMULUS ADVICE LANDSCAPE

The rat–cookie experiment is a laboratory model, which shows that repellents and pain failed to protect Oreo® cookies from being eaten by rats when a noxious or painful stimulus was no longer present. Taste aversion conditioning succeeded in protecting the cookies. In applied settings, the failure to understand the evolutionary relationship between aversive stimuli and organized species-specific defense responding accounts for many of the failures or inappropriate recommendations concerning the use of aversive peripheral stimuli in managing animal behavior problems. The failure to make this distinction is illustrated by analysis of two files found on a commercial computer service. The files contained questions asked and advice given to service users about canine and feline behavior problems. One file covered the period July 1, 1990 to May 31, 1991 and the second file covered the period June 1, 1994 to June 15, 1994. The advice

TABLE II
Number of Questions Asked and Recommendations Given According to Species and Type of Behavior

Species	*Behavior Category of Question*	*Questions*	*Recommendations*
Cats	Ingestive		
	Eating	4	8
	Chewing	3	6
	Eating and chewing	0	0
	Coprophagia	0	0
	Noningestive	16	25
	Cats (total)	23	39
Dogs	Ingestive		
	Eating	11	24
	Chewing	0	0
	Eating and chewing	4	11
	Coprophagia	6	15
	Noningestive	88	157
	Dogs (total)	109	207
Cats and dogs combined	Ingestive		
	Eating	15	32
	Chewing	3	6
	Eating and chewing	4	11
	Coprophagia	6	15
	Noningestive	104	182
	Combined (total)	132	246

was offered by veterinarians, veterinary technicians, and obedience trainers. The questions were categorized according to species (cat and dog) and type of behavior (ingestive and noningestive). Table II shows the frequencies of different questions and recommendations according to species and type of behavior.

The 246 recommendations of those offering advice were categorized into six major categories:

- Targeted behavioral approaches consisting of the subcategories appetitive contingency, aversive contingency (including repellents), and other behavior modification techniques (most often systematic desensitization)
- Obedience training
- Establishment of access barriers and/or removal of objects or animals from the situation
- Medical approaches consisting of the subcategories surgery (mostly castration and debarking), drug administration, and alterations in nutrition
- Taste aversion conditioning
- Other approaches consisting of the subcategories unclassified (often political, such as enforcement of leash laws), learning explanations, and ethological or other psychological explanations.

Figure 2 shows the percentage of recommendations in each of the six major categories for each species and type of behavior. Appetitive and aversive contingencies are used here to mean approximately the same as pleasant and unpleasant states of "feeling." The reader may be inclined to substitute the terms *positive reinforcers* and *punishers*, but these terms refer to a specific economic theory proposed by B. F. Skinner and associates and do not address the existence or importance of feeling states. It is my belief that the hedonic qualities of stimuli represent a primary process that drives behavior.

Targeted behavioral approaches using response–contingent procedures dominate the advice landscape. The use of explanations as substitutes for remedial advice is also common for all categories of questions. Establishing access barriers, removing the animal from the situation, and eliminating problem stimuli are frequent prescriptions. Suggesting general obedience training as a problem solution was surprisingly infrequent, and suggesting the use of medical and taste aversion procedures was low.

Figure 3 shows the percentage of recommendations in each of the subcategories of the targeted behavior approaches (appetitive contingency, aversive contin-

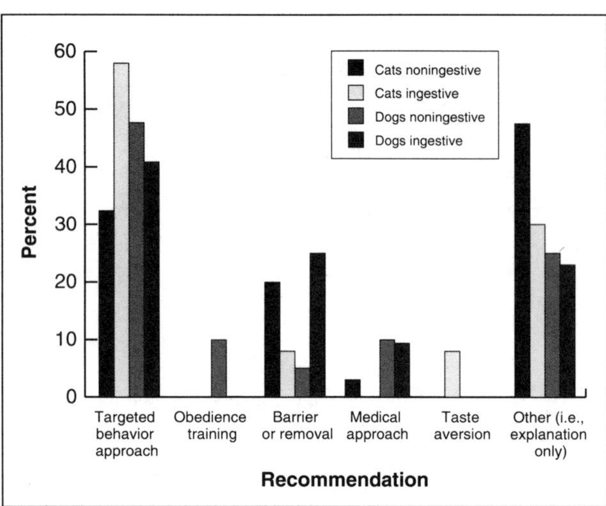

Figure 2—Analysis of two files from a commercial computer service. The files contained 246 recommendations from advisors on animal behavior problems. The figure shows the percentage of recommendations in each of the six major categories of advice, sorted by species and type of behavior.

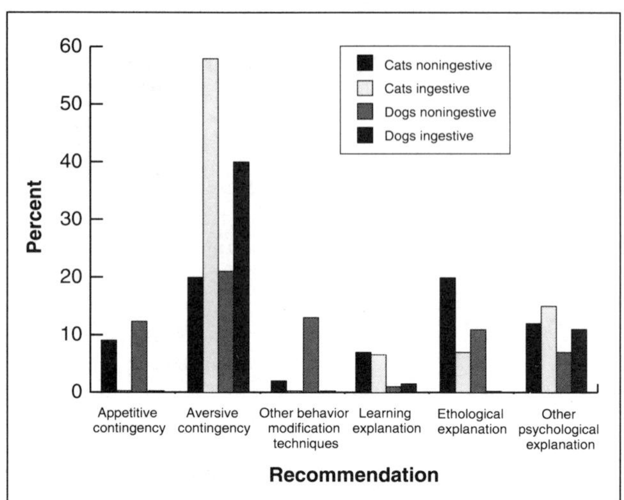

Figure 3—Analysis of two files from a commercial computer service. The percentage of recommendations (total = 246) that involved targeted behavior approaches and explanations for each species and type of behavior.

gency, other behavior modification techniques) and explanations (learning, ethological, and other psychological) for each species and type of behavior. It is very clear that pet owners seeking advice for pet problems involving ingestion were told to present an aversive peripheral stimulus to the animal upon eating or chewing of the item. Among the proposed aversive stimuli were shocks; swats; loud vocalizations; other loud noises; bad tastes (e.g., hot sauces and powders, lemon, vinegar, or bitter substances); leash jerks; streams of water from squirt bottles; corrections and mild corrections; and snap-type mousetraps set under sheets of newspaper. It is of note that the use of aversive contingencies was more often recommended for ingestive than for noningestive behavior problems and that appetitive contingencies were more often recommended for noningestive behaviors than for ingestive behaviors.

Recommending the use of such behavior modification techniques as systematic desensitization was limited to instances of excessive agitation associated with thunderstorms, fireworks displays, and riding in vehicles. Behavior strategies based (correctly or incorrectly) on learning principles were most commonly recommended, but explanations of behavior based on learning principles were the least frequently offered. The responses may reflect the nature of the medium of communication (a commercial computer service), a lack of familiarity with the details of sophisticated behavior analysis, an attempt to respond with "explanations" as treatment when action-based recommendations were not obvious to the advisor, or an assumption that behavior methods are not useful in modifying instinct-based or trauma-based behaviors. The use of gustatory repellents was inaccurately described as taste aversion conditioning on six occasions. Although an appropriate suggestion to use taste aversion conditioning was made on one occasion, details on how to implement the procedure were not given. Rather, a vague reference was made to "some veterinarians use the method." The use of appetitive contingencies was also infrequently suggested.

All recommendations concerning the presentation of aversive peripheral stimulation had the intention of terminating unwanted activity (the appropriate and expected outcome of a sufficiently intense repellent or operant punishment procedure). It also was expected that the recommended procedure would diminish the probability of the unwanted response on future occasions in different settings or when repellents were no longer present (an outcome that is not strongly supported by laboratory findings). Of the 246 recommendations, 85 involved the use of aversive procedures. On only one occasion concerning ingestive behaviors was taste aversion conditioning suggested. The presentation of an aversive peripheral stimulus or an explanation for the behavior was offered as advice on 84 occasions. From a practical point of view, the most common advice given by the advisors concerned the use of aversive procedures. On 84 of 85 occasions, advice was either inappropriate for the behavior involved or included expectations that are not consistent with the laboratory findings upon which the advice was originally founded. Regarding the use of aversive procedures, the advice given was in error 98.8% of the time.

Although I would expect the patterns identified in the described computer files to be representative of the advice given to pet owners in other situations, these computer files are not a scientific sample of advice given to pet owners by experts. One should be careful not to make generalizations concerning these data. The information is presented here to serve as a platform for discussion and comparison of the discrepancies between the scientific psychological literature and commonly given behavior advice.

SPECIES-SPECIFIC DEFENSIVE BEHAVIOR AS A CONSTRAINT ON IF–THEN ASSOCIATIONS

The only responses readily trained with the use of aversive peripheral stimuli are responses that are consistent with the evolutionary-based defensive actions of an animal. In the laboratory, it has been shown that pigeons cannot be trained to peck (an eating response) at a mechanical disk to avoid electric shock; however, pigeons can be trained to fly from perch to perch or step on a mechanical lever to escape or avoid shock. For the pigeon, flying and stepping seem to be consistent with actions associated with fleeing.[3,4] In the laboratory, it has been shown that rats placed in the center of a box can be trained to run to the edge of the box to escape and then avoid foot shock; however, rats placed near the edge of the box and required to move to the open center of the box to escape or avoid foot shock will persist in running around the edge of the box while a shock is given. When threatened or attacked, wild rats run to edges of, for example, rocks, walls, or bushes. In the artificial setting of the laboratory, rats use the wisdom of their evolutionary heritage. Jumping toward an edge of the shock box is consistent with the evolved defensive activities of the species, whereas jumping into an open area is not consistent.[5] In the laboratory, some rats can be trained to avoid foot shock by pressing a lever—but only after hundreds and sometimes thousands of shocks.

When rats are trained to press a bar for food reward, the form of the bar press response resembles the activities of eating. When the rat presses a bar to escape or avoid foot shock, the bar-pressing activities of the rat resemble fighting and attacking behaviors. Apparently, some rats can figure out how to incorporate portions of their natural eating and defensive behaviors into actions necessary to manipulate a laboratory lever, but more rats succeed in doing so when working for food than when defending themselves from pain.

When correcting behavior problems and trying to establish appropriate behaviors, an often subtle but extremely critical factor is selecting the form of the response the animal has to learn. Shock collars may be used successfully to train dogs to turn right or left for field trials because training the direction of running is consistent and compatible with the use of running to flee from an aversive tactile stimulus. Using shock to suppress attacking and biting in a dog is likely to elicit continued, if not exacerbated, attacking and biting unless the intensity of the aversive stimulus is sufficient to change the defensive behavior of the dog from attacking to immobilization or fleeing.

The Oreo® cookie experiment shows that training procedures using painful or noxious stimuli are extremely sensitive to contextual cues and extremely limited in the types of behavior that can be trained—this is a critical concept that is usually not presented clearly in the popular literature or in the classroom. Aversive peripheral conditioning works best with behaviors relevant to the skin defense system, such as freezing, escaping, avoidance, and attacking. Otherwise, the use of peripheral aversive stimuli is often unsuccessful. When using shock as a training stimulus in relatively simple avoidance procedures in the laboratory, some responses can be trained only after hundreds of shocks (occasionally thousands) have been presented. When the avoidance response is inconsistent with the evolved defensive activities of the species, animals may never learn to perform the required responses. Several excellent sources of information on this topic are available.[6–8]

APPLIED USE OF TASTE AVERSION CONDITIONING

Unlike the if–then associations of the skin defense system, in which target items remain desirable and associated stimuli serve as predictive cues indicating an aversive event, in the gut defense system, the associated stimulus (taste) changes from appetitive to aversive. The animal does not treat gut defense associations in temporal and spatial terms, but rather, the hedonic quality of the flavor associated with illness changes from desirable to undesirable, from good to bad, from pleasant to unpleasant, from palatable to unpalatable, from consumable to repellent and repugnant. After conditioning, a flavor paired with illness becomes an aversive stimulus; it does not predict one. After conditioning, the now aversive flavor terminates eating and elicits appropriate coping behavior. With additional experience, the animal learns to treat the items with this flavor as nonfood. For example, sheep in a flock can be protected from predators by using repellents. Each sheep in a flock must be sprayed with the repellent to provide protection from predators. Any sheep not treated with the repellent is unprotected. If a coyote becomes ill after consumption of a sheep-flavored meal, all sheep are protected from that coyote. By using conditioned taste aversion, the flavor of the food paired with illness becomes repellent. Only sheep not having the flavor of the aversion meal (sheep) are unprotected.

Garcia at al[9] reviewed the critical parameters for establishing conditioned taste aversions:

- The flavor of the food, and only the flavor of the food, is the effective stimulus for association with illness. Neither the color, shape, mobility, or location of the food is a relevant factor in determining aversion establishment or performance.
- The onset of illness may occur hours after a meal. A delay in the onset of illness of approximately 20 minutes seems to be optimal. Beyond the optimal 20-minute delay, the longer illness is delayed, the more the association is diminished. Caution: The rate of change for this modulatory process is a matter of hours, not seconds and minutes. In applied contexts, concern about delays in the onset of illness (unless the delays are more than several hours) probably are inappropriate. Illness can be an effective training stimulus even while an animal sleeps.
- Flavor–illness associations are readily established with one or two pairings of sufficient illness with a flavor.
- Conditioned taste aversions are established to stronger or more concentrated flavors better than to weaker or less concentrated flavors.
- Illness that is more intense or longer lasting establishes stronger conditioned taste aversions than illness that is less intense or of shorter duration.
- When a meal consists of two flavors (equivalent in preference, strength, and concentration) paired with illness, the less familiar flavor will become aversive. Caution: This finding has been frequently misinterpreted to suggest that familiar flavors cannot become aversive and only novel flavors are effective in taste aversion conditioning. Conditioned taste aversions can be readily established using flavors with lifelong exposure and familiarity.

In 1973, Gustavson et al[10] demonstrated that coyotes fed a sheep-flavored meal followed by illness induced with lithium chloride will not only refuse to eat sheep-flavored meals but will also refuse to kill sheep. Since then, several researchers have shown similar effects in a wide variety of predator and prey species.[11–14]

In captivity, coyotes have been trained to avoid eating dog food, rabbits, and chickens in addition to sheep. Wolves have been trained to stop killing and to avoid eating sheep. Red-tailed hawks were trained to stop killing and to avoid eating mice. Opossums, ferrets and rats were trained to stop killing and to avoid eating mice. Mountain lions were trained to avoid eating deer meat. Dingoes and New Guinea wild dogs were trained to avoid eating minced lamb. Magpies were trained to avoid eating chicken meat and owls trained to not eat chicks. Blackbirds were trained to avoid eating sunflower seeds. Pheasants were trained to avoid eating and thus to stop damage to sprouting corn. Blue jays were trained not to kill and eat butterflies. Dolphins and sea lions have been trained not to eat specific fish. Cattle, sheep, goats, and deer have been trained not to eat specific plants. Chemotherapy agents have been shown to establish conditioned taste aversions in persons suffering from cancer.[15]

In the field, taste aversion conditioning has been used to stop coyotes from killing sheep, turkeys, and antelope fawns; to stop wolves from killing cattle; to stop ravens and crows from damaging and eating the eggs of cranes; to stop endangered foxes from eating bait in traps intended for feral cats that were killing and eating endangered shrikes; to stop raccoons from killing chickens; to stop coyotes and bears from eating garbage in and around campgrounds and villages; and to stop baboons from eating growing corn. Recently, plant toxicosis has been shown to change the diet selection of free-ranging ungulates.[16]

Some investigators have reported results of experiments using procedures that were not consistent with the procedures outlined earlier for establishing conditioned taste aversion. Burns[17] adulterated the flavor of the aversion meal (chicken) by using a concentration of lithium chloride between 11.25 and 22.5 times stronger than the concentration advocated by Gustavson and Nicolaus.[14] When the flavor of the aversion meal was adulterated with lithium chloride in this manner, coyotes that ate the meal and subsequently became ill refused to eat salty chickens but continued to kill and eat unadulterated chickens. Bourne and Dorrance[18] attempted to protect free-ranging sheep by distributing baits containing lithium chloride throughout the range. Unlike the baits advocated by Gustavson et al,[19] which consisted of 1 kilogram of sheep meat and hide and 2 to 4 grams of lithium chloride, the baits used by Bourne and Dorrance were made of sheep meat, sheep hide, and lithium chloride as well as the strong petroleum-flavored chemical rhodamine B. Because the baits used to train conditioned taste aversions to free-ranging coyotes tasted different from the free-ranging sheep, only minimal field protection was obtained.

Conover[20] attempted to use ipecac as an illness-producing agent to protect waterfowl eggs from free-ranging raccoons; however, ipecac caused regurgitation, which Conover confused with illness, and in addition, ipecac adulterated the flavor of the eggs.

Many events will cause animals to regurgitate (e.g., food begging by offspring, sudden loud noises, and mechanical stomach irritation) but fail to support taste aversion conditioning. Many chemicals (e.g., thiabendazole, ketamine hydrochloride, and estrogen) will support taste aversion conditioning but fail to reliably cause regurgitation. Some chemicals (e.g., lithium chloride, digitalis, and copper sulfate) reliably produce regurgitation (in those animals that are able to regurgitate) and also produce strong conditioned taste aversions. Ipecac reliably produces regurgitation but fails to establish strong conditioned taste aversions.[21,22]

Lehner and Horn[23,24] automatically administered painful, concentrated lithium chloride injections into the abdomens of coyotes as they approached a rabbit but before they could kill and eat the rabbit. The injections took place in a confined funnel-shaped space. When tested several days later (after recovering from illness), the coyotes avoided the funnel-shaped space but readily killed and ate rabbits presented away from the funnel-shaped environment. Lehner and Horn[23,24] used skin-defense procedures and obtained the expected results—spatial avoidance. They were disappointed that they did not obtain gut-defense results—rabbit avoidance.

Rafe[25] advocated that dog owners who wanted to prevent their dogs from interacting with snakes, especially poisonous snakes, should present a nonpoisonous species to the dog just before the expected onset of signs produced from the administration of a dose of ipecac or a sublethal dose of some other malaise-inducing chemical. This procedure has four fundamental flaws. Ipecac is a poor choice of illness-inducing drug. The dog does not eat the snake before illness is induced. The procedure implies that it is important to pair the presentation of the snake close in time with the onset of the signs of illness. Finally, even if an appropriate drug were administered after consumption of the nonpoisonous snake and a conditioned taste aversion occurred to this nonpoisonous snake, it is very unlikely that the dog would avoid other (including poisonous) species of snakes.

Taste aversion conditioning holds great promise for treating a variety of ingestion-related behavior problems in companion animals and livestock. Dogs, cats, and other pets as well as livestock frequently attack, eat, and chew a variety of undesirable items, such as feces, dirt, clothes, plants, squirrels, and rabbits. Yelling, slapping, or squirting the animal with water does not prevent future misbehavior; however, allowing the animal to eat the undesirable item on one occasion followed by a controlled and safe illness will probably prevent future consumption or destruction. The illness can be induced after consumption of the item by two methods: the illness-producing agent can be injected or it can be included in the aversion meal. The agent must be safe and must not alter the flavor of the meal.

Sometimes, pets cannot be allowed to eat a particular item that needs to be protected or the use of repellents would render the item unusable. To address this problem, a neutral flavor that has no odor can be used in a conditioned taste aversion procedure. After an aversion is established to the neutral flavor, the flavor can be used as a gustatory repellent, which is repugnant only to the conditioned animal. This conditioned repellent approach suffers from the same difficulties associated with the use of any other highly effective gustatory repellent, that is, it must be present to be effective. When animals need to be protected from poisonous sources that are not themselves noxious (for example, household chemicals, antifreeze, swimming pool chemicals, and recently "shocked" pool water), conditioned taste aversions could be established to otherwise nonoffensive odorants and flavorants. These now repugnant flavors can be added to toxic but sometimes attractive sources.

Finally, illnesses, worming agents, chemotherapy, and hormone treatments can change an animal's food habits and appetite through taste aversion conditioning. By understanding the processes of taste aversion conditioning, the negative aspects of illness and medical procedures can be prevented or minimized and the animal's normal food consumption can be restored.

CONCLUSION

When companion animals misbehave, the advice commonly given is to present an aversive peripheral stimulus concurrently or immediately after the occurrence of the unwanted behavior. For noningestive behaviors, this suggestion may be appropriate if termination of the unwanted behavior is the only acceptable outcome and the defensive behaviors (e.g., cowering, running and staying away, biting, and attacking) elicited by the aversive peripheral stimulus do not become problems in and of themselves.

Considering the limited effectiveness, the high probability of eliciting additional unwanted responses, and the intense vigilance required when using an aversive peripheral stimulus to control the occurrence of unwanted noningestive behavior, the use of aversive peripheral stimulation is probably overadvised. Such usage, however, is widespread and broadly advised because users and advisors are deceived by common sense into believing that future occurrences of the unwanted behavior will be diminished. To the contrary, laboratory experiments have shown that although conditioning with aversive peripheral stimuli usually stops a particular behavior, the use of aversive peripheral consequences to establish avoidance behavior is extremely difficult.

Methods other than the use of aversive peripheral stimuli are often more efficacious for eliminating noningestive behavior problems but may be perceived by owners as requiring greater effort, creativity, self-discipline, time, and cost. The burden lies with pet advisors to convince pet owners that alternative methods are a better long-term investment.

Behavior advisors are more likely to prescribe the use of aversive peripheral stimuli when companion animals engage in unwanted pursuit, killing, chewing, or eating as opposed to when the problem involves noningestive behavior (Figure 3). When the unwanted behavior is in the ingestive behavioral domain, the use of aversive peripheral stimuli is appropriate only for termination of the currently occurring behavior. Following this prescription, however, will have no inhibiting effect on future occurrences of the unwanted ingestion. For eliminating unwanted ingestive behaviors in companion animals, the use of taste aversion conditioning, the process by which animals naturally learn to avoid foods that lead to illness, holds greater promise of efficacy than does the application of aversive peripheral stimuli.

REFERENCES

1. Garcia J: Learning without memory. *J Cognitive Neurosci* 2(4):287–305, 1990.
2. Gustavson CR, Gustavson JC: Food avoidance in rats: The differential effects of shock, illness, and repellents. *Appetite* 3:335–340, 1982/83.
3. Moore BR: The role of directed Pavlovian reactions in simple instrumental learning in the pigeon, in Hinde RA, Stevenson-Hinde J (eds): *Constraints on Learning.* London, Academic Press, 1973.
4. Smith RF, Gustavson CR, Gregor GL: Incompatibility between pigeon's unconditioned response to shock and the conditioned key-peck response. *J Exp Anal Behav* 18:147–153, 1972.
5. Kelly MJ, Grossen NE: Differential acquisition of an avoidance response as a function of the rat's species-specific behavior. Paper presented at the annual meetings of the Western Psychological Association, San Francisco, California, April 1971.
6. Bolles RC: Species-specific defense reactions and avoidance learning. *Psychol Rev* 77:32–48, 1970.
7. Bolles RC, Fanslow MS: Endorphins and behavior. *Annu Rev Psychol* 33:87–1001, 1982.
8. Seligman MEP, Hager JL: *Biological Boundaries of Learning.* Engelwood Cliffs, New Jersey, Prentice-Hall Inc, 1972.
9. Garcia J, Hankins WG, Rusiniak KW: Behavioral regulation of the milieu interne in man and rat. *Science* 185:824–831, 1974.
10. Gustavson CR, Garcia J, Hankins WG, Rusiniak KW: Coyote predation control by aversive conditioning. *Science* 184:581–583, 1974.
11. Gustavson CR: Comparative and field aspects of learned food aversions, in Barker LM, Best MR, Domjan M (eds): *Learning Mechanisms in Food Selection.* Waco, Texas, Baylor University Press, 1977, pp 23–43.
12. Gustavson CR, Brett LR, Garcia J, Kelly DJ: A working model and experimental solutions to the control of predatory behavior, in Markowitz H, Stevens V (eds): *Behavior of Captive Wild Animals.* Chicago, Nelson-Hall, 1979, pp 21–66.
13. Gustavson CR, Gustavson JC: Predation control using conditioned food aversion methodology: Theory, practice and implications, in Braveman N, Bronstein P (eds): *Experimental Assessments and Clinical Applications of Conditioned Food Aversions. Ann N Y Acad Sci* 348–356, 1985.
14. Gustavson CR, Nicolaus LK: Taste aversion conditioning in wolves, coyotes and other canids: Retrospect and prospect, in Frank H (ed): *Man and Wolf.* Dordrecht, The Netherlands, Dr. W. Junk Publishers, 1987, pp 169–203.
15. Bernstein IL: What does learning have to do with weight loss and cancer? *Science and Public Policy Seminars.* Washington, DC, Federation of Behavioral Psychological and Cognitive Sciences, 1988.
16. Provenza FD: Postingestive feedback as an elementary determinant of food selection and intake in ruminants. *J Range Manage* 48:2, 1995.
17. Burns RJ: Evaluation of conditioned predation aversion for controlling coyote predation. *J Wildl Manage* 44:938–942, 1980.
18. Bourne J, Dorrance MJ: A field test of lithium chloride to reduce coyote predation on domestic sheep. *J Wildl Manage* 46:235–239, 1982.
19. Gustavson CR, Jowsey JR, Milligan DN: A three year evaluation of taste aversion coyote control in Saskatchewan. *J Range Manage* 35:57–59, 1982.
20. Conover MR: Reducing mammalian predation on eggs by using a conditioned taste aversion to deceive predators. *J Wildl Manage* 54(2):360, 1990.
21. Barker LM, Best MR, Domjan M (eds): *Learning Mechanisms in Food Selection.* Waco, Texas, Baylor University Press, 1977.
22. Braveman N, Bronstein P (eds): *Experimental Assessments and Clinical Application of Conditioned Food Aversions. Ann N Y Acad Sci*:1985.
23. Lehner PN, Horn SW: Effectiveness of physiological aversive agents in suppressing predation on rabbits and domestic sheep by coyotes. Research proposal submitted to the U.S. Fish and Wildlife Service, Denver Wildlife Research Center, Denver, Colorado, 1975.
24. Lehner PN, Horn SW: Effectiveness of physiological aversive agents in suppressing predation on rabbits and domestic sheep by coyotes. Final research report submitted to the U.S. Fish and Wildlife Service, Denver Wildlife Research Center, Denver, Colorado, 1977.
25. Rafe SC: Snakeproof dogs without risk. *Quail Unlimited.* July 1989, pp 30–31.

PSYCHOPHARMACOLOGY AND ENDOCRINOLOGY

Behavioral Pharmacotherapy

Diplomate, American College of Veterinary Behaviorists
The Veterinary Behavior Clinic
Southern Pines, North Carolina
Barbara S. Simpson, PhD, DVM

Diplomate, American Board of Psychiatry and Neurology
Department of Psychiatry
Moore Regional Hospital
Pinehurst, North Carolina
Dale M. Simpson, MD, PhD

Behavioral pharmacotherapy (the use of drugs to treat behavioral problems) is rapidly expanding in veterinary medicine. Many recent applications in veterinary behavioral pharmacotherapy involve drugs already widely used by psychiatrists to treat psychiatric illness in humans, and it seems certain that the use of these drugs to treat behavior problems in animals will continue to expand. This review summarizes the psychopharmacologic agents used most often for humans and considers their applications in veterinary medicine. Although the human psychiatric literature serves as a useful reference, drugs tested and labeled for human psychiatric problems may have different effects, side effect profiles, or toxicities in companion animals.

NEUROTRANSMITTERS

All psychoactive drugs are thought to produce their behavioral effects through their actions on neurotransmitters in the brain. The following five neurotransmitters are particularly relevant to the action and side effect profiles of psychopharmacologic agents: acetylcholine, dopamine, norepinephrine, serotonin, and γ-aminobutyric acid (GABA).[1] Dopamine, serotonin, and norepinephrine are related by their chemical structure and are called monoamine neurotransmitters. Monoamines are especially concentrated in the midbrain, the hypothalamus, and the limbic system (a part of the brain thought to be central to the control and expression of emotions).

Acetylcholine, perhaps the most widely distributed neurotransmitter in the brain and the body, is produced from choline and is rapidly inactivated by acetylcholinesterase. Cholinergic receptors are widely distributed and have numerous physiological and behavioral effects. Many psychiatric drugs produce important anticholinergic side effects, similar to those of atropine. These include dry mouth and eyes, retention of urine and feces, dilated pupils, and cardiogenic effects.

Dopamine is produced from L-dopa in the presynaptic vesicles; L-dopa is produced from dietary tyrosine. Dopaminergic neurotransmission is complex, with dopamine activating at least five subtypes of dopamine receptors located presynaptically and postsynaptically.[2] Dopaminergic antagonists produce behavioral quieting without cortical depression.

Norepinephrine or noradrenaline is the precursor to epinephrine and also acts centrally as a neurotransmitter. Norepinephrine is formed from the hydroxylation of dopamine. Norepinephrine agonists are typically behaviorally stimulating, increasing arousal through activation of the reticular activating system and other mechanisms.

Serotonin is the common name for 5-hydroxytryptamine (5-HT). It is produced in the brain from the amino acid tryptophan. There are at least nine serotonin receptor subtypes,[3] each of which has different anatomical distributions and different behavioral roles. Regulation of the serotonin receptor system is complex and can occur at presynaptic or postsynaptic sites. Serotonin is thought to play an important role in the control of sleep, pain, aggression, sexual behavior, thermoregulation, and food intake.

γ-Aminobutyric acid (GABA) is the prototypic inhibitory neurotransmitter.[4] It is produced from glutamate, is widely distributed in the brain, and is thought to produce numerous metabolic and behavioral effects that are, as yet, not well characterized.

PSYCHOPHARMACOLOGIC AGENTS

The most common human psychopharmacologic agents are classified into four major groups based on their distinguishing behavioral effects in humans. These are antipsychotic drugs, antidepressants, anxiolytics, and mood stabilizers.[5] Note that the behavioral effects of these drugs are complex and that the four classifications represent an oversimplification of their behavioral actions.

ANTIPSYCHOTICS

Antipsychotics, also called neuroleptics or major tranquilizers, have been used for almost 30 years to treat psychotic disorders in humans.[2] Typically, psychotic disorders involve a severe disturbance of brain function characterized by disruption of thought and speech and the presence of hallucinations or delusions. Severely agitated or nonresponsive behavior occurs. Although psychotic disorders are not presented in veterinary medicine, antipsychotic drugs are commonly used for tranquilization and for the brief treatment of states of behavioral arousal. All drugs in this class have similar therapeutic effectiveness in the treatment of psychosis in humans and differ primarily in their side effect profile.

Antipsychotics are generally divided into two main groups to distinguish differing side effect profiles. These have been called the low-potency and the high-potency drugs because changes in potency seem to parallel changes in side effect profile. Additionally, antipsychotics may be classified according to their structural similarity. The largest structural class, the phenothiazines, has been derived from the earliest prototype antipsychotic, chlorpromazine. Because the phenothiazines include both low- and high-potency antipsychotics, this structural classification tells little about the characteristics of any specific drug.

Compared with their high-potency counterparts, low-potency antipsychotics require larger doses (typically 1 to 3 mg/kg) and produce more sedation, anticholinergic side effects, and cardiovascular effects. They also produce fewer extrapyramidal side effects of the sort discussed below. Low-potency antipsychotic drugs familiar to veterinarians include the phenothiazines acepromazine maleate, chlorpromazine, and thioridazine. High-potency antipsychotic drugs are effective in smaller doses (typically 0.5 to 1 mg/kg), produce less sedation and anticholinergic side effects, but have a greater incidence of extrapyramidal side effects. The most commonly used high-potency antipsychotics are haloperidol, fluphenazine, trifluoperazine hydrochloride, prochlorperazine, and thiothixene.

Pharmacology and Mode of Action

All antipsychotic medications act as dopamine antagonists. They block dopamine receptors in the basal nuclei and limbic system and produce behavioral quieting without necessarily producing sedation, although sedation is a side effect of the low-potency antipsychotics. Antipsychotics produce ataraxia—a state of decreased emotional reactivity and relative indifference to stressful situations. Antipsychotics also tend to suppress spontaneous movements and other complex behaviors but spare spinal reflexes and unconditioned pain reflexes.

Antipsychotics are significantly metabolized in the liver by first-pass metabolism and are highly protein bound with no renal excretion. Many have active metabolites. In humans, these drugs have fairly long half-lives of between 10 to 30 hours. Two antipsychotics, haloperidol and fluphenazine, are available in long-acting depot forms that are given to humans as a single injection every 2 to 4 weeks. These depot antipsychotics could simplify treatment in animals that have responded to oral haloperidol or fluphenazine. Once injected, however, these depot medications take days to weeks to clear completely, so tolerance and response to these specific agents is best first established with their oral forms.

Side Effects

Antipsychotics produce many side effects. Sedation, anticholinergic effects, and α-adrenergic blocking effects (producing orthostatic hypotension) occur commonly, especially with low-potency antipsychotics. Since dopamine inhibits prolactin secretion by the anterior pituitary, most antipsychotics increase serum prolactin levels by blocking this action.

In humans taking high-potency antipsychotics, extrapyramidal motor side effects, including parkinsonism, dystonic reactions, and akathisia, are relatively common. Parkinsonism is characterized by difficulty in initiating movement; motor stiffness; resting tremor; stiff, shuffling gait; and reduced facial movement (masklike facies). Dystonic reactions are characterized by abrupt-onset spasms typically of the muscles of the eyes, tongue, or back. These spasms produce involuntary eye-rolling, tongue protrusion, or opisthotonos.

Akathisia or the "restless leg syndrome" is characterized by motor restlessness, particularly in the legs, with difficulty in remaining still. Akathisia, exhibited by motor restlessness, pacing, and agitation, occurs at therapeutic doses in certain animals. In humans, extrapyramidal side effects are treated with such antiparkinsonian agents as benztropine mesylate, trihexyphenidyl, diphenhyradramine, or amantadine hydrochloride. In humans, propranolol and other ß-blockers are effective in treating akathisia.

Neuroleptic malignant syndrome is a serious but rare and idiosyncratic side effect of therapeutic levels of antipsychotics. Signs include severe muscle rigidity, autonomic instability including hyperthermia and tachycardia, and changing levels of consciousness.

Tardive dyskinesia is a slowly developing side effect produced by long-term treatment with antipsychotics. It is characterized by facial grimacing and abnormal movements, especially of the mouth and tongue. The presence of tardive dyskinesia as a potential side effect of antipsychotic treatment has greatly limited the long-term use of these drugs in humans, as it may not disappear with termination of drug treatment and the afflicted individual may be left with a permanent disability. High levels of antipsychotics can cause catalepsy, a syndrome characterized by immobility, increased muscle tone, and abnormal postures.

Applications

The antipsychotics chlorpromazine and acepromazine maleate have long been used by veterinarians as tranquilizers for chemical restraint and to modify behavior.[6–8] In general, the antipsychotics produce behavioral quieting, nonspecifically reducing responsiveness but sparing spinal reflexes. Hart[6] suggested the use of phenothiazines to reduce aggression unrelated to fear and to reduce excitement and hyperactivity.[7] When given to laboratory mongrels, chlorpromazine (1.5 mg/kg orally) in comparison with a placebo significantly decreased measures of excitability.[6] Normal laboratory beagles conditioned to a task showed a significant decline in the operant response rate when given any one of six antipsychotic drugs; this change was attributed to loss of motivation, not loss of motor ability.[9] It is well-known among veterinarians that dogs apparently tranquilized with antipsychotic medications can still respond aggressively to painful stimuli.[10]

Chlorpromazine has been used in veterinary medicine to modify behavior for many years but is no longer marketed for veterinary use.[8] In one case, chlorpromazine in combination with behavior modification therapy was effective in the treatment of interdog aggression.[7] At high doses in cats, chlorpromazine has been reported to cause extrapyramidal effects, including tremors, shivering, rigidity, and loss of the righting reflexes.[8]

Acepromazine maleate is available as an injectable or oral medication in the pharmacy of most veterinary hospitals. It is approved for use in dogs and cats as a preanesthetic and as an aid in controlling intractable animals. It is often used to treat infrequent anxiety states. As a component of behavioral therapy, acepromazine maleate was recommended in one case to manage interdog aggression and in another case to tranquilize a dominant aggressive dog, thus allowing the owners to assert control.[7] Some animals exhibit signs of hyperactivity (possibly akathisia) when given acepromazine maleate.[10] An aggressive cat given acepromazine maleate (10 mg orally) demonstrated increased agitation and irritability.[11]

Thioridazine (1.1 mg/kg) has been used in one case to treat a dog for aberrant motor behavior, including erratic episodes of tail chewing, growling and snapping, and barking.[12] The drug controlled the abnormal behavior, but the behavior resumed when the drug was withdrawn. At a higher dose (2.2 mg/kg), mild side effects of tachycardia and firm feces reportedly occurred.[12]

In summary, the antipsychotics have marked effects on behavior, nonspecifically reducing responsiveness, and have a low toxicity profile. Although they are not selective antianxiety agents, like some of the drugs discussed below, they can be effective in episodic anxiety-producing situations by reducing responsiveness in general. Beyond these general indications, the use of these agents in veterinary medicine does not reflect their wide use in the treatment of human psychiatric illness. There are no common behavioral illnesses in animals similar to the common psychosis-producing illnesses in humans. At present, there are few veterinary indications for their long-term use.

ANTIDEPRESSANTS

As a group, the antidepressants have a much wider and more heterogeneous range of behavioral effects than any other group of behaviorally active drugs. They are used extensively in human psychiatry for many behavioral illnesses extending far beyond the treatment of depression. In the following section, the most commonly used antidepressants are organized into three classes: tricyclic antidepressants, selective serotonin-reuptake inhibitors, and atypical antidepressants. As was the case with the antipsychotics, all antidepressants appear equally effective for the treatment of depression in humans. They clearly differ in their effects on central neurotransmitters and in their side effect profile. In contrast to the antipsychotics, the various antidepressants do appear to differ in their effectiveness for treating behavioral problems other than depression, probably reflecting the wide heterogeneity of neurochemical effects that occur within this group.

Tricyclic Antidepressants

The tricyclic antidepressants represent a class of structurally similar compounds used in psychiatry for over 30 years.[1] The most commonly used of these drugs in psychiatric practice are the tertiary amines imipramine hydrochloride, amitriptyline hydrochloride, and doxepin hydrochloride and the secondary amine desipramine hydrochloride. Desipramine is actually a metabolite of imipramine and is produced in significant amounts by the hepatic metabolism of imipramine.

Compared with other tricyclic antidepressants, the tertiary amines are the most sedating and have more anticholinergic and cardiovascular side effects. They are available in generic forms and are inexpensive. Several other available tricyclic antidepressants, such as nortriptyline, trimipramine maleate, and protriptyline hydrochloride, have somewhat different side effect profiles but have no clear advantage to veterinarians over the less expensive drugs mentioned above. Also, to date no veterinary literature demonstrates their clinical application in animals.

Clomipramine hydrochloride (Anafranil®, Basel Pharmaceuticals) is an effective antidepressant with a side effect profile resembling that of the other, more sedating tricyclics. Clomipramine hydrochloride is distinguished neurochemically from the other tricyclics by its potent serotonin reuptake–inhibiting properties, which resemble those of the selective serotonin-reuptake inhibitors discussed in the next section.

All tricyclic antidepressants are well absorbed by the gastrointestinal tract, and most of their metabolism occurs in the liver through demethylation, aromatic hydroxylation, and glucuronide conjugation.[1] The long half-lives of the tricyclic antidepressants permit the convenience of once-a-day dosing, a particular advantage in animals that are difficult to medicate. Note that tricyclic antidepressants have a narrow therapeutic index, with considerable risk of toxicity in overdose. A one-week supply may be fatal to a person or animal if taken all at once, primarily because of quinidinelike effects on cardiac conduction and central nervous system toxicity.[13,14] There are no antidotes for tricyclic antidepressant toxicity.

Pharmacology and Mode of Action

Tricyclic antidepressants appear to block both norepinephrine and serotonin presynaptic neurotransmitter receptors in the brain, acutely reducing norepinephrine and serotonin turnover and effectively increasing the action of these neurotransmitters to varying degrees. For example, desipramine hydrochloride appears to affect norepinephrine receptors to a much greater degree than it affects serotonin receptors; clomipramine hydrochloride affects serotonin receptors preferentially.

Although the effects of all antidepressant drugs on central neurotransmitter receptors are immediate, they all require several weeks to reliably produce changes in depressive symptoms in humans. When used for behavioral illness other than depression, the therapeutic effects may appear much more rapidly, thus suggesting that the multiple behavioral effects of these drugs are probably mediated through various mechanisms. Intermediary mechanisms, such as alterations in receptor sensitivity or number, have been suggested for at least some of the behavioral effects.[1]

Side Effects

Tricyclic antidepressants have numerous side effects, as predicted by their action at various receptors.[1] The most clinically important side effects are cardiovascular, anticholinergic, antihistaminic, and sedative. Cardiovascular side effects include elevated heart rate and orthostatic hypotension. Direct effects on the heart include antiarrhythmic properties and slowing of cardiac conduction. Although not a problem in healthy individuals, this slowed cardiac conduction in patients with preexisting heart blocks can result in complete heart block.[15] This slowing of cardiac conduction produces widening of the QRS complex, which has been used as a bioindicator of overdose.

Anticholinergic effects caused by blockade of muscarinic receptors are a common result of the use of tricyclic antidepressants. These effects include mydriasis, dry mouth, reduced tear production, urine retention, and constipation. Generally, these pose few problems for young, healthy individuals but may lead to serious complications in compromised individuals. Even the seemingly benign side effect of dry mouth can, with long-term treatment, produce serious dental pathology.[1]

Sedative effects are significant with the tertiary amines amitriptyline hydrochloride, imipramine hydrochloride, and doxepin hydrochloride and are probably secondary to anticholinergic and antihistaminic activity. Sedative effects are reduced with desipramine hydrochloride and nortriptyline. Doxepin hydrochloride has the strongest antihistaminic effects and is actually one of the most potent antihistaminic drugs available. Weight gain is a common effect of the use of tricyclic antidepressants in humans, especially with amitriptyline hydrochloride and doxepin hydrochloride.[1]

Applications

In human psychiatry, tricyclic antidepressants are used to treat numerous clinical conditions, most notably depression.[1] All are labeled for use in humans as antidepressants, but most of them reportedly have additional applications. Of special interest is the use of tricyclic antidepressants to treat panic disorder and agoraphobia (fear of open spaces). Although imipramine hydrochloride and nortriptyline are not labeled for that use, controlled studies confirm the efficacy of both for the treatment of panic disorder, generally starting at a low dose and increasing the dose gradually.[1]

Desipramine hydrochloride, a tricyclic that blocks norepinephrine reuptake but has little serotonergic activity, has been used to treat hyperactivity in children.[16] Other uses of tricyclics in humans include treatment of narcolepsy, chronic pain, peptic ulcer, and pruritus.[1] In human psychiatry, tricyclic antidepressants remain a mainstay in the treatment of depression, producing clinically significant improvement in about 70% to 75% of patients with depression, compared with the 20% to 35% of patients whose condition improves with placebo.[17]

There are few controlled clinical trials on the use of tricyclic antidepressants in veterinary medicine. An exception is a series of studies by Rapoport and associates comparing the atypical tricyclic clomipramine hydrochloride with other medications.[18,19]

Amitriptyline hydrochloride is the most commonly used tricyclic antidepressant in veterinary practice,[10] although there is an acute shortage of clinical studies on its use. It reportedly has anxiolytic effects as well as sedative effects. When administered to normal dogs, amitriptyline hydrochloride (0.5 to 1.5 mg/kg orally) had no discernible effect on behavioral measures of friendliness, excitability, or fearfulness.[6] Amitriptyline hydrochloride has been used to treat urine marking in cats (a daily oral dose of 5 to 10 mg) and separation anxiety in dogs (a daily oral dose of 2.2 to 4.4 mg/kg).[20,21]

In a single case of feline hypervocalization, amitriptyline hydrochloride (5 mg per day orally) reduced the frequency of excessive vocalization.[22] In a case of a dog with diagnosed fearful aggression and inappropriate elimination related to anxiety, amitriptyline hydrochloride (25 mg orally every 12 hours for a 21 kg dog) and a behavior modification program apparently reduced the dog's anxiety.[23] Imipramine hydrochloride, which is used to reduce nocturnal enuresis in children, is reportedly effective (2.2 to 4.4 mg/kg every 8 to 12 hours orally) in combination with nonpharmacologic behavior techniques to control canine submissive urination and excitement-induced urination.[10]

Unlike the other tricyclic antidepressants, clomipramine hydrochloride has been shown to produce significant therapeutic effects in the treatment of human obsessive-compulsive disorder (OCD), probably because of its potent effects on serotonergic neurotransmission.[1] Other pathologic compulsive conditions (e.g., trichotillomania [hair pulling] and onychophagia [nail biting]) that are possibly related to obsessive-compulsive disorder have also responded to clomipramine hydrochloride therapy.[24,25]

In a single-blind crossover study of canine lick granuloma, Goldberger and Rapoport[18] reported reduced licking with clomipramine hydrochloride but not desipramine hydrochloride in six of nine dogs. Rapoport and coworkers[19] reported that clomipramine hydrochloride (up to 3 mg/kg daily orally) but not desipramine hydrochloride (up to 3 mg/kg daily orally) reduced licking behavior by 50% in 6 of 13 patients with canine acral lick granuloma. Short-term side effects of clomipramine hydrochloride reportedly occurred in 5 of 13 dogs and included lethargy, anorexia, diarrhea, and growling. A 6-month follow-up of six improved dogs revealed that only two dogs continued clomipramine hydrochloride therapy. For the other owners, the cost of the medication was prohibitive.[19]

Currently, there is a great deal of interest in the comparison between human obsessive-compulsive disorder and canine stereotypies, such as excessive tail chasing, fly biting, flank-sucking, and the licking associated with canine lick granuloma.[26] Whether these aberrant canine behaviors represent a model of human obsessive-compulsive disorder is beyond the scope of this paper (see the article by Hewson and

Luescher elsewhere in this book). Supporting evidence comes from the fact that drugs, including clomipramine hydrochloride, used to treat human obsessive-compulsive disorder are often effective in dogs exhibiting compulsive disorder.[18,19]

Overall[27] has documented in three dogs the success of clomipramine hydrochloride (1 mg/kg every 12 hours orally increased to 3 mg/kg every 12 hours orally) but not amitriptyline hydrochloride concomitant with behavior modification techniques to treat stereotypies, including circling. When clomipramine hydrochloride was withdrawn, the problematic behaviors tended to recur.[27] In another case, a dog self-traumatized as a result of circling and tail chasing was successfully treated with clomipramine hydrochloride (0.5 mg/kg every 12 hours orally increased to 3 mg/kg per day orally) but not diazepam, naloxone hydrochloride, or phenobarbital.[28] After 14 months, the dog was weaned from the medication and only occasionally exhibited the behavior.

Selective Serotonin-Reuptake Inhibitors

The selective serotonin-reuptake inhibitors have been hailed as an important advance in the pharmacotherapy of depression in humans as well as several other behavioral illnesses.[3] Their effectiveness for a wide range of clinical conditions and their relatively low incidence of undesirable side effects have made drugs in this class extremely popular in human psychiatry. In addition to the treatment of depression, selective serotonin-reuptake inhibitors are used to treat obsessive-compulsive disorder, eating disorders, panic disorders, and various other conditions.

Four of these drugs are currently available for clinical use: fluoxetine hydrochloride (Prozac®, Dista Products), paroxetine hydrochloride (Paxil™, SmithKline Beecham Pharmaceuticals), sertraline hydrochloride (Zoloft®, Roerig), and fluvoxamine maleate (Luvox™, Solvay Pharmaceuticals). They are all structurally distinct and differ in their pharmacokinetics and to a small degree in their side effect profile. None is available as a generic, and all are relatively expensive.

For treating depression, a single daily dose is the standard regimen, although higher doses are frequently used and appear to be well tolerated. Many of these drugs may produce some inhibition of hepatic enzymes, thus elevating blood levels of other drugs; so when they are used concurrently with other drugs, a standard reference source[29,30] should be consulted.

Pharmacology and Mode of Action

As the name implies, selective serotonin-reuptake inhibitors enhance central serotonin by blocking presynaptic neuronal input at serotonin receptors. These agents may also act through increased output or increased postsynaptic receptor sensitivity.[3] Several other antidepressants affect serotonergic neurotransmission. These include, most notably, clomipramine hydrochloride but also amitriptyline hydrochloride and other tricyclic antidepressants. The selective serotonin-reuptake inhibitors act more selectively, with little effect on other neurotransmitter systems.[3]

Side Effects

Clinically, the selective serotonin-reuptake inhibitors have a distinct side effect profile that differs greatly from that of the tricyclic antidepressants. The former rarely produce significant sedation, although fluvoxamine appears to be the most sedating, followed by paroxetine hydrochloride, then sertraline hydrochloride, and then fluoxetine hydrochloride. Since the selective serotonin-reuptake inhibitors have little affinity for cholinergic, adrenergic, or histamine receptors, they have fewer side effects than tricyclic antidepressants.[3]

The most common side effects in humans and in companion animals given selective serotonin-reuptake inhibitors are gastrointestinal symptoms and signs, including anorexia, nausea, and diarrhea. In humans, these gastrointestinal side effects can affect up to 25% of patients treated with these drugs. These effects can be reduced by starting at a low dose and increasing the dose to therapeutic levels to allow the development of tolerance, which usually occurs rapidly.[3] Other common side effects in humans and possibly other animals include anxiety, agitation, and insomnia.

Applications

Numerous studies have demonstrated the effectiveness of selective serotonin-reuptake inhibitors over placebo in the treatment of human depression.[3] In crossover trials, selective serotonin-reuptake inhibitors and tricyclic antidepressants produce similar improvement in depression. The latency and quality of response to a selective serotonin-reuptake inhibitor, however, is highly individualized. No clear relationship has been shown between clinical response and plasma concentration with fluoxetine hydrochloride or paroxetine hydrochloride.[3] Selective serotonin-reuptake inhibitors are used to treat several disparate disorders in human psychiatry, including obsessive-compulsive disorder, panic disorder, anxiety, and eating disorders, such as bulimia and anorexia.[3]

Fluoxetine hydrochloride is currently the most prescribed antidepressant in the world.[a] It is labeled for use in humans for treatment of depression, obsessive-

[a]Murphy V: Personal communication, Eli Lilly Corp., Indianapolis, September 11, 1995.

compulsive disorder,[31] and the eating disorder bulimia. Fluoxetine hydrochloride has at least one important active metabolite, norfluoxetine, which has a half-life of 4 to 16 days. This long half-life may reduce any effect of missed doses and may allow dosing at greater than 24-hour intervals but has the disadvantage that significant blood levels of the drug may remain long after discontinuation of the parent drug, thus creating problems in the case of adverse reaction or drug failure.

Paroxetine hydrochloride differs from fluoxetine hydrochloride in that it has no active metabolites and has a much shorter elimination half-life of approximately 20 hours.[3] Steady-state systemic levels of paroxetine hydrochloride are attained in approximately 10 days, although clinical improvement in human depression may not be evident for 2 to 3 weeks. Sertraline hydrochloride, like the previously discussed selective serotonin-reuptake inhibitors, is labeled for use as an antidepressant but is also effective in the treatment of obsessive-compulsive disorder.[32] The elimination half-life of the parent compound is approximately 25 hours.[3] A weak metabolite, desmethylsertraline, exceeds the half-life of its parent (65 versus 25 hours).

Fluvoxamine maleate is the most recently introduced selective serotonin-reuptake inhibitor and currently is approved for the treatment of obsessive-compulsive disorder only, although it appears to be as effective as others in this group for the treatment of depression, panic attacks, and generalized anxiety in humans.[3] Fluvoxamine has no active metabolites, and its elimination half-life is 15 hours, the shortest of the four selective serotonin-reuptake inhibitors.

The veterinary success of fluoxetine hydrochloride has been chronicled in the popular press, including innumerable newspaper, magazine, and television features as well as in the veterinary press.[33,34] In general, the reports have been based on anecdotal cases treated by veterinary behaviorists, dermatologists, and others. Unfortunately, few case-controlled studies are available to support these testimonials.[34,35] An exception is a study of the effect of fluoxetine hydrochloride and other medications on licking behavior of dogs with diagnosed acral lick granuloma.

Rapoport and coworkers[19] compared the effect of clomipramine hydrochloride, two selective serotonin-reuptake inhibitors (fluoxetine hydrochloride and sertraline hydrochloride), the tricyclic antidepressant desipramine hydrochloride, and a serotonin-releasing agent (fenfluramine hydrochloride) in three double-blind crossover drug comparisons. Fluoxetine hydrochloride was more effective than placebo, desipramine hydrochloride, fenfluramine hydrochloride, or sertraline hydrochloride at reducing licking behavior. In the fluoxetine hydrochloride trial, two dogs showed complete remission of symptoms. Four of fourteen dogs treated with fluoxetine hydrochloride showed side effects of lethargy, anorexia, and hyperactivity. At follow-up, a third of the fluoxetine hydrochloride responders continued favorable response without medication.[19]

In an open trial of 65 dogs with diagnosed psychogenic pruritus (n = 31), acral lick granulomas (n = 14), tail mutilation (n = 6), separation anxiety (n = 6), and miscellaneous behavioral problems (n = 8) and treated with fluoxetine hydrochloride (approximately 1 mg/kg per day orally), Melman[36] reported average onset of efficacy from 5 to 16 days. Mild side effects in some animals included "mellow" demeanor, lethargy, hyperactivity, polydipsia, diarrhea, decreased appetite, and increased appetite. Dogs with psychogenic pruritus showed varying degrees of improvement, graded on an equivocally defined scale. Ten of fourteen dogs were successfully treated for acral lick dermatitis with fluoxetine hydrochloride in addition to other therapies. Five of six dogs diagnosed with tail mutilation disorder and six of six with diagnosed separation anxiety recovered totally. In this study, other therapeutic modalities, such as antibiotics and allergy vaccines, were pursued concomitantly, confounding interpretation of the results. Diagnostic criteria and quantitative measures of treatment success were not provided.[36]

Others[10,37,38] have suggested the use of fluoxetine hydrochloride (0.5 to 1 mg/kg daily orally) for treatment of canine compulsive disorder. Successful fluoxetine hydrochloride treatment for psychogenetic alopecia in a cat has been reported (1 to 1.5 mg daily orally for a 4.1 kg cat).[39] Over 18 months, patches of alopecia recurred when the therapy was discontinued for a few weeks or when the dose was lowered (< 0.5 mg per day). Others[37] suggest higher doses of fluoxetine hydrochloride (0.5 to 1 mg/kg daily orally) for cats.

Dodman and Mertens[40] presented data on the use of fluoxetine hydrochloride for treatment of dominance-related aggression in nine dogs. In an incomplete-crossover design, dogs were given placebo for 1 week, then treated for 4 weeks with fluoxetine hydrochloride (1 mg/kg daily orally) for a total of 5 weeks. No other therapy was used, and owners were blind to the control week. During weeks 4 and 5 (after 3 and 4 weeks of fluoxetine hydrochloride therapy, respectively), owners reported significantly fewer aggressive responses (e.g., growl, snap, or bite) than during the placebo week. Side effects reported by owners included fatigue, lethargy, and decreased appetite.[40] In a case report of a behaviorally complex dog exhibiting dominance aggression, fearfulness, and circling, Overall[23] reported that fluoxetine hydrochloride (20 mg daily orally, later 10 mg daily orally) reduced aggression.

The use of other selective serotonin-reuptake inhibitors has been largely unexplored in the veterinary literature, although distinguishing features, such as their availability in tablet form and differences in pharmacokinetics, should stimulate controlled clinical trials of their use.

Atypical Antidepressants

This class contains several different drugs that are classified as antidepressants but are neither classic tricyclic antidepressants nor selective serotonin-reuptake inhibitors. These drugs are used frequently in human psychiatry and are just finding their way into veterinary medicine. The two atypical antidepressants reviewed here are trazodone hydrochloride and nefazodone hydrochloride (Serzone®, Bristol-Myers, Squibb). There are several other atypical antidepressants, such as bupropion hydrochloride (Wellbutrin®, Burroughs Wellcome) and venlafaxine (Effexor®, Wyeth Ayerst Laboratories), which apparently have unusual neurochemical mechanisms of action and side effect profiles but at present have no clear veterinary applications.

The monoamine oxidase inhibitors represent an additional class of atypical antidepressant drugs but are not considered here. They have multiple side effects and many problematic drug interactions and at present have limited application in veterinary or human behavioral medicine. One exception is the use of the monoamine oxidase B inhibitor, selegiline hydrochloride (also known as L-deprenyl), for treatment of "cognitive dysfunction."[41] Further information on any of these drugs is available in a standard reference on human psychopharmacology.[5]

Trazodone Hydrochloride and Nefazodone Hydrochloride

Trazodone hydrochloride and nefazodone hydrochloride are structurally related compounds that are unique in several respects. Both are inactive in standard animal tests for antidepressant drugs but inhibit painful and conditioned emotional responses.[42] Trazodone hydrochloride is a relatively weak inhibitor of serotonin reuptake; but the active metabolite, *m*-chlorophenylpiperazine, is a potent direct serotonin agonist. It is considered a mixed serotonergic agonist and antagonist with minimal effects on norepinephrine or dopamine reuptake.

Nefazodone hydrochloride has been made available only recently in psychiatry and was developed from trazodone with the goal of increasing serotonergic effects and reducing side effects.[42] Nefazodone hydrochloride blocks serotonin reuptake more specifically and also acts as a serotonin$_2$ receptor antagonist. The metabolite, *m*-chlorophenylpiperazine, again acts as a potent direct serotonin agonist.

The most common side effects of trazodone hydrochloride are sedation, dizziness, orthostatic hypotension, and headache. Although rare (approximately 1 in 6000 cases), trazodone hydrochloride can cause priapism (persistent penile erection) in human males, which may cause permanent loss of erectile function.[42] Nefazodone hydrochloride appears to be safe and well-tolerated clinically, with fewer side effects than trazodone hydrochloride. Reported side effects, such as dry mouth, constipation, sweating, and tremor, occur more frequently than in controls but less frequently than with imipramine hydrochloride. In several reported overdoses, patients recovered uneventfully.

In open and double-blind trials, the efficacy of trazodone hydrochloride and nefazodone hydrochloride are comparable to that of the tricyclic antidepressants in the treatment of depression. Trazodone hydrochloride and nefazodone hydrochloride are reported to have antianxiety effects as well.[42] Because of its relative safety and its sedating and antianxiety properties, trazodone hydrochloride has been used for the treatment of behavioral agitation and hostility in geriatric patients with dementia.[43] It is also frequently used as a single nighttime dose in combination with other antidepressants, such as the selective serotonin-reuptake inhibitors, to aid in sleep induction. Both trazodone hydrochloride and nefazodone hydrochloride represent unusual behaviorally active agents that, because of their antianxiety and behavioral calming effects and their benign side effect profile, may have an increasing role in the treatment of animal behavioral problems.

ANXIOLYTICS

Drugs with antianxiety properties include benzodiazepines, azapirones, barbiturates, and antihistamines. The benzodiazepines have become the standard drug class of choice for the treatment of anxiety in humans. A specific azapirone, buspirone hydrochloride (BuSpar®, MeadJohnson Pharmaceuticals), offers a pharmacologically distinct alternative to these drugs. The anxiolytic effect of the barbiturates is associated with a significant degree of cortical depression; and the anxiolytic effects of the antihistamines, although real, appear to be a nonspecific effect associated with the sedating properties of some of these drugs. Only the benzodiazepines and buspirone hydrochloride are considered in the sections below. Other previously discussed drugs, such as certain tricyclic antidepressants and the atypical antidepressants trazodone hydrochloride and nefazodone hydrochloride, may also be useful in the treatment of some anxiety states.

Benzodiazepines

The benzodiazepines have become synonymous with the term anxiolytics, and the vast majority of pharmacologic treatments for anxiety in humans involve benzodiazepines.[4] The most commonly used benzodiazepines are diazepam, lorazepam, alprazolam, chlorazepate dipotassium, oxazepam, chlordiazepoxide, and clonazepam. Several others (e.g., temazepam and flurazepam hydrochloride) are labeled only as sleeping pills. All benzodiazepines have great structural similarity and appear to work through the same mechanisms but differ widely with regard to pharmacodynamics and pharmacokinetics. With the exception of clonazepam (Klonopin®, Roche Laboratories), all commonly used benzodiazepines are available in generic form and are inexpensive. They are extremely safe in overdosage and have a limited number of side effects.[4]

Pharmacology and Mode of Action

Benzodiazepines have been available since chlordiazepoxide was introduced in 1960.[4] There are now at least 39 benzodiazepines on the market.[4] These drugs are thought to act by binding to GABA receptors in the central nervous system, thus promoting the inhibitory activity of GABA. Their anxiolytic properties seem to result from GABAergic activity at the levels of the cerebral cortex and limbic system. GABA receptors are widely distributed, so effects at other sites explain the side effects of benzodiazepines. For example, the sedating effects result from GABA activity in the reticular activating system, muscle-relaxing effects result from GABA effects at the level of the spinal cord, and anticonvulsant effects are mediated by GABA circuits at multiple levels of the central nervous system.[4]

The benzodiazepines are lipid soluble and are thus well absorbed from the gastrointestinal tract and pass easily into the brain. In humans, they reach peak plasma levels from 0.5 to 6 hours, depending on the absorption rate of each particular drug and other variables, such as the amount of food in the patient's stomach. After a single dose of a benzodiazepine, its behavioral action is reduced primarily by drug distribution. Diazepam, for example, is rapidly absorbed and rapidly distributed following a single oral dose. Even though diazepam has a long elimination half-life, its duration of behavioral action is relatively brief.[44]

In human psychopharmacology, these drugs are classified according to whether the elimination half-lives of the parent compound and active metabolites in humans are short (< 6 hours), intermediate (6 to 20 hours) or long (> 20 hours).[4] The half-life determines the time required to achieve a steady-state concentration, approximately four half-lives. Until a drug reaches steady state, it is accumulating and the optimal dosage for a particular patient cannot be determined.[44] Drugs with elimination half-lives greater than 24 hours may be administered once a day. Drugs with shorter half-lives should be given more often.

The intermediate–half-life benzodiazepines lorazepam, oxazepam, and temazepam are metabolized by conjugation to inactive glucuronides, sulfates, and acetylated substances. They have no active metabolites and are often used for elderly patients or those with compromised hepatic function. Alprazolam, another intermediate–half-life benzodiazepine, undergoes oxidation but has no clinically important metabolites. Only lorazepam and midazolam are reliably absorbed after intramuscular injection.[44]

Long–half-life benzodiazepines include diazepam, chlordiazepoxide, clorazepate dipotassium, and clonazepam. All are biotransformed by hepatic oxidative reactions to at least one active metabolite, desmethyldiazepam (nordiazepam). Clorazepate dipotassium requires biotransformation to an active form in the gastrointestinal tract. The fastest absorption occurs in an acid environment, thus clorazepate dipotassium is most effective when administered on an empty stomach and may be poorly absorbed in states of high gastric pH.[29]

Empirically, the duration of action of benzodiazepines in small animals, especially dogs, appears shorter than that in humans.[45] Thus, the timetable for elimination half-lives of benzodiazepines in humans cannot reliably be applied to other species. Other characteristics of the pharmacokinetics appear to differ between humans and dogs. For example, in humans but not in dogs, clorazepate dipotassium sustained delivery (Tranxene SD®, Abbott Laboratories) produces prolonged serum concentrations compared to the regular formulation (Tranxene®, Abbott Laboratories).[45]

Side Effects

Benzodiazepines have few side effects beyond the previously mentioned sedation, cortical depression, and muscle relaxation. They produce little respiratory depression even in overdoses, although they can potentiate the respiratory depression produced by other sedative hypnotics, such as the barbiturates. They have little effect on the cardiovascular system and apparently raise seizure threshold, although benzodiazepine withdrawal can be associated with seizures.

In humans, intravenous use produces amnesia. Oral dosing, particularly of the rapidly absorbed lipid-soluble benzodiazepines, may also produce amnesia and interferes with learning conditioned responses.[44] Chlordiazepoxide administration, however, greatly facilitates operant conditioning of nervous pointer dogs.[46] Use of a benzodiazepine may therefore reduce the effectiveness of a behavior modification program for some patients[21] but may facilitate associative learning by extremely anxious patients.

Partial tolerance to all the aforementioned side effects may develop with time. In general, tolerance develops to the nonspecific sedative effects of a benzodiazepine during long-term therapy, but tolerance to its anxiolytic effects is less common.[47] Gradual withdrawal is recommended after long-term use of benzodiazepines in humans and animals to avoid the discontinuation syndrome. In one regimen for withdrawal, the daily dose is reduced by 25% per week.[47]

In cats, there are reports of fatal idiopathic hepatic necrosis in response to short-term oral administration of benzodiazepines, especially diazepam.[48,49] This rare phenomenon has been confirmed as a result of oral generic diazepam and at least four proprietary formulations.[49] Although questions about the dosage given to affected cats and about their underlying medical status have been raised,[50] the phenomenon does not appear to be dose related and has occurred in cats with normal pretreatment hepatic enzyme assays.[48,49] There have been speculations that some cats produce toxic metabolites from diazepam, possibly related to species-specific differences in hepatic metabolism.

Applications

All benzodiazepines are more effective than placebos for the short-term treatment of generalized anxiety disorder in human patients.[47] Trials comparing different benzodiazepines have failed to demonstrate consistent differences in efficacy among various drugs of this class.[47] Several benzodiazepines have other applications. Alprazolam is labeled for the treatment of panic disorder. Clonazepam is labeled for treatment of epilepsy and is also used in the treatment of panic disorder. Flurazepam hydrochloride and temazepam are marketed for insomnia—likely more a marketing decision than a reflection of unique pharmacologic profile. Because they are biotransformed by conjugation, not oxidation, and have nonactive metabolites, lorazepam and oxazepam are used with success for agitated patients with compromised hepatic function.[29]

Despite the widespread use of benzodiazepines in veterinary practice, there is a shortage of scientific data for rational guidance of their use in animals. Hart[6] evaluated the effect of diazepam (0.7 mg/kg orally) compared to other drugs and placebo in dogs. In 7 of 10 dogs, treated animals (compared with controls) mildly increased behavioral measures of excitability and decreased fearful responses. Clordiazepoxide (1 mg/kg orally) given to a few dogs showed little effect on these measures.[6] Numerous reviews have suggested the benzodiazepines to be effective at reducing the intensity of fearfulness and its associated anxiety in dogs,[10,20,21] although few studies have tested these effects (see the Voith and Borchelt articles on fears and phobias and separation anxiety elsewhere in this book). Voith[21] recommends diazepam (0.55 to 2.2 mg/kg orally) for thunderstorm anxiety in dogs. Beaver[51] recommends chlordiazepoxide (0.5 to 5.0 mg orally) or diazepam (0.5 to 2.0 mg orally) for situations involving "frustration" or social anxiety in cats.

The effects of benzodiazepines on aggressive behaviors in animals are variable, possibly reflecting that different brain centers mediate different types of aggression.[51] Chlordiazepoxide reportedly has "taming" effects on wild cats and other zoo animals.[6,52,53] Marder[10] reported improvement with diazepam treatment in a 1-year-old dog exhibiting unpredictable aggression toward its owners as well as flank-chewing behavior. Review articles generally indicate that benzodiazepine can be used effectively to reduce fear-induced aggressive behavior of cats.[10,21] Schwartz,[11] however, reported the unsuccessful use of diazepam (0.5 to 1.0 mg orally every 12 hours) in two cases of feline aggression. These cats were successfully treated with the mood stabilizer carbamazepine.[11]

Benzodiazepines have been recommended for treatment of fear-related behaviors, including fear-induced aggressive behavior.[6,21] Several veterinary behaviorists, however, report that benzodiazepines can cause a paradoxical increase in some types of aggression in animals.[10,20,21] Diazepam may escalate predation in cats.[54] Benzodiazepines may actually disinhibit normal agonistic responses, thus increasing the likelihood of aggression, a particular problem in fear aggression in dogs and cats.

It is useful to think of benzodiazepines as being "disinhibitors" of suppressed behavior.[47] If aggression is a manifestation of underlying fear or anxiety, reduction in fear or anxiety will reduce aggression. In contrast, if fear or anxiety is restraining the expression of aggression, then reduction in fear or anxiety will release the expression of aggression. Because these underlying states may be difficult to differentiate, benzodiazepines should be used with caution for aggressive patients.

Several clinical studies have evaluated the efficacy of diazepam for the treatment of urine spraying. Urine spraying is a highly motivated behavior influenced by social and sexual factors.[55] Since urine spraying is generally resistant to other therapeutic modalities, such as behavior modification, pharmacotherapy is often important in the management of this behavior problem (see the article on cat elimination problems by Borchelt and Voith). In a prospective trial, Marder[10] treated 23 cats with this diagnosis with diazepam (1 mg orally every 12 hours; if no response, increased to 2 mg orally every 12 hours). Treatment was continued at the effective dose for 1 month, then tapered over 1 month. When assessed later, 43% of the cats had stopped spraying completely and 74% of owners expressed satisfaction, defined as a 75% or greater improvement in the problem.

Eleven of the cats that had previously been given synthetic progestins without success were effectively treated with diazepam. Owners reported side effects of increased affection, lethargy, appetite increase, and ataxia. Of the cats that responded, 75% resumed the behavior after the medication was discontinued.[10]

Cooper and Hart[56] tested the clinical effect of diazepam in an open trial on 20 surgically sterilized cats presented for urine spraying. Eleven of the cats had previously been treated unsuccessfully with progestins. Diazepam (1 to 4 mg orally every 12 hours) was given for 2 to 20 weeks in an experimental design contingent on each cat's response. Nine cats ceased spraying and two others showed marked reduction in spraying for an overall response rate of 55%. All responders except one (10 of 11) resumed spraying when the drug was discontinued after 6 to 8 weeks of treatment. Reported side effects included ataxia (8 of 20), which usually abated as cats became tolerant to the drug; increased sleepiness (10 of 20), increased appetite (3 of 20), weight gain (2 of 20), reduced intercat aggression (1 of 20), and increased affection toward people (3 of 20).[56]

A case presentation of urine spraying and reclusive behavior by a cat from a multicat household highlights common social interactions that complicate treatment in these cases.[57] The cat was successfully treated with diazepam (1 mg orally every 12 hours, later increased to 2 mg orally every 12 hours) until it relapsed—over a year later when successful treatment with the atypical anxiolytic buspirone hydrochloride was initiated.[57]

Various other uses for benzodiazepines have been reported. Diazepam has been used as an appetite stimulant.[20,51] Clonazepam has been used to correct REM sleep disorders in dogs, including violent movement during sleep.[58]

Buspirone Hydrochloride

Buspirone hydrochloride (BuSpar®, MeadJohnson Pharmaceuticals) was the first nonsedating antianxiety drug to be developed and marketed. Its antianxiety effects differ markedly from those of the benzodiazepines and other anxiolytics in that buspirone hydrochloride has no immediate antianxiety effects and requires a week or more of daily treatment to produce measurable antianxiety effects in humans.

Pharmacology and Mode of Action

Buspirone hydrochloride is in a class of drugs called azapirones. Its antianxiety effect is attributed to blocking serotonin$_{1A}$ presynaptic and postsynaptic receptors. It appears that the presynaptic serotonin$_{1A}$ blockade increases serotonergic activity when such activity is low but reduces it by blocking postsynaptic receptors when serotonergic activity is high.[59] Buspirone hydrochloride causes downregulation of serotonin$_2$ receptors and additionally acts as a dopamine agonist throughout the brain. Unlike the benzodiazepines, it has no direct effect on GABA receptors, although it does enhance benzodiazepine binding.

In large-scale, double-blind clinical studies, buspirone hydrochloride has been shown to be as effective as benzodiazepines in the treatment of generalized anxiety disorder or other anxiety states.[59] In smaller, short-term trials, particularly in crossover experimental designs with benzodiazepines, buspirone hydrochloride often fails to show efficacy.[60] This may be because of the slow onset of action of buspirone hydrochloride (in comparison with the benzodiazepines) or because it appears to be less effective in patients that had previously received benzodiazepines than in naive patients.[59] Overall, many clinicians and patients have found the effectiveness of buspirone hydrochloride for the treatment of anxiety to be disappointing.

Buspirone hydrochloride is ineffective in the control of panic disorder, thus emphasizing that panic disorder should be differentiated from generalized anxiety disorder in presentation and efficacy of drug treatment. Evidence suggests that patients who suffered panic attacks or social phobia showed preferential response to medications that were not "anxiolytic" per se, such as tricyclic antidepressants. Other drugs (such as buspirone hydrochloride and most benzodiazepines) that are effective in the treatment of generalized anxiety disorder are ineffective in the treatment of panic attacks.[59] This suggests that the pathophysiology of these states may not be the same. It may be useful to classify fearful behavior in animals similarly.

Buspirone hydrochloride lacks sedative, muscle relaxant, or anticonvulsant actions and does not impair performance on motor tasks.[59] The drug has a short elimination half-life, from 1 to 10 hours in humans. Recommended dosing in humans is 5 to 15 mg three times a day. Unlike benzodiazepines, buspirone hydrochloride can be abruptly withdrawn without complication.

Side Effects

In humans, buspirone hydrochloride has a propensity to elevate prolactin and growth hormone at high doses but not at standard therapeutic doses.[59] Despite its effect on dopamine receptors, buspirone hydrochloride lacks antipsychotic effects. Buspirone hydrochloride is relatively free of serious side effects and drug interactions, and it lacks abuse potential.

Applications

In humans, buspirone hydrochloride is used for the treatment of anxiety and little else. Although there have been some reports of the use of buspirone hydrochloride to augment the effectiveness of the selective serotonin-reuptake inhibitors in the treatment of depression and obsessive-compulsive disorder, well-controlled studies have failed to show increased efficacy over selective serotonin-reuptake inhibitors alone.[61]

Buspirone hydrochloride has been used by veterinary behaviorists, although the doses are significantly higher than those used for depressed human patients. Buspirone hydrochloride has the disadvantages of multiple daily dosing, relatively high cost, and extended latency to effect. It has the advantages of being safe and well tolerated.

In a prospective open trial, Hart and coworkers[62] recorded owner assessment of the effect of buspirone hydrochloride on the incidence of urine spraying and marking in 62 surgically sterilized cats. The presence or absence of concomitant therapy, such as environmental control, was unspecified. Buspirone hydrochloride doses were increased if the cat failed to respond. The dose range (2.5 to 7.5 mg orally every 12 hours) was only slightly less than doses used in adult humans.[29] Results indicated buspirone hydrochloride to be effective in approximately 55% of cats, similar to results of diazepam treatment by the same group,[56] although Marder[10] reported diazepam to be effective in 74% of cats that sprayed urine. Approximately half of the responders resumed spraying after buspirone hydrochloride was withdrawn, in contrast to a similar study with diazepam in which 91%[56] or 75%[10] of the subjects resumed spraying, although the analysis depends on how the operational terms are defined.[57] Side effects included sedation (4 of 62), increased affection to owners (12 of 62), and agitation after pill administration (5 of 62).

Overall[57] presented a case report of a cat whose problems of urine spraying and social withdrawal recurred with long-term diazepam therapy. With buspirone hydrochloride therapy (2.5 mg orally every 12 hours), the cat stopped spraying and exhibited more exploratory behavior, although the behavior problems recurred when attempts were made to withdraw the medication. In a complex case of a dog exhibiting dominance-related aggression and fearfulness with circling, Overall[23] used buspirone hydrochloride, in addition to behavior therapy, to reduce fearfulness and generalized anxiety, although it had no effect on circling behavior.

MOOD STABILIZERS

Lithium, carbamazepine, and valproic acid are dissimilar compounds used in human psychiatry as mood stabilizers in the treatment of bipolar disorder and to reduce impulsivity, emotional reactivity, and aggression. In humans, bipolar disorder is characterized by episodic mood swings, typically with periods of mania and depression. Carbamazepine and valproic acid also are used as anticonvulsants, but this is not clearly related to their mood-stabilizing effects.

Pharmacology and Mode of Action

Lithium has been used therapeutically for more than 150 years.[63] Lithium is an element and is not metabolized but is excreted by the kidneys. Its elimination half-life in humans is from 18 to 24 hours and it is sensitive to renal function. In general, lithium enhances serotonin neurotransmission and may also affect dopamine, norepinephrine, and acetylcholine pathways in certain brain regions.

Carbamazepine has a tricyclic structure similar to that of the tricyclic antidepressant imipramine hydrochloride. As an antiepileptic, it acts on neuronal ion channels to reduce high-frequency repetitive firing of action potentials; and it acts pre- and postsynaptically on multiple neurotransmitter systems, including norepinephrine, dopamine, serotonin, acetylcholine, GABA, glutamate, and others.[63] The specific mechanism for carbamazepine's psychotropic actions are unknown.

Valproic acid has multiple effects, but the mechanisms of action are not well understood. It has been suggested that valproic acid alters the metabolism of GABA.[64]

Side Effects

Lithium has a narrow therapeutic index with a recommended therapeutic serum concentrations of 0.8 to 1.2 mEq/L.[63] Careful monitoring of serum lithium levels is necessary to reduce risk of side effects. Toxicity becomes more evident as serum concentrations exceed the recommended range. Side effects commonly reported include polyuria, polydipsia, memory problems, weight gain, and diarrhea. At elevated serum concentrations, central nervous system toxicity is manifest by cognitive impairment, restlessness, fatigue, and irritability, progressing to delirium, seizures, coma, and death. Clinicians should consider lithium's other side effects and many potential drug interactions.[29,30,63] In human patients receiving lithium therapy, serum lithium levels and measures of kidney function (usually serum creatinine) are monitored regularly. The side effect profile of carbamazepine is more favorable than that of lithium, although 33% to 50% of patients taking carbamazepine experience some side effects.[64] Those most commonly reported are transient dizziness, nausea, fatigue, and blurred vision. Sometimes liver enzymes are elevated transiently. More serious but rare are blood dyscrasias, including agranulocytosis. Carbamazepine can interact with various other drugs.[64] It is generally recommended that complete blood counts and liver chemistry profiles of patients receiving carbamazepine therapy be routinely monitored.

Valproic acid has a better side-effect profile than lithium, other antiepileptics, and antipsychotics.[64] It is less likely to cause cognitive impairment and rarely causes renal, thyroid, cardiac, or allergic effects. Dose-related side effects include gastrointestinal distress, tremor and sedation, and elevation of hepatic enzymes. Most of these abnormalities subside with time or reduced dosage. Irreversible hepatic failure is rare and idiosyncratic. It is recommended that hepatic and hematologic values be monitored periodically.

Applications

Lithium maintenance therapy has been demonstrated to reduce mood swings in bipolar disorder in humans.[64] Of particular interest to veterinarians is the reported antiaggression effect of lithium therapy.[63] Although most studies have been conducted in institutions, lithium has been shown to exert antiaggression effects on psychiatric patients, children with behavior problems, mentally retarded individuals, and prisoners with rage outbursts.[65]

Both carbamazepine and valproic acid have been shown to have similar mood-stabilizing effects. Both drugs are frequently used as an alternative to lithium therapy for the treatment of bipolar disorder in human patients intolerant of or nonresponsive to lithium. Both drugs have been shown in double-blind studies to be as effective as lithium and antipsychotics in the short-term treatment of acute mania and the long-term treatment of bipolar disorder.[64] In cases and small trials, carbamazepine has been used successfully in aggressive patients refractory to conventional therapies. Several of these individuals had abnormal electroencephalograms in the absence of seizures, thus suggesting organic disease.[66]

The possible value of mood-stabilizing agents in the treatment of aggression makes them potentially of interest to veterinary behaviorists[66]; however, these drugs have received little attention so far. Lithium's side-effect profile and requirements for blood-level monitoring make its use difficult for most practicing veterinarians. Reisner[67] reported the use of lithium in one dog in the treatment of dominance-related aggression.

Although the short elimination half-life (1 to 2 hours) of carbamazepine in dogs[68] makes it an unsatisfactory choice for seizure control, it may be useful in the treatment of certain behavior problems. When diazepam failed to control aggression in two cats, Schwartz[11] reported attenuation of aggression with carbamazepine (25 mg orally every 12 to 24 hours). Valproic acid (30 to 180 mg/kg daily in divided oral doses every 8 hours) has been recommended in large-breed dogs with idiopathic epilepsy resistant to more-conventional therapy.[68] This use of valproic acid to treat seizures in dogs suggests that the drug may be beneficial for certain behavior problems.

OTHER DRUGS

This review has focused on human psychotropic drugs and summarizes the studies to date on their use in the treatment of veterinary behavior problems. Omitted from this discussion are drugs that historically have played an important role in veterinary behavior therapy because of their documented psychoactive effects. Such drugs as barbiturates, antihistamines, narcotic antagonists, and progestin hormones[10,20,21,38,69] are familiar to most veterinarians. They have confirmed behavioral effects but are rarely used therapeutically in human psychiatry. Reasons for this include their nonspecific effects (compared with those of other drugs) or risks associated with their use, including fatal overdose or abuse potential.

Other drugs (e.g., such narcotic antagonists as naloxone hydrochloride) are of particular interest for their role in the treatment of compulsive disorders in animals.[70–72] ß-Blockers, such as propranolol or newer agents, such as naldolol or pindolol, reduce signs of fearfulness in humans and may have a role in the treatment of aggression.[66] Methylphenidate hydrochloride or dextroamphetamine may be used to treat a specific diagnosis of hyperactivity.[38,73] The reader is referred to reviews in the veterinary literature for a discussion of these agents.[10,20,21,38,62] Of additional interest are discussions of specific animal behavior disorders, their applicability as models of human behavior conditions, and their treatment.[26,35,74]

PRACTICE OF BEHAVIORAL PHARMACOTHERAPY

In veterinary psychopharmacology, as in human psychopharmacology, treatment is best accomplished by initially identifying a set of problem behaviors targeted for pharmacologic modification. These behaviors should be quantifiable and should occur often enough that the owner can notice changes in their frequency or severity within a reasonable time frame (e.g., before the next visit to the veterinarian). An important part of this process is the involvement of the pet owner as behavior observer. Rarely do behavioral problems present themselves in the office, where the treating veterinarian can effectively assess frequency and severity of the behaviors in the absence of the owner. The pet owner needs to understand and agree with the selection of behavioral targets and must be able to identify and report on any changes in the behavioral targets attributable to the drug treatment. Further, the owner needs to understand that behavioral problems are rarely cured but frequently improved. This process of establishing a realistic groundwork of understanding and expectations is central to the effective practice of veterinary behavioral pharmacotherapy. A solid basis for clear communication between veterinarian and owner about the frequency and severity of target behaviors both before and during drug treatment must be established from the start.

Once target behaviors are agreed upon, the decision about which drug to use then becomes a question of which drug is most likely to have a beneficial influence on these target behaviors. This selection of a drug begins with an understanding of the various types of available drugs, their general behavioral actions, side effects, and previously reported therapeutic effects in animals. Many of these drugs have similar behavioral effects and differ only in pharmacokinetics or side-effect profile. A thorough working knowledge of a few different behaviorally active drugs will therefore be more beneficial to the practicing veterinarian than a less complete knowledge of a large number of similar drugs.

Drug selection in veterinary behavioral pharmacotherapy has in the past been guided by personal experience, extrapolation from the human literature, a few case reports in animals, and a "best guess" approach applied case by case. Recently, there have been more attempts to generate scientific studies of psychopharmacologic treatment of animals to help the clinician in the selection of a therapeutic agent. Despite this trend, finding a drug that actually improves the target behaviors often involves repeated therapeutic trials with different agents.

Once a drug is selected, the expected behavioral effects and possible side effects are discussed with the owner, and the therapeutic trial begins. If the initial drug trial is unsuccessful either because of lack of positive behavioral effect or intolerable side effects, then an alternative drug should be chosen. In the case of lack of desirable behavioral effect, pharmacologically different agents should reasonably be tried on successive therapeutic trials. Where side effects caused problems but positive behavioral effects were achieved, similar drugs with different side effect profiles should be tried.

Veterinarians have available to them an increasingly large number of behaviorally active drugs. Many of these drugs are relatively new in human psychiatry and are just beginning to be applied in veterinary medicine. Caution should be exercised in applying these drugs to species, such as dogs or cats, for which they have not been formally tested. Veterinarians should obtain all available information on the effectiveness and safety of the agent in the relevant species and discuss with their clients the agent and its documented use in veterinary medicine. Then, the veterinarian can expect to improve or ameliorate an increasingly large number of their patients' behavioral problems.

Acknowledgments

We thank Sukey Jacobsen, Donna Witt, and Jacqueline Gadison for assistance with library research.

REFERENCES

1. Potter WZ, Manji HK, Rudorfer MV: Tricyclics and tetracyclics, in Schatzberg AF, Nemeroff CB (eds): *Textbook of Psychopharmacology.* Washington DC, American Psychiatric Press, 1995, pp 141–160.
2. Marder SR, Van Putten T: Antipsychotic medications, in Schatzberg AF, Nemeroff CB (eds): *Textbook of Psychopharmacology.* Washington DC, American Psychiatric Press, 1995, pp 247–262.
3. Tollefson GD: Selective serotonin reuptake inhibitors, in Schatzberg AF, Nemeroff CB (eds): *Textbook of Psychopharmacology.* Washington DC, American Psychiatric Press, 1995, pp 161–182.
4. Ballenger JC: Benzodiazepines, in Schatzberg AF, Nemeroff CB (eds): *Textbook of Psychopharmacology.* Washington DC, American Psychiatric Press, 1995, pp 215–230.
5. Schatzberg AF, Nemeroff CB (eds): *Textbook of Psychopharmacology.* Washington DC, American Psychiatric Press, 1995.
6. Hart BL: Behavioral indications for phenothiazine and benzodiazepine tranquilizers in dogs. *JAVMA* 186:1192–1194, 1985.
7. Hart BL, Hart LA: *Canine and Feline Behavioral Therapy.* Philadelphia, Lea & Febiger, 1985.
8. Plumb DC: *Veterinary Drug Handbook*, ed 2. Ames, Iowa State University Press, 1995.
9. Bruhwyler J, Chleide E: Comparative study of the behavioral, neurophysiological, and motor effects of psychotropic drugs in the dog. *Biol Psychiatry* 27:1264–1278, 1990.
10. Marder AR: Psychotropic drugs and behavioral therapy. *Vet Clin North Am [Small Anim Pract]* 21:329–342, 1991.
11. Schwartz S: Carbamazepine in the control of aggressive behavior in cats. *JAAHA* 30:515–519, 1994.
12. Jones RD: Use of thioridazine in the treatment of aberrant motor behavior in a dog. *JAVMA* 191:89–90, 1987.

13. Johnson LR: Tricyclic antidepressant toxicosis. *Vet Clin North Am [Small Anim Pract]* 20:393–403, 1990.
14. Rudorfer MV, Robbins E: Amitriptyline overdose: Clinical effects of tricyclic antidepressant plasma levels. *J Clin Psychiatry* 43:457–460, 1982.
15. Glassman AH, Roose SP, Giardina EGV, et al: Cardiovascular effects of tricyclic antidepressants, in Meltzer HY (ed): *Psychopharmacology: The Third Generation of Progress.* New York, Raven Press, 1987, pp 1437–1442.
16. Donnelly M, Zametkin AJ, Rapoport JL, et al: Treatment of childhood hyperactivity with desipramine: Plasma drug concentration, cardiovascular effects, plasma and urinary catecholamine levels, and clinical response. *Clin Pharmacol Ther* 39:72–81, 1986.
17. Montgomery SA: Venlafaxine: A new dimension in antidepressant pharmacotherapy. *J Clin Psychiatry* 54:119–126, 1993.
18. Goldberger E, Rapoport JL: Canine acral lick dermatitis: Response to the antiobsessional drug clomipramine. *JAAHA* 27:179–182, 1991.
19. Rapoport JL, Ryland DH, Kriete M: Drug treatment of canine acral lick: An animal model of obsessive-compulsive disorder. *Arch Gen Psychiatry* 49:517–521, 1992.
20. Houpt KA, Reisner IR: Behavioral disorders, in Ettinger SJ, Feldman EC (eds): *Textbook of Veterinary Internal Medicine,* ed 4. Philadelphia, WB Saunders Co, 1995, pp 179–187.
21. Voith VL: Behavioral disorders, in Ettinger S (ed): *Textbook of Veterinary Internal Medicine,* ed 3. Philadelphia, WB Saunders Co, 1989, pp 227–238.
22. Houpt KA: Animal behavior case of the month. *JAVMA* 204:1751–1752, 1994.
23. Overall KL: Animal behavior case of the month. *JAVMA* 206:629–632, 1995.
24. Leonard HL, Lenane MC, Swedo SE, et al: A double-blind comparison of clomipramine and desipramine treatment of severe onychophagia (nail biting). *Arch Gen Psychiatry* 48:821–827, 1991.
25. Swedo SE, Leonard HL, Rapoport MD, et al: A double-blind comparison of clomipramine and desipramine in the treatment of trichotillomania (hair pulling). *N Engl J Med* 321:497–501, 1989.
26. Rapoport JL: Animal models of obsessive-compulsive disorder. *Clin Neuropharmacol* 15(Suppl 1):261A–271A, 1992.
27. Overall KL: Use of clomipramine to treat ritualistic stereotyped motor behavior in three dogs. *JAVMA* 205:1733–1741, 1994.
28. Thornton LA: Animal behavior case of the month. *JAVMA* 206:1868–1870, 1995.
29. American Hospital Formulary Service: *Drug Information.* Bethesda, MD, American Society of Hospital Pharmacists, 1994.
30. *Physicians' Desk Reference,* ed 48. Montvale, NJ, Medical Economics Data Production Co, 1994.
31. Tollefson GD, Rampey AH, Potvin JH: A multicenter investigation of fixed-dose fluoxetine in the treatment of obsessive-compulsive disorder. *Arch Gen Psychiatry* 51:559–567, 1994.
32. Murdoch D, McTavish D: Sertraline: A review of its pharmacodynamic and pharmacokinetic properties, and therapeutic potential in depression and obsessive-compulsive disorder. *Drugs* 44:604–624, 1992.
33. Kauffman S: Problem pets may now get Prozac. *Raleigh News-Observer* August 1:1B–5B, 1994.
34. Marder A: The promise of Prozac. *Vet Product News.* May/June:1, 45, 1995.
35. Karel R: Fluoxetine use in dogs provides animal model for human mental disorders. *Psychiatric News* September 16, 1994; p 12.
36. Melman SA: Use of Prozac in animals for selected dermatological and behavioral conditions. *Vet Forum* 12(8): 19–27, 1995.
37. McKeown DB, Luescher UA, Halip J: Stereotypies in companion animals and obsessive-compulsive disorder, in *Purina Specialty Review: Behavioral Problems in Small Animals.* St. Louis, MO, Ralston Purina Company, 1992, pp 30–35.
38. Overall KL: Practical pharmacological approaches to behavior problems, in *Purina Specialty Review: Behavioral Problems in Small Animals.* St. Louis, MO, Ralston Purina Company, 1992, pp 36–51.
39. Hartmann L: Cats as possible obsessive-compulsive disorder and medication models (letter). *Am J Psychiatry* 152:1236, 1995.
40. Dodman NH, Mertens PA: Fluoxetine (Prozac) for the treatment of dominance-related aggression in dogs (abstract). *Newsletter Am Vet Soc Anim Behav* 17(2):3, 1995.
41. Ruehl WW: L-deprenyl for treatment of behavioral and cognitive problems in dogs: Preliminary report of an open label trial (abstract). *Appl Anim Behav Sci* 39:191, 1994.
42. Golden RN, Bebchuk JM, Leatherman ME: Trazodone and other antidepressants, in Schatzberg AF, Nemeroff CB (eds): *Textbook of Psychopharmacology.* Washington DC, American Psychiatric Press, 1995, pp 195–213.
43. Simpson DM, Foster DL: Improvement in organically disturbed behavior with trazodone treatment. *Clin Psychiatry* 47:191–193, 1986.
44. Rosenbaum JF, Gelenberg AJ: Anxiety, in Gelenberg AJ, Bassuk EL, Schoonover SC (eds): *The Practitioner's Guide to Psychoactive Drugs,* ed 3. New York, Plenum Medical Book Co, 1991, pp 179–218.
45. Brown SA, Forrester SD: Serum disposition of oral clorazepate from regular-release and sustained-delivery tablets in dogs. *J Vet Pharmacol Ther* 14:426–429, 1991.
46. Reese WG: A dog model for human psychopathology. *Am J Psychiatry* 136:1168–1172, 1979.
47. Shader RI, Greenblatt DJ: Use of benzodiazepines in anxiety disorders. *N Engl J Med* 328:1398–1405, 1993.
48. Center SA, Elston TH, Rowland PH, et al: Hepatotoxicity associated with oral diazepam in 12 cats. *Proc ACVIM* 13:1009, 1995.
49. Levy J, Cullen JM, Bunch SE, et al: Adverse reaction to diazepam in cats (letter). *JAVMA* 205:156–157, 1994.
50. Horwitz D: Adverse reactions to diazepam (letter). *JAVMA* 205:966, 1994.
51. Beaver AV: *Feline Behavior: A Guide For Veterinarians.* Philadelphia, WB Saunders Co, 1992.
52. Yen HCY, Krop S, Mendez HC, et al: Effects of some psychoactive drugs on experimental "neurotic" (conflict induced) behavior in cats. *Pharmacol* 3:32–40, 1970.
53. Heuschele WP: Chlordiazepoxide for calming zoo animals. *JAVMA* 136:996–998, 1961.
54. Pellis SM, O'Brien DP, Pellis VC, et al: Escalation of feline predation along a gradient from avoidance through "play" to killing. *Behav Neurosci* 102:760, 1988.
55. Hart BL, Cooper L: Factors relating to urine spraying and fighting in prepubertally gonadectomized cats. *JAVMA* 184:1255–1258, 1984.
56. Cooper L, Hart BL: Comparison of diazepam with progestin for effectiveness in suppression of urine spraying behavior in cats. *JAVMA* 200:797–801, 1992.
57. Overall KL: Animal behavior case of the month. *JAVMA* 205:694–696, 1994.
58. Hendricks JC, Lager A, O'Brien, et al: Movement disorder

during sleep in cats and dogs. *JAVMA* 194:686–689, 1989.

59. Cole JO, Yonkers KA: Nonbenzodiazepine anxiolytics, in Schatzberg AF, Nemeroff CB (eds): *Textbook of Psychopharmacology.* Washington DC, American Psychiatric Press, 1995, pp 231–244.
60. Olajide D, Lader M: A comparison of buspirone, diazepam, and placebo in patients with chronic anxiety states. *J Clin Psychopharmacol* 7:148–152, 1987.
61. Grady TA, Pigott TA, L'Heureux F, et al: Double-blind study of adjuvant buspirone for fluoxetine-treated patients with obsessive-compulsive disorder. *Am J Psychiatry* 150:819–821, 1993.
62. Hart BL, Eckstein RA, Powell KL, Dodman NH: Effectiveness of buspirone on urine spraying and inappropriate urination in cats. *JAVMA* 203:254–258, 1993.
63. Lenox RH, Manji HK: Lithium, in Schatzberg AF, Nemeroff CB (eds): *Textbook of Psychopharmacology.* Washington DC, American Psychiatric Press, 1995, pp 303–349.
64. McElroy SL, Keck PE: Antiepileptic drugs, in Schatzberg AF, Nemeroff CB (eds): *Textbook of Psychopharmacology.* Washington DC, American Psychiatric Press, 1995, pp 351–375.
65. Nilsson A: The anti-aggressive actions of lithium. *Rev Contemp Pharmacother* 4:269–285, 1993.
66. Yudofsky SC, Silver JM, Schneider SE: Pharmacologic treatment of aggression. *Psychiatric Ann* 17:397–406, 1987.
67. Reisner I: Use of lithium for treatment of canine dominance-related aggression (abstract). *Appl Anim Behav Sci* 39:183–192, 1994.
68. De LaHunta A: *Veterinary Neuroanatomy and Clinical Neurology,* ed 2. Philadelphia, WB Saunders Co, 1983.
69. Hunthausen WL, Landsberg GM: *A Practitioner's Guide to Pet Behavior Problems.* Denver, CO, American Animal Hospital Association, 1995.
70. Brown SA, Crowell-Davis S, Malcolm T, Edwards P: Naloxone-responsive compulsive tail chasing in a dog. *JAVMA* 190:884–886, 1987.
71. Dodman NH, Shuster L, White SD, et al: Use of narcotic antagonists to modify stereotypic self-licking, self-chewing, and scratching behavior in dogs. *JAVMA* 193:815–819, 1988.
72. White SD: Naltrexone for treatment of acral lick dermatitis in dogs. *JAVMA* 196:1073–1076, 1990.
73. Luescher UA: Hyperkinesis in dogs: Six case reports. *Can Vet J* 34:368–370, 1993.
74. Stein DJ, Dodman NH, Borchelt P, et al: Behavioral disorders in veterinary practice: Relevance to psychiatry. *Comp Psychiatry* 35:275–285, 1994.

Endocrinopathies That Affect the Central Nervous System of Cats and Dogs

Virginia Tech
Todd L. Towell, DVM Linda G. Shell, DVM

KEY FACTS

- **Clinical signs that indicate dysfunction of the central nervous system include seizure activity and abnormalities in behavior, mentation, head posture, coordination, and cranial nerve function.**
- **Endocrinopathies that affect the central nervous system include diabetes mellitus, hypothyroidism, hypoparathyroidism, hyperparathyroidism, hypoadrenocorticism, hyperadrenocorticism, and insulinoma.**
- **Diabetes mellitus, hypothyroidism, and insulinoma are reported to cause coma via different mechanisms.**
- **Seizure activity in patients has been associated with hypothyroidism, hypoparathyroidism, hyperparathyroidism, and insulinoma.**
- **Patients with endocrinopathies that affect the central nervous system may present with severe, insidious, or vague neurologic signs.**

Metabolic derangements that result from endocrine disorders can adversely affect the central nervous system. Clinical signs indicating involvement of the central nervous system include seizure activity and abnormalities in behavior, mentation, head posture, coordination, and cranial nerve function.[1] Atypical vital signs may also indicate neurologic involvement. Patients should be carefully evaluated to determine the presence or absence of such signs; careful assessment of clinical signs helps practitioners to localize lesions and develop a prognosis for recovery. Endocrinopathies that have been reported to cause disturbances of the central nervous system include diabetes mellitus, hypothyroidism, hypoparathyroidism, hyperparathyroidism, hypoadrenocorticism, hyperadrenocorticism, and insulinoma (Table I).

DIABETES MELLITUS

Diabetic ketoacidosis and diabetic hyperosmolar nonketotic syndrome are complications of diabetes mellitus that may produce neurologic dysfunction. Both conditions are rarely reported in veterinary medicine.[1–3] Diabetic ketoacidosis is the most common cause of death in children and teenagers with diabetes mellitus.[4] One study documented that subclinical swelling of the brain is commonly found during the treatment of children with ketoacidosis.[5] Hyperosmolar nonketotic syndrome is reported to represent 5% to 10% of hyperglycemic comas; the condition occurs primarily in the elderly and is associated with a much higher mortality rate than is diabetic ketoacidosis.[6]

Diabetic ketoacidosis results from an absolute or relative deficiency of insulin and a relative excess of glucagon and other counterregulatory hormones (i.e., cortisol, growth hormone, and catecholamine).[2] The increased ratio of counterregulatory hormones to insulin promotes hepatic glyco-

TABLE I
Neurologic Signs Associated with Endocrinopathies

Endocrinopathy	*Neurologic Signs*
Diabetes mellitus[1–9]	Depression, coma
Hypothyroidism[11–15]	Myxedema stupor, coma Hypothermia Nonpitting edema Muscle weakness
	Cerebral artery atherosclerosis Disorientation Circling Blindness Seizures Head tilt
Hypoparathyroidism[17–19]	Seizures Behavioral changes Muscle tremors Stiff gait
Hyperparathyroidism[22]	Depression Seizures
Hypoadrenocorticism[24]	Seizures
Hyperadrenocorticism[25,26]	Behavioral changes Aggression Depression Anxiousness Increased motor activity
Insulinoma[27,28]	Seizures Depression, coma Visual deficits Tremors Irritability Weakness

genolysis and gluconeogenesis. Hyperglycemia occurs when the production of glucose exceeds the removal of glucose by peripheral tissues and renal excretion. The increased ratio also stimulates the hormone-sensitive lipase which, in turn, causes the release of free fatty acids from adipose stores. Because of the increased glucagon-to-insulin ratio, free fatty acids in the liver are oxidized to ketones, acetoacetate, and β-hydroxybutyrate dehydrogenase instead of being esterified into triglycerides. Hydrogen ions are released from the ketones and accumulate in the extracellular fluid. Metabolic acidosis ensues as soon as the accumulation overwhelms compensatory mechanisms, the bicarbonate buffer system, and renal excretion of hydrogen ions. These metabolic derangements eventually result in hyperglycemia, ketoacidosis, hypovolemia (from osmotic diuresis), and total body electrolyte (particularly sodium and potassium) deficits.[2]

Patients are considered to have hyperosmolar nonketotic syndrome if plasma glucose is greater than 600 mg/dl, serum osmolality is greater than 330 mOsm/kg, and ketones are absent or minimal in the serum or urine.[7] There is some controversy regarding the pathophysiology of hyperosmolar nonketotic syndrome. Some believe that there is possibly a continuum of decompensated diabetes, with diabetic ketoacidosis and hyperosmolar nonketotic syndrome representing the two extremes. Most patients would be presented because of clinical signs between these two conditions.[7,8] The following commonly accepted mechanisms may account for the differences between diabetic ketoacidosis and hyperosmolar nonketotic syndrome: (1) free fatty acid and counterregulatory hormone levels are lower in patients with hyperosmolar nonketotic syndrome; (2) pancreatic insulin secretion, measured by C-peptide levels, is higher in patients with hyperosmolar nonketotic syndrome and prevents lipolysis; and (3) the hyperosmolar state alone inhibits lipolysis.[8]

Serum hyperosmolality seems to be the one metabolic derangement most likely to cause neurologic dysfunction in patients with hyperosmolar nonketotic syndrome. Dehydration that results from long-term osmotic diuresis and renal insufficiency (which promotes hyperglycemia) may be significant in the development of hyperosmolar nonketotic syndrome.[7] Coma is believed to result from dehydration of brain cells secondary to chronic hypovolemia, osmotic diuresis, and shifts in fluid balance between the intracellular and extracellular compartments.

Diabetic ketoacidosis and hyperosmolar nonketotic syndrome commonly occur in patients that are not diabetic. The onset of both conditions is generally insidious, and the diagnosis is based on the results of blood chemistry panels and urine chemistry tests. Determination of serum osmolality, directly or by calculation, aids the diagnosis. One study of hyperosmolar nonketotic syndrome in humans indicates that corrected sodium levels (measured sodium mmol/L + 1.6 mmol/L for every 100 mg/dl increment in the serum glucose concentration) may be the most sensitive index of the state of cerebral cellular hydration, with corrected hypernatremia indicating cerebral dehydration.[8] Neurologic abnormalities that have been documented to result from a gradual increase in serum osmolality include restlessness, ataxia, nystagmus, muscle twitching, convulsions, and hyperthermia.[9] Death from respiratory failure has also been noted.[9] Veterinary patients are more likely to present with vague signs, such as lethargy, depression, anorexia, and possibly stupor. Classic clinical signs of diabetes mellitus (i.e., polydipsia, polyuria, polyphagia, and weight loss) may not be reported by the owner.

Treatment regimens for patients diagnosed with diabetic ketoacidosis or hyperosmolar nonketotic syndrome are similar. The goals of therapy are to treat circulatory collapse; replace sodium, potassium, and fluid deficits; and correct hyperglycemia slowly without rapid decreases in glucose or hypoglycemia. Excessive fluctuations in glucose, osmolarity, and pH should be avoided in order to prevent cerebral edema. Strict monitoring of fluid balance, blood glucose, electrolyte concentration, and urine production are essential. Insulin therapy should not commence until initial fluid deficits have been corrected.

HYPOTHYROIDISM

Myxedema stupor and coma, the most severe complications of clinical hypothyroidism, are caused by diminished metabolic processes related to thyroid hormone deficiency.[10] Of four cases reported in the veterinary literature,[11,12] none of the dogs was previously diagnosed as having hypothyroidism. Characteristic clinical signs at presentation were hypothermia, nonpitting edema, mental dullness, and muscle weakness.[11,12] In addition, the dogs had dermatologic, reproductive, and hematologic signs consistent with hypothyroidism. The most peculiar feature of myxedema stupor and coma in canine[11,12] and human[10] patients is hypothermia in the absence of shivering. This condition is attributed to impaired hypothalamic temperature regulation or impaired calorigenesis.[13]

Atherosclerosis is a focal thickening of the inner portion of the artery wall that is associated with lipid deposits.[14] In dogs, cerebral artery atherosclerosis, which is a rare complication of hypothyroidism, has been reported to cause clinical signs that result from abnormalities of the central nervous system, such as disorientation, circling, blindness, seizures, head tilt, and ataxia.[14,15] These signs may be present intermittently for months or may occur acutely.[15] Cerebral hypoxia, secondary to cerebral artery atherosclerosis, is the likely cause of these neurologic abnormalities.[15] Electroencephalogram patterns from dogs with anomalies of the central nervous system exhibit diffuse flattening in all leads with periodic spike and slow-wave discharges.[15] These discharges are consistent with patterns observed after prolonged cerebral hypoxia and with the presence of an epileptic focus, respectively.[15] The same pattern of diffuse flattening has also been documented in dogs diagnosed as having hypothyroidism; however, these dogs did not have clinical signs of atherosclerosis.[16] Patients with hypothyroidism that are fed high-cholesterol diets seem to be at risk for developing this condition.[14]

Secondary hypothyroidism can be caused by pituitary neoplasia. Dogs with this condition are commonly presented with neurologic abnormalities[15] caused by the compressive effects of a tumor.[17] Clinical signs may include ataxia, depression, seizures, and head pressing. Presentation of patients with secondary hypothyroidism varies because clinical signs often depend on the location and extent of the tumor.[17]

HYPOPARATHYROIDISM

Dogs with hypoparathyroidism are most commonly presented because of an acute onset of neurologic or neuromuscular disturbances that result from abnormal plasma calcium concentrations.[18] Disturbances of the central nervous system include seizures, changes in behavior (e.g., aggression, increased excitement, and disorientation), and gait abnormalities.[18] Seizures begin as muscle tremors in one limb that gradually spread to other parts of the body and culminate in a violent generalized seizure.[18] Clinical signs relating to neuromuscular disturbances have been reported to range from focal muscle twitching and generalized muscle tremors to a stiff gait resulting from tetanic muscle contractions.

Clinical signs during the early stages of hypoparathyroidism may be subtle. Animals may seem to be in pain or may be abnormally tense, nervous, irritable, or nonplayful.[19] In all reports, neurologic disorders were originally episodic in nature; intervals of clinically normal behavior lasted from minutes to more than a week.[18,19]

Latent tetany occurs in humans when calcium concentrations fall to an abnormal level but are not low enough to cause obvious tetany.[20] Overt tetanic contractions, however, can be elicited by hyperventilation and the subsequent alkalinization of body fluids that increases the irritability of nerves.[20] A similar syndrome seems to occur in hypocalcemic dogs. Owners report that sudden excitement, activity, or petting occasionally causes clinical signs, such as muscle cramping, pain, irritability, or aggression.[19]

The pathophysiology of these neurologic signs is related to the lack of calcium and not to a direct effect of decreased levels of parathyroid hormone.[19] Calcium acts to stabilize the membrane of virtually all cells. As blood calcium concentrations decrease, neuronal membrane permeability, and therefore excitability,[19] progressively increases. The increase in excitability is possibly a mechanism for the development of seizures and muscle twitches or tetany. Ataxia, stiff gait, and extensor muscle rigidity may be related to cerebellar or extrapyramidal motor system dysfunction in the presence of hypocalcemia.[18,21]

HYPERPARATHYROIDISM

Primary hyperparathyroidism and hypercalcemia of malignancy (pseudohyperparathyroidism) can cause disturbances of the central nervous system. Pseudohyperparathyroidism is caused by malignant tumors of nonparathyroid origin that secrete parathyroid hormone–like proteins. Primary hyperparathyroidism is most often caused by a functional adenoma of the chief cells.[19] Pseudohyperparathyroidism is more commonly reported in veterinary literature than is primary hyperparathyroidism.

Neurologic disturbances associated with hyperparathyroidism are related to increased calcium concentrations, not to hormone or hormonelike substances. Patients diagnosed as having hyperparathyroidism are often presented because of an insidious onset of nonspecific signs; however, the most common complaint is polyuria and polydipsia.[19] Common clinical signs involving the central nervous system include depressed mentation and seizures.[19,22] Increased calcium concentrations depress the excitability of the neurologic system and subsequently result in depressed mental activity. Although the mechanism involved in the development of seizures is unclear, one theory suggests that calcium ions selectively depress the zone of inhibitory neurons surrounding a seizure focus, thus allowing the seizure impulse to spread.[22] Selective suppression of inhibitory neurons has occurred in patients with hypoxia and other metabolic abnormalities and is possible in patients with calcium abnormalities.[23] Seizure activity in one dog with primary hyperparathyroidism abated as soon as normocalcemia was established.[22]

Because primary hyperparathyroidism and pseudohyperparathyroidism usually result from malignant tumors, metastasis to the central nervous system should be considered as a possible cause of any neurologic abnormality. The clinical presentation of animals with metastatic lesions, therefore, would vary greatly.

HYPOADRENOCORTICISM

Hypoadrenocorticism, an endocrine disorder uncommonly diagnosed in dogs, is rare in cats.[24] Onset of the disorder is insidious; clinical signs are often nonspecific and wax and wane. The major problem results from decreased concentrations of glucocorticoids and mineralocorticoids that cause electrolyte imbalances (i.e., hypercalcemia, hyponatremia, hypochloremia, and hyperkalemia). These imbalances may be mild or severe, and clinical signs vary accordingly.

Although low glucocorticoid concentrations are responsible for mild hypoglycemia in 10% to 30% of patients with hypoadrenocorticism,[24] glucose concentrations are not usually low enough to produce neurologic signs.[19] Clinical signs of hypoglycemia are confounded by regular feeding intervals and a relative decrease in insulin secretion due to low blood glucose levels.[19] Patients with hypoadrenocorticism are rarely presented because of seizure activity; however, hypoglycemic-induced seizures should be considered as a potential complication, and patients should be managed accordingly.

HYPERADRENOCORTICISM

Excessive concentrations of adrenal corticosteroids may be the result of primary or iatrogenic hyperadrenocorticism or accidental toxicosis. Side effects of increased corticosteroid levels are numerous and include effects on the neurologic system. In humans, chronic use of corticosteroids has been associated with vertigo and behavioral disturbances.[25] Insomnia, nervousness, irritability, and excessive motor activity are manifestations of acute toxicity from ingestion of large doses of corticosteroids.[25] Psychotic experiences and convulsions have also been reported in humans.[25]

Some veterinary patients that receive high doses of glucocorticoids have side effects ranging from anxious, aggressive, or psychotic behavior to depression.[25] Other potential pharmacologic effects of corticosteroids on the neurologic system include decreased seizure threshold, decreased response to pyrogens, somnolence, and stimulation of the appetite.[26] Some of these behaviors are believed to be caused by a depletion of γ-aminobutyric acid—an inhibitory neurotransmitter in the brain.[26]

INSULINOMA

Insulinoma is a functional pancreatic β-cell tumor most commonly found in the endocrine pancreas in dogs.[27] Adenocarcinomas are more common in dogs than adenomas; more than 75% of adenocarcinomas are malignant.[27] Clinical signs of insulinoma are associated with inappropriate secretion of biologically active insulin.[27] Resultant hypoglycemia is directly responsible for neurologic manifestations of the syndrome.[27]

The brain requires glucose as a primary energy source but only has minimal glycogen reserves; therefore, a continuous supply of glucose is essential. Blood glucose homeostasis results from complex interactions of intake and adsorption of carbohydrates, hepatic production, and peripheral utilization.[27] Progressive hypoglycemia decreases the ability of the brain to extract oxygen from blood; therefore, neurologic signs that result from hypoglycemia are similar to signs caused by cerebral hypoxia.[27]

Clinical signs in dogs with insulinoma may be episodic or incited by feeding, fasting, or exercise.[27,28]

If glucose concentrations rapidly decrease, hypothalamic glucoreceptors respond by increasing the tone of the sympathetic nervous system.[28] Clinical signs associated with insulinoma include tachycardia, tremors, nervousness, irritability, and intense hunger.[28] Gradual reductions in glucose concentrations only cause neuroglucopenic signs, such as hypothermia, visual disturbances, mental dullness, confusion, seizures, and coma.[28] Anticonvulsant therapy, instituted for seizure control, subsequently fails to control seizures induced by hypoglycemia.[27]

Diagnosing insulinoma is often difficult. Physical examination, hematologic testing, and radiographic findings are often equivocal. The only consistent metabolic abnormality is hypoglycemia, but it may be present intermittently.[28] Several blood glucose determinations should be evaluated in every middle-aged to old dog that is presented because of seizures. Levels should be determined after the patient has fasted 24 to 48 hours and, if possible, assessed during a seizure. Analysis of serum insulin concentrations and calculations of the amended glucose–insulin ratio may aid the diagnosis. As soon as hyperinsulinism has been established, celiotomy is indicated.[28] Surgical resection of adenomas or adenocarcinomas that have not metastasized may be curative.[27] Other options include dietary management and medical therapy. The prognosis for patients with insulinoma is fair to poor; survival times of a year or more have been reported.[27]

CONCLUSION

Cats and dogs with manifestations of endocrinopathies that affect the central nervous system may present with dramatic clinical signs; often, patients are in an endocrine crisis and must be managed accordingly. Clinical signs, however, may also be vague and insidious. Veterinarians should consider endocrine disorders as a possible cause of seizures, behavioral changes, cranial nerve defects, and various other neurologic signs because appropriate and prompt treatment often results in clinical improvement.

About the Authors

Drs. Towell and Shell are affiliated with the Department of Small Animal Clinical Sciences, Virginia–Maryland Regional College of Veterinary Medicine, Virginia Tech, Blacksburg, Virginia.

REFERENCES

1. Chrisman CL: *Problems in Small Animal Neurology*. Philadelphia, Lea & Febiger, 1982, pp 115–153.
2. Wheeler S: Emergency management of the diabetic patient. *Semin Vet Med Surg (Small Anim)* 3(4):265–273, 1988.
3. Schaer M: Diabetic hyperosmolar nonketotic syndrome in a cat. *JAAHA* 11:42–46, 1975.
4. Schaer M, Scott R, Wilkins R, et al: Hyperosmolar syndrome in the nonketotic diabetic dog. *JAAHA* 10:357–361, 1974.
5. Greene SA, Hefferson IF, Baum JD: Cerebral oedema complicating diabetic ketoacidosis. *Dev Med Child Neurol* 32: 633–644, 1990.
6. Krane EJ, Rockoff MA, Wallman JK, Wolfsdorf JI: Subclinical brain swelling in children during treatment of diabetic ketoacidosis. *N Engl J Med* 316:857–859, 1987.
7. Alberi KGMM: Diabetic emergencies. *Br Med Bull* 45(1): 242–263, 1989.
8. Ellis EN: Concepts of fluid therapy in diabetic ketoacidosis and hyperosmolar hyperglycemic nonketotic coma. *Pediatr Clin North Am* 37(2):313–321, 1990.
9. Daugirdas JT, Kronfol NO, Tzamaloukas AH, Ing TS: Hyperosmolar coma: Cellular dehydration and the serum sodium concentration. *Ann Intern Med* 110(11):855–857, 1989.
10. Rosenberg I: Hypothyroidism and coma. *Surg Clin North Am* 48(2):353–360, 1968.
11. Chastain C, Graham C, Riley M: Myxedema coma in two dogs. *Canine Pract* 9(4):20–34, 1982.
12. Kelly M, Hill J: Canine myxedema stupor and coma. *Compend Contin Educ Pract Vet* 6(12):1049–1055, 1984.
13. Forester C: Coma in myxedema. *Arch Intern Med* III:734–743, 1963.
14. Sikwang L, Tilley L, Tappe J, Fox D: Clinical and pathologic findings in dogs with atherosclerosis: 21 cases (1970–1983). *JAVMA* 189(2):227–232, 1986.
15. Patterson J, Rusley M, Zachary J: Neurologic manifestation of cerebrovascular atherosclerosis associated with primary hypothyroidism in a dog. *JAVMA* 186(5):499–503, 1985.
16. Redding R: Canine electroencephalography, in Hoerlein B (ed): *Canine Neurology*, ed 3. Philadelphia, WB Saunders Co, 1978, p 150.
17. Chastain C, Riedesel DH, Graham CL: Secondary hypothyroidism in a dog. *Canine Pract* 6(5):59–64, 1979.
18. Sherding R, Meuten D, Chew D, Knaack K: Primary hypoparathyroidism in the dog. *JAVMA* 176(5):439–444, 1980.
19. Feldman E, Nelson R: *Canine and Feline Endocrinology and Reproduction*. Philadelphia, WB Saunders Co, 1987, pp 328–355.
20. Rassinussen H: Parathyroid hormone, calcitonin and the calciferols, in Williams R (ed): *Textbook of Endocrinology*. Philadelphia, WB Saunders Co, 1974, p 724.
21. Capen C: Parathyroid hormone, calcitonin and cholecalciferol: The calcium regulating hormones, in McDonald L (ed): *Veterinary Endocrinology and Reproduction*, ed 3. Philadelphia, Lea & Febiger, 1988, p 60.
22. Ihle S, Nelson R, Cook J: Seizures as a manifestation of primary hyperparathyroidism in a dog. *JAVMA* 192(1):71, 1988.
23. Adams R, Victor M: Epilepsy and other seizure disorders, in Adams R (ed): *Principles in Neurology*, ed 3. St. Louis, McGraw-Hill Book Co, 1985, p 233.
24. Hardy R: Hypoadrenocorticism: An internist delight/nightmare. *Proc 8th ACVIM Forum*:65–68, 1990.
25. Knecht C, Henderson B, Richardson R: Central nervous system depression associated with glucocorticoid ingestion in a dog. *JAVMA* 173(1):91–92, 1978.
26. Chastain C: Use of corticosteroids, in Ettinger S (ed): *Textbook of Veterinary Internal Medicine*. Philadelphia, WB Saunders Co, 1989, pp 413–428.

27. Eckersley G, Fockema A, Williams J, et al: An insulinoma causing hypoglycemia and seizures in a dog: Case report and literature review. *J S Afr Vet Assoc* 58(4):187–192, 1987.

28. Leifer C, Peterson M, Matus R: Insulin secreting tumor: Diagnosis and medical and surgical management in 55 dogs. *JAVMA* 188(1):60–64, 1986.

ANXIETIES, FEARS, AND STEREOTYPIES

Separation Anxiety in Dogs

Victoria L. Voith, DVM, PhD
Department of Clinical Studies
School of Veterinary Medicine
University of Pennsylvania
Philadelphia, Pennsylvania

Peter L. Borchelt, PhD
Animal Behavior Consultants, Inc.
Forest Hills, New York
The Animal Medical Center
New York, New York

One of the most common complaints heard from the owners of companion dogs is that the dogs engage in disruptive behavior when left alone. A dog may urinate, defecate, bark, howl, chew, or dig. The owner often describes the dog as "spiteful" or as "being angry." Such explanations, however, have no basis in the scientific study of animal behavior. It is more accurate to describe the dog's disruptive behavior when left alone as a distress response to separation from the person or persons to whom it is attached.

Attachment and attachment behaviors are essential for animals whose survival is benefited by sociability. Attachment, whether defined as an emotion or strictly in terms of the behavior an animal exhibits when separated from a companion, serves to keep individuals together and as such is a mechanism for social cohesion. It helps maintain social contacts and bonds between adults and between parent and infant. The mother-infant bond is one of the strongest forms of attachment and has been studied frequently in an attempt to understand the attachment phenomenon.[1-5]

Many social animals engage in distress responses when separated from their companions. Although behavioral differences exist between species and individuals in the same circumstances, in general, an infant's response to separation from its mother is increased activity and vocalization. The result usually is a quick reuniting of the infant and its mother. If the high activity phase fails, however, the infant may withdraw to a hypoactive or depressed attitude, perhaps to conserve energy and draw no attention to itself until the mother returns. These behaviors are not unlike the responses of older animals when separated from individuals to whom they are attached. For instance, when a loved one dies, humans normally cry and display short-term depressions lasting several weeks or months. Because dogs are highly social creatures and become attached to the people with whom they live, it is not surprising that they engage in separation distress responses.[6-8] In addition to separation anxiety, many other reasons exist that might cause a dog to eliminate in the home, vocalize excessively, chew, and/or dig. Just as the presenting sign of polyuria has numerous possible medical causes, a behavioral complaint can stem from many possible causes. A differential diagnosis must be made before appropriate treatment for separation anxiety problems can be started.

Characteristics of Canine Separation Anxiety

Separation anxiety behaviors are exhibited by male and female dogs with equal frequency and are not breed or age related (except at the time of weaning).

Separation anxiety can be manifested in several ways. An individual dog can show one or a combination of signs (Table I). Usually dogs vocalize, eliminate, or engage in destructive behavior. Conversely, they may manifest behavioral depressions, such as absence of play, immobility, or anorexia when left alone. Owners often remark that even when they leave an appealing tidbit for the dog as they depart, it is still there when they return. Then, after the dog greets the owner, it runs over and eagerly eats the tidbit that has been there for the entire period of absence. Frequently, dogs left in boarding kennels do not eat, remain inactive, and look "depressed" until the owner returns.

Dogs with separation anxiety usually engage in exaggerated greeting behaviors, which are more intense and prolonged than exhibited by most dogs. In fact, some owners present these dogs for hyperactivity. When the owners are at home, their dogs continuously seek to maintain contact with them, following the owners from room to room, frequently leaning against them or climbing onto their laps. These dogs rarely spend any time outdoors by themselves. When let out to urinate or defecate, they immediately want to come back in unless the owner is outside with them. Occasionally, psychosomatic responses, such as diarrhea, bloody stools, or pulling or chewing hair, can occur as a result of separation anxiety.

Sometimes a dog with separation anxiety is presented with the complaint of aggression exhibited toward the owner at the time of departure. The dog is usually described as trying to prevent the owner from leaving. A detailed history usually reveals, however, that the dog bit or snapped at the owner as a response to physical restraint implemented by the owner at the point of exit. Complete histories disclose that dogs who exhibit aggression in this circumstance are also aggressive in other situations that are compatible with dominance aggression.[9,10] Dominant dogs usually resist physical restraint and heavy tactile pressure by owners. If a clinician is presented with the complaint of canine aggression related to the owner's departure, the possibility that the dog has both separation anxiety and dominance aggression problems should be explored.

TABLE I

Signs of Separation Anxiety

Vocalization—howling, barking, whining
Elimination—urination, defecation
Destructive behavior—chewing, digging
Anorexia/"depression" or inactivity
Psychosomatic/medical consequences—diarrhea, vomiting, excessive licking of haircoat
Overactivity—excessive greeting behavior, constant pestering of owners

The most obvious feature of separation anxiety in dogs is that the problem behaviors occur only when a dog is separated from the individual(s) to whom it is attached. Usually separation occurs when the owner leaves the home; however, the behaviors may also occur when the owner is home but the dog is prevented access to the person—such as when the dog is confined in a room, kept outside, or put in a cage. Separation anxiety also may occur when the owner is out of the home and the dog is left with other dogs, cats, or even with other humans.

Dogs that exhibit separation anxiety are usually model pets when the owner is at home; they are housebroken and not destructive. They never or rarely engage in excessive barking or howling and may have been to obedience school where they performed well. Anxiety behaviors are not the result of disobedience but are a distress response.

One of the key differentiating features between separation anxiety and other behavioral disorders with similar signs is that, for the former, the dog engages in separation responses within a short time after being left alone—often within minutes (Figure 1). The owner may discover this on returning to the home to pick up a forgotten item a few minutes after leaving and finding that elimination or destructive behavior has occurred. The owner (or neighbors) might hear the dog vocalize shortly after being left alone.

Usually, dogs that experience separation anxiety begin exhibiting distress signs before the owner's departure. They follow the owner closely, often panting, pacing, salivating, and sometimes trembling as the owner prepares to depart. Some dogs become extremely excited and try to leave with the owner—resisting attempts to be restrained. Occasionally, dogs are described as looking "sad" or "depressed" when the owner leaves; the head, ears, and tail are lowered and the dogs are very quiet. They usually watch the owner constantly and are hypoactive, although they may tremble.

Typical Histories of Dogs with Separation Anxiety

Prolonged constant contact between dog and owners is a common history. Usually these are dogs that have grown up in a household where a person has always been present, such as a nonworking spouse or an elderly relative. Rather than being left alone, the dogs always accompanied a person. The problem often coincides with resumption of the school year or the beginning of employment of all the household residents. For the first time, suddenly, the dog is left alone for long intervals.

Occasionally a dog that has demonstrated tolerance for being alone will suddenly begin exhibiting separation anxiety. These sudden onsets are often preceded by periods of constant contact with the owner, such as a month of vacation or a convalescent period during which

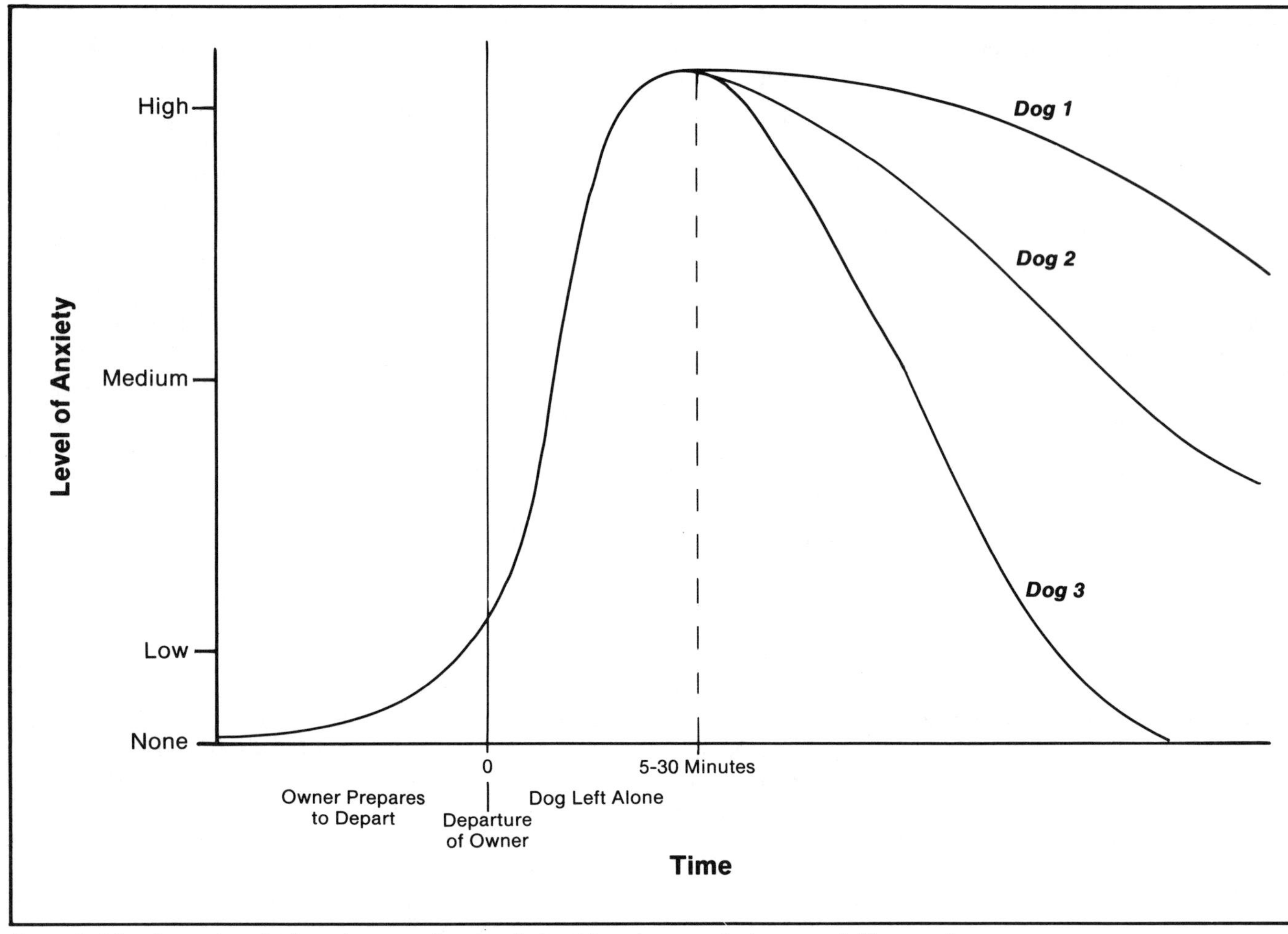

Figure 1—The change in a dog's anxiety over time in relationship to its owner's departure is illustrated.

the owner never left the home. When the owner resumes leaving the dog alone, the pet may engage in separation distress behaviors.

Separation anxiety is sometimes preceded by a stay at a boarding kennel or other long separation from the owner. When the owner and dog are reunited and the owner then leaves the dog alone, the dog becomes distressed. Another factor that has been implicated in the onset of separation anxiety is a dramatic change in the constituency of the household involving a permanent or long-term absence of a person to whom the dog is attached. This can be the consequence of divorce, a student leaving for college, or death of one of the individuals in the household. Dogs that fear thunderstorms sometimes develop separation anxiety behaviors. If these dogs are traumatized by a storm while alone, they often are afraid to be alone subsequently, regardless of whether or not a storm is imminent. Not surprisingly, a dog left in a strange place or a new apartment may show distress responses even though it was fine in its prior home.

It is interesting that many dogs adopted from humane societies or found as strays exhibit separation distress behaviors when left alone. Perhaps it was separation behavior that led to their being abandoned, but it is also possible that this period of "abandonment" predisposes dogs to form very strong attachments when the opportunity presents itself. When subsequently separated from a new companion, the dog becomes exceedingly distressed.

Differential Diagnosis

The following is an outline of the most common differentials for the presenting complaint of eliminative, destructive, or excessive vocalization behaviors:

- I. Elimination behaviors
  1. Separation anxiety
  2. Housebreaking
  3. Lack of opportunity
  4. Fear of excitement
  5. Submission, greeting, and marking
- II. Destructive behaviors
  1. Separation anxiety
  2. Playful behavior
  3. Puppy chewing
  4. Fear responses
  5. Reaction to arousing stimuli
  6. Overactivity
- III. Excessive vocalization
  1. Separation anxiety

2. Reaction to arousing stimuli
3. Socially facilitated
4. Play
5. Aggression
6. Fear responses

Elimination Behaviors

Urination and/or Defecation

When a dog is presented because it is urinating and/or defecating in the home, pathophysiologic disorders should be considered. Appropriate physical examination and laboratory tests and careful determination of the dog's history should either confirm or exclude a disease process.

If the dog is presented for both urinating and defecating in the house, the problem is unlikely to be disease related (unless the dog is geriatric), as few diseases affect both urogenital and gastrointestinal systems. The major behavioral differentials when a dog is presented for urinating and defecating in the home are separation anxiety, ineffective housebreaking, and lack of opportunity to eliminate in an acceptable location. The key features of separation anxiety have already been described and should be compared with the characteristic features of other problems that are accompanied by similar presenting signs.

Housebreaking. A dog with housebreaking problems will generally eliminate in the house regardless of whether the owner is home or away. Housebreaking problems are related to lack of inhibition about eliminating in the household and to the length of time since a dog last urinated and/or defecated. Housebreaking problems rarely occur shortly after an owner's departure if a dog has had an opportunity to eliminate outdoors before its owner leaves. "Accidents" usually occur several hours after the dog has been left alone. Conversely, a dog with separation anxiety usually urinates and/or defecates within minutes of the owner's departure regardless of whether it has just eliminated. Occasionally, dogs with housebreaking problems will eliminate only when the owner is gone. Such dogs are let out frequently when the owner is home, have learned to signal their owners when they must go out, or have learned to inhibit their eliminative behavior only when the owners are present. When the owner is not home, however, these dogs have little inhibition about eliminating in the house.

In summary, the key characteristics of housebreaking problems are that the incidents occur whether or not the owner is present, do not occur immediately after the owner's departure, and are related to the last time the dog eliminated.

Lack of Opportunity. Some owners appear to be ignorant of the fact that dogs may have to urinate or defecate several times a day. Although some dogs can accommodate 12- and 16-hour stints in the home, others must relieve themselves during that long a time span despite their effort not to eliminate in the house. These dogs may be housebroken most of the time, but if the owner leaves them for an extended period they will eliminate in the house. Gradually these dogs may become "unhousebroken" as they lose all inhibition about relieving themselves within the home.

Fear or Excitement. Defecation often accompanies excitement and fear responses. Dogs that are afraid of loud noises, such as thunder and exploding firecrackers, may defecate in response to those stimuli. Because its owner's presence may offer some reassurance to a dog, its responses to loud noises may be less intense when the owner is home. The dog may simply shiver, shake, hide, or seek out the owner. When the owner is not home, however, the dog is more frightened and may eliminate as well as engage in destructive (escape) behaviors. The history of a dog's responses to specific stimuli should help determine if defecating in the home when alone is related to a fear response. These dogs routinely show signs of fear to certain stimuli. They usually do not eliminate or act destructively when left alone; they only eliminate in the home when the owner is gone and the fear-eliciting stimuli occur.

Submission, Greeting, and Marking. Submissive and greeting urination behaviors are most common in puppies in the presence of a human or another animal and typically occur in a squatting posture or in lateral recumbency with one leg lifted. Urination of this type is often elicited by the return of the owner to the household, reaching toward the dog, and petting or disciplining the dog.

Urine marking is generally performed in a leg-lifting posture by male dogs after they reach puberty. The dog generally deposits small amounts of urine regardless of the owner's presence, although some will only urine mark in the owner's absence. Urine marking in these cases is usually unrelated to the time of the owner's departure. It can, however, occur in response to specific stimuli, such as a mail carrier making a delivery or another dog outside when the owner is absent.

Destructive Behavior

When an animal is presented for chewing and digging behaviors, the most likely behavioral causes are separation distress, play behavior, puppy "chewing," or fear responses. Other possible causes are responses to arousing stimuli or overactivity.

Separation Anxiety

Destructive behavior related to separation anxiety is usually directed at a point of exit from the home, such as a door, windows, or the floor around doors and windows. A destructive dog may also chew objects that frequently contact the owner's body, such as pillows, bedspreads, couches, or chairs. Destructive behavior usually occurs shortly after the owner's departure and does not occur when the dog has access to the owner. The behavioral profile and history of the animal also fit the profile and history described for separation anxiety.

Play Behavior

Play behavior is generally exhibited by young animals and is unrelated to the departure of the owner, although a young dog may engage in more destructive play when the owner is gone because there is no one to prevent it. Play behavior is often directed toward objects that are "fun" to shred or toss—sponges, foam rubber, pillows, toilet paper. Although dogs may engage in more destructive play when alone, they usually engage in the same behaviors when the owner is present. Most may not do so directly in front of the owner because they have been previously punished for doing so. Some dogs actually look at the owner as they grab objects and attempt to engage the owner in play.

Puppy "Chewing"

Puppies are notorious for chewing on a variety of objects. Gnawing on hard objects, such as table legs, probably is related to tooth development. Chewing on other objects may be related to exploratory behavior. Most young mammals, including humans, investigate novel objects by placing them in the mouth.

Fear Responses

Fear responses, particularly to loud noises, such as thunder and fireworks, may involve destructive behaviors. Phobic responses of dogs include jumping through glass windows as well as digging or chewing through floors, walls, and doors. These dogs show fear responses to the stimuli regardless of the owner's presence or absence, although when the owner is home the dog's responses may be restricted simply to shivering, shaking, or leaning against the owner. In the owner's absence, the dog will engage in what appear to be escape behaviors that result in destruction of objects.

Reaction to Arousing Stimuli

Dogs may react to arousing stimuli by engaging in behaviors that are disruptive to the household. For example, some dogs attack the door whenever mail is delivered, or they lunge and scratch at windows when another dog or pedestrian passes by. Usually these dogs exhibit at least a modified form of such behaviors when the owners are at home. It is possible that a dog may have learned not to respond to these stimuli in the presence of the owner and will only do so when left alone. These dogs, however, would not have the same behavioral profiles or histories as dogs with separation distress responses.

Overactivity

Very active dogs often cause havoc in a household. These are generally young dogs or breeds that have been selected for high activity levels (e.g., some hunting and running dogs). Their destructive behavior is often the result of knocking objects over as they run through the house. Overactivity is related to the amount of exercise the animal gets and is not related to the departure of the owner. These dogs usually engage in overactive behaviors when the owner is home as well as when the owner is away. The objects and nature of their destruction are not the same as those occurring with separation anxiety nor are the dogs' behavioral profiles and histories compatible with separation anxiety.

Excessive Vocalization

Excessive barking and howling may be related to separation distress, reaction to external stimuli, social facilitation by other dogs, play behavior, or aggression or may occur as part of a fear response.

A dog that is vocalizing as part of a separation distress response will do so at the time of the owner's departure or very shortly thereafter. Rarely does separation anxiety cause a dog to bark and howl or whine hours after the owner's departure. The behavioral profile of the dog before and at the time of the owner's departure as well as how the dog interacts with the owner when she or he is at home should confirm a separation anxiety response.

Some dogs bark excessively when reacting to such stimuli as passing pedestrians, loud noises, or mail carriers. These vocalizations are usually not related to the absence of the owner. Generally, dogs that vocalize in response to fearful stimuli or bark in response to the barking of other dogs also do so when the owner is present. Vocalizations related to play and aggression would occur in the contexts of those activities. A detailed history should allow a clinician to make a differential diagnosis involving barking behavior.

Treatment of Separation Anxiety

The principle underlying all treatment techniques for phobias, fears, or anxieties is to enable an animal to experience situations that elicit fear or anxiety without being afraid or anxious. In treating dogs with separation anxiety, this is generally achieved by gradually getting the dog used to being alone starting with many short separations that do not produce anxiety and then gradually increasing the duration of the separation periods. Sometimes these graduated separations need to be supplemented with counterconditioning to predeparture cues that elicit anxiety even before the owner leaves. Antianxiety medication may also suppress separation anxiety responses and can be used in conjunction with behavioral treatment procedures. If the dog's responses are mild, drug therapy alone can be used effectively.

On the surface it seems puzzling that a dog would not learn to relax when the owner leaves because she or he invariably returns, usually in 8 to 10 hours. This is probably because the dog experiences a species-typical ("innate") response to separation and has become conditioned (both classically and operantly) to experiencing intense emotional and behavioral responses to the owner's departure and events associated with it. Even if dogs theoretically have the ability to "think ahead" and associate events that occur hours apart, it is unlikely that a dog experiencing separation anxiety could do so because of the high state of arousal and anxiety it experiences at the time of departure. It is well known from the literature on animal behavior and psychology that high anxiety states inhibit learning.[11] What is required is that a dog experience separation in a nonanxious state.

Many dogs that exhibit separation anxiety when left alone at home are perfectly well behaved when left alone in the owner's car, perhaps because most owners gradually get their dogs accustomed to being alone in the car. Initially the owner leaves the dog only for a few minutes, for instance, as he or she makes a quick purchase at a fast-food store or mails a letter. Few people leave a dog alone in a vehicle for eight hours the first time. Contrast this with initial departures from the home, which are usually for several hours. The dog's "apprehension" is justified—the owner does not appear to be returning. Car absences also differ from home absences in that the duration of separation involving a car are usually short even though the dog is occasionally safely left in the car for long time periods. Conversely, absences from the home are almost always long. This observation suggests that gradually getting the dog accustomed to being alone in the home might be an efficacious approach to treating separation anxiety.

It is also interesting that dogs "know" when their owners are preparing to depart. Some owners report that their dogs act differently on weekend mornings compared with workday mornings. Obviously such dogs attend to stimuli or behaviors of their owners, which predict the departures. Dogs that learn to associate specific cues with departures from the house for short-term absences are not anxious during those periods. For example, one client presented her dog because of its destructive behavior (chewing) whenever left alone. On one occasion when this owner reentered her house for forgotten items only minutes after having left for work, she discovered the dog had already begun chewing on the woodwork. Detailed questioning of this owner revealed that the dog was never destructive when the woman left to do her laundry in the basement of the building. Departures to do laundry meant relatively short absences—sometimes only 5 to 10 minutes—in comparison with work departures. Definite cues (laundry items) indicated a short absence to the dog. Consequently, the dog was not anxious before or during the owner's absences to do laundry. This observation suggests that teaching a dog to associate specific cues with departures leading to "safe" short absences might be helpful in getting it accustomed to being alone and not being anxious.

Getting Ready to Practice Graduated Departures

Before starting graduated departures, it is helpful to countercondition a dog's anxiety responses to its owner's movement around the exit door and to other predeparture cues. For instance, the dog can be taught to sit and stay and to associate rewards of food or praise with these activities. At first, the dog is required to sit as the owner merely steps away and then back. Gradually, the owner increases the distance and length of time away from the dog. Additionally, the owner adds departure-related cues, such as picking up keys, rattling them, turning the door knob, and putting on a coat, while the dog sits and stays and receives a tidbit or praise for doing so. Next the owner might step out of view with the door open before reappearing and returning to the dog. Then the owner might leave, close the door, stay outside for a second or two, reopen the door, and return to the dog. If the dog is extremely food oriented, it should not be rewarded with food on return of the owner, because in its anticipation of food the dog may become anxious or excited.

Besides counterconditioning a dog's responses to departure cues, the owner can habituate the dog to these cues by exposing the dog to them numerous times a day without leaving. The owner might pick up and rattle the car keys, then put them down; put on a coat, then take it off; open the door, then shut it. These counterconditioning and habituation procedures reduce a dog's anxiety responses to predeparture cues. If a dog is less anxious before its owner's departure, it is more likely to tolerate the absence, especially if it is *very short.*

Graduated Departures

When an owner begins therapeutic departures, the dog should be left alone only for very short periods of time—shorter than it takes for the dog to engage in an anxiety response. This may mean absences of only one or two seconds in some cases or one or two minutes in other cases. Most dogs will tolerate absences of a few minutes if the owner has spent some time counterconditioning the dog's predeparture responses before beginning the graduated departures.

Because dogs can associate cues in the environment (e.g., laundry items, being left in the car) with "safe" or "probably safe" departures, some behavioral therapists suggest that the owner consistently present a "safety cue" to associate with the short graduated departures.[6,12,13] After the dog has become accustomed to being alone in the house without being anxious, the safety cues can be eliminated. One such cue might be a consistent sentence stated by the owner, such as "Be good, I'll be right back." Other cues that can be used to give a constant signal while the owner is gone are leaving a radio or television on, a loud ticking clock, or availability of an attractive chew toy or bone. It is important to note that these "safety cues" are not likely to work by themselves. Most owners of dogs with separation anxiety problems *do* present one or more of the preceding cues before they leave, but the dogs still engage in separation anxiety behavior. A dog must *learn* to associate the cue and the separation with a nonanxious state.

Obviously, the "safety" cue should not be something the dog has already associated with anxiety. If the owner has previously left a radio on while the dog engaged in separation anxiety responses, using a radio would not be a good cue to try to associate with safe departures because the dog already associates anxiety with the radio. Few people leave a television set on while they are gone from a house, and it can serve effectively as a safety cue. It is important to note that safety cues will not have to be used indefinitely.

A dog that manifests distress by chewing can be given a palatable bone (that cannot be splintered) or an attractive chew toy both as a signal that the departure is for a short, nonanxiety-provoking absence and as a target for the chewing behavior. If a bone or toy is incorporated into practice departures, the item should not be available at other times. This makes the item more interesting when it appears, thus the dog is more likely to direct chewing behavior toward it. The novel item should not be given to the dog when the owner leaves for a period longer than the dog is likely to tolerate because the item will lose its value as a safety signal if it is associated with anxiety.

No controlled clinical studies have been done to prove how efficacious the use of "safety cues" is in treating separation anxiety in dogs. Some animal behavior therapists report success without their use. If cues are used, however, they should be presented just before the owner's departure. The owner should present the cue, leave, and then return before the dog becomes anxious. The dog should be greeted in a low-key manner and the safety cue removed or terminated.

Many dogs become slightly excited by even a short absence. If this happens, the owner should wait until the dog is relaxed before repeating a short absence. The waiting period may be several minutes. When the owner is ready to depart, the safety cue can be presented. Then the owner leaves and again returns within the prescribed interval. The owner's greeting should be low-key, the cue (if used) removed; and the owner should again wait until the dog is relaxed before practicing another short absence. This last, seemingly minor point, is important. The authors have found that owners, naturally motivated to solve a departure problem as quickly as possible, often attempt to expose a dog to practice departures that are spaced too closely together. If a departure leads to an anxiety response (indicated by the dog's behavior on the owner's return) and the next departure occurs before the dog's level of anxiety/greeting behavior has diminished, then the next departure will lead to a higher level of anxiety. Departures performed too closely together can thus increase anxiety, and the problem will get worse rather than better. Practice makes perfect—but only if practiced perfectly!

How to Assess Progress

The goal of the procedure is a gradual increase in the duration of time a dog is left alone, as quickly as possible without eliciting anxiety. The owner must judge at what point a dog is able to tolerate an increase in the length of separation. Because each dog reacts differently, it is not helpful to set a rigid time interval for an owner to follow. A written time schedule can be used as a guide, but only if the owner understands that she or he must assess the dog's response to the previous practice departure every time to determine when and how long the next departure and absence should be. The danger of a fixed time schedule is that the owner will not pay attention to the dog's behavior and will inadvertently present the dog with durations of absence that are too long for the dog to tolerate, thereby worsening the problem. In some circumstances, separation anxiety occurs only when one member of the family leaves. Thus, other family members could serve as observers of the dog's anxiety level during graduated separations from the person to whom the dog is attached. In the majority of the cases, however, no one would be in the home to observe the dog. In cases where no direct observation is possible, the dog's level of anxiety related to a separation can be assumed by observing the dog's behavior just before the owner's departure (predeparture anxiety level) and after the owner returns (arrival elation). The assumption is that predeparture anxiety and arrival elation reflect the dog's anxiety level during the separation.

An owner should increase the duration of separation only after two criteria are met: (1) The dog exhibits no predeparture anxiety, and (2) the dog does not exhibit prolonged or exaggerated greeting behavior when the owner returns. For instance, if the dog tolerates repeated 5-minute absences of the owner, then the owner might extend the duration of separation to 10 minutes. If the dog begins to display predeparture anxiety or too high a greeting response to 10-minute absences, then the owner should extend the time only to, perhaps, 6 minutes.

As progress occurs, the separation durations become longer, for instance, 1 minute, 2 minutes, 3 minutes, or 5 minutes—up to 10 minutes—up to 15 minutes. The durations should not extend steadily (e.g., 5, 10, 15, 20 minutes) but should increase according to a variable schedule (e.g., 1, 2, 1, 2, 1, 3, 2, 4, 1, 4, 3, 5 . . . minutes). Regardless of the duration, it is critical to remember that the dog must remain nonanxious.

The owner should be aware that too rapid an exposure to separations that are too long (i.e., leading to anxiety) will make the problem worse. For instance, if the dog can tolerate 1-minute separations and the owner attempts to extend the duration of absences quickly to 10 minutes, the dog will probably become anxious during the 10-minute separations. Subsequently, the dog will resume anxiety at 1-minute separations because it again has associated anxiety rather than a nonanxious state with the owner leaving.

Fortunately, an owner does not have to increase absences gradually minute by minute until 8 to 10 hours of separation. Most distress responses occur within 30 minutes of an owner's departure; the owner will spend much more time getting the dog accustomed to being alone for the first 30 minutes of separation compared with later, longer time intervals. How rapidly the program progresses varies among individuals. As a guideline, the owner should expect to do many short absences, gradually increasing in length, under 30 minutes. Then, instead of increasing in 1-minute to 5-minute increments, 15-minute increments can be initiated, with 5- and 10-minute absences introduced periodically. After the dog can accommodate a 30-minute departure, the owner might leave it for 45 minutes, then 10, then 30, 45, 15, 60, 15, 30, 5, 75 minutes and so on. If the dog can be left alone for 90 minutes, it probably will remain relaxed for 4 to 8 hours.

As a rule of thumb, a dog that can tolerate being alone for an hour and a half will probably tolerate being alone for an entire day. It is best, however, to leave the dog alone for three hours to four hours before leaving a dog for eight hours.

Most owners can accomplish these procedures within a few weeks. This typically means several practice sessions on weekends, before going to work, and in the evenings. These practice sessions can be scheduled in several ways, depending on the owners' available time. One approach is to schedule one long session per day, initially comprising many (10 to 30) short absences and later during the course of treatment several (1 to 5) longer absences. Another approach is to conduct several sessions a day consisting of fewer departures per session.

Depending on the severity of the separation anxiety, the owner may initially have to make the practice departures "safer" by avoiding cues that the dog has already associated with "unsafe" or anxious departures. For example, the owner might wear weekend clothes instead of workday clothes, avoid rattling car keys, and not take a briefcase. Eventually such cues must be incorporated into the practice departures, initially with short and then gradually longer separations.

What to Suggest While the Owner Is at Work

After an owner implements therapeutic departures, optimum progress is achieved if the dog does not experience any traumatic separations until it is "cured." Each distressful separation erases some of the improvement that has been achieved by the owner. Ideally, the dog should not be left alone for any period of time longer than it can tolerate. Owners may elect to take the dog to a friend's house, hire a dog sitter, board the dog while they are working during the day, or take the dog with them if possible.

If a dog must be left, however, its distress might be alleviated with antianxiety medication. The dog can be left alone for long intervals under the influence of the antianxiety drug while the owner works on increasing the amounts of time the dog can be alone in the unmedicated state. If the dog must be left alone in an unmedicated state for longer than the owner predicts it will tolerate, the safety cues (e.g., a radio) should *not* be presented because to do so would ruin the association between the cues and safe, nonanxious departures.

Antianxiety Medications

Antianxiety drugs are potentially helpful in treating separation anxiety in dogs. Such drugs can be used to suppress the distress response if a dog must be left at home for an interval of time longer than the owner has practiced. Some dogs are so distraught at any departure of the owner that even practice departures for short absences cannot be implemented without the help of antianxiety medication. Sometimes dogs with mild separation anxiety can become accustomed to being left alone under the influence of a drug without concurrent behavior modification techniques; as the dose is gradually reduced over repeated separations, the dog gets used to its surroundings.

A good antianxiety drug does not sedate the dog; in fact, sedation might interfere with the dog's learning abilities. For separation anxiety responses, an antianxiety drug does not need to have a long effect; all that appears necessary is to reduce the dog's anxiety at the time of departure. There may be individual differences in responses to antianxiety drugs and a dog may respond to some antianxiety drugs and not others. If the owner and clinician elect to try antianxiety medication, the drug should first be given while the owner is at home to determine if there are any side effects. If no apparent side effects occur, the drug should be administered one hour before the owner's actual or practice departures.

When the dog has demonstrated that it can experience a sufficiently long unmedicated practice departure, the antianxiety drug can be discontinued. If the dog is being medicated without concurrent behavior modification techniques, it should probably experience 10 to 14 departures at the initial dose, 10 to 14 departures at one half the initial dose, then 10 to 14 departures at one quarter the initial dose before the medication is discontinued. Depending on the case, the dosage may be reduced more quickly or gradually. Regardless of the precise dosage schedule, the drug should be gradually decreased. Drugs given for separation anxiety problems should only be used temporarily.

Drugs that are potentially useful in suppressing separation anxiety responses in dogs are tricyclic antidepressants, progestins, barbiturates, phenothiazines, and benzodiazepines. Dogs' responses to antianxiety drugs vary greatly between individuals. The goal is to reduce anxiety without inducing sedation that could interfere with learning.

The tricyclic antidepressants, such as amitriptyline hydrochloride and imipramine hydrochloride have been reported efficacious in reducing anxiety and depression in humans as well as separation distress in dogs.[3, 14] Although the exact physiologic mechanism of action is unknown, the drugs appear to functionally reduce distress related to separation and loss. The authors believe amitriptyline hydrochloride effectively reduces separation anxiety in dogs at a dose of 1 to 2 mg/kg. If no side effects occur and the drug proves efficacious for an individual, the authors use the medication according to the guidelines in this section. Although the literature indicates that the tricyclic antidepressants may require two to three weeks of administration before achieving therapeutic effects in treating human depression, the authors' clinical impression is that these drugs immediately reduce anxiety in some dogs.

The *Physician's Desk Reference*[a] cautions against the use of tricyclic antidepressants in human patients with a history of seizures, urinary retention, glaucoma, and cardiovascular disorders. The tricyclic antidepressant drugs, such as amitriptyline hydrochloride, have been reported to produce arrhythmias, sinus tachycardia, and prolonged conduction time. Other side effects that may occur are hypertension; states of excitement; such anticholinergic effects as a dry mouth, blurred vision, and constipation; allergic responses; and gastrointestinal changes. Clinicians should become familiar with contraindications for use, the signs of toxicity, and potential side effects as described in the *Physician's Desk Reference* or package inserts supplied with the drugs. Again, the drug should always be given first when the owners are home for an extended period of time to determine whether or not side effects are produced.

The synthetic progestins have a marked antianxiety and calming effect in some animals. Megestrol acetate at 1 to 2 mg/kg orally may produce tranquilizing effects without rendering the animal ataxic. Clinicians should be aware of potential side effects of synthetic progestins and consider them in selecting these drugs for treatment. Most dogs exhibit an increase in appetite and consequent weight gain while on progestin therapy. Occasionally elevated glucose levels and mammary gland hyperplasia occur. Progestins also suppress production of luteinizing hormone, testosterone production, and spermatogenesis; these functions should return to normal after medication is stopped but should be taken into consideration particularly if an animal is intended for breeding. Prolonged administration or idiosyncratic responses to the drug may result in hyperglycemia; elevated insulin and growth hormone levels; acromegaly; glucose intolerance/diabetes mellitus; polyphagia, polyuria, and polydypsia unaccompanied by hyperglycemia; lowered packed cell volume; reduced exercise tolerance; thickening of the gallbladder with a possible sequela of jaundice; mammary gland hyperplasia and development of mammary gland nodules; suppression of adrenal function; and, in the intact female, cystic endometrial hyperplasia-pyometra complex. The older the animal, the more likely the occurrence of glucose intolerance/ diabetes mellitus.[15-18] A wise precaution is to determine a dog's blood glucose level before medicating it with progestins and then monitor serum or urine glucose levels periodically while the dog is taking medication. If there are no contraindications or side effects, progestins can be used according to the guidelines described at the beginning of this section.

The phenothiazines, such as promazine hydrochloride, acepromazine maleate, and chlorpromazine hydrochloride, are tranquilizers that are frequently used in veterinary medicine. The authors believe, however, that in dogs these drugs usually must be given at such a high level to suppress destructive behavior and vocalization stemming from separation anxiety that the dog is ataxic. Moreover, it is unclear if the dog's anxiety is reduced or if the animal is simply rendered immobile.

The benzodiazepines, such as diazepam or chlordiazepoxide hydrochloride are frequently used with great success in reducing anxiety in humans and, hence, would seem likely candidates for treating anxieties in dogs. Diazepam and chlordiazepoxide hydrochloride can have a paradoxical effect in dogs, however. Some dogs become more excited, hyperactive, and consequently more destructive when treated with this group of drugs. Benzodiazepines are not general central nervous system depressants and can excite certain regions of the brain.

Punishment

Punishment is not an effective means of treating separation anxiety. Generally, punishment increases anxiety. Even if punishment were not contraindicated in principle, it would be difficult for most owners to meet the criteria for effective punishment. To be effective, punishment must be applied while the dog is engaging in the behavior and administered each time it engages in the behavior without producing more anxiety and without leading to inappropriate escape and avoidance responses.

The commonly suggested use of mousetraps and balloons to keep dogs away from furniture and the use of bitter-tasting substances applied to items that will probably be chewed *are* likely to "catch the dog in the act," but such techniques usually work only on play-related or "teething"-related destructive behavior, if they work at all. These techniques usually fail to decrease separation-related destruction because they are not directed at the underlying problem—distress. At best a dog that experiences separation anxiety may learn to stay away from the booby-trapped items but will continue to display anxiety in other ways. Likewise, the commonly suggested technique of having the owner hide inside the home or wait outside the exit door for the dog to vocalize and then rapidly return to yell at the dog, throw a can filled with pennies, or give the dog a leash correction might meet the criteria of punishing a misbehavior during or soon (a second or two) after its occurrence, but those approaches do not directly address the problem of separation anxiety and additionally have the isidious side effect of associating the owner with the punishment. Those techniques usually lead to the dog simply delaying the onset of barking or displaying anxiety through some other means.

The occasionally suggested technique of the owner reentering the home and rolling or forcing a dog over on its back to place it in a submissive posture is irrelevant

[a]Medical Economics Company, Oradell, New Jersey

because separation anxiety is not related to dominance. Moreover, the rollover technique may not actually be aversive to many dogs and thus is not a punishment.

Punishing a dog long after it has engaged in a behavior (such as when the owner returns) is useless in trying to prevent separation anxiety behaviors from recurring. Any punishment technique in which the dog is presented an aversive stimulus long after (more than a second or two) it has engaged in a behavior is ineffective. Many dogs do, however, learn to associate "punishment" with the return of the owner and may develop anxiety that is related to the owner's return and urinate or defecate just before the owner's reentry.

The most important criticism of the preceding punishment techniques is that a dog does *not* learn a non-anxious response in the stimulative context of being left alone.

Despite evidence supporting the ineffectiveness of punishment for separation behaviors, owners typically insist that their dogs can remember that they have "done something wrong" because, on the owners' return, the dogs "look guilty" only when they have done something wrong. Owners conclude, therefore, that punishment must be effective. They usually persist in this belief even though they have repeatedly disciplined and punished their dogs for misbehaviors without any effect on the problem.

The "guilty look" is actually a combination of submissive and fear postures, elicited by the anticipation of punishment. A number of simpler explanations than "guilt" exist as to why a dog exhibits these expressions only when it has done something wrong. The owner may be selectively remembering the times when disruptive behaviors and the dog's expressions did coincide and forgetting the times when events and expressions did not coincide.The dog may actually be reacting to subtle changes in the expression of the owner as she or he detects the mishap. Another explanation is that the dog has learned that when two stimuli are present simultaneously (return of the owner and signs of recent destruction or elimination), it is punished. Therefore, when both stimuli coincide, the dog anticipates punishment and "looks guilty." It is not reflecting on what it did several hours before but is merely responding to two presently occurring stimuli: returning owner and destroyed items or elimination.

Each of the authors has had several cases similar to the following story, which supports the hypothesis that dogs probably associate the onset of punishment with the presence of two stimuli. An owner housebroke his dog by taking her out frequently as a puppy. He rarely caught her in the act of eliminating in the house, but when he discovered that she had done so, he would discipline her and "point out her mistake." When he returned home, she only looked "guilty" when she had "done something wrong." The dog became housebroken (because she had been taken out frequently and developed a location preference for eliminating outdoors) and was a model pet for seven years. Then the owner acquired a puppy, which was not yet housebroken. The character of the puppy's feces and the volume of its urine was sufficiently different from the character and amount of the urine and feces of the adult dog that it was easy to tell which dog had eliminated. When the owner returned home from work and the *puppy* had eliminated in the house, the *adult* dog greeted him looking "guilty."

Ancillary Measures

In addition to departure techniques, counterconditioning procedures, and drug therapy, some ancillary measures might help reduce the excessive attachment of a dog to an owner. While the owner is home, she or he might begin by requiring the dog to stay in a room by itself for a few minutes before following the owner. A dog might be prevented from sleeping on the owner's bed, sitting on laps, and leaning against the owner. Because most owners enjoy letting a dog behave in these ways, they often are reluctant to implement the suggestions. The owner can be assured that the measures are temporary and after the dog no longer shows signs of separation anxiety, she or he can probably resume interacting with the dog as before.

Use of Crates

Although use of a crate might be effective in housebreaking a puppy or preventing puppy chewing, for treating separation anxiety its use is usually counterproductive. When confined, a dog may simply continue to engage in distress responses in the crate. It may still eliminate, howl, whine, salivate, or attempt to escape (consequently injuring itself as it tries to do so).

Sometimes crates can be used if owners get their dogs accustomed to the crates gradually, before leaving them confined alone for a long period. The owners may spend a few weeks feeding the dog in the crate, giving it tidbits in the crate, and confining it for a few minutes at a time until the animal is comfortable in the crate when the owner is home. After the dog is accustomed to the crate on that basis, it still must become gradually accustomed to being left alone at home in the crate without the owner present. The owner might start getting the dog used to crate confinement by leaving the room for just a few minutes, then gradually lengthening these periods of absence. After the dog tolerates short in-house separations, the owner can actually leave the house for a few minutes. Now the owner can essentially repeat the behavior modification program outlined in this article, but with the dog in a crate. The next step would be to accustom the dog to being loose in the house while the owner is gone.

Even if a crate can be used safely, there are some undesirable consequences to its use. Most owners do not like the thought of their companion animals being caged all day. Dogs in crates are less effective watchdogs. Undoubtedly, a dog can learn to associate a tidbit with entering a crate and may move in and out of an open crate on its own, also using the crate as a resting place; but a question certainly exists about the comfort of so small an area for confinement 8 to 12 hours a day. Although a wolf may use a den to raise cubs or to escape from inclement weather, there are no reports in the literature to indicate that canids routinely spend entire days in small enclosures.

Obedience School

Obedience training does not directly influence separation anxiety. Many dogs with this problem have been to obedience school and have done well. Separation anxiety is not the result of disobedience or lack of training; it is a distress response.

Preventive Measures

One of the greatest services a clinician can offer clients is instruction in ways to prevent separation anxiety from developing in their dogs. During the first examination of a puppy, it would be beneficial to explain that the dog should not always accompany the owner, that it should gradually become accustomed to being alone, and thereafter should periodically be left alone for several hours at a time. Although dogs with separation anxiety are usually treatable, it is much easier to prevent the problem than to cure it.

When to Consider Referral

The behavior modification techniques outlined in this article are based on principles of learning. This means many practice sessions and many departures per session are required, with careful attention to the relevant stimuli for and responses of the individual dog. Done properly, progress is usually seen within a few days or weeks. If the practitioner and the owner follow the preceding recommendations without success, they should not give up. It is likely that some small, but critically important, detail has been overlooked. Referral to an animal behaviorist with extensive knowledge of these techniques is likely to lead to a successful resolution of the problem. The prognosis for successful treatment of separation anxiety is excellent.

REFERENCES

1. Harlow HF: The development of affectional patterns in infant monkeys, in Foss BM (ed): *Determinants of Infant Behavior*. London, Methuen, 1961.
2. Hess EH: Imprinting in birds. *Science* 146:1128-1139, 1964.
3. Scott JP, Stewart JM, DeGhett, VJ: Separation in infant dogs: Emotional response and motivational consequences, in Scott JP, Senay E (eds): *Separation and Depression: Clinical and Research Aspects* (AAAS No. 94). Washington, DC, American Association for the Advancement of Science, 1973.
4. Bowlby J: *Attachment*. New York, Basic Books, 1969.
5. Bowlby J: *Separation*. New York, Basic Books, 1973.
6. Voith VL: Destructive behaviour in the owner's absence, Parts I and II. *Canine Pract* 2(3):11; 2(4):8, 1975.
7. Voith VL: Attachment between people and their pets: Behavior problems of pets that arise from relationships between pets and people, in Fogle B (ed): *Interrelations Between People and Pets*. Springfield, IL, Charles C Thomas, 1981.
8. Borchelt PL, Voith VL: Classification of animal behavior problems. *Vet Clin North Am* [*Small Anim Pract*] 12:511-585, 1982.
9. Voith VL, Borchelt PL: Diagnosis and treatment of dominance aggression in dogs, in Voith VL, Borchelt PL (eds): *Vet Clin North Am* [*Small Anim Pract*] 12(4):655-663, 1982.
10. Line S, Voith VL: Profile of dominant aggressive dogs. *Applied Anim Ethol*, manuscript submitted, 1984.
11. Spielberger CD (ed): *Anxiety and Behavior*. New York, Academic Press, 1966.
12. Tuber DS, Hothersall D, Voith VL: Animal clinical psychology: A modest proposal. *Am Psychol* 29:762-766, 1974.
13. Tuber DS, Hothersall D, Peters MF: Treatment of fears and phobias. *Vet Clin North Am* 12:607, 1982.
14. Lapras M: Modern psychotrope therapy in the dog. *Proc 6th World Cong, World Sm Anim Vet Assoc*. Post Acad. onderwijs Publikatie, No.8., 1977, p 129.
15. Eigenmann JE: Diabetes mellitus in elderly female dogs: Recent findings on pathogenesis and clinical implication. *JAAHA* 17:805-812, 1981.
16. Eigenmann JE, Eigenmann RY: Influence of medroxyprogesterone acetate (Provera) on plasma growth hormone levels and on carbohydrate metabolism I and II. *Acta Endocrinol* 98:599-602, 602-608, 1981.
17. Eigenmann JE, Venker-Van Haagen AJ: Progestagen-induced and spontaneous canine acromegaly due to reversible growth hormone overproduction: Clinical picture and pathogenesis. *JAAHA* 17:813-822, 1981.
18. Chastain CB, Graham CL, Michols CE: Adrenocortical suppression in cats given megestrol acetate. *Am J Vet Res* 42(12):2029-2035, 1981.

UPDATE

CLINICAL INCIDENCE

Separation anxiety accounts for 20% to 40% of the canine cases in widely distributed behavior practices across the United States and Europe. One of us (V.V.) and coworkers[1] reported that 21% of 489 canine cases seen at the Animal Behavior Clinic of the Veterinary Hospital of the University of Pennsylvania (ABC-VHUP) in Philadelphia from March 1984 to June 1987 were separation related; separation anxiety was the most common diagnosis (13/26) for older dogs presented for behavior problems at ABC-VHUP during the same period.[2] Borchelt indicated that 39% of the canine cases in a New York City–based practice were separation anxiety.[3] Wright and Nesselrote[4] indicated that 26.7% (28/105) of canine cases presented to a Georgia behavior practice were separation-related problems.

Landsberg[5] reported that destructive behaviors were the third most commonly seen behavior problem in three behavioral practices in North America: Toronto (N = 176), Cornell University (N = 397), and Kansas City (N = 170). Separation anxiety was diagnosed as the cause of the destructive behaviors in 46%, 70%, and 26% of the cases, respectively. O'Farrell[6] in Edinburgh, Scotland, wrote that the most common type of excitable behavior presented to behavioral therapists was destructive behavior and that the dog's agitation at being left alone was by far the commonest cause of destruction in the owner's absence. Van der Borg and coworkers[7] stated that aggression and separation

Figure 1–Adult horses develop strong attachments to conspecifics and often manifest distress responses of vocalization, hyperactivity, and defecation when separated from a companion.

anxiety were the most frequently encountered reasons dogs were returned to animal shelters in the Netherlands.

LABORATORY STUDIES

Separation distress is an interesting and complex phenomenon exhibited by many species, most often during infancy.[8] In some species, such as horses (Figure 1) and monogamous titi monkeys,[9] adults also manifest acute distress responses when separated from their companions. Separation distress responses in dogs have been viewed as a model for attachment, anxiety, depression, and drug addiction in humans.[8,10–12]

The separation distress response in puppies is a species-typical response that requires no learning. Separation-distress vocalizations are manifested the first time a puppy of the appropriate age is separated from its littermates and mother. The sounds are characteristic for all puppies and vary little except in volume. Separation-distress vocalizations arise when the puppies are about 3 weeks of age; peak when the puppy is 6 to 7 weeks old; and then occur in a novel environment but with declining frequency until the puppy is 12 to 14 weeks old.[8,13,14] The decline in vocalizations is attributed to maturation and the appearance of fear responses to novel stimuli.[8,10] Separation-distress vocalizations are independent of other causes of distress vocalization (e.g., hunger or cold, which can motivate a puppy to emit care-soliciting calls at birth).[15]

Because of the increasing likelihood of dogs reacting fearfully to novel stimuli as they mature, most laboratory separation anxiety studies are done with young puppies. Little information exists about the development of canine separation-distress reactions as the animals mature. Although the characteristic distress vocalization in novel locations is greatly diminished by 16 weeks, the literature indicates that older dogs still vocalize when isolated, albeit much less than puppies, and they exhibit a greater variation in their response.[8,15] Older animals reportedly spend more time trying to escape.[8,15] The Coppingers have observed that mother dogs retrieve puppies or tape recorders emitting puppy distress vocalizations for only a few weeks after parturition (see their article elsewhere in this book).

Laboratory studies indicate that young puppies vocalize more when isolated in a strange place than when they are alone at home.[8] The presence of a littermate reduces distress responses at home and in novel environments.[16] These studies also report that increases in distress vocalizations are usually accompanied by an increase in the intensity of the vocalizations and an increase in general activity. Separation distress in puppies is sometimes accompanied by urination, defecation, and destructive behaviors; but the occurrence of these behaviors is more variable. Reports differ regarding the effect of "restriction of activity" or "confinement" on the increase of the distress responses of puppies.[14,16,17] These conditions are usually produced by putting the puppy inside a box with one or more wire-mesh sides; it is therefore difficult to decide how much of any increase in vocalization is due to restricted space and how much to the animal's perception of being in a strange environment.

The increase in magnitude of a puppy's response in a novel environment has been hypothesized to reflect a summation of separation distress reactions to the absence of familiar conspecifics and the absence of familiar environment.[18] As the puppy matures, a different pattern emerges in response to being placed in a novel environment—that of being quiet and attempting to escape.[8,10] When placed alone in a novel environment, an older puppy may be reacting to two separate phenomena: the absence of familiarity and the fear of novelty. The separation evokes vocalization (a signal for help), and the fear response to something novel motivates the dog to be quiet. For this reason, as the animal gets older, it becomes quieter when alone in a novel environment.

Tuber and coworkers[19] clearly demonstrated that a novel environment has a significant effect on adult dogs and that the responses of mature dogs when isolated in a novel area are different from those manifested when alone in familiar surroundings. These investigators showed that vocalizations of mature adult laboratory dogs *increased* when they were alone in a *familiar* environment and were *suppressed* when they were alone in a *novel* environment. The latter is the opposite of what is observed in very young puppies and compatible with the trend of decreasing vocalizations observed in isolation studies in novel environments in puppies after 8 weeks of age.

In the Tuber and coworkers study, the frequency of barking of these dogs was measured at intervals over a 4-hour time span in the following conditions: together in the home kennel, alone in the home kennel, together in a novel room, alone in a novel room, and in a novel room with a familiar person. During the first hour, some barking sporadically occurred in the control condition (dogs together in the home kennel), there was a marked increase in vocalizations by dogs left alone in the home kennel, and virtually no vocalizations were made in the novel environment whether the dogs were alone, with a conspecific, or with a person (Figure 2).

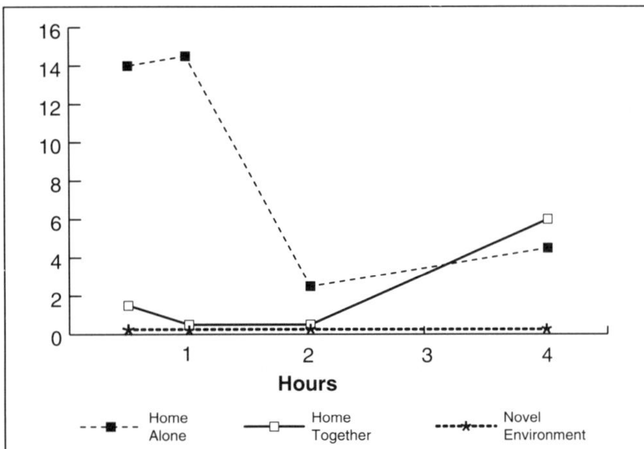

Figure 2—The median number of vocalizations of adult laboratory dogs at specific times over 4 hours under the following test conditions: in the home kennel together, in the home kennel alone, and in a novel environment with or without kennelmate. (Data from Tuber DS, Hennessy MB, Sanders S, Miller JA: Behavioral and glucocorticoid responses of adult domestic dogs (*Canis familiaris*) to companionship and social situation. *J Comp Psychol*, in press.)

David Tuber and colleagues also looked at the glucocorticoid (cortisol and corticosterone) responses of these eight dogs, which had been maintained in pairs for many years. The mean glucocorticoid levels in samples taken at the end of the 4-hour tests showed marked and statistically significant differences among the conditions. In ascending order, the mean glucocorticoid levels were: together in the home kennel, with a person in a novel environment, alone in the home kennel, with a kennelmate in a novel environment, and alone in a novel environment (Figure 3). The analysis of variance (ANOVA) of the mean glucocorticoid levels indicated significant differences between the dogs that were with a kennelmate at home compared with those with a kennelmate in a novel environment as well as between those with a kennelmate in the home environment and those left alone in the novel environment. Clearly, being in a novel environment has an effect in addition to that of being separated from familiar conspecifics. Interestingly, the glucocorticoid levels were significantly lower in the novel environment when a dog was with a person than when it was with its canine kennelmate.

EFFECT OF A PERSON

Several studies have demonstrated the stress-alleviating effect of a person on a dog or puppy in novel or aversive situations. Lynch and McCarthy[20] demonstrated that petting by a person reduced the elevated heart rate of a dog's reaction to both conditioned fear responses and unconditioned pain responses to mild electric shocks to the forepaw. Pettijohn and coworkers[21] showed that distress vocalizations of a puppy in a novel room were reduced more by contact with a person than by the presence of the puppy's mother or an unfamiliar adult dog. Tuber and coworkers[19] showed that the glucocorticoid response of a dog in a novel room was lowered more by the presence of a person than by the presence of its conspecific kennelmate.

We and other applied animal behaviorists are aware that companion dogs often engage in separation-distress responses when left by their owners, even though another household dog or an unfamiliar person is present. Sometimes, separation anxiety is manifested even in the presence of other family members when a dog is left by the person to whom it is most attached.

CLINICAL PROFILES

Retrospective Studies

Two overlapping series of separation-anxiety cases (N = 42 and N = 104) seen at the ABC-VHUP (1983–1987) indicated that 26% and 20%, respectively, of the dogs had been acquired from an animal shelter or similar source. Only 8% of the dogs owned by people filling out questionnaires while waiting for medical or surgical veterinary care at VHUP in 1981 had come from shelters.

In the above two series of dogs with separation anxiety, 64% and 65% were males. The percentages of males with diagnosed separation anxiety were significantly greater ($P < .01$) than the percentage of males seen in the VHUP medical/surgical population (N = 11,466; 51.5%). Male rhesus monkeys and humans may be more adversely affected by separation anxiety than are females of these species.[22]

In another series of canine behavior cases (N = 489: 158 intact males, 158 castrated males, 55 intact females, 118 spayed females) seen between March 1984 and June 1987 at ABC-VHUP were analyzed for relationships between the source of the dog, diagnosed behavior problem, sex of the dog, and breed.[1,23] The six most common sources of the dogs were breeders (37%), friends (15%), animal shelters (14%), pet shops (10%), advertisements (9%), and found (8%). The most prevalent diagnostic categories were dominance aggression (37%), territorial aggression (22%), separation anxiety (21%), fear-induced aggression (16%), interdog aggression (13%), and housebreaking problems (9%). Some dogs had more than one diagnosis.

Chi-squared analysis of diagnosed behavior problem by source of dog yielded several significant differences ($P<.05$). Dogs originating from shelters were significantly more likely to receive the diagnosis of separation anxiety than were dogs from other sources; when purebreds and mixed breeds were considered separately, this effect was shown to be due entirely to purebreds. Analyzed separately, purebred males and females were both more likely to receive the diagnosis of separation anxiety than were mixed-breeds, but the effect was significant for males alone. Dogs obtained from breeders were less likely to be presented for separation anxiety. Of the 39 dogs that had been found as strays by their owners, 13 were given the diagnosis of separation anxiety; the likelihood of dogs found as strays being given the diagnosis of separation anxiety compared with dogs from all other sources combined approached significance ($P = .054$).

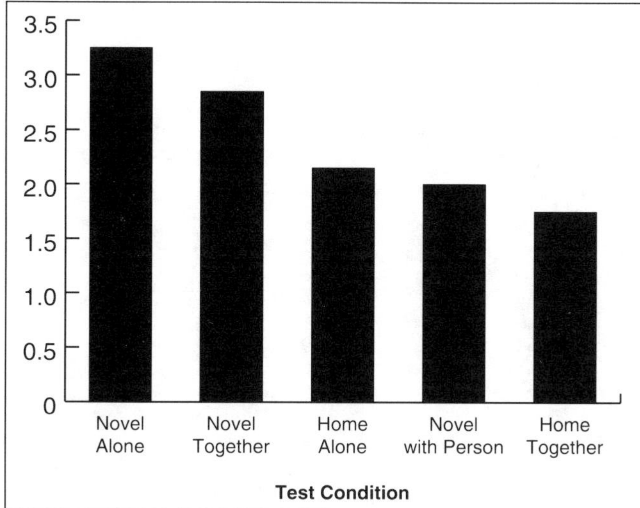

Figure 3–Histogram depicting differences in glucocorticoid levels (μg/100 ml) of dogs under five conditions. (Data from Tuber DS, Hennessy MB, Sanders S, Miller JA: Behavioral and glucocorticoid responses of adult domestic dogs (*Canis familiaris*) to companionship and social separation. *J Comp Psychol*, in press.)

Prospective Study

In a prospective study of dogs presented to ABC-VHUP, the dogs were specifically examined and the owners interviewed about separation-anxiety behaviors regardless of the presenting complaint.[24] Before being interviewed, the owners completed a questionnaire about their dogs' behaviors. This information was entered into a raw data set along with the diagnosis and additional information obtained during the behavioral interview. In a series of 103 dogs presented consecutively, 36 were given the diagnosis of separation anxiety, 64 were determined not to be exhibiting separation anxiety, and in 3 cases it was ambiguous whether the animals were manifesting the syndrome. Chi-squared analyses were used to compare the profiles of the 100 dogs with a definitive diagnosis regarding the presence or absence of separation anxiety.

Dogs were given the clinical diagnosis of separation anxiety if they exhibited an active distress response (e.g., vocalization, destruction, or elimination) within an hour after being separated from all people in the household and if an active distress response occurred most of the times the dog was left. If the onset of the distress response was unknown, separation anxiety was diagnosed (N = 4) if the dog showed signs of anxiety prior to the owner's departure, if destructive behaviors were directed to exit points, and if the dog always or almost always exhibited the problem behavior when left alone but rarely or never when with people.

In descending order, the most common *presenting complaints* of owners of dogs with separation anxiety were destruction, vocalization, elimination, aggression, and overactivity; several other behaviors were only mentioned once. Detailed interviews of the owners of dogs with separation anxiety indicated that 90% vocalized when alone, 80% were destructive, and 55% eliminated inappropriately. Most of these dogs rarely or never engaged in these behaviors when someone was with the dog. Several of the dogs with separation anxiety also exhibited other behavior problems with similar signs (e.g., destructive play behaviors, other problem elimination behaviors, or brontophobia). Not all behaviors exhibited when a dog is left alone are necessarily related to separation distress.[25,26]

Of the 36 dogs with a definitive diagnosis of separation anxiety, 26 had been presented with a primary complaint that was related to separation anxiety, 6 dogs showed separation anxiety as an ancillary problem contributing to the primary complaint, and 4 dogs exhibited significant signs of separation anxiety that were unrelated to any of the presenting complaints.

Chi-squared analyses revealed no significant differences between dogs with and without separation anxiety as to sex and age of dog at presentation; the age of the dog at acquisition; whether the dog had been acquired from an animal shelter; whether the owners had moved with the dog one or more times; whether the dog lived in a household with one adult, two adults, or other number of people of various ages; or how long the dog had been owned by the people presenting it. In this population of dogs with behavior problems, mixed-breeds were more likely to be presented for separation anxiety than were purebreds.

Dogs with separation anxiety were significantly more likely to greet their owners by barking at them, jumping on them, or "running around a lot" (Figure 4). Dogs with diagnosed separation anxiety were more likely to follow the owner at home, but the difference was not statistically significant (Figure 5). There was no statistical difference between the dogs regarding how well they behaved when left alone in their owners' cars; most of the dogs in both groups behaved well in the car.

Monitoring of Behavior

David Tuber's audiotapes of hundreds of dogs at home while the owners were gone indicated that dogs with separation anxiety were often active for several hours after the owners left. The dogs could be heard intermittently pacing, vocalizing, or scratching at the door for many hours. Our subsequent studies corroborated that dogs can manifest active distress responses for over an hour after being left at home by their owners. Sometimes Tuber had a small bell attached to the dog's collar to help monitor its behavior and to distinguish its behaviors from those of other dogs in the household.[27] Tuber used the audiotapes to help differentiate behaviors related to separation anxiety from other causes, such as object play, teething, and barking at exciting stimuli. Audiocassettes can also be used to monitor the effectiveness of treatment or degree of distress the dog may show while the owner is gone.

BEHAVIORAL TREATMENT

Most behavioral treatment techniques involve reducing a dog's arousal or excitement at the time the owner leaves or

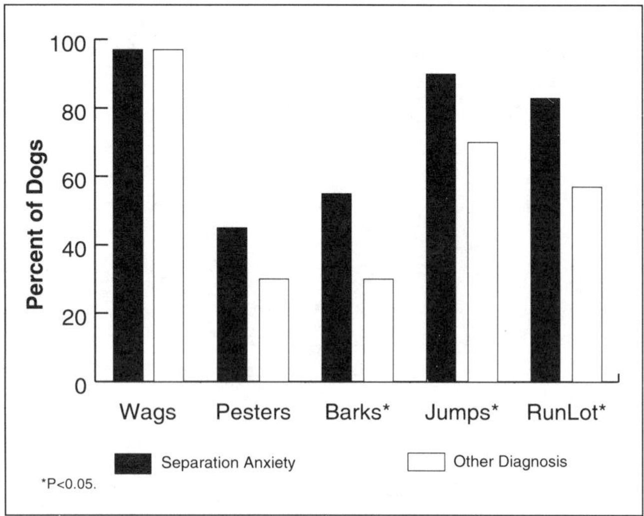

Figure 4—Comparison of frequency of greeting behaviors of dogs with and without clinical separation anxiety (n = 100).

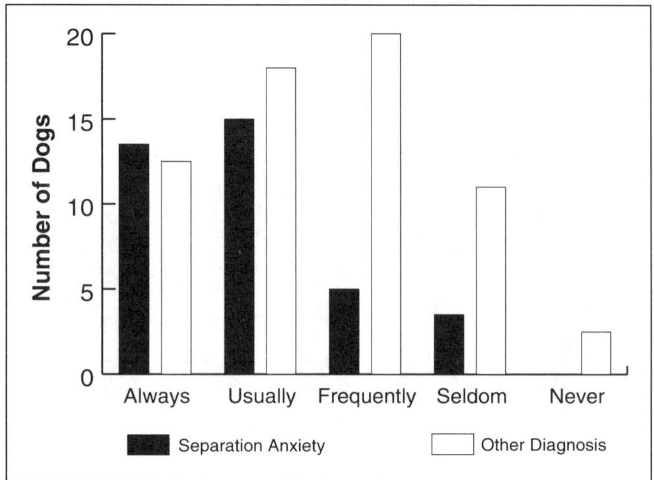

Figure 5—Graph of following behaviors of dogs with and without separation anxiety (n = 100).

inducing a competing emotional state (e.g., eating or relaxation).[25] Tuber often used a massage technique he called the "soft exercise."[27] See the update for the article on fears and phobias in companion animals by Voith and Borchelt for a description of this technique.

The high incidence of dogs returned to animal shelters prompted Tuber and his colleagues to devise a means of easing the dog's transition from the shelter environment to its new home. They found that if a dog was *slowly and gradually acclimated, over several days while at the shelter,* to staying in a large, comfortable airline crate and then sent home with that crate and instructions on its proper use, the likelihood of the person keeping the dog was greatly increased. The safe, familiar environment of the airline crate helped the animal make a successful transition to its adoptive home—and may have also helped in housetraining the dog.

Perhaps the reason that shelter dogs can become accustomed to an airline crate in only a few days is because the crate is a positive contrast to the rest of the shelter environment. Tuber and other behaviorists have found that at least 2 weeks are usually required to accustom a dog to a crate when the dog is already exhibiting separation anxiety in the home. Sudden confinement to a crate from a comfortable home environment may result in a negative contrast to the comfortable home environment and increase the distress and escape responses of the dog. This effect is further compounded if the crate is put in an unfamiliar area. While the dog is being acclimated to the crate, be it in a shelter or at home, it is critical that the dog is never left in the crate for a length of time that will induce distress.

The concern about successful adoptions has encouraged the development and assessment of objective tests to determine the likelihood of the development of specific behavior problems, such as separation anxiety. Van der Borg and colleagues[7] reported that distress vocalizations of dogs during a 5 minute isolation in a *strange* car was fairly predictive of behaviors compatible with separation anxiety after the dog was adopted.

Figure 6—Dogs with separation distress at home are often content when left alone in the owner's car.

PHARMACOLOGIC TREATMENTS

Scott and colleagues[18] tested a range of doses of several drugs on the separation-anxiety response of puppies. They found that imipramine suppressed distress vocalizations in puppies without any observable side effects. Chlorpromazine, reserpine, meprobamate, diazepam, alcohol, sodium pentobarbital, and amphetamine did not reduce distress vocalizations in young puppies except at toxic or sedating doses. Several years later, Panksepp and coworkers examined the effect of drugs with different mechanisms of action on separation-distress responses of animals.[28] They found that the opioid system is involved in attachment and separation-distress behaviors.[11,12,28,29] Low levels of morphine suppressed separation distress responses of young puppies, guinea pigs, and chicks without observable side effects. Moreover, this suppression could be blocked by the opioid antagonist naloxone. Naloxone also increased separation-distress vocalizations in young guinea pigs and chicks, but its effect on puppies was inconsistent. Naloxone did increase solicitous behaviors in dogs (e.g., tail wagging and licking of people's faces) and increased the time spent near a person.[29]

The responses of a distressed household pet when the owners leave is not identical to that of isolated young puppies. The household pet exhibits a conditioned anxiety prior to the owner's departure and a wide range of vocalizations and various other behaviors when left alone. Adult dogs in the home pant, whine, grunt, bark, howl, yip, and scream shortly after the owners leave. They may vocalize, salivate, pant, or attempt to escape and sometimes chew personal objects of the owner, eliminate, groom excessively, self-mutilate, shake, shed hair profusely, sweat through their feet, pace, watch the door or through a window, and do not eat.

The separation anxiety behavior of the household dog may be due to one or more of the following: a species-typical distress response, an unconditioned generalized anxiety, a panic attack, or a conditioned fear/panic/distress response to the aversive experience of being left alone. The observation that separation-anxiety responses in a dog can be environmentally contingent (e.g., anxious when left at home but not when left in the car) suggest that much of the response may be conditioned. A dog may exhibit one or a combination of these types of responses. Depending on the physiologic mechanisms of these types of response, drugs representing a variety of categories might affect the response.

To date, the most commonly used medication for separation anxiety in companion dogs appears to be the tricyclic antidepressant amitriptyline. Some dogs improve somewhat when treated with progestins, tricyclic or atypical antidepressants, or acepromazine maleate. In addition to showing that morphine alleviates separation-distress vocalizations, Panksepp and coworkers also indicated that the serotoninergic system is involved in distress vocalizations.[28] Drugs that blocked serotonin facilitated distress vocalizations, and serotonin receptor agonists decreased distress vocalizations in chicks.

Although Panksepp and coworkers showed species differences in the effects of psychotropic drugs on separation-distress vocalizations, it would be interesting to assess the effects of the serotoninergic drugs currently available on the separation distress responses of household dogs. We hope that within the next few years, well-designed clinical studies that evaluate the efficacy of drugs on the prevalent problem of separation anxiety in dogs will be implemented and published. See the Simpson and Simpson article elsewhere in this book for more information about veterinary psychopharmacology.

REFERENCES

1. Voith VL, Goodloe L, Chapman B, Marder AR: Comparison of dogs presented for behavior problems by source of dog. Paper presented at AVMA meeting, Seattle, Washington, July 18, 1993.
2. Chapman BL, Voith VL: Behavioral problems in old dogs: 26 cases (1984–1987). *JAVMA* 196:944–946, 1990.
3. Borchelt PL: Separation-elicited behavior problems in dogs, in Katcher AH, Beck AM (eds): *New Perspectives on Our Lives with Companion Animals*. Philadelphia, University of Pennsylvania Press, 1983.
4. Wright JC, Nesselrote MS: Classification of behavior problems in dogs: Distribution by age, breed, sex, and reproductive status. *Appl Anim Behav Sci* 19:169–178, 1987.
5. Landsberg G: The distribution of canine behavior cases at three behavior referral practices. *Vet Med* 86:1011–1018, 1991.
6. O'Farrell V: Destructiveness in the owner's absence, in *Manual of Canine Behaviour*. West Sussex, England, British Small Animal Veterinary Association, 1986, pp 104–108.
7. Van der Borg JAM, Netto WJ, Planta DJU: Behavioural testing of dogs in animal shelters to predict problem behaviour. *Appl Anim Behav Sci* 32:237–251, 1991.
8. Scott JP, Stewart JM, DeGhett VJ: Separation in infant dogs, in Scott JP, Senay EC (eds): *Separation and Depression: Clinical and Research Aspects*, publ. no. 94. Washington DC, American Association for the Advancement of Science, 1973, pp 3–33.
9. Cited in Tuber DS, Hennessy MB, Sanders S, Miller JA: Behavioral and glucocorticoid responses of adult domestic dogs (*Canis familiaris*) to companionship and social separation. *J Comp Psychol*, in press.
10. Scott JP: The domestic dog: A case of multiple identities, in Roy MA (ed): *Species Identity and Attachment*. New York, Garland SRPM Press, 1980, pp 129–144.
11. Panksepp J, Herman B, Conner R, et al: The biology of social attachments: Opiates alleviate separation distress. *Biol Psychiatr* 13:607–618, 1978.
12. Panksepp J, Herman BH, Vilberg T, et al: Endogenous opioids and social behavior. *Neurosci Biobehav Rev* 4:473–487, 1980.
13. Scott JP, Fredericson E, Fuller JL: Experimental exploration of the critical period hypothesis. *Personality* 1:162–183, 1951.
14. Elliot O, Scott JP: The development of emotional distress reactions to separation in puppies. *J Genet Psychol* 99:3–22, 1961.
15. Scott JP, Bronson FH: Experimental exploration of the et-epimeletic or care-soliciting behavioral system, in Leiderman PH, Shapiro D (eds): *Psychobiological Approaches to Social Behavior*. Stanford, CA, Stanford University Press, 1964, pp 174–193.
16. Fredericson E: Perceptual homeostasis and distress vocalizations in puppies. *J Personality* 20:472–477, 1952.
17. Ross S, Scott JP, Cherner M, Denenberg VH: Effects of restraint and isolation of yelping puppies. *Anim Behav* 8:1–5, 1960.
18. Scott JP: Effects of psychotropic drugs on separation distress in dogs. *Proc IX Congr Neuropsychopharmacol*, Excerpta Medica International Congress Series 359:735–745, 1974.
19. Tuber DS, Hennessy MB, Sanders S, Miller JA: Behavioral and glucorticoid responses of adult domestic dogs (*Canis familiaris*) to companionship and social separation. *J Comp Psychol*, in press.
20. Lynch JJ, McCarthy JF: The effect of petting on a classically conditioned emotional response. *Behav Res Ther* 5:55–62, 1967.
21. Pettijohn TF, Wong TW, Ebert PD, Scott JP: Alleviation of separation distress in 3 breeds of young dogs. *Dev Psychobiol* 10(40): 373–381, 1977.
22. Mineka S, Suomi SJ: Social separation in monkeys. *Psychol Bull* 85(6):1376–1400, 1978.
23. Voith VL, Ganster D: Separation anxiety: Review of 42 cases (abstract). *Appl Anim Behav Sci* 37:84–85, 1993.
24. Voith VL: Profiles of dogs with separation anxiety and treatment approaches. Paper presented at AVMA meeting, San Francisco, July 10, 1994.
25. Tuber DS, Hothersall, Peters MF: Treatment of fears and phobias in dogs. *Vet Clin North Am Small Anim Pract* 12(4):607–623, 1982.
26. McCrave EA: Diagnostic criteria for separation anxiety in the dog. *Vet Clin North Am Small Anim Pract* 21(2):247–255, 1991.
27. Tuber DS: The soft exercise. *Anim Behav Consultants Newsletter* 3(1):2, 1986.
28. Panksepp J, Meeker R, Bean NJ: The neurochemical control of crying. *Pharmacol Behav* 12:437–443, 1979.
29. Panksepp J, Comer R, Forster PK, et al: Opioid effects on social behavior of kennel dogs. *Appl Anim Ethol* 10:63–74, 1983.

Fears and Phobias in Companion Animals

Victoria L. Voith, DVM, PhD
Department of Clinical Studies
School of Veterinary Medicine
University of Pennsylvania
Philadelphia, Pennsylvania

Peter L. Borchelt, PhD
Animal Behavior Consultants Inc.
Forest Hills, New York
The Animal Medical Center
New York, New York

Fears in companion animals are common. Veterinary clinicians frequently encounter dogs that are afraid of loud noises (such as thunder or exploding firecrackers), sudden movements, unfamiliar people, or novel environments. Extensive literature exists in the fields of animal behavior, experimental psychology, and human clinical psychology/psychiatry on the topic of fears and phobias. Such literature contains pertinent information for the treatment of behavior problems in animals. This article provides an overview of fear in general with a focus on companion animals and the treatment options available for fears and phobias in dogs.

What Is Fear?

Fear is best conceptualized as a complex system involving the interaction of behavioral, emotional, and physiological components. Each of these components is influenced by the environment and by a myriad of variables, such as species, age, and experience. Fear can neither be defined simply with reference to any one of these components nor merely as a specific behavior. Research on the physiology of fear[1] clearly shows that various measures, such as heart rate, blood pressure, and endocrine changes, do not always correlate with each other or with specific fear behaviors. Different fear-evoking situations can produce different types of fear responses.[2] For example, an animal seeing or hearing a predator at a distance may *freeze*. If the predator approaches, the animal *flees*. If escape attempts are unsuccessful and the animal is cornered, it may *fight*. When grasped and held by a predator, many species show a prolonged *immobility* response. Finally, individual animals can differ in response to the same potentially fear-provoking situation. For instance, one dog in a home might become phobic about thunderstorms or firecrackers and the other dogs in the home ignore them.

Despite the complexity of fear and difficulty of a precise definition, there is a "common sense" aspect to *fear* as the term is used in everyday speech. The authors believe that owners of companion animals are usually accurate when labeling a pet's behavior as fearful. An owner is familiar with his or her animal, its history, and the particular circumstances related to the fear and usually accurately infers the animal's emotional state. Owners are aware of *both* the specific circumstances or situations that evoke the fear and the sequence of behaviors that the animal exhibits.

Functional Analysis of Fear Behavior

Viewed in terms of function, fear behavior is conceptualized as an adaptive sequence of responses that removes or protects an animal from immediate or impending dangers or noxious stimuli. The earliest responses to evolve were simple *withdrawal* responses. Simple animals, such as amoebas, withdraw when contact receptors are stimulated by touch, certain chemicals, vibration, or extremes of temperature. As animals evolved distance receptors, they could show more complicated behaviors, such as *startle* and *escape*. Eventually, as the ability to learn evolved, animals became capable of *avoiding* stimuli previously associated with danger or discomfort.

The stimuli that naturally evoke fear responses[3] include:

- *Predators,* which have fear-eliciting stimuli including proximity, specific and sudden movements, visual features, and olfactory and/or auditory cues
- Physical and *environmental* dangers, such as heights, noise, or extremes of temperature (e.g., fires)
- Situations or events *associated* with risk of predation or danger, such as novel or unfamiliar places or objects
- Behaviors of *companions* or *conspecifics,* for instance, threats, intrusions of personal space, or the arrival of a stranger.

Fears, Experience, and Learning

In the real world, there often is no chance for an animal to learn to withdraw from, escape, or avoid a dangerous situation. That is, the first experience with a dangerous stimulus may be the last unless the animal does the right behavior immediately. Consequently, unlearned species-typical or unconditioned behaviors have evolved that prepare animals to respond appropriately and effectively to the typical dangers they encounter in their usual environment.[4,5]

Experience and learning also play an important role in coping with potential dangers. If an animal encounters the same potentially dangerous situation repeatedly and if that situation is perceived as very frightening, then the animal will continue to be afraid of it and probably begin to show fear responses to stimuli associated with the situation. For instance, if an attack by a predator (from which it escaped successfully) was preceded by a particular sound, then the animal will rapidly learn to escape more quickly from that sound and will eventually learn to avoid other stimuli associated with the sound. This process of learning involves classical conditioning and instrumental (operant) conditioning, both of which have been well researched in laboratory and field settings. Classical conditioning processes often allow an animal to anticipate, predict, or prepare for an eventually dangerous stimulus; operant processes allow the animal to learn the most effective escape and/or avoidance responses.

If an animal encounters a possibly dangerous situation over and over but nothing happens, then eventually that situation is not perceived as dangerous and the animal will probably ultimately lose its fear. The fear response is said to habituate to that particular stimulus. For instance, an animal startled by a novel sound in the distance may freeze; but, if the sound is repeated frequently and gets no closer, the animal will soon learn to ignore it. Habituation is adaptive because it prevents the animal from responding repeatedly to low-level stimuli that, in fact, are not dangerous. If, however, the sound quality were to change (for instance, different frequency or intensity), the animal would again be alert or startle to it. This is called *dishabituation* and is adaptive in that a change in the stimulus might be a signal of a real danger.

Learning permits an animal flexibility in perceiving and appropriately responding to the inevitable and variable dangers in its environment. The basic principles derived from studies of learning (particularly laboratory studies) are useful for understanding how fears and phobias develop and, more important, for treating them.

Treatment of Fears and Phobias

Although there is value in knowing what events caused a fear to develop, such information is unnecessary to the treatment of fears.[6,7] What is critical, however, is being able to *identify* and *control* all (or most) of the stimuli that presently evoke the fear.

Essentially, the treatment of fears and phobias involves experiencing the fear-eliciting stimulus without being afraid. This can be done in a gradual manner (by desensitization and counterconditioning) or by forcing the individual to experience an intense stimulus until there is no longer any fear (flooding). There are suggestions that anxiolytic drugs might also be helpful, but some laboratory research is contradictory and the majority of clinical information is based on case reports with few well-controlled studies.

Historical Perspectives

In the course of history, many cultures have independently developed similar approaches to the treatment of fears, either intuitively or by trial and error. Marks[8] describes two historical accounts of treatment techniques. More than 300 years ago, Locke described the technique presently known as *desensitization*. He wrote:

> "(if) your child shrieks and runs away at the sight of a frog, let another catch it, and lay it down at a good distance from him; at first accustom him to look at it; when he can do that to come nearer to it, and see it leap without emotion; then to touch it lightly, when it is held fast in another's hand; and so on, until he can come to handle it as confidently as a butterfly or sparrow."

Over 500 years ago, an eminent Chinese physician reportedly treated a woman with a noise phobia by holding her down in a chair while iron gongs were struck nearby until she became accustomed to the noise. Today this technique is called *flooding*.

A more recent classical example involves the treatment of children with phobias.[9,10] In these cases, the feared objects were brought progressively closer to the children while they were eating their mid-morning lunch. The entire procedure was done without evoking fear at any time. One child's fear of small animals and related objects was treated by gradual exposure to a rabbit as the child was playing or eating candy. These are examples of counterconditioning.

The effectiveness of counterconditioning has also been demonstrated in animals. Masserman and Wolpe[6] observed that cats made "neurotic" by being shocked in a small cage would lose their fear when they were induced to eat in those cages. Wolpe also observed that the cats' phobic responses to these cages and the testing rooms generalized to similar environments. The intensity of the cats' fearful responses were in proportion to the degree of similarity of these environments to the settings where the cats had been shocked. Wolpe used the phenomena of stimulus generalization and response generalization to his advantage in treating the cats' fears. Although the cats were too frightened to eat in the original cages and in similar-appearing cages, they would eat in places where they were only slightly anxious. Wolpe first fed the cats in the locations that were similar to the original. When the cats were no longer anxious in these locations, they were moved to settings slightly more like those where they had originally been frightened. Each cat was again fed. Gradually the cats were advanced along a continuum until they were eating and were no longer afraid in the original cages.

Behavior Modification

Laboratory experiments indicate that general principles of learning appear to apply, at least in many cases, to the development, spread, and amelioration of fears. Wolpe[6] defines behavioral therapy or conditional therapy as the use of experimentally established principles of learning for the purpose of changing nonadaptive behaviors. In the past 30 years, the field of behavior modification for humans has developed rapidly. Therapeutic techniques have been subjected to well-designed clinical trials, and several journals that are primarily devoted to behavioral therapy are now published. Many of the techniques that have been developed by behavior therapists for treating humans can be applied to dogs and cats.[11,12]

Terminology

The box on the next page highlights some general concepts of learning and behavior modification and the associated terminology. Use of these terms enables more efficient descriptions of the therapeutic techniques that are applicable in treating canine behavioral problems.

Drug Therapy Literature

The literature about the influence of drugs on the amelioration of fear responses is (at least on the surface) ambiguous. Many studies appear contradictory. This may result from differences in the intensity and type of fear involved, drug dosages, or environmental variables that are not always noted. Perhaps at the present time the value of this literature is in knowing that at least sometimes a particular type of drug can have a certain effect. Information concerning the use of drugs in the treatment of fears and phobias in humans has largely been based on individual cases. Only in recent years have controlled, double-blind, clinical studies been conducted.

Some general conclusions drawn from laboratory and clinical studies[7,8,13-15,17,23,25] are:

1. Although some drugs, such as the β-blockers, reduce autonomic signs associated with fear, the emotional and behavioral avoidance responses can remain.
2. If drugs do reduce fear when the individual is exposed to the fearful stimulus, the effect may not transfer to the nondrugged state.
3. The positive effects of drug therapy are more likely to transfer to the nondrugged state if the drug dosage is gradually reduced.
4. Lasting positive effects of drug therapy are more likely to occur if the individual is *continuously* protected from any phobic response during the course of therapy.
5. Different classes of drugs (barbiturates, phenothiazines, benzodiazepines, β-blockers, etc.) may affect different aspects of the fear response or different types of fears (e.g., social phobias, noise phobias).
6. There are likely species, breed, and individual differences in response to these drugs.

Fears in Companion Animals

Most of the fear behaviors displayed by companion animals involve such responses as freezing, trembling, panting, increased heart rate, digging, hiding, or fleeing if possible. Aggressive or fighting responses, such as growling, hissing, or biting, may occur if the fear-evoking stimulus is close and moving and there is no avenue

BEHAVIOR MODIFICATION TERMINOLOGY	
Phobia	A maladaptive fear response that is out of proportion to the real threat of the stimulus.
Stimulus and Response Generalization	After an animal acquires a response (whether an operantly or classically conditioned response) to a specific stimulus, the animal tends to engage in similar responses to similar stimuli. The quality and intensity of these responses to other stimuli are proportional to the degree of similarity the stimuli have to the original stimulus.
Desensitization	Systematic desensitization is a technique used to reduce anxiety and fear responses in a stepwise fashion by exposing the individual to weak fear-eliciting stimuli or to non-fearful stimuli. Gradually, the animal is exposed to increasing intensity of the fearful stimuli without them evoking a fearful response.
Counterconditioning	Counterconditioning is a technique by which an animal is conditioned to acquire responses that are physiologically and behaviorally incompatible with an undesirable response. Counterconditioning is usually performed simultaneously with desensitization.
Flooding	Flooding is a technique whereby an individual is continuously exposed to an anxiety-evoking or fear-evoking stimulus until the individual stops exhibiting the fearful behavior. The stimulus is not withdrawn until sometime after the animal has significantly or completely relaxed. Hence, at the end of the session, the animal is experiencing the stimulus in a nonanxious state.
Habituation	Habituation is the decrease or loss of a response to a stimulus simply as the result of repeated exposure to that stimulus without any pleasant or aversive associations.

for escape. Immobility responses to close and prolonged contact to a predator or a predatorlike stimulus do not typically occur as clinical cases.

Fears and phobic responses make up a large proportion of the case loads of animal behavior therapists. Dogs are most often presented because they fear loud noises, humans, traffic, and specific environments, such as dog-show rings. Cats are usually presented because they fear unfamiliar humans or other cats.

Tuber et al[12] reviewed the literature regarding the incidence of fearful behavior in dogs in selected populations. Of 118 dogs seen in an animal behavior practice in Columbus, Ohio, 20% were treated for fear of loud noises and 8% for fear of strangers or strange environments. Of over 600 dogs (German shepherds, golden retrievers, and Labrador retrievers) evaluated by the Guide Dog Program, 19% were frightened by loud noises, 15% by cars and farm machinery, 12% by other animals, and 4% by bicycles. Of 400 German shepherds in a specific breeding program evaluated by Project Fortunate Fields, over 11% were judged to be oversensitive to auditory stimuli. In a series of 41 behavior problems of dogs seen in Berkeley, California, four involved fears: one of noises, three of humans.[18] Melzack[19] documented that many household dogs are afraid of novel stimuli, such as open umbrellas, electric trains, 12-inch balloons, and masked familiar humans.

Fears and phobias can develop at any age. Sometimes their origins can be traced to a specific frightening event. For example, a five-month-old female cocker spaniel began exhibiting fear of strangers, particularly men, after a man in the neighborhood had a psychotic episode during which he ran up and down an alley nearby screaming and banging trash-can lids against the fences.[a] A golden retriever that was very attached to its owners became afraid of thunderstorms after it had been left alone, tied in a strange location during an intense thunderstorm.[a]

Many times, however, owners are unaware of any precipitating event that has led to the development of a fear. Many dogs seem to gradually develop extreme fear of thunderstorms or show rings after one year of age. Anecdotal responses from dog breeders suggest that some dogs become frightened or phobic at about one year of age. This stage may be "outgrown" if no traumatic fear-eliciting circumstances occur. Interestingly, many fears and phobias in humans do not appear to be related to specific traumatic origins.[8]

Apparently no sex predilection exists for fears and phobias in dogs, and phobias can occur in any breed. In fact, wolves, the presumed ancestors of dogs, are noted for being highly neophobic, i.e., afraid of novel stimuli.

[a]Clinical cases seen at the Veterinary Hospital, University of Pennsylvania, Philadelphia, Pennsylvania.

There may be a genetic predisposition for the development of some fearful behaviors. Numerous anecdotal reports are passed among dog breeders about certain breeds or strains within a breed that are more likely to exhibit fear responses to novelty. Murphree and colleagues[20,21] have bred two strains of German shorthaired pointers, one that exhibits extreme phobias to noise, novelty, and humans (even familiar humans) and another strain that does not exhibit such fears. The fearful behaviors develop at about three months of age, occur independently of adverse conditioning experiences, and are quite resistant to change.

Early Experience

The early experience of an animal can have a profound effect on the development of fear. Dogs that have restricted experience with humans or novel environments from an early age show fear when exposed to them later. Clark et al[22] reared Scottish terriers in a 3 × 6-ft cage with little human contact. When placed in an unfamiliar room or handled by humans for the first time at seven and one-half months, the dogs froze, flattening themselves on the floor with front legs out and ears back. Repeated experiences with human handling resulted in some decrease in fear within a week, however, and little indication of fear after six months of human contact.

The relationship between fear and early experience at different ages is nicely illustrated in a study done by Freedman, King, and Elliott.[23] These investigators raised six groups of puppies with their mothers in one-acre fields without any human contact until the puppies reached various ages. Individual groups of puppies were first exposed to periods of daily handling indoors for one week starting at either two, three, five, seven, or nine weeks of age; another group was not handled at all during the first 14 weeks. During handling tests, the two-week-old puppies were too immature to do much but eat, sleep, or crawl about randomly. Puppies handled at three weeks of age were immediately attracted to humans; five-week-old puppies were initially wary but quickly became friendly; seven-week-old puppies were frightened and wary for the first two days of handling; and pups nine weeks old were frightened and wary for the first three days of handling. At 14 weeks of age, all of the groups were removed from the field environment and experienced daily contact with humans indoors for two weeks. The group that had not been handled at all was extremely frightened of humans, and their attraction to humans improved little during the two weeks of handling. One of these puppies was subsequently handled daily for three months with only a slight decrease in fear. The group that had been handled at two weeks of age showed an initially lower attraction to humans than did the groups handled at three, five, seven, and nine weeks; but after two weeks of handling, all of these groups were much less fearful and about equally attracted to humans.

These studies indicate that early experiences with humans and novel environments influence the development of fear and that there apparently is an age range (sensitive or critical period) at which this experience has a maximum effect. It is also clear (and usually ignored in discussion of these studies) that fear resulting from early restriction of experiences can be reduced by later experiences. For example, three Scottish terriers that had been raised without human contact, except for brief daily feeding and cleaning periods, were initially more likely to avoid humans and freeze in novel environments than were littermates raised with extensive handling. After one week of exposure, however, some of the peculiarities of the deprived dogs were greatly diminished and six months later could only be elicited under certain conditions.[22] Another experiment comparing Scottish terriers raised in a restricted environment versus a free environment found that the experimental dogs were initially more likely to avoid "friendly" and less likely to avoid "bold" men than were puppies raised in a free environment. A year later, however, there were no statistical differences between the two groups regarding behavior.[24]

Another study reported that after two weeks of testing, no differences were evident in the attraction to humans by groups of 16-week-old puppies that received one and one-half hours of daily socialization for a period of a week at two weeks of age, three weeks of age, five weeks of age, seven weeks of age, or nine weeks of age. Those not exposed to humans until 14 weeks of age were less attracted to them but *doubled* in their attraction after the two weeks of testing. Yet another study in which the amount of exposure that puppies between 4 and 16 weeks of age had to humans concluded that delaying experience to the age of 16 weeks did *not* reduce the number of positive responses to humans (although the vigor of the advances was less) and that relatively brief periods of experience permitted essentially normal development of social and manipulative behavior.[25]

Few studies have investigated the relationship between early experiences and fear in cats, but it appears[26] that kittens handled early in life (3 to 14 weeks) are initially less frightened of humans than kittens handled later (7 to 14 weeks).

Thunderstorm Phobias in Dogs

Fear of thunderstorms is very common in dogs and provides a good example of a specific learning principle as well as serves as a clear model of behavior modification techniques for the treatment of phobias. Dogs with thunderstorm phobias generally display a gradient of fear intensity that is proportional to the intensity of the storm. Low-level fear responses, such as pacing, panting, looking wary, staying close to the owner, and so forth, occur well before the actual onset of the storm is obvious to the owner. Thunderstorms consist of loud noises (thunder) that have frequencies both above (ultrasounds) and below (infrasounds) the range of human hearing. Infrasounds carry for hundreds of miles and are probably used by dogs and other animals to predict the likelihood of thunderstorms many hours before their occurrence and perhaps "in the wild" serve as a signal to find shelter.

The stimulus of a thunderstorm can be composed of a number of features, any of which individually or in combination might be the eliciting stimuli for a fear response. Sounds of thunder, rain, and wind; lightning; and changes in atmospheric pressure, ionization, illumination, and odors are all stimuli that can occur during thunderstorms. Although it is unknown which of these potential stimuli plays a role in thunderstorm phobias, the most overpowering stimulus and the easiest to replicate for treatment is the auditory component.

As the intensity of a storm increases, the dog exhibits a greater intensity of fear. Pacing and panting may give way to hiding, sitting on the owner, or frantically trying to escape the storm by attempting to dig out of the house or, if outside, attempting to get in. Dogs have been known to jump through glass windows during storms, dig their way through floors, and even collapse into a paralyzed state. During the height of a storm, a phobic dog is unresponsive to food, play, praise, or petting, much the same as Wolpe's cat was unresponsive to food in the cage in which it had experienced an extremely frightening event (shock).

It is impossible to use the stimulus of a real storm to implement a behavior modification procedure. The onset of a storm is too rapid to allow desensitization and counterconditioning techniques, which require that the stimulus not be presented at such an intensity that it evokes fear. Storms evidently do not last long enough, or are too intense, for a flooding technique to be applicable. Phobic dogs give no evidence of habituating to storms. In order for flooding techniques to work, the subject must experience a substantial reduction or total amelioration of fear while being exposed to the stimulus. This does not happen during the course of an actual thunderstorm.

Although it is impossible to use the real stimulus of a thunderstorm, it is feasible to construct a facsimile of a thunderstorm, the intensity of which can be controlled. In order for the facsimile to be successfully used, it must be perceived by the dog to be similar to a real thunderstorm. The dog's fear of the facsimile is treated, then the treated unanxious attitude to the facsimile should generalize to the real storm. The phenomenon of stimulus and response generalization should now work in the dog's favor.

An "artificial" controllable thunderstorm can be constructed with the use of stereo recordings of thunderstorms, quality stereo equipment, and, if need be, such ancillary stimuli as a darkened room, strobe lights (to replicate lightning), and the sound of rain on the roof or against the windows (use of a sprinkler or hose). Certainly it is impossible to duplicate accurately all of the stimuli of a real thunderstorm. All that is usually needed, however, is to use enough stimuli to evoke a response that indicates the dog perceives the situation to be similar to a real storm. Often the sounds of a thunderstorm are all that is needed.

The actual desensitization and counterconditioning procedures are quite straightforward and have been described in detail.[12,27] The dog is exposed, without experiencing fear, to gradually increasing intensities of the artificial storm; however, this must be done correctly or it will not work. Simultaneously, the dog is counterconditioned to associate pleasant experiences (such as eating delicious tidbits) with the gradual presentation of the storm stimuli. The dog may also be counterconditioned to behave differently, e.g., sit and stay or lie down instead of pacing.

First, a recording—or a combination of a recording and other stimuli—must be found that causes the dog to react in a fearful manner. As soon as it has been determined that the dog is afraid of these stimuli (increased panting, restlessness, salivation, etc.), the stimuli should be terminated. The idea is not to evoke a complete fear response.

Next, when the dog is totally relaxed, the behavior modification procedures can begin. The recording is always played below the threshold level that would evoke an anxious or fearful behavior. Initially, this may be below an audible level. After about five minutes, the sound is increased slightly and remains at this level for several minutes. While the dog is listening in an unanxious state, it should periodically be fed (especially after a thunderclap) "mouth-watering" food tidbits or petted. The volume is gradually increased.

If a mild fearful reaction is evoked (and it will be mild if the intensity of the stimulus has been gradually increased), there are two alternatives: Immediately turn the volume down, or wait and see if the dog habituates to that level of intensity. If the response is mild, the authors usually wait for the dog to habituate; as soon as the dog is relaxed (e.g., respiratory rate returns to baseline), the dog is given some tidbit and/or petted. The intensity of the stimulus is kept at this level for a considerable period of time before advancing further. If the dog does not habituate within a few minutes, the intensity of the stimulus should be decreased until the dog relaxes.

The most common mistake owners make is to play a recording too loudly or to advance it too quickly and thus evoke a fear response that does not habituate during the session. Great care must be taken not to evoke any fear.

The increments of improvement in desensitization and counterconditioning procedures take disproportionately longer in the beginning of a program than later. If meticulous care is taken in the first two or three treatment sessions, subsequent sessions usually proceed much more rapidly. In the treatment for fear of thunderstorms, it often takes only three to five treatment sessions to surpass the initial threshold level at which the dog was tested and reacted fearfully.

After the dog is ignoring loud thunderstorm recordings, the volume should be reduced and other stimuli added, such as darkening the room or flashes of light. Gradually all of the stimuli, together or independently, are increased in intensity. The desensitization and counterconditioning procedures should also be implemented in a variety of locations. Each time an aspect of the program is changed (e.g., a stimulus is added or there is a change in the room), the intensity of the stimulus complex should be reduced temporarily.

Eventually, the facsimile storm should be presented at a low-level intensity at times other than formal practice sessions—perhaps initially during the dog's dinnertime or during play sessions. The intensity must be low enough that the dog is not inhibited from eating or playing. Later, the stimuli should be presented randomly at other times while the owner is home. Finally, the recording can be set on a timer to play at low levels for very short periods of time (initially) while the owner is out of the house.

Factors To Consider in the Treatment of Phobias

The principles are the same for treating any fear or phobia, whether it results from noises produced by firecrackers, doors slamming, or traffic or such visual stimuli as humans, dog-show rings, or other dogs.

How quickly the dog can be successfully cured of its phobia depends on the severity of the phobia and the frequency of treatment sessions. The authors generally recommend several sessions a week, daily if possible and more often when time permits (for instance, weekends). There is evidence that long desensitization and counterconditioning sessions are more effective than multiple short ones, even if the total amount of exposure is the same.[28] The authors recommend 30- to 40-minute treatment sessions. Moderate phobias can be successfully treated in a few weeks and severe ones in one to one and one-half months.

Theoretically, even if an owner only practices once a week, eventually the dog could be successfully treated. The dog will not forget what it has learned. If very long periods of time (months) elapse between sessions, however, the dog is likely to experience an exacerbation of its fear response. Likewise, if a real thunderstorm occurs between practice sessions, much of the progress that has been made will be lost. Therefore, it is advantageous to treat the dog in as short a time period as possible.

Spontaneous recovery of a fear may occur after an animal has been successfully treated for a phobia, if a long period of time elapses between occurrence of the stimuli and the present. Because thunderstorms are seasonal, many months can elapse between episodes. Therefore, to prevent spontaneous recovery, dogs should periodically be exposed to the facsimile that was used during treatment. Such special precautions are not necessary to prevent spontaneous recovery from many other phobias if the animal is frequently exposed during the course of its regular activities to the relevant stimulus after treatment. For example, if the dog is frequently exposed to humans, show rings, or rides in cars after treatment, there is no need to set up specific structured sessions subsequently.

An important feature in any learning process is the contextual situation in which the learning occurs. For instance, specific locations, the ambient temperature, odors, and the presence or absence of humans are all stimuli that could have been heeded at the time the animal learned to become frightened of a particular stimulus. Via classical conditioning, these associated stimuli can play a role in eliciting the phobic response. Thus, in a desensitization or flooding procedure it is ideal to know what contextual stimuli occur during the phobic responses, and these should be incorporated in the treatment procedures. Likewise, if there is a substantial difference in the context of the treatment procedures versus the real situations, there may be little generalization from the treatment sessions to the natural situations.

Flooding is a commonly used technique for the treatment of fears and phobias in humans but has not been used extensively by animal behavior therapists.[28] Because of the dangerous consequences of many animals' phobic responses, flooding is probably more readily applicable in the treatment of moderate fears than in treating intense phobias. Dog-show fanciers and breeders frequently take dogs to crowded shopping centers if they want to ensure that their dogs get used to crowds. If the dog is already frightened by crowds, this procedure will only work if the dog is kept at the shopping center until its fearful behavior substantially subsides or, ideally, disappears completely. If the dog is taken home before it experiences a reduction in fear, the procedure will not work and the dog may get worse. Because its fearful behaviors are still occurring while it is removed from the situation, nothing prevents the dog from associating its fearful behavior with the termination of the stimulus. The dog may perceive that its fearful behavior is what leads to the termination of the stimuli.

A variation on the flooding technique that might work, however, is to present graduated levels of stimulus intensity that evoke only a mild fearful response. The dog can be exposed to that level of intensity until it habituates, even though the time required may be several hours. In treating for fear of thunderstorms, if the owners are home on a weekend, they can play a thunderstorm record at a level that merely leads to increased respiration by the dog. The record is left playing until the dog ignores it. This session may last several hours. The next session would involve playing the record at a slightly higher level of intensity. If this does work, considerable progress can be made in a relatively short period of time. If the starting level of the record is too high and the dog does not habituate, however, the dog's phobia may actually increase in severity. Great care must be taken in determining at what level to begin a flooding technique; and, if flooding is implemented, the session must be carried to completion.

Drug Therapy in Companion Animals

Veterinarians frequently use anxiolytic drugs, primarily such phenothiazines as acetylpromazine acetate, to facilitate restraint or to calm a frightened animal. Often these drugs have the desired effect of enhancing management of the animal, either by actually reducing fear or by rendering the animal disoriented, ataxic, and less able to move. Drugs are warranted if at least some of the animal's phobic responses can be suppressed to reduce self-inflicted injuries, damage to property, and perhaps further reinforcement of the fear.

Phenothiazines and phenobarbital are frequently prescribed for dogs with thunderstorm phobias. Owners report that sometimes these drugs help suppress the dogs' frantic responses, but only if dosed at such a level that the animals are disoriented and ataxic. Usually, the dog still appears frightened.

No evidence is available to support that the use of a phenothiazine during thunderstorms has ever cured a dog's phobia. This may be because: (1) the drug does not reduce the dog's fear during the storm, (2) the drug does suppress the fear during the storm but also inhibits the dog's learning that the storm is not frightening, and (3) the drug is only used sporadically and the dog frequently experiences phobic episodes between effective drug therapy.

Clinically, it would be advantageous to treat an animal before it begins exhibiting any fear. The appropriate dosage and drug may have to be reached by trial and error. One of the authors (Voith) has successfully used diazepam (0.11 to 0.45 mg/kg orally) to suppress some dogs' phobic responses to thunderstorms. Owners report that the drug does appear to reduce the fear to the storm and not simply render the dog disoriented and ataxic. Diazepam also has the advantage of lasting only 3 to 4 hours compared with phenothiazines, which often last 12 to 24 hours. There is, however, a potential side effect of hyperactivity with the benzodiazepines; the first time any drug is administered, the owner should be home to monitor for side effects.

Another successful use of benzodiazepines involved the continuous medication of a five-year-old, spayed, female German shepherd for a noise phobia that generalized to any loud sound, such as a car backfiring, pop guns going off, and even car doors slamming. During the summer months when the windows were open, the dog lived in a constant state of agitation. The animal did not respond to stereo recordings of similar noises, and therefore behavior modification techniques could not effectively be implemented. The owner was considering euthanasia because the animal was so distraught. The dog was treated daily with 22.5 mg of clorazepate dipotassium (Tranxene®—Abbott Laboratories) throughout the entire summer and gradually withdrawn from the drug in the fall. This long-acting benzodiazepine ameliorated the dog's fears of the outside noises. During thunderstorms the dog was medicated with additional clorazepate dipotassium or diazepam. The next spring and entire summer, the dog tolerated most outside noises in an undrugged state, although it was still afraid of thunderstorms.[a]

The authors recommend medicating a phobic animal whenever exposure to a fearful stimulus cannot be avoided. When possible, the animal should be treated before it begins manifesting a fear response. If chronic drug therapy has been successfully implemented, the dose should gradually be reduced (while the animal is still exposed to the stimulus) before discontinuing drug therapy. The clinician may have to try several drugs at different doses before finding an effective treatment. The owner should always be home for several hours following the first administration of a drug.

At this time, the authors do not recommend the use of drugs during desensitization and counterconditioning procedures. There is no evidence that drug therapy significantly improves the long-term efficacy of behavior modification techniques, and there is the possibility that drugs may inhibit learning.

Summary

Some general guidelines apply to the treatment of fears, although factors related to specific stimuli, intensity of fears, species, and breed may have to be taken into consideration.

- The relevant stimuli, contextual cues, and circumstances that are associated with a fear must be identified and arranged along a continuum.
- The relevant stimuli must be controllable so that they can be presented to the animal at the appropriate intensity.
- In desensitization and counterconditioning techniques, the stimuli are presented gradually **without evoking any anxiety or fear**. The entire stimulus complex might be presented at a reduced intensity or for very brief periods of time, or individual components of the stimulus complex might be presented and later gradually combined.
- If flooding techniques are used, the animal must remain in the fear-eliciting situation until the fear ceases (ideally) or is greatly reduced.
- The animal must be protected from experiencing phobic responses during the course of therapy, or treatment will be retarded.
- If during the course of treatment, exposure to fear-eliciting stimuli cannot be avoided, the animal should be medicated with the most efficacious fear-suppressing drug.
- If the animal is on continuous medication, because of constant, unavoidable exposure to fearful stimuli, the drug should be withdrawn gradually.
- After an animal is successfully treated, it should be

periodically exposed to the (previously) fearful stimuli to avoid spontaneous recovery of the fear.

- If the eliciting stimuli can be identified and controlled, treatment of most fears and phobias is highly successful.

REFERENCES

1. Mayes A: The physiology of fear and anxiety, in Sluckin W (ed): *Fear in Animals and Man.* New York, Van Nostrand Reinhold Co, 1979, pp 24-55.
2. Ratner SC: Animals' defenses: Fighting in predator-prey relations, in Pliner P, Alloway TM, Krames L (eds): *Advances in the Study of Communication and Affect: II. Nonverbal Communication of Aggression.* New York, Plenum, 1976.
3. Russell PA: Fear-evoking stimuli, in Sluckin W (ed): *Fear in Animals and Man,* New York, Van Nostrand Reinhold Co, 1979, pp 86-124.
4. Archer J: Behavioral aspects of fear, in Sluckin W (ed): *Fear in Animals and Man.* New York, Van Nostrand Reinhold Co, 1979, pp 56-85.
5. Bolles RC: Species-specific defensive reactions and avoidance learning. *Psychol Rev* 77:32-48, 1970.
6. Wolpe J: *Psychotherapy by Reciprocal Inhibition.* Stanford, CA, Stanford University Press, 1958.
7. Wolpe J: *The Practice of Behavior Therapy.* New York, Pergamon Press Inc, 1973.
8. Marks I: *Cure and Care of Neuroses.* New York, John Wiley & Sons, 1981.
9. Jones MC: Elimination of children's fears. *J Exp Psychol* 7:382, 1924.
10. Jones MC: A laboratory study of fear: The case of Peter. *J Genet Psychol* 31:308-315, 1924.
11. Tuber DS, Hothersall D, Voith VL: Animal clinical psychology: A modest proposal. *Am Psychol* 29:762-766, 1974.
12. Tuber DS, Hothersall D, Peters M: Treatment of fears and phobias in dogs. *Vet Clin North Am [Small Anim Pract]* 12:607-623, 1982.
13. Miller NE: The analysis of motivational effects illustrated by experiments on amylobarbitone sodium, in Steinberg H (ed): *Animal Behavior and Drug Action.* Boston, Little, Brown & Co, 1964, pp 1-22.
14. Miller NE, Murphy JV, Mirsky IA: Persistent effects of chlorpromazine on extinction of an avoidance response. *Arch Neurol Psychiat* 78:526-530, 1957.
15. Sherman AR: Therapy of maladaptive fear-motivated behaviour in the rat by the systematic gradual withdrawal of a fear reducing drug. *Behav Res Ther* 5:121-129, 1967.
16. Fuller JL, Clark LD: Genetic and treatment factors modifying the postisolation syndrome in dogs. *Physiol Psychol* 61:251-257, 1966.
17. Yeung DPH: Diazepam for treatment of phobias. *Lancet* 1:475, 1968.
18. Voith VL: Clinical animal behavior. *Cal Vet* June:21-25, 1979.
19. Melzack R: Irrational fears in the dog. *Can J Psychol* 6:141-147, 1952.
20. Murphree OD: Reduction of anxiety in genetically timid dogs: Drug-induced schizokinesis and autokinesis. *Cond Refl* 7(3):170-176, 1972.
21. Dykman RA, Murphree OD, Reese WG: Familial anthropophobia in pointer dogs? *Arch Gen Psychiatry* 36:988-993, 1979.
22. Clark RS, Heron W, Fetherstonhaugh ML: Individual differences in dogs: Preliminary report on the effects of early experience. *Can J Psychol* 5(4):150-156, 1951.
23. Freedman DG, King JA, Elliot O: Critical period in the social development of dogs. *Science* 133:1016-1017, 1961.
24. Melzack R, Thompson WR: Effects of early experience on social behavior. *Can J Psychol* 10:82-90, 1956.
25. Fuller JL: Effects of experiential deprivation upon behaviour in animals. *World Congress Psychiatry* 3:223-227, 1963.
26. Karsh EB: The effects of early handling on the development of social bonds between cats and people, in Katcher AH, Beck AM (eds): *New Perspectives on Our Lives with Companion Animals.* Philadelphia, University of Pennsylvania Press, 1983, pp 22-28.
27. Voith VL, Borchelt PL: *Fear of Thunderstorms and Other Loud Noises* (pamphlet). Lawrenceville, NJ, Veterinary Learning Systems, Co, Inc, in press, 1985.
28. Young MS: Treatment of fear-induced aggression in dogs. *Vet Clin North [Small Animal Pract]* 12(4):645-653, 1982.

UPDATE

INCIDENCE

Recent data corroborate the high incidence of fears and phobias in dogs. In a self-selected survey (1993–1994)[1] completed by people who owned dogs representative of a multitude of breeds and mixed breeds, 38% of the 2018 respondents indicated that their dog sometimes, often, or always shook, trembled, tucked its tail between its legs, panted and paced, or tried to hide when hearing loud noises, such as thunder or firecrackers. Of the respondents, 22% indicated that their dog sometimes, often, or always avoided or was fearful of unfamiliar adults; 33% indicated that their dog sometimes, often, or always avoided or was fearful of unfamiliar children. Fourteen percent indicated that their dog was similarly frightened by nonthreatening unfamiliar dogs.

Fearful or phobic behaviors were the primary complaint of 10% of 1718 consecutive calls about dog behavior problems made to the ABC-VHUP[2]; the frequency of male and female dogs in this sample was in proportion to that seen for medical and surgical reasons at VHUP and to those registered by local licensing agencies. Thirty percent of the patients presented to the Clinical Behavioral Service, College of Veterinary Medicine, at the University of Tennessee had fear-related problems. The sex ratios of 30 of the dogs presented for noise phobias did not indicate a sex predisposition for noise phobias.[3]

BEHAVIOR MODIFICATION

Two difficulties that frequently arise in the treatment of noise phobias are finding a realistic stimulus to be used during treatment sessions and generalization of the dog's responses from the treatment stimulus to the real situation. Some of the critical factors to consider when creating a treatment and testing environment have been described in detail elsewhere.[3,4] For dogs with a history of noise phobias but that do not outwardly react to artificial stimuli, behaviorists have used desensitization and counterconditioning techniques and have had some positive effect on the dogs during the real situations.

Perhaps Tuber succeeded in transferring training sessions to the home environment because he always worked with the clients and the dogs for the first several sessions and because he always established a conditioned stimulus during the training sessions that was then transferred to the home environment. In his behavior-modification protocols, Tuber always associated the counterconditioned state in the training environment with a specific stimulus (cue) that is then transferred to the dog's normal environment. Such a cue serves several purposes. It is a discriminative stimulus that differentiates the environment now from the environment in which the phobia occurred. The new environment has a weaker association with the phobic experiences. The cue is also a conditioned stimulus associating a positive appetitive state and the phobic stimulus; this association further weakens the phobic response in the home environment.

During the treatment sessions for a noise phobia, for example, the dog lay on a specific rug (perhaps further distinguished from other rugs by the addition of a salient novel odor—e.g., oil of cloves). The rug is used during treatment sessions at the clinician's office and then at home where the sessions are replicated.

Eating palatable foods is not the only pleasant emotional state that can be used in counterconditioning procedures. Studies have shown that petting an aroused dog lowers its heart rate[5] and mutual grooming by horses lowers their heart rates.[6] Grooming and play have been used to reintroduce fearfully aggressive cats to each other.[7]

In the early 1970s, David Tuber introduced a massage technique he called the *soft exercise*.[8] He used it in counterconditioning procedures to treat fears and anxieties and as a way of easing a dog's transition from an aroused emotional state to a calm one (e.g., after the dog had become excited during play and as the owner prepared to leave). The dog is placed on its side and petted and massaged until it is so relaxed that it is limp. The technique is practiced in quiet surroundings until the owner can get up and walk away while the dog remains recumbent and relaxed.

A specific word (e.g., *soft*) is used during the exercises. Later, the word is used as a discriminative stimulus to signal to the dog that the soft exercise is about to begin and also as a conditioned stimulus associated with a relaxed state. After the dog has reliably and readily learned to relax, the technique is applied in mildly exciting situations, such as after a walk. Later, the technique is applied in arousing or potentially arousing situations, such as when visitors and children are present.

To use this technique most effectively, it is important to do the following:

- Use the same word or phrase (which becomes a classically conditioned cue) during the practice and applied situations
- Practice the exercise until the dog readily becomes totally relaxed before the exercise is tried when the dog is in arousing situations
- Continue to practice the exercise in quiet situations even after it is implemented in exciting situations.

This last point provides reinforcement of the word cue with the relaxed state.

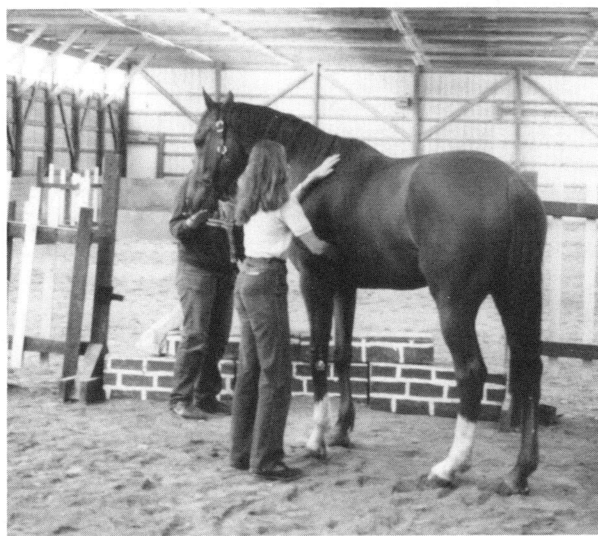

Figure 1—The horse and therapists in the treatment setting, an indoor arena with a public-address system installed specifically for these sessions. A pedometer is attached to the horse's front leg.

TREATMENT OF A NOISE PHOBIA IN A HORSE

The following example illustrates the applicability of desensitization and counterconditioning procedures across species as well as how effectively and quickly such procedures can work. While at the racetrack, a Thoroughbred gelding had acquired phobic responses to the sound of crowds and of broadcasts over public-address (PA) systems. The horse reacted to these sounds with an increase in cardiac rate and vigor (a rider could feel the horse's heart pound against her leg), trembling, sweating, and flight. The current owner was unable to ride the horse in any environment where such sounds occurred.

Baseline and treatment sessions were conducted twice a week for five consecutive weeks. There were two half-hour pretreatment/baseline sessions, a 3-minute test session conducted at the beginning and the end of the last baseline session, and eight subsequent half-hour treatment sessions. Heart rate and respiratory rate were monitored at the beginning and at 5-minute intervals during the 30-minute sessions and at the end of the 3-minute test session when the validity of the test stimulus (a tape recording of racetrack and basketball game sounds) was confirmed. Other parameters recorded were a pedometer reading at the end of each session, incidence of defecation, and the occurrence of any behaviors different from those observed during the baseline sessions. All but the last two treatment sessions took place in an indoor arena. A portable PA system was installed specifically for the sessions. The horse was fed grain by hand during the sessions (Figure 1).

During the baseline sessions, the horse did not defecate, exhibit muscle tremors, or move about much; his heart and respiratory rate fluctuated little and were within the normal range. During the 3-minute test session, the horse exhibited muscle tremors, increased locomotor activity, and elevated

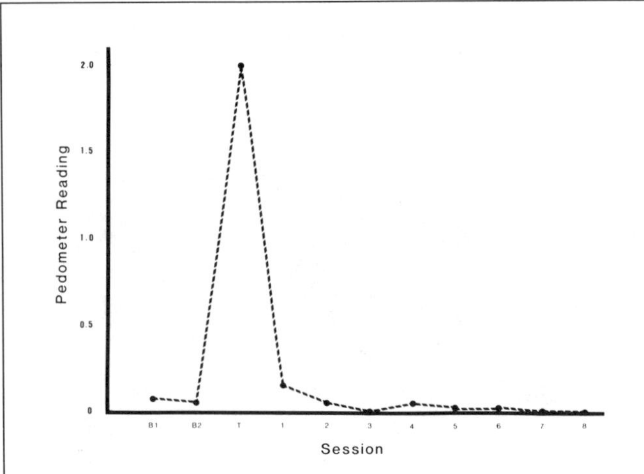

Figure 2—Pedometer readings at the end of each of the sessions. *B1* and *B2* are baseline sessions, *T* is the 3-minute test session, and *1* through *8* are the treatment sessions.

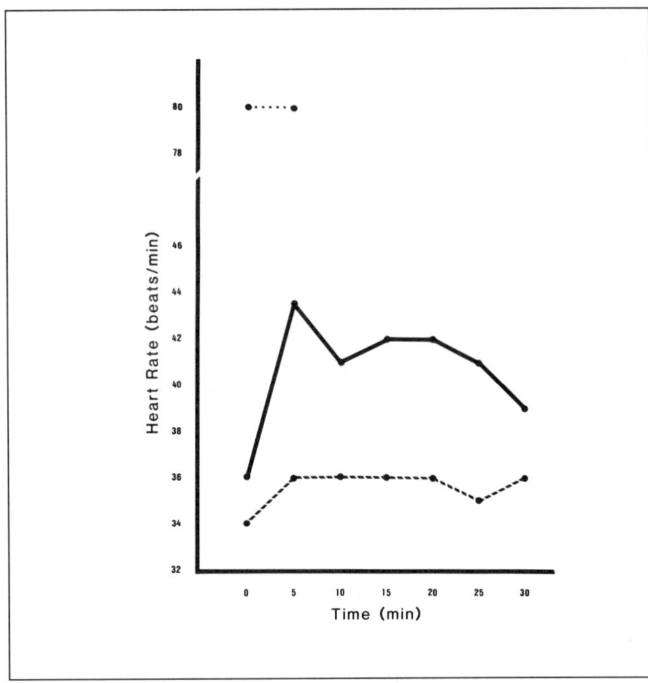

Figure 3—Heart rate during three conditions: Average of the two baseline sessions *(solid line)*, during the 3-minute test session *(dotted line)*, and average of the last two treatment sessions *(broken line)*.

head carriage. His heart rate was twice that of baseline, and the pedometer reading increased twentyfold (Figure 2). His respiratory rate was not elevated, nor did he defecate. The horse's fear response persisted after the recording was turned off. The tape recording clearly elicited a fearful response similar to that in the "real" situations.

Ideally, fearful reactions should be avoided during desensitization/counterconditioning procedures. In this case, however, the lowest setting on the equipment still elicited mild fear responses. More than three of the eight treatment sessions were conducted at the lowest possible volume setting (an even lower setting would have been preferable) before the volume was increased. During each session, the horse became habituated to the sounds; and at each succeeding session, his initial phobic reaction was less intense and the horse became habituated sooner.

During the fourth session, the sound was gradually increased. Periodically as the volume was increased (the equipment being used was not a finely tuned acoustical system), the horse exhibited temporary increases in heart rate, which quickly diminished. During the fourth session, the horse continued to eat but appeared restless, held his head low, shifted his head from side to side, and yawned. The volume was not increased unless the heart rate was at baseline. By the end of the fourth session, the horse's heart rate was at baseline and he appeared relaxed.

Over the next four sessions, the horse either accommodated or quickly became habituated to increases in the intensity of the sounds. The horse appeared relaxed and interested in food. During the sixth treatment session, the volume at which the horse was originally tested was exceeded. The last two treatment sessions were conducted outdoors instead of in the indoor arena where the other exposures had taken place. During the sixth, seventh, and eighth session, the horse's heart rate was below baseline (Figure 3).

There are several important points in this example. None of the sessions ended while the animal was exhibiting signs of fear. The volume was kept at the lowest setting for slightly over three sessions before being increased. Then within the next three sessions, the volume could be increased threefold, exceeding the volume at which the horse was originally tested. The initial sessions of a behavior modification program are almost always disproportionately long. But once an association is established between a positive appetitive state and the phobic stimulus, the techniques usually proceed rapidly. Improper implementation often causes desensitization and counterconditioning procedures to fail. Patience, the ability to recognize a low-intensity response, and adherence to fundamentally correct guidelines are essential.

There are some interesting caveats in this case. Sometimes, the horse looked calm even though his heart rate had increased. The lack of correlation, especially in the early phase of some of the treatment sessions, between heart rate and observable fearful behaviors has also been noticed in the treatment of humans with phobias.[9] Sometimes the horse ate grain while in a moderately fearful state (somewhat elevated heart rate, vigilant attitude, or some muscle tremors). Eating might have helped reduce the fear.[7] At some point on the continuum of fear, the horse would not eat (e.g., during the initial 3-minute exposure to the test stimulus). Disinterest in food is often an indication of the degree of anxiety or fear an animal is experiencing. After the 3-minute test session, the horse continued to exhibit anxious behaviors for some time

Figure 4—Licking behavior or licking-intention movements are a common but often unrecognized early sign of uneasiness in dogs.

Figure 5—Horses often yawn when their movement is restricted or when they are prohibited access to food.

after being returned to his stall. His respiratory rate was relatively unaffected throughout the sessions; the horse never defecated during any of the sessions.

A few months after the last treatment session, the owner took the horse to a small nearby outdoor horse show, a situation in which the horse previously had trembled, attempted to bolt, and was led away. It had been too dangerous for the rider to remain seated. This time, the horse was vigilant, held his head high, and would not stand still when he arrived at the show. The horse's activity, however, could be restricted to a walk with the owner astride. The horse was allowed to walk until he became habituated to the situation. In about an hour, the horse could stand still and tolerate the activities. Eventually, he was relaxed enough to stand in a resting position with one foot cocked. At no time did the horse pose a danger to himself, his rider, bystanders, or property. Although the treatment with the tape recording did not generalize completely to the real circumstances, the horse's fear responses had been reduced to a low enough level to safely allow flooding or habituation to the real situation.

SIGNS OF CONFLICT, ANXIETY, AND FEAR

Species-specific expressions and body postures have been correlated with aggressive offensive and defensive states, dominant and submissive attitudes, anxious and fearful states, and states of ambivalence.[11,12] An often unrecognized but common early sign of uneasiness in dogs is licking behavior or licking-intention movements (Figure 4). Sometimes the dog's tongue protrudes only slightly before being retracted. At other times, the tongue is extended; occasionally, the dog licks its nose. It has been hypothesized that this behavior has evolved from the greeting and food-soliciting behaviors of puppyhood.[12] This behavior is one of the first exhibited by a fearful dog as someone slowly approaches it.

Yawning and sleep are also conflict-associated behaviors for various species.[13] For example, dogs often yawn while they wait at the door to be let out or while they await their turn to engage in an activity they enjoy. A dominant-aggressive dog may yawn after the owner releases its muzzle. Horses yawn when their movement is restricted (e.g., in cross-ties) or when they are prohibited access to food in view (Figure 5). Humans also often yawn while they are waiting for something.

Some animals that are frightened, are in a quiet setting, and have been stationary for some time appear to become sleepy and to fall asleep involuntarily. Early in her career, one of us (V.V.) noticed that dogs often slept in the behavior consulting room during the interviews. The sessions were long, the dog may have traveled far, and the room was quiet. Initially, she assumed that the dogs were simply tired. She noticed, however, that some dogs, usually those with fear and phobia problems, fought falling asleep. One could see them struggling to keep their eyes open while they were standing, sitting, or in sternal recumbency with head erect (Figure 6). Finally, their eyes would close and the dogs would lose all muscle tone and begin to collapse. As they began to fall, they would awaken with a jerk and resume their erect posture.

When a dog fell asleep in a sitting or lying posture, it sometimes didn't wake up until its chin hit the floor. It is as though the dog suddenly went into REM sleep; but unlike narcoleptic attacks, these sudden involuntary onsets of

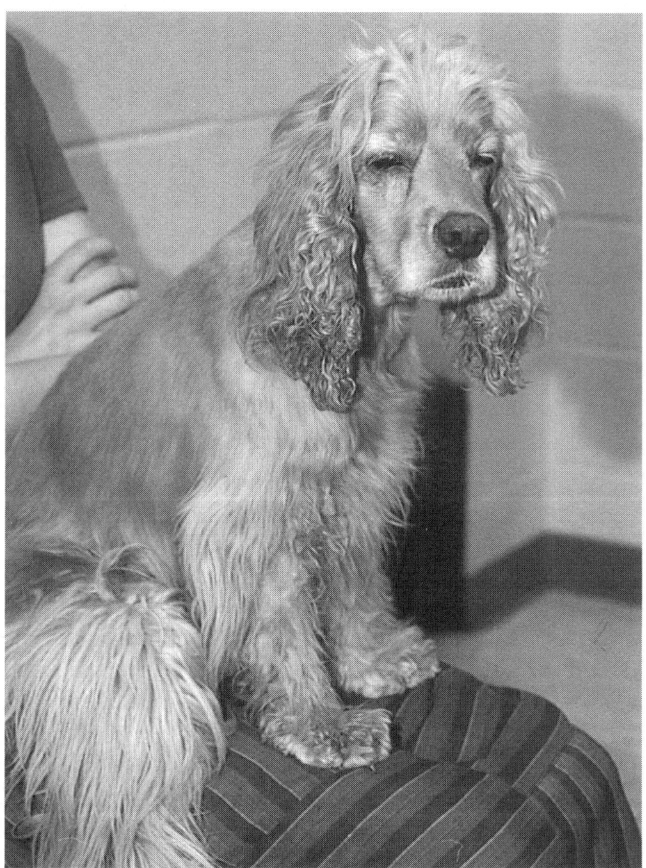

Figure 6—Some dogs with phobia problems struggle to keep awake while sitting or standing.

sleep occurred in quiet, instead of highly arousing, situations and the dogs immediately awoke as soon as they perceived they were falling. Sometimes, one sees horses struggle to keep their eyes open in uncomfortable but quiet situations from which they cannot leave.

Grooming is probably the most common behavior seen in conflict-eliciting and ambivalent situations.[11,14] Birds preen; dogs scratch themselves and lick their forelimbs; cats lick their sides; and humans chew their fingernails, twist their hair, and rub their skin. Sometimes these behaviors result in pathologies (e.g., feather-picking in birds or acral lick granulomas in dogs).

PHARMACOLOGIC TREATMENT

At the 1994 AVMA meeting, Shull summarized her observations of phobic and nonphobic dogs before, during, and after exposure to thunderstorm recordings in a laboratory setting. Using audiocassette stereo recordings of thunderstorms,[15] she could elicit phobic responses by all dogs with a history of brontophobia. She noted that the initial responses of all dogs (those known to be brontophobic and those known *not* to be brontophobic) were similar. Dogs with histories of no fear of thunderstorms became habituated to the recordings, and the others got progressively worse. Shull also observed that fearful dogs continued to exhibit phobic behaviors for some time after cessation of the fear-eliciting noise (a recording of a thunderstorm). Sometimes, medications that did not suppress the concurrent phobic response to the noise did shorten the dogs' fearful behaviors after the noise stimulus had been removed.

Shull also reported on the effects of several antianxiety medications on noises phobias in 16 clinical canine cases. Effective drugs and doses varied among animals; some dogs were refractory to all treatments that were tried. Overall, acepromazine maleate had the greatest effect in suppressing observable phobic behaviors of dogs, but the owners generally found the side effects (prolonged ataxia and lethargy) unacceptable. We have also found that owners are concerned about the side effects of acepromazine maleate and that they believe that the dog still "looks afraid" even though it is not engaging in disruptive behavior. As a class, the benzodiazepines were relatively effective. Propranolol was only effective for dogs with mild phobic responses; in some dogs, it shortened the recovery phase after the storm.

The oral doses of acepromazine maleate and propranolol were within the ranges specified on products labeled for veterinary use. The dose of alprazolam ranged between 0.02 and 0.22 mg/kg orally; chlorazepate dipotassium 0.5 to 43 mg orally; and imipramine 2.2 mg/kg orally. See the Simpson and Simpson article for review of published papers on pharmacologic treatment of fears in animals.

REFERENCES

1. Goodloe L, Borchelt PL: Unpublished data, University of Pennsylvania, Philadelphia, Pennsylvania, 1995.
2. Borchelt PL, Voith VL: Aggressive behavior in dogs and cats. *Compend Contin Educ Pract Vet* 7(11):949–957, 1985.
3. Shull-Selcer EA: Advances in the understanding and treatment of noise phobias. *Vet Clin North Am Small Anim Pract* 21(2):353–367, 1991.
4. Tuber DS, Hothersall D, Peters MF: Treatment of fears and phobias in dogs. *Vet Clin North Am Small Anim Pract* 12(4):607–623, 1982.
5. Gantt WH, Newton Jeo, Royer FL, Stephens JH: Effect of person. *Conditional Reflex* 1:18–35, 1966.
6. Feh C, De Mazieres J: Grooming at a preferred site reduces heart rate in horses. *Anim Behav* 46:1191–1194, 1993.
7. Borchelt PL, Voith VL: Diagnosis and treatment of aggression problems in cats. *Vet Clin North Am Small Anim Pract* 12:665–671, 1982.
8. Tuber DS: The soft exercise. *Anim Behav Consultants Newsletter* 3(1): 1986.
9. Leitenberg H, Agras S, Butz R, Wincze J: Relationship between heart rate and behavioral change during the treatment of phobias. *J Abnorm Psychol* 78:59–68, 1971.
10. Wolpe J: *Psychotherapy by Reciprocal Inhibition.* Palo Alto, CA, Stanford University Press, 1958.
11. Eibl-eibesfeldt I: *Ethology: The Biology of Behavior*, ed 2. New York, Holt, 1975.
12. Fox MW: *Behavior of Wolves, Dogs and Related Canids.* New York, Harper & Row, 1971.
13. Voith VL, McCrave E, Marder AR: Yawning, "licking," and sleep behaviors in dogs in relationship to conflict, anxiety, and fear (poster presentation). Presented at the Animal Behavior Society annual meeting, Williamsburg, MA, 1987.
14. Borchelt PL: Care of the body surface (COBS) in Denny MR (ed): *Comparative Psychology: An Evolutionary Analysis of Animal Behavior*. New York, John Wiley & Sons, 1980.
15. Shull EA: Analysis and treatment of noise phobias. Presented at the American Veterinary Medical Association meeting, San Francisco, July 9, 1994.

Compulsive Disorder in Dogs

Department of Population Medicine
Ontario Veterinary College
University of Guelph
Caroline J. Hewson, DVM
U. A. Luescher, DVM, PhD

Dogs and other species are known to suffer from a syndrome of repetitive behavior.[1–6] The syndrome has not been precisely defined, nor have diagnostic criteria been determined. Various authors have called the syndrome *stereotypy*,[4,5] *obsessive-compulsive disorder*,[2,3] or *compulsive disorder*.[7]

DEFINITION

Ethologists define *stereotypy* as a repetitive and invariant pattern of behavior that serves no obvious purpose in the context in which it is performed.[8–10] It is often implied that such behavior is exhibited only by captive animals in suboptimal environments,[5] by subjects under the influence of psychotropic drugs,[11] and by humans with psychiatric disorders.[10] Stereotypies, however, vary in cause and appearance.[10] One hypothesis is that stereotypies might provide some benefit to the individual performing them (e.g., by reducing stress, arousal, or awareness of surroundings).[12,13] An alternative hypothesis is that stereotypies serve no such function[14]; they merely indicate neuropathology.[15] Mason[10] and Lawrence and Rushen[16] have given comprehensive reviews of stereotypic behavior in animals.

Obsessive-compulsive disorder (OCD) is a human psychiatric condition that is defined in the American Psychiatric Association's Diagnostic and Statistical Manual of Mental Disorders (DSM-IV).[17] Obsessions are persistent intrusive thoughts that cause extreme anxiety and that the patient tries to suppress or ignore.[17] Compulsions are repetitive behaviors that are aimed at preventing or reducing this anxiety and that the person feels driven to perform.[17] The behavior syndrome in animals might not be wholly analogous to obsessive-compulsive disorder in humans. Because no one knows whether animals have obsessive thoughts, we prefer to omit the word *obsessive* and use the term *compulsive disorder* to describe the behavioral syndrome in animals.

We propose the following *working* definition of compulsive disorder (CD): behaviors that are usually brought on by conflict (see Situations of Conflict or Frustration on page 154) but that are subsequently shown outside the original context. The behaviors might share a similar pathophysiol-

Situations of Conflict or Frustration

Physical restraint
- Chaining
- Close confinement

Social conflict
- Changes in social group
- Competition for status or resources
- Separation

Lack of appropriate releasing stimuli (target objects) for normal behavior
- Lack of social or sexual partner (e.g., a dog kept in isolation)—Dogs are social animals and are motivated to interact with others; if kept in isolation, a dog has no outlet for this highly motivated behavior.
- Sucking deprivation (puppies)
- No object for prey-catching behavior

Motivational conflict
- Two equally strong, conflicting motivations (e.g., approach–withdrawal)—A dog that is at once territorial (approach) and fearful (withdrawal) may be in conflict when a stranger approaches the property; similarly, many dogs are both sociable (approach) and fearful (withdrawal) at the veterinary clinic.

Unpredictable or uncontrollable environment
- Novel situation
- Inconsistent environment (e.g., inappropriate use of punishment or lack of training or ineffective training methods)
- Inability to control or avoid aversive stimuli (e.g., thunderstorm or punishment)

Conditioned conflict
- Inconsistent reward or punishment
- Reward or punishment not contingent on behavior
- Biological constraints on learning

ogy (e.g., changes in serotonin, dopamine, and β-endorphin systems). Compulsive behaviors seem abnormal because they are displayed out of context and are often repetitive, exaggerated, or sustained (e.g. persistent tail-chasing or freezing in one position).

CLINICAL SIGNS AND PATHOGENESIS

There has been little formal research into canine compulsive disorder, and much remains to be learned. The pathophysiology is unknown, although some authors have proposed that it may be similar to that of obsessive-compulsive disorder in humans.[18,19]

Figure 1—Compulsive circling by a dog with no opportunity for exercise and no social contact.

The prevalence of compulsive disorder among dogs in North America is unknown. The lifetime prevalence of obsessive-compulsive disorder in humans in North America is 2.5%.[20] The prevalence of compulsive behaviors in Thoroughbred horses has been estimated at 2%.[21] At the Ontario Veterinary College, compulsive disorder constitutes 6% of the canine behavior caseload. Canine compulsive disorder is probably underdiagnosed because few veterinary schools give their students thorough training in clinical ethology. Hence, compulsive disorder may go unrecognized or misdiagnosed. Furthermore, pet owners may not always report the behavior to the veterinarian.

Compulsive disorder may begin as a normal, adaptive response to situations of conflict or frustration.[22,23] Behaviors of dogs in conflict situations include licking the nose and lips, shaking the head, yawning, circling (see Figure 1), pacing, tail-chasing, self-mutilation, snapping at the air, excessive grooming, and rhythmic barking. In our experience, some conflict behaviors are seen more frequently in certain breeds[24] (Table I).

After varying periods of expression, the conflict behavior may become emancipated from the original context and may be performed in other situations when arousal exceeds a critical threshold.[25,26] We hypothesize that at this point, neurochemical changes have occurred in the central nervous system, thus producing a pathological condition. The hypothesized changes in dogs have not been documented; but clinical studies in dogs,[18,27–31] horses,[32,33] and humans[34,35] and laboratory studies in other species[36–43] suggest that canine compulsive disorder may involve β-endorphins, serotonin, and dopamine. Neuropeptides, such as vasopressin and corticotropin (ACTH), may also play a role.[35] Note also that compulsive behaviors usually evoke a response from the owner and may be unwittingly reinforced.

TABLE I
Breed Predispositions for Conflict Behaviors

Breed	*Conflict Behavior*
Large breeds (e.g., Doberman pinscher, golden retriever, German shepherd, Labrador retriever)	Persistent licking of one area, usually carpus or tarsus, causing a lesion (lick granuloma)
English bullterrier	Whirling; freezing in one position, often with head in closet or in bushes
German shepherd	Tail chasing with or without tail mutilation
Miniature schnauzer	Turning and checking the rear or the floor, with or without scratching at the floor
Doberman pinscher	Flank-sucking

HISTORY AND PRESENTING SIGNS

Canine compulsive disorder commonly presents as grooming, flank-sucking, pacing (e.g, in a circle or figure eight), whirling, tail-chasing, snapping at the air, and chewing. Dogs may be presented at different stages of development of the behavior, depending on how observant the owners are and on how much the behavior upsets them. Dogs that mutilate themselves may be presented quite early, whereas dogs whose behavior is not injurious (e.g., flank-sucking) may never be presented for it. Dogs whose behavior is physically disruptive (e.g., whirling English bullterriers who knock over furniture and their owners) may also be presented early.

We have found that dogs suffering from compulsive disorder typically start to show the behavior in situations of conflict or frustration. Later, the behavior becomes emancipated from the initiating context and is shown in several other situations, typically those that excite the dog (e.g., when the owner returns from work or when the leash is put on for a walk). The behavior is also shown when the dog is not in contact with the owner (e.g., when the dog is alone outside). Dogs differ in the number of eliciting contexts and in the duration of each episode of the behavior. Some dogs show the behavior in only two or three contexts, but other dogs show it in ten or more contexts. An episode of the behavior may be brief (10 to 30 seconds) or prolonged (in extreme cases, the dog shows the behavior persistently, stopping only to eat or sleep). The dog is fully conscious while performing the behavior and can be interrupted by a sufficiently distracting stimulus (e.g., a loud noise). Similarly, the dog is attentive to stimuli as soon as an episode of behavior has ended—there is no disorientation as is usual following an epileptiform seizure.

It is not always possible to identify a source of initial conflict. The owner may report that the dog seems always to have shown the behavior or that the behavior started without a known inciting stimulus.

In taking the history, the veterinarian should focus on the following:

- A detailed description or observation of the behavior
- The breed
- The contexts in which the behavior was shown initially and in which it is currently shown (e.g., where and when the behavior appears and who is present)
- The behavior of the dog immediately before and after a typical episode
- The owner's reaction to the behavior
- The dog's temperament (excitable or fearful dogs may be more likely to develop compulsive disorder)
- How the owner interacts with the dog (e.g., method of training and type of punishment)
- Concurrent behavior problems that may indicate a conflict (e.g., aggression between dogs in the same household).

Affected dogs may not show the behavior in the waiting room, clinic, or hospital kennel; so a videotape recording of the dog engaging in the behavior elsewhere is useful. A 10- to 15-minute recording should be obtained showing the following:

- Performance of the behavior in two or more typical contexts
- Typical duration of an uninterrupted episode of the behavior
- Ease with which the dog can be interrupted while performing the behavior.

DIAGNOSIS

There are no generally accepted diagnostic criteria for canine compulsive disorder. The diagnosis is made from the history, the presenting signs, and the breed. In cases where the developmental history of the behavior is unknown or where a strong genetic predisposition is believed to exist, we sometimes base the diagnosis on presenting signs and breed alone.

At a minimum, a physical examination (including a neurologic examination) with complete blood count, chemistry profile, and urinalysis should be done to rule out easily diagnosed clinical or subclinical conditions. Additional tests may be used for differential diagnosis, depending on the presenting signs and the owner's resources.

Diagnostic differentials include normal behaviors (e.g., responses to acute conflict[23] and operantly reinforced behaviors), neurologic disorders (e.g., epilepsy, tumors, or hydrocephalus), dermatologic conditions, systemic disease (e.g., hepatic encephalopathy), and hyperkinesis. Note that a dog may suffer concurrently from compulsive disorder and other pathologic conditions. The latter might provide an internal stressor from which the dog cannot escape, thus producing compulsive disorder; however, compulsive disorder is more likely to stem from an external source of conflict or frustration.

TREATMENT

There is no proved treatment for canine compulsive disorder. No randomized, controlled clinical trials have been reported except for studies of lick granuloma.[18,30] A combination of behavioral techniques and medication is reportedly the most effective treatment regimen for humans with obsessive-compulsive disorder.[44] Such a combination is also used successfully for canine patients at the Ontario Veterinary College.

The behavioral techniques we use include removing the sources of conflict whenever possible. If this is impossible, dogs may be systematically desensitized to the eliciting stimuli. Daily exercise and reward-based training are also advocated. We have found that inconsistent or unpredictable actions by the owner are likely to contribute to compulsive disorder. Therefore, we recommend that, for 2 to 4 weeks, the owner's interaction with the dog be limited to daily exercise and to daily reward-based training (e.g., "Come!" and "Fetch!"). These prescribed interactions eliminate inconsistent signals from the owner. Inconsistent signals may be a source of conflict for the dog. The interactions also reduce the likelihood that the owner will inadvertently reinforce the behavior. Reward-based training provides the owner with the opportunity to interact with the dog in a predictable and consistent way. Training also enables the owner to control the dog in situations in which the owner might otherwise resort to punishment.

We find that a choke collar is difficult to use effectively and consistently and can produce unpredictable neck pain for the dog. Choke collars are therefore *not* recommended for dogs with compulsive disorder. A flat collar or head halter is an effective alternative. Owners are advised to be careful *not* to reinforce the compulsive behavior inadvertently.

We advocate that the owners *never* rebuke or punish the dog for engaging in the behavior. Inconsistent, unpredictable punishment is likely to worsen the behavior. Aversive treatment has reportedly stopped some compulsive behavior[45]; however, the technique was used according to a particular protocol, something most owners cannot do.

If the above procedures combined with the drug treatment are ineffective, the dog can be trained to perform an alternative, acceptable behavior in lieu of the compulsive behavior. The dog should first be trained to perform a behavior that is incompatible with the compulsive behavior. Then, as soon as the dog shows intentions to perform the compulsive behavior, it should be distracted using a novel sound (e.g., duck-call or shrill whistle) and commanded to perform the alternative behavior; it should then be rewarded for doing so. To be successful, this technique must be used every time the dog is about to perform the compulsive behavior. When supervision is impossible, the dog should be put in a situation in which it cannot or is unlikely to perform the behavior (e.g., an Elizabethan collar can be used).

A defined 6-week treatment period is useful to increase compliance. Owners should be encouraged to keep a diary during this time, with weekly goals. Weekly follow-up calls are also useful to increase compliance and provide support.

DRUGS

Drugs may be a necessary adjunct to treatment. For example, β-endorphin receptor blockers have been used experimentally and have temporarily stopped the performance of some compulsive behaviors in dogs[27–29] and horses.[32,33] The drug with the most clinical relevance is naltrexone.[46] In one reported case, however, the use of naltrexone in a dog was associated with pruritus.[47]

The drugs used most often for compulsive disorder are serotonin-reuptake inhibitors, such as clomipramine and fluoxetine. Both of these drugs are effective in the treatment of obsessive-compulsive disorder in humans[48] and have shown promise for the treatment of lick granulomas in dogs.[18,30]

It is recommended that humans with obsessive-compulsive disorder continue taking a specific serotonin-reuptake inhibitor for 10 to 12 weeks before the drug is considered ineffective.[48] In dogs, a minimum 5-week period of treatment may be indicated.[18]

We advise that after a compulsive behavior has been ameliorated by drug therapy and behavioral management, the dog should be weaned off the drug gradually. This is done by reducing the dose (but not the dose frequency) over three weeks.

If drugs are effective, they should *not* be regarded as the sole long-term solution but as an adjunct to behavioral therapy. Veterinarians and owners should never rely on medication alone to correct the underlying dysfunction.[48] Furthermore, owners must be advised that a dog may have a genetic predisposition to engage in compulsive behavior and that behavioral management may have to be continued throughout the dog's life.

REFERENCES

1. Fox MW: Spontaneous displacement activities, compulsive behaviour and abnormal social behaviour in the dog. *Vet Rec* 76:840–843, 1964.
2. Luescher UA, McKeown DB, Halip J: Stereotypic or obsessive-compulsive disorders in dogs and cats. *Vet Clin North Am [Small Anim Pract]* 21(2):401–413, 1991.
3. Overall KL: Recognition, diagnosis and management of obsessive-compulsive disorders, part 1. *Canine Pract* 17(2): 40–44, 1992.
4. Houpt KA, McDonnell SM: Equine stereotypies. *Compend Contin Educ Pract Vet* 15(9):1265–1271, 1993.
5. Fraser AF, Broom DM: *Farm Animal Behaviour and Welfare.* London, Bailliere Tindall, 1990, pp 305–317, 363.
6. Carlstead K, Seidensticker J, Baldwin R: Environmental enrichment for zoo bears. *Zoo Biol* 10:3–16, 1991.
7. Luescher UA: Conflict, stereotypic and compulsive behavior, in *Official Convention Program, AVMA 131st Annual Meeting.* Schaumberg, IL, American Veterinary Medical Association, 1994.
8. Odberg F: Abnormal behaviours (stereotypies), in *Proceedings of the First World Congress on Ethology Applied to Zootechnics.* Madrid, Editorial Garsi, Industrias Graficas España, 1978, pp 475–480.
9. Immelmann K, Beer C: *A Dictionary of Ethology.* Cambridge, MA, Harvard University Press, 1989, p 291.
10. Mason GJ: Stereotypies: A critical review. *Anim Behav* 41: 1015–1037, 1991.
11. Nymark M: Apomorphine provoked stereotypy in the dog. *Psychopharm (Berl)* 26:361–368, 1972.
12. Dantzer R, Mormede P: De-arousal properties of stereotyped behaviour: Evidence from pituitary-adrenal correlates in pigs. *Appl Anim Ethol* 10:233–244, 1983.
13. Wiepkema PR: Abnormal behaviours in farm animals: Ethological implications. *Neth J Zool* 35(1,2):279–299, 1985.
14. Odberg FO: Future research directions, in Lawrence AB, Rushen J (eds): *Stereotypic Animal Behaviour: Fundamentals and Applications to Welfare.* Tucson, CAB International, 1993, pp 180–187.
15. Dantzer R: Behavioral, physiological and functional aspects of stereotyped behavior: A review and a reinterpretation. *J Anim Sci* 62:1776–1786, 1986.
16. Lawrence AB, Rushen J (eds): *Stereotypic Animal Behaviour: Fundamentals and Applications to Welfare.* Tucson, CAB International, 1993.
17. American Psychiatric Association: *Diagnostic and Statistical Manual of Mental Disorders,* ed 4. Washington DC, American Psychiatric Association, 1994, pp 422–423.
18. Rapoport JL, Ryland DH, Kriete M: Drug treatment of canine acral lick: An animal model of obsessive compulsive disorder. *Arch Gen Psychiatry* 49:517–521, 1992.
19. Stein DJ, Dodman NH, Borchelt P, et al: Behavioral disorders in veterinary practice: Relevance to psychiatry. *Compr Psychiatry* 35(4):275–285, 1994.
20. Karno M, Golding JM, Sorensen SB, et al: The epidemiology of obsessive-compulsive disorder in five U.S. communities. *Arch Gen Psychiatry* 45:1094–1098, 1988.
21. Vecchiotti GG, Galanti R: Evidence of heredity of cribbing, weaving and stall-walking in Thoroughbred horses. *Livestock Production Sci* 14:91–95, 1986.
22. Wood-Gush DGM: *Elements of Ethology: A Textbook for Agricultural and Veterinary Students.* London, Chapman and Hall, 1983, pp 134–144.
23. Hinde RA: *Animal Behavior,* ed 2. New York, McGraw Hill, 1970, pp 396–421.
24. Blackshaw JK, Sutton RH, Boyhan MA: Tail chasing or circling behavior in dogs. *Canine Pract* 19(3):7–11, 1994.
25. Cooper JJ, Odberg F: The emancipation of stereotypies with age, in Appleby MC, et al (eds): *Applied Animal Behaviour: Past, Present and Future. Proceedings of the International Congress.* Potters Bar, England, Universities Federation for Animal Welfare, 1991, p 142.
26. Ridley RM, Baker HF: Stereotypy in monkeys and humans. *Psychol Med* 12:61–72, 1982.
27. Brown SA, Crowell-Davis S, Malcolm T, et al: Naloxone-responsive compulsive tail-chasing in a dog. *JAVMA* 190:884–886, 1987.
28. Dodman NH, Shuster L, White SD, et al: Use of narcotic agonists to modify stereotypic self-licking, self-chewing and scratching behavior in dogs. *JAVMA* 193:815–819, 1988.
29. White SD: Naltrexone for treatment of acral lick dermatitis in dogs. *JAVMA* 196(7):1073–1076, 1990.
30. Goldberger E, Rapoport JL: Canine acral lick dermatitis: Response to the anti-obsessional drug clomipramine. *JAAHA* 27(2):179–182, 1991.
31. Overall KL: Use of clomipramine to treat ritualistic stereotypic motor behavior in three dogs. *JAVMA* 205(12):1733–1741, 1994.
32. Dodman NH, Shuster L, Court MH, et al: Investigation into the use of narcotic antagonists in the treatment of stereotypic behavior pattern (crib-biting) in the horse. *Am J Vet Res* 48:311–319, 1987.
33. Dodman NH, Shuster L, Court MH, et al: The use of narcotic antagonist (nalmefene) to suppress self-mutilation behavior in a stallion. *JAVMA* 192:1585–1586, 1988.
34. Goodman WK, Dougle CJM, Price LH: Pharmacotherapy of obsessive compulsive disorder. *J Clin Psychiatry* 53(Suppl 4):29–37, 1992.
35. Altemus M, Swedo S, Leonard H, et al: Changes in cerebrospinal fluid neurochemistry during treatment of obsessive compulsive disorder with clomipramine. *Arch Gen Psychiatry* 51:794–803, 1994.
36. Cooper SJ, Dourish CT: *Neurobiology of Stereotyped Behavior.* New York, Clarendon Press, 1990.
37. Cabib S: Neurobiological basis of stereotypies, in Lawrence AB, Rushen J (eds): *Stereotypic Animal Behaviour: Fundamentals and Applications to Welfare.* Tucson, CAB International, 1993, pp 119–145.
38. Cronin GM, Wiepkema PR, van Ree JM: Endorphins implicated in stereotypies of tethered sows. *Experientia* 42:198–199, 1986.
39. Schouten W, Rushen J: Effects of naloxone on stereotypic and normal behaviour of tethered and loose-housed sows. *Appl Anim Behav Sci* 33:17–26, 1992.
40. von Borell E, Hurnik JF: The effect of haloperidol on the performance of stereotyped behavior in sows. *Life Sci* 49: 309–314, 1991.
41. Kennes D, Odberg FO, Bouquet Y, et al: Changes in naloxone and haloperidol effects during the development of captivity-induced jumping stereotypy in bank voles. *Eur J Pharmacol* 153:19–24, 1988.
42. Vandebroek I, Odberg FO, Caemaert J: Microdialysis study

of the caudate nucleus of stereotyping and non-stereotyping bank voles, in Rutter SM, Rushen J, Randle HD, et al (eds): *Proceedings of the 29th International Congress of the International Society for Applied Ethology*, Potters Bar, England, Universities Federation for Animal Welfare, 1995, pp 245–246.

43. Cools AR, Van Rossum JM: Caudal dopamine and stereotypy behavior of cats. *Arch Int Pharmacodyn Ther* 187:163-173, 1970.
44. Greist JH: An integrated approach to treatment of obsessive compulsive disorder. *J Clin Psychiatry* 53(4, suppl):38–41, 1992.
45. Eckstein R: Use of electronic stimulation to treat acral lick dermatitis, in *Official Convention Program, AVMA 131st Annual Meeting*. Schaumberg, IL, American Veterinary Medical Association, 1994.
46. Dodman NH, Shuster L: Pharmacologic approaches to managing behavior problems in small animals. *Vet Med* 89:960–969, 1994.
47. Schwartz S: Naltrexone-induced pruritus in a dog with tail-chasing behavior. *JAVMA* 202(2):278–280, 1993.
48. Rasmussen SA, Eisen JL, Pato MT: Current issues in the pharmacologic management of obsessive compulsive disorder. *J Clin Psychiatry* 54(6, suppl):4–9, 1993.

Equine Stereotypies

Cornell University
Katherine A. Houpt, VMD, PhD

University of Pennsylvania
Susan M. McDonnell, PhD

Rutgers University
Sarah Ralston, VMD, PhD

KEY FACTS

- **Stereotypical behavior may be a protective mechanism in a stressful environment.**
- **Although cribbing and wood chewing are different behaviors, both tend to begin when roughage is reduced in the diet.**
- **There is no evidence that cribbing is learned, but there is evidence that it is inherited.**
- **Providing stall toys is usually ineffective in reducing stereotypical behavior.**
- **Managing horses on pasture with other horses is currently the most reliable means of reducing locomotor stereotypies, such as weaving and stall walking.**

The behavior problems discussed in this article are often referred to as vices. We avoid this term because (1) it is anthropomorphic, implying that horses are making moral decisions to act in an evil manner, and (2) the behaviors reflect a horse's response to a stress and may actually help to alleviate the stress of confinement. Veterinarians should approach these problems not as equine misdeeds that should be punished but as medical problems that can be understood on the basis of physiologic processes and treated via pharmacologic and management techniques.

Stereotypies are defined as stylized, repetitive, apparently functionless motor responses or sequences.[1] They occur in domesticated and captive wild species as well as in humans. Stereotypies in horses can be classified as locomotor stereotypies (which include weaving, stall circling, fence pacing, head bobbing, pawing, and wall kicking) and oral stereotypies (e.g., cribbing, wood chewing, and flank biting). Equine stereotypies vary considerably in the percentage of the horse's time occupied by the activity and in the vigor and persistence with which the behavior is performed.

Horses that exhibit stereotypies are often unthrifty as a result of the increased energy utilization or reduced feed intake or because the stereotypy is performed to the exclusion of normal eating or grazing. In some individuals, episodes of a stereotypy occur at predictable times, either seemingly unprovoked or in association with particular environmental events. In other individuals, the episodes are sporadic but obviously triggered by environmental events. Some horses with stereotypies seem nervous and prone to panic; others are relatively even-tempered and apparently well-adjusted animals.

There are numerous hypotheses regarding the origin of stereotypies in domesticated and captive wild animals. The most common scientific interpretation is that stereotypies begin as displacement activities, vacuum activities, intention movements, or mimicry. Each of these is presumed to be normal and adaptive in wild animals.

Displacement activities are behavioral sequences that appear in an unusual context. In birds, the classic example of a displacement activity is feeding or grooming in the middle of a fight sequence. Displacement responses are believed to occur with inherent motivational conflicts or when goal-directed activity is thwarted. The animal is apparently anxious or frustrated. A stabled horse may pace, weave, or paw as the feed cart advances down the aisle; the horse is anticipating the grain but unable to reach it. Similarly, stall walking or fence walking and weaving often begin when a horse is held behind in a stall or pasture when others are turned out or taken away. Stallions that can see or hear mares being teased but cannot reach them tend to paw, weave, pace, or circle. It is typical for a horse experiencing pain, fear, or stress to paw, yawn, circle, weave, eat wood or dirt, or bite itself (flanks, chest, knees, or shoulders). Most displacement behaviors subside as the stressful environmental conditions resolve.

Vacuum activities are behaviors that occur in the absence of a stimulus. A classic example is elaborate nest-building behaviors in caged birds with no access to nesting materials. In horses, head shaking and head bobbing have been interpreted as vacuum activities aimed at nonexistent insects. Habitual vacuum activities are behaviors that were initially evoked by a meaningful stimulus (e.g., skin parasites in the case of flank biting) but that persist after the stimulus ceases.

Intention movements are abbreviated initial portions of behavior sequences. Stall circling or fence walking may initially result from thwarted attempts to escape confinement. Over time or even in the same escape episode, the distance traveled while circling or stall walking may gradually decrease; the horse just paces one side of the stall or just back and forth at the pasture gate. Eventually, the horse may only weave; weaving thus is apparently the final condensation of the pacing or circling and can be interpreted as an intention movement of walking or pacing. In a hungry horse, pawing may be an intention movement of the sequence of uncovering forage or of movement during grazing.

Mimicry involves stereotypies that may evolve by observational learning. Abundant but inconclusive anecdotal evidence supports the belief that stereotypies tend to develop in association with exposure to other horses that engage in the activities.

Finally, some stereotypies can be induced by drugs. This fact suggests possible mechanisms of development. Apomorphine is an opiate that acts as a dopaminergic agonist and stimulates the emetic center. Although horses do not vomit after receiving apomorphine, locomotor stereotypies and yawning are exhibited. Most normal horses that are confined to a stall will circle, pace, and/or weave when given apomorphine.

Regardless of how stereotypies arise and to what degree they are initially normal, they lead to well-established and aberrant behavior in domesticated or captive wild animals. The stereotypical behavior may occupy the major portion of the animal's time, often excluding or diminishing normal maintenance behaviors. Many stereotypies are maintained by inadvertent reinforcement, such as feeding a horse that paws as the feed cart comes down the aisle. A stallion's daily pacing at the stall door as he anticipates breeding time is reliably reinforced by breeding. Changing management to eliminate reinforcement or, alternatively, rewarding cessation of the behavior can effectively eliminate some stereotypies.

Isolation and stimulus deprivation increase the likelihood of stereotypies. Because noisy stereotypies often lead to attention from humans, stereotypies that produce rhythmic sound may be inherently rewarding to isolated or stimulus-deprived animals. By means of 24-hour video surveillance, I (Dr. McDonnell) observed that several horses performed stereotypies only when humans were in the barn; the handlers had the erroneous impression that the stereotypical behavior was continuous. My preliminary findings indicate that repetitious motor behavior may reinforce itself via the release of endogenous opiates.

In humans, stereotypies may be one feature of a more complex psychophysiologic disorder, such as obsessive-compulsive disorder. Stereotypies in animals may represent a similar disorder[2]; however, a specific diagnosis of obsessive-compulsive disorder in animals is premature in the absence of a better understanding of the neurochemical basis of such complex disorders.

Researchers determined that 15% of horses in Ontario engage in some form of stereotypical behavior (including masturbation). Until recently, spontaneous erection and masturbation were considered equine stereotypies.[3] Close observation of confined, pastured, and free-ranging horses has confirmed that all normal stallions exhibit spontaneous erection and masturbation.[4–6] Ejaculation actually occurs rarely. These behaviors are evident from birth through adulthood and are not affected by age or sociosexual environment. Even geldings exhibit spontaneous erection and masturbation, with reduced frequency and duration. Masturbation thus should not be considered a stereotypy in equine species.

Equine stereotypies are undesirable because they can lead to considerable damage to the animal and property. The remainder of this article considers some of the common stereotypies in horses and presents possible therapies. Further discussions of animal stereotypies, including those common in horses, have been published.[1,7,8] Table I lists types of equine stereotypical behavior and Table II outlines the breed distribution of the various problems presented to a behavior clinic. McGreevey et al found that 5% of British Thoroughbreds were wood chewing, 3.3% were weaving, and 1% were stall walking.[8a] When all anomalous behavior, including cribbing, was considered, 11.9% of horses engaged in some stereotypic behavior. The factors that predispose to stereotypic behavior are less than 6.8 kg of forage per day, fewer than 75 horses on the farm, feeding more than three times a day, hay as the only forage, and bedding material other than straw.

TABLE I
Types of Equine Stereotypical Behavior[a]

Behavior	*Number of Cases*
Self-mutilation	11
Stall walking	11
Cribbing	8
Wood chewing	2
Tongue play	2
Miscellaneous	3

[a]Presented to the Animal Behavior Clinic, New York State College of Veterinary Medicine, Cornell University.

ORAL STEREOTYPIES

Cribbing

Cribbing is an oral behavior in which the horse grasps a surface (e.g., the rim of a bucket or the rail of a fence) with its incisors and then simultaneously flexes its neck and was thought to swallow air. Recently, McGreevy et al have found that a cribbing horse does not swallow the air.[8b] The role of contraction of the strap muscles in cribbing is to create a pressure gradient in the soft tissues surrounding the esophagus, which provides movement of air into the esophagus. This air produces a transient dilation of the upper esophagus. The characteristic noise of wind sucking coincides with an in-rush of air through the cricopharynx. There is no deglutition. The esophageal dilation is relieved when the air returns to the pharynx. Only small amounts pass caudally to the stomach. Air reaches the stomach only if a bolus of food pushes it down. Some horses grasp vertical surfaces; some press their necks against a horizontal object without using their teeth. Less commonly, a horse may extend the neck and gulp without grasping or touching an object. Generally, one to two hours per day are consumed by cribbing.

Two European studies indicate that approximately 2.5% of Thoroughbreds crib.[9,10] Some data suggest that cribbing is inherited.[9,11] The behavior apparently develops more readily in confined horses (usually near mealtime) but also occurs in pastured horses. Anecdotally, cribbing has been reported immediately after ingestion of concentrates or before a meal (if feeding time is delayed). Cribbing has been associated with colic, and swallowed air has been incriminated as a possible cause for the pain. Figure 1 depicts a related behavior that, like cribbing, occurred after the horse ate grain.

There are substantial anecdotal reports (but no experimental, objective evidence) that cribbing is modified by learning. A horse generally first cribs during exposure to a cribbing horse. Particular methods of cribbing are apparently learned from other horses.

TABLE II
Breed Distribution of Equine Stereotypies and Aggressive Behavior[a]

Breed	*Number of Cases Involving Stereotypies*	*Number of Cases Involving Aggression*
Quarter horse	15	11
Thoroughbred	8	14
Arabian	5	0
Appaloosa	0	5
Morgan	4	3
Standardbred	2	3
Other breeds	3	7
Donkey	1	0

[a]Cases presented to the Animal Behavior Clinic, New York State College of Veterinary Medicine, Cornell University.

Because they are very oral and are teething, young horses may be more likely than adults to learn cribbing from other horses. Learning to crib may be contingent on genetic predisposition—a horse that has inherited the disposition to crib will begin the behavior when it sees another horse cribbing. Alternatively, the environment that causes one horse to crib may elicit cribbing in other horses.

There are various general approaches to the treatment of cribbing. The following approaches are discussed here: elimination of surfaces to grasp, punishment, surgery, and pharmacologic treatment.

Elimination of Inviting Surfaces

Cribbing may be reduced by painting surfaces with an unpalatable substance or electrifying fence rails and other horizontal surfaces. A cribbing muzzle that allows grazing and drinking but prevents grasping of a surface with the incisors usually inhibits cribbing; aerophagia (excessive swallowing of air) may continue.

Punishment

The most common method used to reduce cribbing is a strap around the throat that exerts pressure when the horse arches its neck and attempts to swallow air. A strap with spikes may punish more effectively. A horse with a cribbing strap may continue to grasp objects, but aerophagia is reduced. Shock has been used to punish cribbing.[12] Commercially available electronic dog-training collars can be adapted to fit the equine neck and can be remotely controlled so the horse does not associate punishment with the presence of a human. I (Dr. McDonnell) have observed that the use of dummy collars before and after training may prevent the association of punishment with the collar itself.

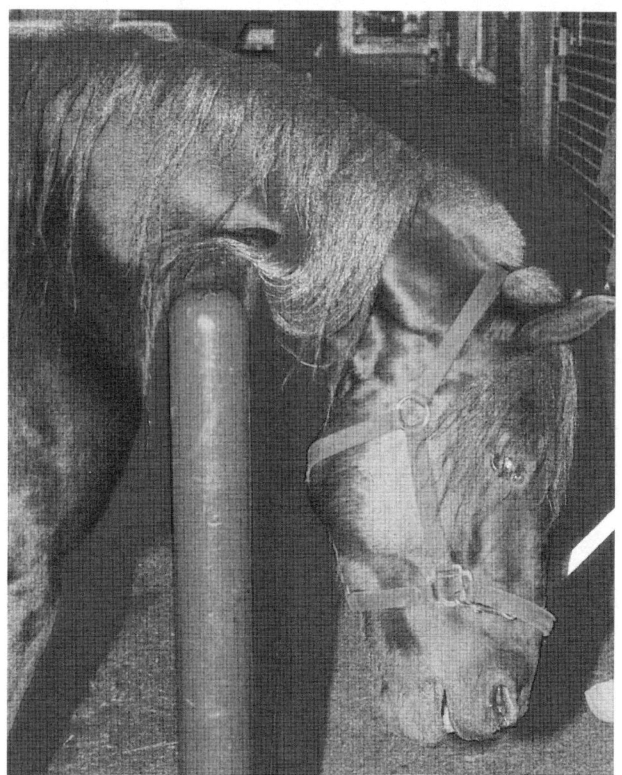

Figure 1—A novel type of stereotypy involving a horse pressing its neck around a horizontal bar.

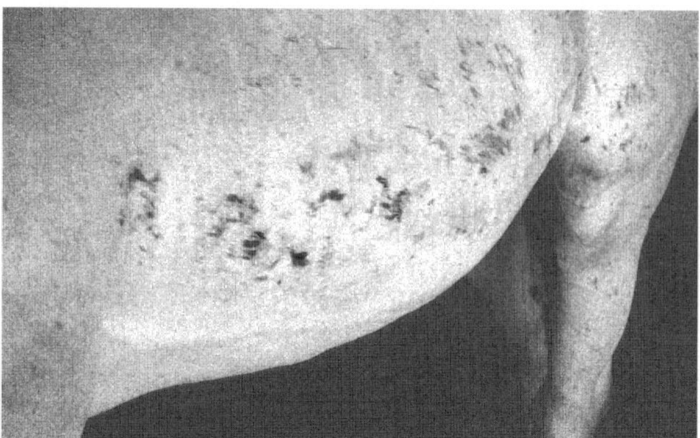

Figure 2—Lesions produced by self-mutilation.

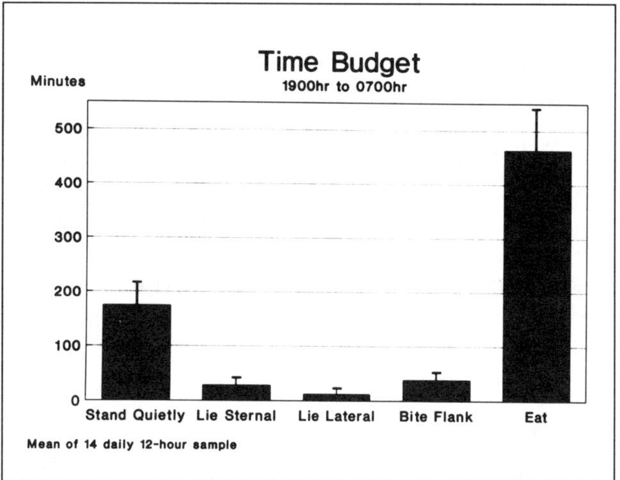

Figure 3—Time budget of a flank-biting stallion, based on 14 consecutive days of observation. Illustrated is the mean time spent in each activity per 12-hour sample (7 PM to 7 AM).

Surgical Approaches

The various surgical treatments for cribbing include buccostomy, spinal accessory neurectomy (11th cranial nerve), myotomy or myectomy of the ventral neck muscles, or a combination of partial myectomy and spinal accessory neurectomy. The success rates of these treatments vary from 0% to 70%.[13–19] Cribbing often recurs if the environment is not changed.

Pharmacologic Treatment

Theoretically, horses must derive some pleasure from cribbing; otherwise the behavior would not be sustained. Gillham et al have found that baseline levels of endogenous opiates are lower in cribbers than in noncribbers.[19a] Cribbing may lead to the release of endogenous opiates and thus produce pleasure. Opiate receptor blockers (e.g., naloxone hydrochloride at a dose of 0.02 mg/kg) stop horses from cribbing for 20 minutes following a 20-minute latency period. Intravenous naltrexone (0.04 mg/kg), another opiate blocker, suppressed cribbing for two to seven hours.[20] Cribbing is suppressed only while the opiate antagonists are present; the short half-life and expense of these drugs have made long-term opiate receptor blockade impractical.

Wood Chewing

Wood chewing is different from cribbing in that the wood is actually torn away or ingested and that air is not swallowed. Wood chewing may be a stereotypy or may reflect a normal effort to satisfy nutritional requirements. Horses that chew wood thus cannot be categorically classified as exhibiting a stereotypy. Some cases of wood chewing cease immediately if the diet is changed (e.g., if salt or more hay is made available). Highly concentrated or pelleted diets and infrequent meals increase the incidence of wood chewing.[21,22] Indigestible roughage may play a role in the diet—feral horses as well as well-fed pastured ponies ingest trees and shrubs even when grasses are freely available. Wood chewing increases in cold and wet weather[23,24]; this suggests a survival value in ingesting shrubs in the winter, when little else is available (see Table III for causes).

Traditional methods for reducing wood chewing include eliminating wood surfaces, covering them

TABLE III
Potential Causes of Excessive Wood Chewing

1. **Inadequate volume of feed**—Caloric and nutrient needs fulfilled by one or two feedings per day, which are consumed in a total of 1 to 4 hours as opposed to the normal 10 to 12 hours that the animals would voluntarily spend eating.
2. **Lack of environmental stimulation**—Horses confined to stalls for prolonged periods without companionship or other forms of diversion.[a] This is often exacerbated by a lack of feeding time.
3. **Inadequate protein?**—Amount of wood chewed was reported to decrease when crude protein percentage of a complete feed was increased to >10%.[b]
4. **Decreased cecal pH secondary to feeding complete pelletal feeds?**—Wood chewing slightly decreased in ponies fed pelletal feeds when cecum pH was buffered to 6.8 with $NaHCO_3$.[b] This is obviously not a practical solution to the problem.

[a]Ralston SL, Van den Broek G, Baile CA: Feed intake patterns and associated blood glucose, free fatty acid and insulin changes in ponies. *J Anim Sci* 49(3):838–845, 1979.
[b]Willard J, Willars SC, Baker JP: Dietary influences on feeding behavior in ponies. *J Anim Sci* 37:227, 1973.

TABLE IV
Solutions for Excessive Wood Chewing

1. Provide more chewing time with diet.[a]
   Feed loose hay ad libitum.
   Feed wafered or cubed forages that take longer for the animals to eat.[b]
   Put horses out on adequate pasture.
   Give ration in four to five separate feedings.
2. Provide environmental stimulation.
   Provide a companion animal such as a goat, sheep, or pony.
   Provide a toy such as a plastic jug with stones in it suspended from ceiling,[c] or a small rubber tire, sacks, or balls to bat around the stall.
   Hang a metal mirror[d] at the horse's eye level.
   Provide more exercise.
3. Cover all wooden surfaces with creosote, metal,[e] or aversive flavor (citrus?) or stimulus (electric wire).
4. Add sodium bicarbonate to diet (?)[f]—not proven to be effective.
5. Check if quantity and quality of diet meets 1989 National Research Council requirements. Try increasing percentage of total protein and fiber.[g]

[a]Applicable to most behavior problems.
[b]Haenlin GFW, Holdren RD, Yoon YM: Comparative response of horses and sheep to different physical forms of alfalfa hay. *J Anim Sci* 25:740–743, 1966.
[c]Beware—Some horses chew the plastic and cut themselves on the resultant sharp edges.
[d]Available from army surplus outlets.
[e]Metal covers may cause excessive wear of the incisors if the animals still try to chew.
[f]Willard J, Willars SC, Baker JP: Dietary influences on feeding behavior in ponies. *J Anim Sci* 37:227, 1973.
[g]Schurg WA, Frei DL, et al: Utilization of whole corn plant pellets by horses and rabbits. *J Anim Sci* 45(6):1317–1321, 1977.

with metal or wire, treating them with unpalatable substances, or muzzling the horse with a grazing basket. Providing more roughage and less concentrate in the diet is advisable, as is providing salt and minerals ad libitum. An increase in exercise may reduce wood chewing[25] (see Table IV for solutions).

Self-Mutilation

Self-mutilation is a serious behavior problem that occurs more frequently in stallions. A survey performed by Dodman et al revealed that the behavior occurred more frequently in males (51 of 57 cases).[25a] The most common form involves biting the flanks or (more rarely) the chest and limbs; there may be simultaneous squealing and kicking.[26] The signs mimic those of acute colic but without progression to rolling or depression.[27] Rare forms of self-mutilation include lunging into walls and banging the head against objects. Self-mutilation may originate as a displacement activity related to stress, frustration, or fear. Some stallions begin self-mutilation when moved to stables with other stallions, when moved to isolated stables, or when within sight of (but denied contact with) mares. Figure 2 depicts the effects of self-mutilation. Figure 3 illustrates the time spent in self-mutilation and other activities.

Manipulation of the social environment, even if by trial and error, may reduce self-mutilation. A stallion that self-mutilates in the stable may improve during pasture housing with other horses, particularly mares. A stall companion (e.g., a donkey, goat, or rabbit) may be helpful. Removal of other stallions from the environment is useful. Because any change may be initially stressful, one change should be made and left in place for a minimum of two weeks in order to appraise the effect on the horse's behavior.

Physical restraint, in the form of a muzzle or head cradle, is commonly used to inhibit flank biting. A muzzled horse usually continues to swing its head, thus continuing to inflict injury. Similarly, a horse wearing a cradle usually keeps turning its head as far as possible, inflicting new sores from contact with the restraining device. The horse may kick humans or objects.

Castration usually reduces and sometimes eliminates self-mutilation. Geldings that self-mutilate are often those that exhibit other stallionlike behaviors. Some geldings only self-mutilate in the presence of mares. The opiate antagonists used to manage cribbing reduced self-mutilation in one stallion.[26] Pro-

Figure 4—A horse exhibiting a stereotypical turn at the end of an excursion.

gestin treatment may reduce self-mutilating behavior, particularly in geldings. Typical dosages are as follows: altrenogest, 0.02 ml/kg/day; megestrol acetate, 65 to 86 mg/500-kg horse; and progesterone in oil, 0.4 mg/kg/day. The success rate of progestin treatment has not yet been established. As antiandrogen drugs become available, they may be useful in controlling self-mutilation in stallions.

Other Oral Behaviors

Such oral behaviors as lip smacking, tongue playing, and lolling are largely ignored by owners unless they occur in the show ring. Dropped nose bands, tongue ties, and loose-fitting head stalls that force a horse to hold the bit have been used to control tongue lolling in saddled horses.

LOCOMOTOR STEREOTYPIES

Circling, Weaving, and Pacing

Most normal horses periodically circle in a stall or small paddock and occasionally walk up and down a fence and pace by the gate, especially at feeding time. Circling, fence walking, and gate pacing (Figure 4) are considered to be stereotypies if they become excessive and replace normal periods of resting or eating. Stereotypical walkers wear a trench along the fence or near the gate. Some individuals circle for hours, at a rate of several stall revolutions per minute. Most horses circle in both directions, sometimes forming a figure eight when reversing direction. Serious stall walkers lose condition, and racing or sporting performance may be negatively affected. An increase in dietary roughage may be helpful but is not always effective in reducing the frequency or severity of the problem.

Weaving is a behavior in which the horse remains stationary but shifts its weight from forelimb to forelimb and swings its head from side to side (30 to 90 cycles per minute). Confinement and the presence of

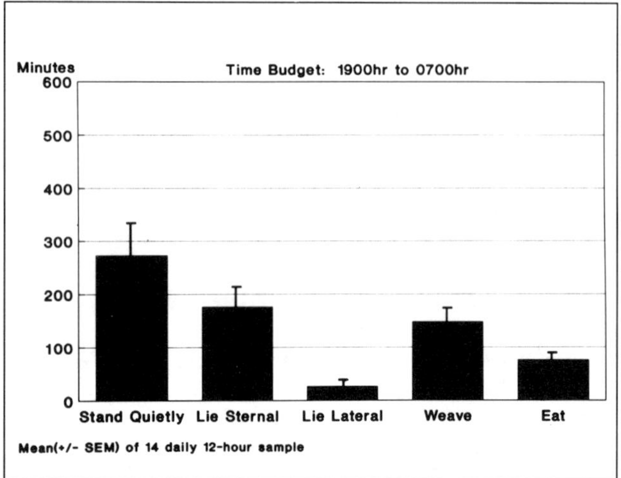

Figure 5—Time budget of a weaving mare, based on 14 consecutive days of observation. Illustrated is the mean time spent in each activity per 12-hour sample (7 PM to 7 AM). *SEM* = standard error of the mean.

a barrier between the horse and a goal (e.g., proximity to other horses, feed, or freedom) are expected causes, at least initially. Horses that are in pain may weave, usually temporarily. Tying a stereotypical circling or pacing horse may lead to weaving for as much as three hours per day. There is some evidence that the tendency to circle, pace, and weave is inherited and possibly related to endorphin production.[9,20] Some owners report that horses circle more frequently before a hunt or show. Figure 5 presents the time budget of a weaving mare; Figure 6 is the time budget of a normal mare for comparison.

Management approaches include providing more exercise, a less stressful environment, and a less concentrated ration. Stall toys are apparently effective only in young, playful horses. Stall size evidently does not influence stall walking behavior—one horse given access to an entire barn continued to circle in one corner.

A 24-hour videotape surveillance of the patient's behavior may help to assess quantitatively and objectively the severity of the problem and suggest management approaches. The diurnal pattern of the stereotypy as well as exacerbating and attenuating environmental conditions may be identified. For example, some stall circlers pause at the hay rack on each circle, periodically stopping to tear a piece of hay and quickly resume circling. One such horse was put on a complete roughage diet, with flakes of hay positioned around the perimeter of the stall. The horse stopped at each flake of hay; eventually the stall walking was transformed into a much slower, relaxed, grazinglike ambulation. Subsequent video surveillance indicated a more than 80% decrease in time spent stall circling.

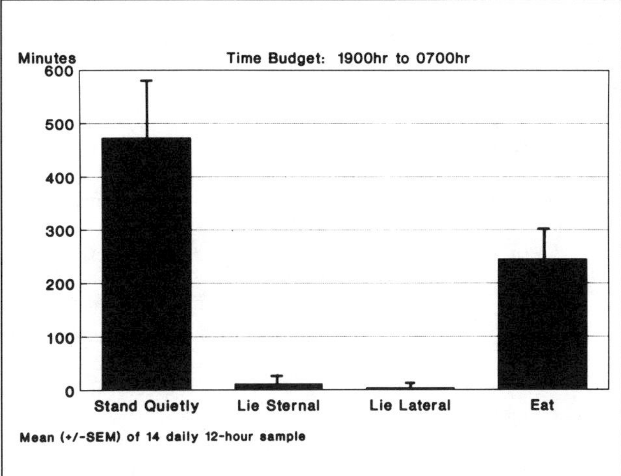

Figure 6—Time budget of a normal mare, based on 14 consecutive days of observation. Illustrated is the mean time spent in each activity per 12-hour sample (7 PM to 7 AM). *SEM* = standard error of the mean.

Tricyclic antidepressants that are used to treat obsessive-compulsive disorder in humans may be of therapeutic value for equine locomotor stereotypies. Preliminary results indicate relatively long-lasting positive effects in the treatment of seven horses that exhibited weaving, stall circling, and fence walking.

Stall Kicking

Stereotypical stall kicking involves repetitive striking of the stall walls with the hooves or hocks or stomping a hindleg against the floor with no target animal. Stereotypical kicking produces detrimental concussion of bones and joints and can damage stall walls. Stall kicking may be a form of self-stimulation or cause satisfaction from the sound made by kicking wood. Stereotypical kicking, like pawing, often begins at feeding time, when the animal is exposed to various auditory, olfactory, and visual cues. Because it is eventually fed, the horse learns that kicking or pawing is rewarded by food and may begin to paw or kick progressively earlier.

One way to extinguish the behavior is to feed the horse frequently and only when it refrains from pawing or kicking. An example of such a regimen is one-half cup of feed when the horse refrains from kicking for two seconds. Gradually, the criterion is raised to no kicking for 5, then 10, then 30 seconds before food is given. The training goes faster if the horse is simultaneously taught a countercommand (e.g., "stand") for a food reward. Alternatively, an animal can be put on a free-choice roughage diet with no specific meal times. Analogous, inadvertently reinforced bad habits include trailer kicking or pawing and stall kicking in stallions waiting to be bred.

Stall kicking that is not clearly associated with external reinforcement is more difficult to eliminate. Management aids include providing more turnout, providing a stall toy, padding the stall walls, and padding the horse's hocks. A rail that prevents the horse from backing into the wall may prevent hock knocking.

Pawing

Pawing (dragging a foot across the ground or motioning in the air) is a normal behavior that becomes a stereotypy when excessive. Horses paw in a variety of situations and with various apparent motivations: if confined, to attain proximity to another horse, to stimulate a recumbent foal to rise, in anticipation of feed, while eating grain or (more rarely) hay, when restricted from moving forward, and if in pain. Confinement can lead to apparent frustration and pawing. While confined in a box stall for training, a Standardbred stallion that had spent most of its life on pasture dug a hole 1.5 meters (four feet) deep. If dirt floors are replaced with concrete, a horse may stop pawing but may cause considerable hoof wear if pawing continues.

Some animals paw while eating grain or hay and apparently not frustrated. In its natural state, a grazing horse continually steps and nibbles. If food is presented in a stationary location (e.g., grain in a bucket or hay in a net), pawing may represent intention movements to perform the normal grazing locomotor sequence. Unless the pawing continues beyond feeding, it may not be appropriately considered a stereotypy. Pawing while restrained from moving is similar to pawing to escape confinement. Pawing in pain may exhibit frustration at not escaping the pain. In horses with colic, such pawing is frequently a prelude to rolling.[28]

Stereotypical pawing may be managed by altering the stall floor surface to prevent holes (concrete floor) or to prevent excessive trauma to the limbs (heavy rubber matting). Providing more turnout time also is helpful.

SUMMARY

Stall walking, weaving, cribbing, pawing, head bobbing, and self-mutilation are examples of stereotypical equine behavior. These behaviors may be responses to isolation, confinement, or deprivation of foraging opportunities. Many horses are subjected to confinement isolation and limited roughage but do not develop stereotypical behavior; there is probably a genetic predisposition to the development of these behaviors. There is some indication that opiate antagonists interrupt cribbing and self-mutilation. The fact that tricyclic antidepressants may decrease weaving and stall walking indicates that the underlying biochemical mechanisms of these behaviors resemble compulsive disorders in humans. The best and most economically feasible treatments involve eliminating the environmental factors that stimulate stereotypical behaviors.

About the Authors

Dr. Houpt is affiliated with the Animal Behavior Clinic, New York State College of Veterinary Medicine, Cornell University, Ithaca, New York. Dr. McDonnell is affiliated with the New Bolton Center, School of Veterinary Medicine, University of Pennsylvania, Kennett Square, Pennsylvania. Dr. Ralston is affiliated with the Department of Animal Sciences, Cook College, Rutgers University, New Brunswick, New Jersey.

REFERENCES

1. Kiley-Worthington M: *Behavioural Problems of Farm Animals.* Stocksfield, England, Oriel Press, 1977, p 134.
2. Goldberger E, Rapoport JL: Canine acral lick dermatitis: Response to the anti-obsessional drug clomipramine. *JAAHA* 27:179–182, 1991.
3. Luescher UA, McKeown DB, Halip J: Reviewing the causes of obsessive-compulsive disorders in horses. *Vet Med* 86: 527–530, 1991.
4. McDonnell SM: Spontaneous erection and masturbation in horses. *Proc 35th Annu Conv AAEP*:567–580, 1989.
5. Wilcox S, Dusza K, Houpt K: The relationship between recumbent rest and masturbation in stallions. *Equine Vet Sci* 11(1):23–26, 1991.
6. McDonnell SM, Henry M, Bristol F: Spontaneous erection and masturbation in equids. *J Reprod Fertil Suppl* 44:664–665, 1992.
7. Leuscher UA, McKeown DB, Halip J: Stereotypic or obsessive-compulsive disorders in dogs and cats. *Vet Clin North Am Small Anim Pract* 21:401–413, 1991.
8. Mason GA: Stereotypies: A critical review. *Anim Behav* 41: 1015–1037, 1991.

8a. McGreevy PD, Cripps PJ, French NP, et al: Management factors associated with steriotypic and redirected behaviour in the Thoroughbred horse. *Eq Vet J* 27:86–91, 1995.

8b. McGreevy PD, Richardson JD, Nicol CJ, Lane JG: Radiographic and endoscopic study of horses performing an oral based stereotypy. *Eq Vet J* 27:92–95, 1995.

9. Vecchiotti GG, Galanti R: Evidence of heredity of cribbing, weaving and stall walking. *Livestock Prod Sci* 14:91–95, 1987.
10. McBane S: *Behavior Problems of Horses.* North Promfret, VT, David and Charles, 1987, p 304.
11. Hosoda T: On the heritability of susceptibility to windsucking in horses. *Jpn J Zootech Sci* 21:25, 1950.
12. Baker GJ, Kear-Colwell J: Aerophagia (windsucking) and aversion therapy in the horse. *Proc AAEP* 20:127–130, 1974.
13. Firth EC: Bilateral ventral accessory neurectomy in windsucking horses. *Vet Rec* 106:30–32, 1980.
14. Forssell G: The new surgical treatment against crib-biting. *Vet J* 82:538–548, 1926.
15. Greet TRC: Windsucking treated by myectomy and neurectomy. *Equine Vet J* 14:299–301, 1982.
16. Hamm D: A new surgical procedure to control crib-biting. *Proc 23rd Annu Meet AAEP*:301–302, 1977.
17. Karlander S, Mansson J, Tufvesson G: Buccostomy as a method of treatment for aerophagia (windsucking) in the horse. *Nord Vet Med* 17:455–458, 1965.
18. Owen RR, McKeating FJ, Jagger DW: Neurectomy in windsucking horses. *Vet Rec* 106:134–135, 1980.
19. Turner AS, White N, Ismay J: Modified Forssell's operation for crib biting in the horse. *JAVMA* 184:309–312, 1984.

19a. Gillham SB, Dodman NH, Shuster L, et al: The effect of diet on cribbing behaviour and plasma B-endorphin in horses. *Appl Anim Behav Sci* 41:147–154, 1994.

20. Dodman NH, Shuster L, Court MH, Dixon R: Investigation into the use of narcotic antagonists in the treatment of a stereotypic behavior pattern (crib-biting) in the horse. *Am J Vet Res* 48:311–319, 1987.
21. Willard JG, Willard JC, Wolfram SA, Baker JP: Effect of diet on cecal pH and feeding behavior of horses. *J Anim Sci* 45:87–93, 1977.
22. Houpt KA, Perry PJ, Hintz HF, Houpt TR: Effect of meal frequency on fluid balance and behavior of ponies. *Physiol Behav* 42:401–407, 1988.
23. Jackson SA, Rich RA, Ralston SL: Feeding behavior and feed efficiency in groups of horses as a function of feeding frequency and use of alfalfa hay cubes. *J Anim Sci* 59(Suppl 1):152–153, 1984.
24. Salter RE, Hudson RJ: Feeding ecology of feral horses in western Alberta. *J Range Manag* 32:221–225, 1979.
25. Krzak WE, Gonyou HW, Lawrence LM: Wood chewing by stabled horses: Diurnal pattern and effects of exercise. *J Anim Sci* 69:1053–1058, 1991.

25a. Dodman NH, Normile JA, Shuster L, Rand W: Equine self-mutilation syndrome (57 cases). *JAVMA* 204:1219–1223, 1994.

26. Dodman NH, Shuster L, Court MH, Patel J: Use of a narcotic antagonist (nalmefene) to suppress self-mutilative behavior in a stallion. *JAVMA* 192:1585–1587, 1988.
27. Murray MJ, Crowell-Davis S: Psychogenic colic in a horse. *JAVMA* 186:381–383, 1985.
28. Odberg FO: An interpretation of pawing by the horse (*Equus caballus* Linnaeus), displacement activity and original functions. *Saugetier Mitteil* 21:1–12, 1973.

Feeding and Elimination

Elimination Behavior and Related Problems in Dogs

Victoria L. Voith, DVM, PhD
Department of Clinical Studies
School of Veterinary Medicine
University of Pennsylvania
Philadelphia, Pennsylvania

Peter L. Borchelt, PhD
Animal Behavior Consultants, Inc.
Forest Hills, New York
The Animal Medical Center
New York, New York

The term *behavior* encompasses everything an animal does. To analyze and understand behavior, it is helpful to classify the types of behavior. One classification scheme is that of behavior systems. Behavior systems are composed of sequences of behavior patterns that serve a common function. Examples of behavior systems include ingestion, elimination, sleep, care of the body surface (e.g., grooming), reproduction, and aggression.

Behavior systems are composed of many types of behaviors. Some of the behaviors are reflexes and are relatively stereotyped; others involve learning and are highly flexible. For example, the ingestive behavior system in canids involves not only eating but also searching for prey, learning how to hunt, catching prey, and perhaps even storage of food items. The actual act of eating is only a small part of the ingestive behavior system of dogs.

The *elimination* behavior system of dogs involves not only urinating and defecating to rid the body of waste but also searching for specific locations and/or surfaces on which to eliminate; preelimination behaviors, such as sniffing, circling, and walking "straddle-legged"; and postelimination behaviors, such as scratching the ground. Although the elimination behavior system has been studied relatively little, it is frequently involved in behavior problems of companion animals. A survey of more than 800 clients who brought their pets to the Veterinary Hospital of the University of Pennsylvania for medical or surgical services showed that 5.5% had dogs with a behavior problem involving elimination.[1] About 20% of the canine cases presented to an animal behaviorist involve elimination behavior.[2,3]

Urine and feces can also be used as part of a communication system between animals. Excretions might be used to mark territories, to communicate to dogs and other animals that another animal has passed by, to identify individuals, to determine the reproductive status of an individual, and perhaps to determine whether or not an individual tends to be dominant or submissive.[4-6] Urine and feces can serve a communicative function primarily or secondarily to ridding the body of waste. Possibly the *act of urinating* has its own communicative function. For example, young dogs often engage in submissive postures, which sometimes involve urinating[7-9]; and dogs frequently urinate in view of each other.

Urination and defecation also occur in contexts related to other behavior systems. Animals may uncontrollably urinate and defecate when extremely frightened and sometimes urinate when excited. Urinating and defecating may accompany states of high anxiety, e.g., when a dog is separated from its owner or when riding in a car. Elimination in such circumstances usually has a degree of control that is completely absent when a dog is extremely frightened. Anxious dogs often deposit urine and feces in specific areas and are able to assume typical elimination postures.

Development of Canine Elimination Behaviors

A bitch stimulates newborn puppies to urinate and defecate by licking their urogenital area while simultaneously ingesting the secretions. After the first week or two of life, puppies may initiate elimination by themselves but the mother still immediately ingests the excretions. Bitches are reported to be "surprisingly quick" to notice a puppy squatting or grunting in relation to elimination, then moving to catch the excretions as they are voided.[10] Urine has been observed being caught by the mother up to five and one-half weeks and feces up to nine and one-half weeks postpartum. If the mother does not catch an excretion, she may still ingest it when found later.

By three weeks of age puppies begin voiding away from the nest box.[11,12] On waking they will walk three to four feet from where they have been sleeping and then urinate and defecate. Although pups initially avoid soiling their sleeping areas, the excretions are deposited in many locations in the pen area; at about five weeks of age puppies have picked a generalized area and by nine weeks of age a relatively specific area in which to eliminate. Interestingly, although these areas are away from the resting site, they are *not* as far away as possible from the feeding areas.[11] In laboratory studies where puppies have been left with the mother, puppies have been observed gradually beginning to urinate and defecate more often in the spot characteristically used by the mother.[10]

Adult dogs in laboratory situations also have specific locations where they eliminate within the pen.[11] Owners of companion dogs report the development of location-specific elimination behaviors in their pets, both in and out of the home. For example, some dogs eliminate only in particular areas of the yard, between parked cars, or on curbs. Others demonstrate very specific surface preferences, such as only on grass, leaves, cement, linoleum, or carpeting. Location and surface preferences for elimination are very common in dogs.

Young puppies usually eliminate immediately after waking from a nap or sleep, a short period of time after eating or drinking, and about every two hours while awake and active. They can, however, sleep for many hours without eliminating.[11,12]

Urinary postures in puppies are sexually dimorphic by two months of age, even though puberty is not reached before five to six months of age. At five to seven weeks postpartum, male laboratory beagles begin to assume a standing and leaning forward posture rather than squatting to eliminate.[13] The full elevated leg-lift postures begin in males between four and six months of age. For about one month thereafter, postures fluctuate between juvenile and adult forms before the dog consistently assumes adult male postures. Neonatally castrated males do not begin showing leg-lift postures until 6 to 12 months of age and then continue to fluctuate frequently between juvenile and adult postures. Administration of testosterone in adulthood does not affect the urinary posture of dogs that were castrated neonatally.[13]

Female puppies squat to urinate, and most continue this posture into adulthood. Androgenized females (those exposed to testosterone prenatally and/or immediately postnatally) may assume masculine urinary postures as adults[13]; however, they squat like female puppies during infancy and the juvenile period. Signs of masculinization do not emerge until later in life. Possibly pet female dogs that begin to assume a leg-lifting posture in young adulthood were exposed in utero to circulating testosterone from adjacent male puppies. Several experiments with mice and rats indicated that female fetuses positioned next to male fetuses will assume more masculine behaviors and postures in adulthood than female fetuses positioned elsewhere.[14-16]

Elimination Postures in Adult Dogs

A wide range of individual variation exists in the elimination postures of dogs, as is true of most animal behavior patterns. The literature previously suggested that normal postures for elimination were squatting for females and leg elevation for males and that variations

Owners of companion dogs report the development of location-specific elimination behaviors in their pets, both in and out of the home.

Figure 1—Elimination postures. (From Sprague RH, Anisko JJ: Elimination patterns in the laboratory beagle. *Behaviour* 47:287, 1973. Reprinted with permission.)

of these were abnormal.[12,17] Both sexes had been reported to defecate in the same squatting posture that a female uses for urination.

More recent detailed observations, however, indicate that both males and females frequently exhibit a variety of postures for elimination (Figure 1; Table I).[18] This study, based on observations of beagles, found that at least 15% of the males sometimes urinated on the ground rather than on vertical objects and 15% urinated using postures other than elevating the rear leg. About 13% of the females urinated at least once on a vertical object and approximately 4% never urinated on the ground; 15% of the females used postures other than the squat or squat-raised leg position.[18] Although the squat position

TABLE I

Percentage of Urinations Performed Using Each Posture

Group	*Number of Urinations*	*Posture*[a,b]								
		Sq	*SqR*	*FR*	*F*	*H*	*L*	*LR*	*R*	*El*
Female	259	68.0 (87)	19.3 (43)	3.1 (9)	0.4 (2)	1.9 (2)	0.4 (2)	0.0 (0)	4.6 (8)	2.3 (6)
Male	963	0.0 (0)	0.6 (2)	0.0 (0)	0.0 (0)	0.0 (0)	0.0 (0)	0.3 (2)	2.1 (13)	97.0 (100)

[a] Posture codes used: *Sq* = squat; *SqR* = squat-raise; *FR* = flex-raise; *F* = flex; *H* = handstand; *L* = lean; *LR* = lean-raise; *R* = raise; *El* = elevate. (Figure 1 illustrates these positions.)

[b] Numbers in parentheses show the percentage of subjects that urinated using each posture. There were 53 female and 60 male subjects that urinated. (From Sprague RH, Anisko JJ: Elimination patterns in the laboratory beagle. *Behaviour* 47:260, 1973. Reprinted with permission.)

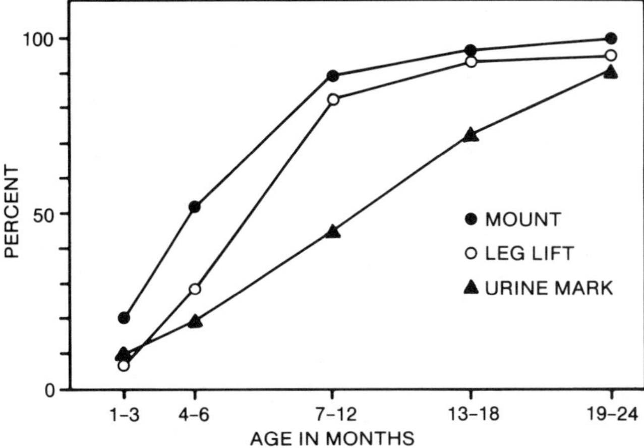

Figure 2—Cumulative percentage of dogs exhibiting mounting, leg-lifting, and urine marking at different ages.

has historically been called the "normal" posture for females, 13% of the female beagles were never observed using it.

Males usually sniffed the ground or an object during urination; females rarely did so, although they spent much of the time in the test areas walking and sniffing the ground. Females appeared to empty their bladders with one or two urinations, but males urinated many times. Males frequently went through the motions of urinating without expelling any urine.

Although females showed greater variability in urination postures, males showed more variation in defecation postures. In addition to arch and handstand postures, males used an arch-raised leg, squat-raised leg, and the full elevated rear-leg posture that is normally associated with urination. Both males and females frequently defecated on vertical objects, such as fences. Males engaged in ground scratching after elimination more frequently than females did. Either the hind feet or all four feet were used to scratch, and scratching was more common after urination than defecation. The scratching did not cover the urine and feces but usually left conspicuous marks on the ground; scratching may be used to disperse the dog's scent over a wider area.

Sexual and Hormonal Influences on Canine Elimination Behavior

The administration of either estrogen or testosterone to adult dogs of either sex increases the frequency of urination but does not affect urinary postures.[13] Castration of adult and juvenile males reduces the frequency of urination but does not alter urinary postures.[13,19]

Females that were implanted with testosterone one to three days postpartum show some masculinization in urinary postures. Females exposed to testosterone in utero and immediately after birth urinate using male postures 50% of the time in adulthood.[13]

Males and females differ in the amount of time they spend investigating the urine of other adult dogs and in the frequency of urinating on the urine deposits of other adult dogs.[20-22] Intact males and anestrous females investigate the urine of estrous females more often than the urine of males or anestrous females. Females in estrus, however, spend more time investigating male urine and urinate two and one-half times more frequently than anestrous females do. Male dogs urinate three times as often as anestrous females do and orient more of their urine toward the urine of other dogs.[21] Males investigate urine from strange males longer and urinate on it more frequently than they do on urine from familiar males or their own urine.[22] Although females investigate the urine of other dogs, they urinate on it less frequently than males do.

Male dogs spend a considerable amount of time investigating and urinating on urine of females in estrus. Evidence has shown that male urine has more of an effect than simply acting as a diluent of female urine and actually appears to mask the odors specific to estrous females. This masking perhaps reduces the probability that other male dogs will detect the estrous female.[23]

Functional Significance of Urine Marking

Many animals communicate through olfactory stimuli. In mammals, the most common sources of chemicals used in scent marking are urine, feces, saliva, body secretions, and scent-gland products.[4] A unique property of olfactory signals is their ability to remain in the environment after the animal has left and thus allow communication between individuals at different points in time. Even humans, not known for olfactory prowess, can detect wolf urine six weeks after it has been deposited.[24]

The term *marking behavior* is often used with the implication that it is territorial. This presumption elicits several problems. In the animal behavior literature, territorial behavior denotes defense of a well demarcated area. The relationship between territoriality and marking in dogs as well as many other animals is unclear. Many animals, particularly dogs, do not limit their marking behavior to their territorial boundaries. They mark multiple locations within the territory as well as areas other than the territory. Additionally, scent marking does not keep other animals out of territories. Dogs typically enter other dogs' territories to urine mark.

Scent marking may provide intruders with a means of identifying and assessing the residents of the area.[6] For example, if an intruder is able to match or compare the specific odor of scent marks in an area with the odor of a specific individual in that area, the stranger might identify the animal as a resident of the territory and act accordingly. In view of this theory, it would be predicted that the owner of a territory will mark frequently throughout its territory, although it might mark more often on the boundaries. The residents allow a stranger to investigate the urine marks as well as the owner itself to enable the stranger to match the odors. In fact, dogs usually urinate in sight of each other and then investigate each other's urine. Perhaps male dogs urine mark over the urine of strange male dogs to mask the other dog's odor so that there is no confusion as to who is the territorial resident. Laboratory studies have demonstrated that, compared with their own urine, male dogs mark twice as often over urine of familiar male dogs and four times more frequently over urine of strange male dogs.[22]

Scent marking in dogs might be used as a sexual attractant.[21,25] Estrous females urinate more frequently than anestrous females, and male dogs spend more time investigating the urine of estrous females than that of anestrous females or urine of other male dogs. When females are in estrus, they spend more time investigating the urine of intact males. In these ways, urine might be used as a means of identifying the sexual status of an individual and perhaps even finding that individual. Roaming behavior increases dramatically in estrous bitches.

Urine marking is also observed during courtship behaviors in dogs and wolves and perhaps serves as a sexual stimulus.[21,24,25] Interestingly, male dogs are more strongly attracted to the urine of estrous females than to their vaginal or anal-sac secretions. Anal-sac secretions elicit no more interest from other dogs than distilled water would.[26]

Anal-sac secretions vary between dogs in consistency, odor, and color; but dogs are unable to identify the sexual or reproductive status of an individual from anal-sac secretions.[27] The individual properties of anal-sac secretions, however, might be used to identify individual dogs. Although the ability to identify one another through anal-sac odors is untested in dogs, human researchers can easily and reliably identify individual dogs from the odor of their anal-sac secretions.[a,b]

Scent marking might also be used to help familiarize an animal with an area and/or to orient the animal within an area. For example, wolves and foxes do not mark food caches that are full but do mark empty food caches or sites that have already been scavenged.[28,29] This behavior provides a "bookkeeping system" that allows more effective scavenging and assessment of food caches with the least expenditure of energy.

In some species, scent marks may allow animals to assess the dominant status of individuals.[30] Whether this determination occurs in dogs is unknown, nor has the relationship between the frequency of urine marking and dominant status in dogs been determined. Even if dominance were related to the frequency of urine marking in dogs, there is no reason to believe that preventing a dog from urine marking alters its dominant status.

Scent marks may serve as alarm signals or warnings to conspecifics. Certainly it is known that the urine, feces, and anal-sac secretions of dogs are expelled when they are frightened. Scent markings might also serve as an indicator of population size or density of a species. This information could influence immigration and emigration rates.[31] Whether these odors serve such a communicative function in dogs is unknown.

Canine Elimination Behavior Problems

Behavior problems that involve urination, defecation, or both in the household generally are from undesirable location and surface preferences (which result in housebreaking problems), separation anxiety, or fear of a particular stimulus (e.g., thunder). The two behavior problems that are most difficult to differentiate are lack of housebreaking and separation anxiety.

Fear Reactions

Some dogs cannot control urination and defecation in response to a frightening stimulus. When fear is diagnosed as the cause of an elimination behavior problem, the stimuli that evoke the fear must be carefully and accurately identified and then the dog can be systematically desensitized and/or counterconditioned. Use of antianxiety drugs to reduce the intensity of the fear response is another approach; however, the use of such drugs in companion animals has not been well studied. Fears and phobias are highly treatable problems in dogs (and cats).[32,33]

Separation Anxiety

Separation anxiety is one of the most common reasons that a dog urinates and defecates in the home. Elimination as a result of separation anxiety occurs when the dog is separated from the owner—either by a closed door or by being left alone in the home. Dogs with a separation-anxiety problem may be well housebroken when the owners are home or when the dogs have access to the owners. These anxious dogs often vocalize and/or become destructive as well as eliminate. The onset of the behaviors is usually close to the time when the owners depart—often within minutes—and occurs almost every time the dogs are left alone. Such dogs also show signs of anxiety prior to an owner's departure. Sometimes they will tolerate daily work departures but become distressed when an owner leaves again in the evening. Separation anxiety is a highly treatable syndrome when treated correctly.[34]

Housebreaking

Urination and defecation stemming from lack of

Dogs with housebreaking problems generally have a history of never having been well housebroken.

[a]Dunbar I: Personal communication, University of California, Berkeley, 1978.

[b]Doty RL: Personal communication, University of Pennsylvania, 1983.

housebreaking usually occur even though a dog has access to the owner. The frequency of the problem may be less when the owner is home because the dog is let outdoors often. Dogs with housebreaking problems generally have a history of never having been well housebroken. Housebreaking problems, however, may follow an illness when the animal could not help eliminating in the house and then does so as a residual habit. Dogs with housebreaking problems develop the habit of eliminating in locations and on surfaces that are undesirable to their owners.

Housebreaking techniques, for young as well as old dogs, involve establishing elimination location and surface preferences that are acceptable both to the owners and the dog.[35] Simultaneously, the tendency to eliminate in inappropriate areas must be reduced; to accomplish this involves not allowing the dog access to the undesirable areas or in some way preventing it from eliminating there. When it is likely that a dog is going to urinate and defecate, the dog should be taken outdoors or to the location where the owner wants the dog to eliminate. Usually the dog needs to eliminate immediately on return of the owner after an extended absence; soon after the dog awakens; 15 to 30 minutes after eating; an individually variable specific period of time after the dog's last elimination (this varies among dogs and puppies); and whenever the dog engages in preeliminatory behaviors, such as sniffing or circling.

When the owner is out of the house for a while and the dog cannot be monitored, it could be kept in a relatively small area where it is likely to inhibit its elimination. If the dog does urinate or defecate in this area, at least it is prevented from establishing a habit of elimination elsewhere in the house. The designated area might be where the dog regularly sleeps, a large crate, or a section of the kitchen. If the owner must be gone for a period of time longer than the dog will inhibit itself, newspapers should be placed in the area where the dog is likely to eliminate. Using newspapers may prevent the dog from establishing a habit of eliminating on inappropriate surfaces, such as rugs.

Punishment after the dog has eliminated is ineffective in teaching it to inhibit its elimination habits. For punishment to be effective, it must occur within a second after the onset of the behavior (0.5 second is optimal); should occur every time the animal engages in the behavior; and should be of appropriate intensity to stop the behavior. Punishment, even a few minutes after a dog has engaged in the behavior, does not meet these criteria. Dogs punished after they have eliminated in the wrong place do not learn to avoid eliminating there in the future. They are likely only to learn to be afraid of their owners, people approaching, and hands reaching toward them. Such dogs sometimes become fear biters.

Owners frequently state that their dog knows it has done something wrong because only when it has urinated and/or defecated in the home does it greet the owner looking "guilty." The guilty look is really a composite of submissive and fearful expressions. The dog has simply learned that when urine and/or feces are present and the owner returns, it gets punished. Therefore, only when urine and feces are present does it manifest submissive/fearful behaviors.

Some dogs develop firmly entrenched habits of eliminating indoors. They will not urinate and defecate outdoors. Even if the original cause of the reluctance to eliminate outdoors (for example, fear of traffic noises) is corrected, the dog may still inhibit elimination outdoors. In such cases, administration of a diuretic and/or laxative may be helpful in overcoming the dog's inhibitions. After it has eliminated outdoors several times, the drugs should be discontinued. The dog should, of course, be watched constantly while indoors and prevented from eliminating inside while it is developing a preference for outdoors.

Urination Problems

Differential diagnosis of urination problems should include disease, pathology, separation anxiety, housebreaking, fear of eliciting stimuli, urine marking, excitement-related urination, and submissive urination.

Urine Marking

Urine provides the most common source of chemicals used in scent marking.[4] On the whole, male mammals engage in more scent marking than females do; this behavior is facilitated by testosterone and decreased by castration. In a retrospective survey of 10 adult dogs presented for urine marking in the home, castration suppressed or greatly reduced urine marking in seven of them.[19] The effect of castration was independent of the age of the dog or the length of time since the dog began urine marking. Progestins also tend to suppress urine marking.[36,37]

Urine marking can be exhibited by an intact male dog that is otherwise well housebroken. The amount of urine involved is usually small and is deposited in a few regular locations, generally on vertical objects. Urine marking might occur in response to visiting dogs, an estrous bitch in the vicinity, or arousing stimuli, such as visitors or new objects in the house.

The age at which dogs begin to urine mark varies. Of dogs presented for urine-marking behavior problems in the home, about 10% began manifesting the behavior by 3 months of age, 20% by 6 months, 40% by 12 months, 70% by 1½ years, and 90% before 2 years of age[38] (Figure 2).

If urine marking occurs in only one or two locations, placing food or water in those locations may inhibit the dog from marking there and it may not eliminate elsewhere. If the stimuli that elicit urine marking can be identified (e.g., new objects in the home or mail carriers) the animal might be counterconditioned, that is, taught to assume a behavior incompatible with urine marking. Because the owner or trainer is usually directly involved in the situation in which this learning occurs, the animal may only learn to inhibit its urine marking in the presence of that person. When unsupervised, it may again urine mark.

The areas that the dog marks should be thoroughly cleaned. Studies of wolves indicate that urine deposits elicit re-marking by the same wolf.[39] In males, the urine of strange male dogs elicits more investigation and marking than the urine of a familiar male dog does; further, the latter evokes a greater response than the male's own urine does.[22] In the studies involving dogs, however, it is unclear whether the male dogs were stimulated to urine mark by the odor of their own urine or were merely re-marking the location per se.

Females in estrus urinate two and one-half times as often as anestrous females. Although they do not use the typical masculine elevated leg-lift posture, they are probably engaging in a form of urine marking. If a bitch tends to urine mark while in estrus, the owner could consider spaying as a means of controlling the behavior. A female that urinates in the house year-round is unlikely to be doing so in relation to estrus and, therefore, spaying is unlikely to resolve the problem.

Although dominant and territory-holding wolves tend to urine mark more than juveniles, subordinates, and lone wolves, no information is available on the frequency of urine marking in dominant and subordinate dogs. Because dogs travel through nonterritorial areas more frequently than wolves do, dogs of all social ranks may urine mark considerably.

Submissive Urination

Young animals and extremely submissive older animals may urinate in response to a returning owner or to someone who moves in a threatening or dominant manner. Usually the dog crouches, assuming a semisquat posture, and may present its inguinal area to the dominant individual or roll over into lateral recumbency, elevating a rear leg. Other concurrent signs indicative of submission are lowering of the head and the neck, flattening the ears against the head, and retracting the lips horizontally into a submissive "grin."[7]

Submissive urination can be treated in several ways. In young animals, sometimes doing nothing, that is, neither punishing or rewarding the dog, is the best approach until the dog gradually matures and is able to control its bladder.

Another approach is to determine the specific stimuli that evoke the behavior, such as the number of pats on the head, a direct stare plus an excited voice, or a stare plus a touch, and avoid the stimuli until the dog matures. The owner can greet the dog but avoid direct eye contact. The owner might also speak softly and only for a very short time.

In submission-greeting urination problems, the owner could encourage the dog to shift out of a greeting "mood" into another competing "mood," such as play or feeding. For example, the owner might immediately throw a ball, engage in a gentle tug-of-war game, or give the dog a biscuit. The dog should then calmly be ushered outside. Once it is outside, play or feeding should cease, and the owner can allow the dog to urinate in a normal manner. When the puppy has an empty bladder, the owner can satisfy the puppy's need to be greeted but reduce threatening or dominant signals by assuming a crouched posture and gently petting the dog as well as talking to it softly. As the dog gets older, the owner can gradually increase the directness, intensity, and immediacy of greeting behaviors and phase out the counterconditioning treatment involving play or feeding.

Punishment is contraindicated; it is likely to increase the tendency or motivation of the dog to indicate submission and hence to urinate more readily. When an owner reaches down to pet and reassure a dog, there are two potential side effects. Looking directly at the dog, reaching, and touching it could be interpreted by the dog as dominant or threatening gestures and thus elicit more submissive urination. Petting and talking to the dog in a friendly manner could also reinforce the urinary behavior problem.

Adult dogs that persist in submissive urination either at greeting times or when reached for at other times may require very specific counterconditioning procedures.[40] For instance, the dog can be taught to assume a different posture (e.g., standing) and to inhibit urination as the eliciting stimuli are presented gradually. This type of treatment technique usually involves the help of a professional animal behaviorist.

Direct physical punishment of dogs with submissive urination problems should be avoided *in all situations* because such discipline is likely to increase submissive behaviors. Methods other than punishment can be developed for treating most behavior problems; pun-

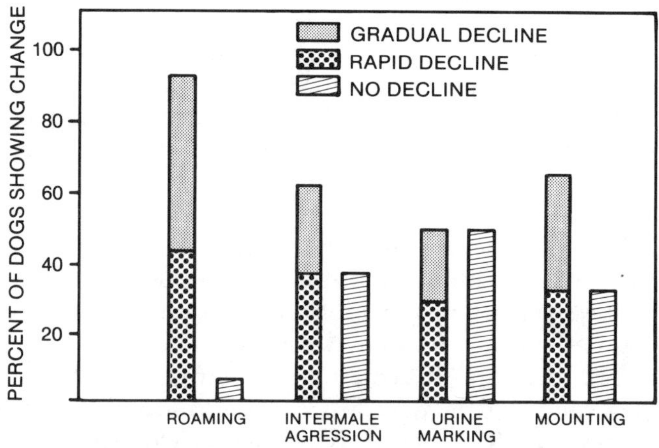

Figure 3—Percentage of dogs experiencing rapid decline, gradual decline, or no change in roaming, aggressive interactions with other males (intermale aggression), urine marking in the house, and mounting of other animals or people after castration. (From Hopkins SG, Schubert TA, Hart BL: Castration of adult male dogs: Effects on roaming, aggression, urine marking, and mounting. *JAVMA* 168:1110, 1976. Reprinted with permission.)

ishment rarely needs to be used. After a submissive urination problem is corrected, punishment techniques—properly implemented—might be used for other problems.

In extreme cases, the stimuli that elicit submissive urination must be identified. Such stimuli might be the owner reaching with an arm, actually touching the dog, or simply looking at the dog. When the stimuli are identified, the dog can be counterconditioned, that is, taught a competing response incompatible with submissive urination. The stimuli must be identified, arranged in an order on a gradient, and presented to the dog in a nonthreatening manner. First, the lowest level stimulus is presented, then the next highest, and so on, gradually metamorphosing into a full-intensity stimulus. A competing response might be standing with the head up and the tail wagging as opposed to the head lowered, ears back, and squatting posture. The dog is first taught to stand and stay for tidbits or some other positive reinforcement. For example, if a dog only submissively urinates when men approach it, then it can first be taught to stand when women approach and reach for it. Next the dog may be exposed to a man who does not look directly at it or try to touch it. Eventually, after many sessions, the dog can be expected to stand as a strange man enters the house, approaches, reaches, touches the dog, and interacts with it. Cases of submissive urination are very treatable.

Urination in Relationship to Excitement

Independent of submissive urination, when young dogs are highly aroused, such as when greeting the owner or engaging in vigorous play, they may be unable to inhibit urination. They may urinate in a standing position or even while walking.

One approach to this problem is to identify the stimuli or situations that elicit the urination and not expose the dog to the conditions when it has a full bladder. Alternatively, the animal can be counterconditioned and taught to relax in these circumstances. Possibly some drugs that decrease the tone of the bladder would raise the threshold of excitement urination.

Punishment techniques are usually unsuccessful because the puppy simply is incapable of inhibiting its bladder when excited. Punishment related to excitement urination may lead to submissive urination or other negative side effects, such as anxiety or aggression.

Perhaps the most practical approach to the problem of excitement urination is to allow the animal to urinate before it engages in vigorous play or greets a returning individual or entering visitor. As a further safety precaution, these activities should be manipulated so that they occur in locations where urination is acceptable and are avoided where urination is unacceptable. Puppies usually outgrow excitement urination problems within a few months.

Defecation Problems

Defecation problems can have the same causes as urination problems, with the exception of marking behaviors. Although dogs sometimes deposit feces on vertical objects (e.g., against trees, fence posts, or on the tops of bushes), fecal marking does not appear to be a common behavior problem in dogs.

Complications

Sometimes a dog urinates and defecates in the house for more than one reason. For example, the dog that urine marks in the home may also begin to lose its housebreaking habits; or dogs that eliminate in the home as a consequence of separation anxiety might eventually begin to mark in the home as well. The odor or presence of urine and/or feces might elicit or facilitate elimination for other reasons, particularly if more than one dog lives in the house.

REFERENCES

1. Voith VL: Human/animal relationships, in Anderson RS (ed): *Nutrition and Behaviour in Dogs and Cats.* Oxford, Pergamon Press, 147-156, 1984.
2. Voith VL: Clinical animal behavior. *California Vet* June:21-25, 1979.
3. Mugford RA: Behaviour problems in dogs, in Anderson RS (ed): *Nutrition and Behaviour in Dogs and Cats.* Oxford, Pergamon Press, 207-214, 1984.
4. Doty RL: Scent marking in mammals, in Denny MR: *Comparative Psychology: An Evolutionary Analysis of Animal Behavior.* New York, John Wiley & Sons, pp 385-399, 1980.
5. Kleiman D: Scent marking in the canidae. *Symp Zool Soc London* 18:167-177, 1966.
6. Gosling LM: A reassessment of the function of scent marking in territories. *Z Tierpsychol* 60:89-118, 1982.
7. Fox MW: *Behavior of Wolves, Dogs and Related Canids.* New York, Harper & Row, 1971.
8. Schenkel R: Expression studies of wolves. *Behaviour* 1:81-129, 1947.
9. Schenkel R: Submission: Its features and functions in the wolf and dog. *Am Zool* 7:319-330, 1967.
10. Rheingold HL: Maternal behavior in the dog, in Rheingold HL (ed): *Maternal Behavior in Mammals.* John Wiley & Sons, New York, 169-202, 1963.
11. Ross S: Some observations on the lair dwelling behavior of dogs. *Behaviour* 2:144-162, 1950.
12. Scott JP, Fuller JL: *Genetics and the Social Behavior of the Dog.* Chicago, The University of Chicago Press, 1965.
13. Beach FA: Effects of gonadal hormones on urinary behavior in dogs. *Physiol Behav* 12:1005-1013, 1974.
14. vom Saal FS: The intrauterine position phenomenon: Effects on physiology, aggressive behavior, and population dynamics in house mice, in Flannelly KJ, Blanchard RJ, Blanchard DC: *Progress in Clinical and Biological Research, 169: Biological Perspectives on Aggression,* 1984.
15. vom Saal FS, Bronson FH: Sexual characteristics of adult female mice are correlated with their blood testosterone levels during prenatal development. *Science* 208(9):597-599, 1980.
16. vom Saal FS: Variation in phenotype due to random intrauterine positioning of male and female fetuses in rodents. *J Reprod Fertil* 62:633-650, 1981.
17. Berg IA: Development of behavior: The micturition pattern in the dog. *J Exp Psychol* 34:343-368, 1944.
18. Sprague RH, Anisko JJ: Elimination patterns in the laboratory

beagle. *Behaviour* 47:257-267, 1973.
19. Hopkins SG, Schubert TA, Hart BL: Castration of adult male dogs: Effects on roaming, aggression, urine marking and mounting. *JAVMA* 168:1108-1110, 1976.
20. Dunbar I, Buehler M, Beach F: Developmental and activational effects of sex hormones on the attractiveness of dog urine. *Physiol Behav* 24:201-204, 1980.
21. Dunbar I: Olfactory preferences in dogs: The response of male and female beagles to conspecific urine. *Biol Behav* 3:273-286, 1978.
22. Dunbar I, Carmichael M: The response of male dogs to urine from other males. *Behav Neural Biol* 31:465-470, 1981.
23. Dunbar I, Buehler M: A masking effect of urine from male dogs. *Appl Anim Ethol* 6:297-301, 1980.
24. Rothman RS, Mech LD: Scent-marking in lone wolves and newly formed pairs. *Anim Behav* 27:750-760, 1979.
25. Beach FA, Gilmore RW: Response of male dogs to urine of females in heat. *J Mammals* 30(4):391-392, 1949.
26. Doty RL, Dunbar I: Attraction of beagles to conspecific urine, vaginal, and anal sac secretion odors. *Physiol Behav* 12:825-833, 1974.
27. Doty RL, Dunbar I: Color, odor, consistency, and secretion rate of anal sac secretions from male, female, and early androgenized female beagles. *Am J Vet Res* 35(5):729-731, 1974.
28. Henry JD: The use of urine marking in the scavenging behavior of the red fox (*Vulpes vulpes*). *Behaviour* 61:82-106, 1977.
29. Harrington FH: Urine-marking and caching behavior in the wolf. *Behaviour* 76:280-288, 1981.
30. Ralls K: Mammalian scent marking. *Science* 171:443-449, 1971.
31. Wynne-Edwards VC: *Animal Dispersion in Relation to Social Behavior*. Edinburgh, Oliver and Boyd, 1962.
32. Voith VL, Borchelt PL: Fears and Phobias in Companion Animals. *Compend Contin Educ Pract Vet* 7(3):209-221, 1985.
33. Voith VL, Borchelt PL: Fear of Thunder and Other Loud Noises. *Vet Tech* 6(4):189-191, 202, 1985.
34. Voith VL, Borchelt PL: Separation Anxiety in Dogs. *Compend Contin Educ Pract Vet* 7(1):42-53, 1985.
35. Voith VL, Borchelt PL: Housebreaking: What Is a New Puppy Owner to Do? *Vet Tech* 6(6):288-290, 1985.
36. Evans JM: Current thoughts concerning the control of hypersexuality in dogs, with particular reference to the role of progestogens. *Proceedings of Refresher Course for Veterinarians*, no. 37. Sydney, Post-Graduate Committee in Veterinary Science, University Press, 1978.
37. Voith VL: Possible pharmacological approaches to treating behavioural problems in animals, in Anderson RS (ed): *Nutrition and Behaviour in Dogs and Cats.* Oxford, Pergamon Press, 1984.
38. Borchelt PL: Development of behaviour of the dog during maturity, in Anderson RS (ed): *Nutrition and Behaviour in Dogs and Cats.* Oxford, Pergamon Press, 1984.
39. Peters RP, Mech LD: Scent-marking in wolves. *Am Sci* 63(2):628-637, 1975.
40. Voith VL: Animal behavior: Submissive urination. *Methods III*(4):5-7, 1980.

UPDATE

Figure 2 in the original article presents in cumulative percentages the number of dogs that exhibited a particular masculine behavior by a particular age divided by the *total number* of dogs that eventually exhibited that behavior. For example, 38 dogs exhibited mounting behaviors directed toward people or other dogs: 20% of them did so by 1 to 3 months of age, 50% by 4 to 6 months of age, 90% by 7 to 12 months of age, and all 38 (100%) had mounted by 24 months of age. In contrast, 51 dogs lifted a leg when urinating (leg-lifting) and about 5% first did so between 1 and 3 months of age. The cumulative percentage of leg-lifting generally parallels that of mounting, although a few dogs that leg-lifted had not yet begun by 24 months of age. The data were obtained over approximately 1 year from interviews of owners of male dogs that exhibited typically masculine behaviors.[1] These retrospective data provide information about the wide variation in age of onset of these typically masculine behaviors.

IMPORTANCE OF HOUSETRAINING TECHNIQUE

Advice on housetraining a dog is the single most valuable kind of advice for helping dog owners prevent myriad behavior problems and promote a satisfying human–animal bond. Housetraining is done by establishing a surface and location preference as soon as possible and by *preventing the dog from eliminating in unacceptable places* (thereby developing an inappropriate surface and location preference). Crating and confinement should be kept to a minimum; dogs isolated from human and canine companionship develop other problems, such as excessive attachment-seeking behaviors (pestering) or too-exuberant play (hyperactivity), when they are released from confinement.

Appropriate housetraining techniques avoid punishment. Direct physical punishment from the owner, even if the dog is "caught in the act" can lead to fearful and defensive behaviors. Punishment unrelated to "the act" results in even more-intense defensive reactions to being approached, reached for, or touched by a person. Although dominance aggression is the most commonly diagnosed behavior problem presented to animal behaviorists, fear-induced aggression is probably the most common type of aggression among pet dogs. Fearfully aggressive housedogs almost invariably have a history of difficulties in housetraining and were inappropriately and unpredictably punished by the owners.

A few months of the "inconvenience" of ensuring that the puppy always goes in the right place pays off in 10 to 15 years of proper elimination behavior by the dog. Interestingly, some of the most quickly and reliably housetrained dogs are those owned by people who know nothing about the theoretical basis for dog training, don't use a crate, and never punish the dog. They are, however, meticulous housekeepers. A fastidious housekeeper will watch a puppy constantly during the day and listen for signs of restlessness at night. Before the puppy has a chance to make a mistake, it's outside (or in a specific location in the home), where it should eliminate. Soon, it inhibits itself until it is in the right location and travels there on its own.

BLADDER AND SPHINCTER CONTROL

Some puppies seem to have delayed maturation of urethral sphincter competence and bladder control. These animals readily urinate when submissive or excited. Implementation of behavioral techniques as mentioned in the above article can usually manage the problem until the puppy can voluntarily control urine flow in these situations. Sometimes short-term use of medications that facilitate contraction of the urinary sphincter or lower the excitatory threshold of the bladder is helpful. Prolonged use of these drugs may mask underlying illness and delay diagnosis of disorders that might be treatable with surgery as opposed to life-long medication.

Phenylpropanolamine is the medication most often prescribed for functional incompetence of the urethral sphincter. Recommended doses include 12.5 to 50.0 mg orally two or three times a day[2] and 1.5 to 3.0 mg/kg orally three times a day.[3] Oxybutynin chloride, an anticholinergic agent with antispasmodic and local anesthetic effects on the bladder, has been used for older puppies and adult dogs with bladder-control problems. The reported dose is approximately 0.2 mg/kg orally two or three times a day.[3] Imipramine and amitriptyline, which are tricyclic antidepressants with anticholinergic properties, have been used with some success for controlling excitement and submissive urination in puppies.[a]

Although these medications may control expulsion of urine, it should be noted that humans taking phenylpropanolamine often report lightheadedness and dizziness, those taking imipramine or amitriptyline may experience hypertension and states of confusion, and those taking oxybutynin chloride report drowsiness and blurred vision. It would be difficult or impossible to detect that a dog is suffering from such side effects; and unless the owner can recognize that a dog is experiencing them, the dog would have to endure them. Behavior modification techniques are usually in the animal's best interest.

HOUSETRAINING PROBLEMS OF PUPPIES AND OLDER DOGS

The urge to urinate frequently despite a relatively small amount of urine in the bladder is often associated with infectious or noninfectious cystitis, small bladder, fibrotic bladder, and space-occupying lesions.[4] The urge to urinate may come on so quickly that the animal cannot inhibit urination until it has access to its normal elimination sites. Housetraining difficulties related to medical causes often stem from urinary tract infections, urinary sphincter incompetence, and bladder dysfunction.[3–6]

Congenital anatomic abnormalities must also be considered when puppies are incontinent; such anomalies as ectopic ureters, abnormal urethral conformation, or abnormal size or position of the bladder occasionally occur. For example, increases in abdominal pressure may be transmitted to an intrapelvic bladder body instead of to the neck of the bladder, thus increasing intravesicular pressure and promoting urine leakage.[5] Acquired anatomic abnormalities can result from inflammation or postsurgical adhesions following such abdominal procedures as ovariohysterectomy, cystotomy, and prostatic surgery.[4]

It is a common misconception that if a puppy or dog doesn't urinate during the night or when kept in a crate but

[a]Amy R. Marder, VMD, personal communication, Cambridge, Massachusetts, 1995.

Figure 1—If a crate or pen is used to facilitate housetraining, it should be large enough that the puppy is comfortable and can move about.

does eliminate when loose during the day then the cause must be "a behavior problem" and unrelated to a medical cause. Urine production can be affected by diurnal and sleep cycles; the animal may drink less because it has no access to water or because it is asleep; the animal is unlikely to be active or to become excited while it is confined and consequently never exceeds the competence of a compromised urethral sphincter or bladder. It is usually abnormal for puppies older than 4 to 5 months of age to urinate every 3 to 4 hours. Dogs and older puppies that urinate frequently during the day, have difficulty controlling urination when excited or submissive, or are incontinent should be examined medically.

Another misconception is that the absence of white blood cells or bacteria from urine sediment rules out bacterial urinary tract infections.[6] Bacteria must reach a certain concentration before they are likely to be detected, and the degree to which white blood cells are present varies with concurrent systemic disease. Detection of urinary tract infection may require a quantitative urine culture. One "negative" urinalysis in the face of persistent urination problems suggests further medical workup.

A minimum workup when a urinary tract problem is suspected includes a physical examination, vaginal examination, urinalysis, urine culture, and neurologic evaluation. Further diagnostic techniques include survey abdominal radiographs, contrast radiographs, and colposcopic procedures. Cystourethroscopic examinations and biopsies can be done. Intravesicular (intrabladder) and urethral pressure profiles can also be determined. Such workups often pinpoint the cause of the problem, and treatment can be specifically directed to the cause. Some of these disorders can be corrected surgically.

Dogs with urination problems should not just be relegated to the backyard. Pollakiuria (frequent urination) or incontinence often indicate that the animal's defenses against urinary infection are compromised.[3] Bacterial urinary tract infections as a result of contamination by fecal or periurethral organisms commonly occur in animals with impaired urinary tract defense mechanisms. Chronic infections of the lower urinary tract often lead to urolith formation and infection of the upper urinary tract and may ultimately result in kidney failure.[6]

INCONTINENCE AMONG GERIATRIC DOGS

Recently, nicergoline hydrochloride and selegiline hydrochloride (also known as L-deprenyl) have been reported to improve urination behavior problems of geriatric dogs. At present, the mechanisms of action of these drugs on behavioral problems of geriatric dogs are speculative.

Nicergoline hydrochloride (Fitergol®, Rhone Merieux) has been approved in the United Kingdom for "improvement of ageing related disorders in dogs, particularly those of behavioural origin."[7] One of the reported benefits of the drug is an improvement in housetraining. Nicergoline hydrochloride blocks α_1- and α_2-adrenergic receptors (thus increasing cerebral vasodilation), blocks serotonin and dopamine receptors, and putatively has neuroprotective actions.

Selegiline hydrochloride (which is marketed for treatment of human Parkinson's disease) reportedly stopped or reduced inappropriate urine elimination by 16 of 19 elderly dogs. Inappropriate urine elimination was defined as complete voiding in inappropriate locations, with or without urine dribbling.[8] Selegiline hydrochloride inhibits monoamine oxidase type B, facilitates dopaminergic activity, and decreases the production of free radicals.[9] Its metabolites include amphetamine and methamphetamine, which have pharmacologic actions of their own.[10] Selegiline hydrochloride is currently being evaluated for a spectrum of canine geriatric disorders, including urinary incontinence.[11]

REFERENCES

1. Borchelt PL: Development of behaviour of the dog during maturity, in Anderson RS (ed): *Nutrition and Behaviour in Dogs and Cats.* Tarrytown, NY, Pergamon Press, 1984, pp 189–197.
2. Marder AR, Voith VL: Canine behavioral disorders, in Morgan RV (ed): *Handbook of Small Animal Practice*, ed 2. New York, Churchill Livingstone, 1992, p 1251.
3. Lane IF, Lappin MR: Urinary incontinence and congenital urogenital anomalies in small animals, in Bonagura JD (ed): *Kirk's Current Veterinary Therapy XII, Small Animal Practice.* Philadelphia, WB Saunders Co, 1995, pp 1022–1026.
4. Moreau PM, Lees GE: Incontinence, enuresis, nocturia, and dysuria, in Ettinger SF, Feldman EC (eds): *Textbook of Veterinary Internal Medicine.* Philadelphia, WB Saunders Co, 1995, pp 164–169.
5. Lane IF: Pediatric urologic problems. *Proc 13 Annu ACVIM Forum*:505–508, 1995.
6. Lulich JP, Osborne CA, Bartges JW, Polzin DJ: Canine lower urinary tract disorders, in Ettinger SF, Feldman EC (eds): *Textbook of Veterinary Internal Medicine.* Philadelphia, WB Saunders Co, 1995, pp 1833–1861.
7. *The Case for Fitergol.* Harlow, Essex, England, Rhone Merieux Ltd, 1995.
8. Ruehl WW, DePaoli A, Bruyette DS: Treatment of geriatric-onset inappropriate urine elimination in elderly dogs (abstract). *J Vet Intern Med* 8(2):178, 1994.
9. Ruehl WW, Bruyette DS, DePaoli A, et al: Canine cognitive dysfunction as a model for human age related cognitive decline, dementia and Alzheimer's disease: Clinical presentation, cognitive testing, pathology and response to l-deprenyl therapy. *Prog Brain Res*, in press, 1995.
10. *PDR: Physicians' Desk Reference*, ed 47. Montvale, NJ, Medical Economics Data, 1993, p 2430.
11. Ruehl WW: Rationale to develop the investigational drug L-deprenyl for use in pet dogs (abstract). *Newsletter Am Vet Soc Anim Behav* 15(1):4–5, 1993.

Elimination Behavior Problems in Cats

Peter L. Borchelt, PhD
Animal Behavior Consultants, Inc.
Forest Hills, New York
The Animal Medical Center
New York, New York

Victoria L. Voith, DVM, PhD
Department of Clinical Studies
School of Veterinary Medicine
University of Pennsylvania
Philadelphia, Pennsylvania

Elimination in inappropriate places is one of the most common behavior problems reported by cat owners. The actual incidence of this type of problem is surprisingly high in view of the reputation cats have for being "clean animals." In a questionnaire,[1] information about behavior problems was obtained from more than 800 owners of pet cats. Forty-seven percent of cat owners reported cats that engaged in one or more behavior problems. Of these owners, 24% reported that their cats eliminated out of the litter box at least some of the time. Thus, based on this sample, a reasonable estimate would be that approximately 10% of pet cats exhibit an elimination behavior problem at some time.

Complaints about elimination behaviors account for about 66% of all telephone calls concerning feline behavior problems (473 of 721 calls) made to the Animal Behavior Clinic at the Veterinary Hospital of the University of Pennsylvania over a period of 24 months.[2] The majority of calls concerned neutered cats. There was no statistical difference between **the percent of males and females with regard to elimination problems and the percent of males and females presented to the Veterinary Hospital of the University of Pennsylvania for general medical and surgical** reasons. Because no data are available on the proportion of each breed in the general feline population, it is impossible to determine if any breeds are more likely to develop elimination behavior problems. The telephone data, however, provide breed-specific information about the percent of calls that were related to elimination and other behavior problems (Table I). For example, 79% of all calls about behavior problems in Persian cats were about elimination problems; while 37% of the calls about Siamese cats involved elimination. These data do not prove that the Persian breed is more likely than other breeds to develop elimination behavior problems, but merely that a telephone call about a Persian cat is highly likely to be about elimination behavior.

TABLE I
Breakdown of Telephone Calls About Specific Behavior Problems in Breeds of Cats[a]

Behavior Problem \ Feline Breed	Domestic Shorthair	Siamese	Persian	Domestic Longhair	Himalayan
Elimination	335 (60.4)	21 (36.8)	37 (78.7)	22 (61.1)	15 (71.4)
Aggression	137 (24.7)	10 (17.5)	3 (6.4)	10 (27.8)	3 (14.3)
Destruction	23 (04.1)	7 (12.3)	2 (4.3)	0 (0)	1 (4.8)
Overactivity	14 (02.5)	1 (01.8)	0 (0)	0 (0)	0 (0)
Vocalization	8 (01.4)	4 (07.0)	0 (0)	1 (2.8)	0 (0)
Fears	8 (01.4)	1 (01.8)	2 (4.3)	0 (0)	0 (0)
Ingestive	5 (01.0)	7 (12.3)	0 (0)	1 (2.8)	0 (0)
Self-inflicted injury/grooming	3 (00.5)	2 (03.5)	2 (4.3)	1 (2.8)	0 (0)
Other	22 (4.0)	4 (07.0)	1 (2.1)	1 (2.8)	2 (9.5)
TOTALS	555	57	47	36	21

[a]Number of calls received is given for each category, followed by percent in parentheses of total calls for that breed. Data were drawn from Reference 2.

This article discusses the three specific behavior problems involving elimination: marking, urination, and defecation. Procedures for diagnosing elimination problems are outlined, and a wide range of possible treatment techniques is described. Additional information can be obtained from other sources in the literature.[3,4] It is important to note that when a cat is presented for an elimination problem, the differential diagnosis should always include disease processes. This article discusses problems for which pathophysiologic causes have been ruled out.

Behavioral Description and Definitions

A cat can void urine in a *standing* or a *squatting* posture. In the standing posture (spraying), the hindlegs are straight, the tail is held upright and typically quivers, and the cat may step alternately with the hindlegs. The spraying posture (Figure 1) is used for urine marking. The urine is sprayed horizontally, usually onto vertical surfaces, such as walls or the sides of furniture. In a squatting posture (Figure 1), the hindlegs are flexed and the urine is deposited vertically onto horizontal surfaces—ideally in the litter box but often on floors, carpets, beds, or couch cushions.

The causative factors of the problem behavior can be determined (or assumed) based on a thorough description and history of the problem as well as the context(s) in which it occurs. *Urine marking*, either in the typical spraying posture or in a squatting posture, often involves expressing urine in amounts smaller than the amount voided during typical urination behavior. Urine marking has a communication function and often takes place in the motivational context of territorial, sexual, or agonistic behavior. Agonistic behavior refers to competitive encounters between animals, usually of the same species. Thus, urine marking can be related to competitive interactions between or among pets within the household, between household cats and outdoor cats, or (rarely) between household cats and people. These competitive interactions typically involve obvious aggressive and/or defensive displays by the cat.

Urination involves voiding the bladder for the purpose of eliminating waste products and not primarily as a communication signal. It occurs in a squatting posture and usually does not occur in the same context as marking does.

Defecation involves voiding feces in a squatting posture. It can serve a marking function in some species, but there is little evidence that it does so for domestic cats. In a study of free-ranging house cats and feral cats in a rural area of Sweden, Liberg[5] observed that cats regularly cover their feces in their home yards or gardens; but in 55% of the defecations observed in the field, feces were left exposed.

Marking

Marking behaviors involve communication between and among animals through the use of odor associated with deposits of urine, feces, or secretions from skin glands and sometimes visual communication signals as well. The exact function (or functions) of urine marking in domestic cats is unclear. In domestic cats, numerous observations[6] clearly indicate that a cat will approach a sprayed object, sniff the mark carefully, and may or may not mark over it. Then the cat usually moves on at a leisurely pace. There is no indication that a cat avoids or retreats from the urine mark of another cat.

Both male and female domestic cats spray, although males are more likely to do so than females. Females in estrus may spray or urine mark in a squatting posture.

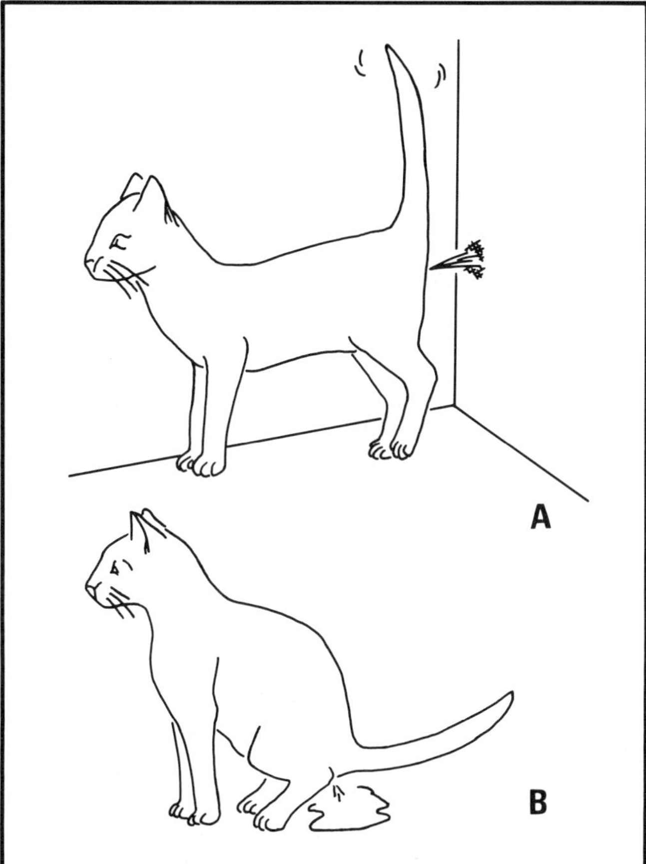

Figure 1—Postures assumed by a cat during elimination of urine are standing and spraying (**A**) and squatting (**B**). (From Borchelt PL, Voith VL: Diagnosis and treatment of elimination behavior problems in cats. *Vet Clin North Am [Small Anim Pract]* 12[4]:674, 1982. Reprinted with permission.)

Spraying as well as other forms of marking behavior are known to be influenced by androgens, and castration reduces the frequency. A retrospective survey of 23 owners who had one- to seven-year-old tomcats castrated for spraying behavior indicated that 78% greatly reduced or stopped spraying within two weeks after the operation.[7] Nine percent gradually reduced spraying between two and six months later. No correlation was found between age of cat and reduction in spraying behavior. There were, however, no data presented concerning duration of problem and effects of castration. A survey of 134 male cats neutered before 10 months of age (and presumably not spraying) indicated that 10% engaged in spraying behavior later in adult life.[8]

The development of urine marking behaviors in domestic cats apparently has not been investigated. Although most house cats never urine mark, our clinical data on spraying problems indicate that onset can occur at any age after sexual maturity and can develop in neutered male and neutered female cats. Individual variation in marking patterns is great, and the amount of urine expressed and frequency of marking may vary among cats. Urine marking generally occurs in several locations in a house.

Stimuli That Elicit Marking

Observations and descriptions of clinical cases of spraying in domestic cats indicate that a variety of causative factors can elicit the behavior. A common eliciting factor for spraying is the presence of one or more cats in the neighborhood outside the home. These outside cats may spray near or on the home (e.g., in bushes, by windows, or on doors or porches), and the odor may elicit spraying by the indoor cat. The indoor cat also may see or hear outside cats and exhibit clear signs of aggressive behavior, such as running to the windows or doors, hissing and growling, or intently watching the outside cat.

It is also possible for behavioral interactions among cats within the home to induce spraying by one or more indoor cats. For instance, the introduction of a new cat in the household may induce spraying by either a resident or the "intruder." Spraying also may be induced by aggressive encounters, for example, following an episode in which one cat chases the other out of specific areas.

It is rare, but possible, that a cat will spray as a result of the introduction or behavior of a person in the household. Spraying may occur if a resident cat is afraid, defensively aggressive, territorial, or highly aroused by a person in the home.

Urinating and Defecating

The elimination behavior system in domestic cats has not been well studied. In fact, to our knowledge, no quantitative observations of urinating and defecating behavior in domestic cats have been published. General observations of various species of cats,[6] our own observations, and behavioral descriptions from clients indicate that elimination in cats involves a sequence of behavior. Domestic cats typically approach a specific surface and/or location, dig with the forepaws to make a depression in the substrate, and then urinate or defecate, followed by sweeping and pawing motions of the forepaws to cover up the urine or feces. Contrary to popular belief, mother cats do not teach their kittens to use litter. Kittens start digging in and using clean, loose material for elimination at about four weeks of age without ever having observed their mother use the material.[a]

As is the case with other species-typical sequences of behavior, large differences exist among individual cats both in the frequency and occurrence of specific components of the sequence. For instance, some owners report that their cats do not typically dig and cover in the litter box when they urinate. Some cats simply walk into the box, squat to urinate, and then walk away. Some cats dig and scratch elsewhere, for instance, on the walls or floor. A cat may dig in the litter before eliminating and not afterward or vice versa. A cat may dig many or only a few times before eliminating and many or only a few times afterward.

[a]Borchelt PL, Voith VL: Personal observations, 1984-1986.

The precise function of "covering-up" behavior in a litterlike substrate is unknown. One hypothesis is that burying the feces reduces the possibility of parasite infestation. There is no evidence to support this theory, and it is unclear why the burying behavior would be more or less restricted to felids.

In the outdoor environment, cats have a variety of substrates or surfaces, such as loose dirt, leaves, or sand, in which to dig, eliminate, and cover. The range of surfaces used by free-living cats and the relative degree of preference for naturally occurring substrates have not been reported. For the domestic cat living in a human household, the range of surfaces in which the cat could dig and cover is extremely limited. Usually, the only substrate provided for the cat to dig in is commercial litter.

Although it is not clear why cats have evolved a sequence of behavior that functions to bury feces and sometimes urine, it clearly is a species-typical behavior. Individual variation in characteristics of species-typical behavior sequences is to be expected; and, thus, it is actually predictable that several factors could be involved in the development and expression of the elimination behavior sequence. The fact that some cats urinate and defecate on surfaces other than the litter provided in the litter box is not surprising. Although no laboratory studies have investigated this behavior, our observations of elimination problems in house cats has begun to reveal some of the factors that can lead to this problem.

Some cats that consistently use a litter box bury the feces in the litter but then continue their burying behavior even though the feces are already covered. These cats continue to dig and scratch in the litter even after the feces have disappeared from view. Other cats engage in much digging and pawing in the litter but leave the litter box without completely covering the feces. Observations of elimination behavior in domestic cats suggest that the digging-eliminating-covering sequence is not controlled or influenced only by visual feedback regarding the sight (or disappearance) of feces. That is, cats do not seem to be digging and covering in litter simply to make the feces disappear from view.

A major factor eliciting and controlling the digging-covering sequence appears to be tactile/kinesthetic feedback from the paws. No direct experimental evidence from either laboratory or field research is available to support this hypothesis, but observations of clinical cases and the efficacy of treatment techniques based on this hypothesis provide some support.

Stimuli That Influence Elimination

In addition to surface features, learned preferences and/or aversions to particular locations can play a role in determining where a cat eliminates. It is important to note that these factors can apply to either urinating or defecating or to both. A cat might, for instance, urinate consistently in the litter box but consistently defecate on another surface.

Surface Aversions and Preferences

Litter Aversion

A common factor in many cases of inappropriate elimination is a cat's aversion to the litter(s) provided. The cat with a litter aversion typically fails to dig, cover, and bury feces or urine and displays behaviors indicating a "dislike" of the litter. Besides not digging in the litter, the cat may stand on the edge of the litter box to avoid touching the litter with its paws, may shake its paws after contact with the litter, or may run quickly out of the litter box. The intensity of litter aversions can vary. A cat with a mild litter aversion may use the litter box consistently but fail to dig and cover in the litter as much as would be expected. Cats with an intense litter aversion rarely if ever use the litter. Such cats eliminate elsewhere, often on carpet, where they do engage in digging and scratching behaviors.

Surface Preference

Cats may learn to associate the tactile/kinesthetic features ("feel") of a surface other than litter with the act of eliminating, which results in the cat eliminating on that surface. For instance, cats that eliminate on carpets frequently have a history of first doing so on a bathmat placed near the litter box in the bathroom. Many owners have observed that initially the cat began to scratch and paw on the bathmat before or after it eliminated in the litter box. Then the cat began to eliminate on the bathmat. At this point, the owner removed the bathmat and the cat then proceeded to eliminate on carpets elsewhere in the house. Apparently, the act of scratching and digging on the bathmat became associated with the act of eliminating and the cat soon learned to prefer, seek out, and eliminate on materials that provided the same tactile/kinesthetic feedback.

Another common history involves a different type of learned surface preference featuring cats that dig and paw at the plastic sides or bottom of a litter box or at the smooth, hard surfaces of a nearby bathtub, toilet, wall, or floor. Some of these cats may then learn to prefer smooth surfaces and eventually eliminate in tubs or sinks or on wood, tile, or linoleum floors.

Occasionally, a cat eliminates on a new surface in the house (usually a new carpet or rug) very soon or immediately after it first scratches on it or even walks on it. This type of problem might be misinterpreted as the cat "marking" a new object in the house but, in fact, may be the result of the cat's preference for the new material, combined with mild aversion to its litter.

Location Preference

Many animals learn to associate specific behaviors with specific locations in the environment. An obvious example is the pet running to its food dish when the refrigerator is opened or the can opener is operated. Cats frequently learn to prefer certain spots or areas for sunning, sleeping, or grooming. A cat that urinates or defecates on carpets usually is reported to do so on only one or a few spots or areas (generally about a few square feet in size). Thus, even though a wall-to-wall carpet provides the cat the same tactile sensation all over, the cat may learn to associate eliminating with only one particular location on that surface.

The intensity of location preferences can vary. Sometimes, a cat's preference for eliminating in one location is so strong that the cat will eliminate there even when the surface features change. Owners have observed cats that have used the litter box consistently but, when the litter box was moved, then eliminated on the floor where the litter box had been located. This behavior indicates a very strong location preference. On the other hand, a cat's location preference may not be strong enough to overcome a surface preference. For example, placing a litter box at the location on a carpet where a cat has been eliminating does not always result in the cat using the box. Instead, the cat may simply eliminate on the carpet near the box.

The fact that cats usually develop a location preference for eliminating is the basis for the common but mistaken notion that the main reason they repeatedly eliminate in a specific place is because of the odor there. Some owners have rotated soiled carpet 180° and found that their cats continue to eliminate in the same location in the room rather than on the spot where the odor has been established. There is an inherent contradiction in the idea that cats prefer to eliminate at locations bearing the odor of urine or feces when they avoid litter boxes because of the odor of urine or feces. Cats that eliminate frequently on carpets are usually reported gradually to enlarge the location used, which may be related to the increasing intensity of odor in one place.

Location Aversions

Several factors can cause the litter box or the location of the box to be aversive to a cat. Obviously, a litter box that is not cleaned frequently enough will become aversive to a cat. An aversion to the odor could become associated with the location or even the box itself. A cat that has been frightened where the litter box is located could begin to eliminate elsewhere as a result. Although uncommon, occasionally a cat fails to use its litter box because it has been "captured" there to be medicated or picked up and petted by a person of whom it is afraid or has been frightened there by a loud noise or chased or threatened by another cat. The common but counterproductive technique of reprimanding a cat for eliminating in the house and then immediately placing it in the litter box can actually make the box aversive. Sometimes, a cover placed on the litter box reduces the attractiveness of the box by not allowing odors to dissipate or interferes with a cat's normal posture in the box or with its typical digging–covering behavior. A cat that soils its paws in the act of digging and covering diarrhea may quickly develop an aversion to the litter box and possibly even to the litter.

Further, a cat that experiences pain in the act of eliminating will often shift locations and/or surfaces, possibly eliminating on a wide variety of spots as if searching for a nonpainful place to eliminate. A cat's use of a variety of surfaces and locations may be a behavioral sign of a pathophysiologic problem.

Emotionally Related Elimination

It is commonly assumed that the major reason cats eliminate outside the litter box is the result of emotional responses to changes in the household. We have found only three circumstances that seem to fit this category: cats that squat to urinate in the context(s) associated with marking behavior, cats that eliminate as a result of the owners' absence, and cats so frightened that they hide in a location that offers no access to the litter box.

Occasionally, cats may squat to deposit small or large amounts of urine in a few or multiple locations in the home in the context of territorial, aggressive, or competitive behavior between cats (inside or outside the home) or people. Marking behaviors involving squatting may be difficult problems to diagnose because some cases involve subtle feline interactions that are difficult to observe or interpret. When a urination problem is refractory to a well thought out and diligent treatment program, it is reasonable to assume that marking behavior could be occurring.

Eliminating as a result of the owners' absence is probably related to separation anxiety, assuming that the litter box is clean. Separation anxiety in cats is much less common than in dogs but generally involves similar behaviors: eliminating, vocalizing, and escalated greeting and play when the owners return. Destructive behaviors, however, are not common in cats that have been left alone. The typical parameters of separation anxiety in cats differ from those of dogs. Intense separation behaviors are not displayed within a few minutes to an hour after the owners' departure as is typical of dogs. Cats generally do not exhibit separation anxiety during an 8- to 10-hour absence but may do so if the owner is absent for 24 hours or longer. Some cats eliminate immediately on the owners' return after a long absence. Stimuli that are consistently paired with an impending absence (e.g., a visible suitcase) may elicit preseparation anxiety and elimination behavior.

The fact that cats with a separation anxiety elimination problem frequently eliminate on an owner's bed, clothing, or favorite furniture is misinterpreted by many owners as "spite." It is more likely that, in a state of anxiety, the cat is attracted to locations that carry an owner's odor.

A cat frightened by a person or animal in the home may hide to avoid the frightening stimulus. If the cat does not come out to use the litter box, it may subsequently develop a new surface/location preference.

Treatment

Figure 2 depicts a systematic approach to diagnosing, deducing the possible cause(s) of, and treatment options

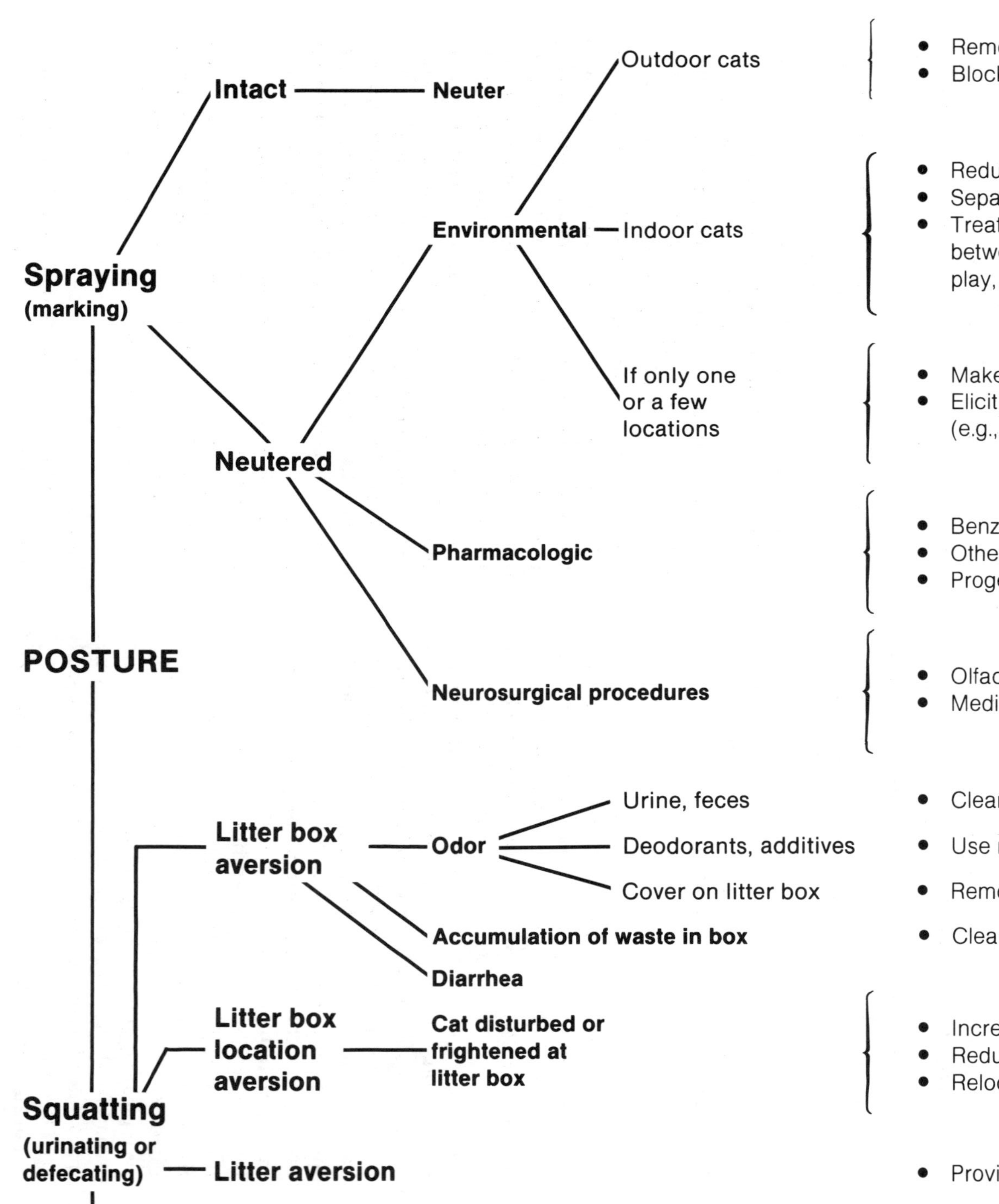

Spraying
(marking)
Intact
Neuter
Neutered
Environmental
Outdoor cats
Indoor cats
If only one or a few locations
Pharmacologic
Neurosurgical procedures
Remove outdoor cats from area
Block indoor view of outdoor cats
Reduce number of indoor cats
Separate (isolate) indoor cats
Treat conflict; change interaction between cats; desensitize fear, redirect play, etc.
Make location(s) aversive
Elicit another behavior in the location(s) (e.g., feeding, play)
Benzodiazepines
Other anxiolytic drugs
Progestins
Drug withdrawal without relapse may require environmental approaches
Olfactory tractotomy
Medial preoptic lesions
POSTURE
Squatting
(urinating or defecating)
Litter box aversion
Odor
Urine, feces
Deodorants, additives
Cover on litter box
Accumulation of waste in box
Diarrhea
Litter box location aversion
Cat disturbed or frightened at litter box
Litter aversion
Clean more frequently
Use nondeodorized litter, no additives
Remove cover
Clean more frequently
Increase number of litter boxes
Reduce stimulus
Relocate litter boxes
Provide alternate materials
May lead to new surface or location preference(s)

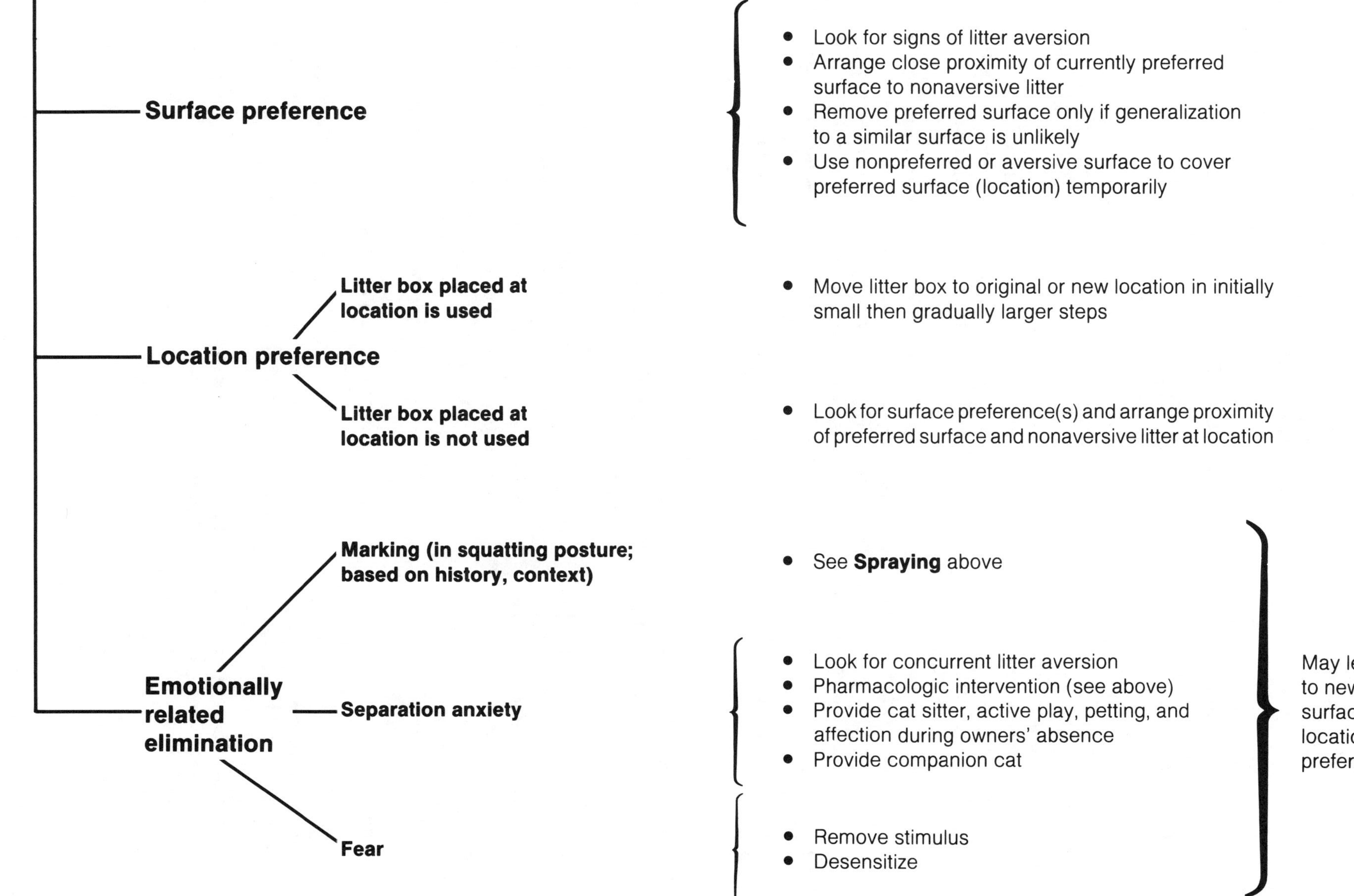

Figure 2—A schematic representation of the first considerations in diagnosis and treatment of feline elimination problems.

for feline elimination problems. This scheme, based on our clinical experience, is a formalized first approach to the process of diagnosis and treatment[9,10] and may change as new data are obtained.

Treatment of Marking Problems

If urine marking is a problem in the intact male or estrous female, neutering is a highly effective treatment. Neutered cats may continue to urine mark, however; and environmental, pharmacologic, and neurosurgical procedures are available.

Environmental

An environmental approach to treatment is the least invasive and, in some instances, is the easiest and quickest solution. The goals are to identify the eliciting stimuli and change the environment so that these stimuli no longer elicit urine marking. For instance, the sight of outdoor cats may be reduced by closing the door to a particular room or by placing translucent material over a portion of the window(s). Sometimes, increasing or decreasing the amount of time a cat spends outdoors affects the frequency of urine marking in the home. An environmental approach to treating urine marking requires close questioning of, and perhaps close observation by, the owner in order to specify the stimuli that are likely to elicit the behavior.

Pharmacologic

Progestins and diazepam have been demonstrated to reduce spraying in cats. Perhaps other benzodiazepines and anxiolytic drugs also may be effective.

Progestins have long been used for the treatment of spraying behavior and the suggested dose and schedule of administration vary.[11-13] Oral megestrol acetate is generally dosed at 5 mg/cat/day for one week and then the dose is systematically lowered to the lowest maintenance dose possible. Repository injectable medroxyprogesterone acetate is usually dosed at 100 mg for a male cat and 50 mg for a female cat and administered as necessary. Care must be exercised about inducing serious side effects if the drug accumulates. Progestins can cause increases in appetite and weight gain, mammary gland hyperplasia, hyperglycemia and adrenocortical suppression.[14] Prolonged treatment with medroxyprogesterone acetate has caused adenocarcinomas in female cats.[15] Progestins in dogs can also result in elevated plasma insulin levels; acromegaly; diabetes mellitus; decreased packed cell volume; reduced exercise tolerance; thickening of the gallbladder epithelium; and, in intact females, chronic endometrial hyperplasia or pyometra.[14,16,17] Use of progestins should not be taken lightly nor should the drugs be dispensed indefinitely or without monitoring the animal. Judiciously used, however, they can be safe and efficacious.

A retrospective survey of neutered cats treated with megestrol acetate or medroxyprogesterone acetate reported that the drugs were effective in about one third of the cases, were more effective at reducing spraying in males and in cats in single-cat households, and that the latter drug induced more lethargy and weight gain than the former. The data indicated that megestrol acetate was more effective than medroxyprogesterone acetate.[13]

A recent study evaluating the effectiveness of diazepam at reducing spraying by neutered cats indicated that the drug completely stopped spraying in 43% of the cases and reduced spraying to less than 25% of the presenting frequency in an additional 32% of the cats.[18] In the study, diazepam stopped spraying in all (11) of the cats that had not responded to progestins. Owners reported some lethargy, temporary mild ataxia, temporary increase in appetite, increased affection toward the owners, and a general reduction in anxiety. The dosage was 1 to 2 mg/cat/orally twice daily for two weeks, then one half of the initial dosage for one month followed by gradual reduction to the lowest possible maintenance dosage.

Periodically (about every two months) drugs used to suppress spraying behavior of cats should be withdrawn gradually. The cat may no longer be motivated to spray and the drug no longer necessary. Cats on long-term drug therapy should be constantly monitored for side effects by the owner and periodically evaluated by the veterinarian who has assumed responsibility for the drug therapy by prescribing the drug.

Neurosurgical Procedures

Lesions of the medial preoptic hypothalamic nuclei eliminate or greatly reduce copulatory behavior in males of most species, including dogs and cats. Lesions in this area also reduce urine marking in male dogs.[19] Medial preoptic lesions placed in nine neutered pet cats that still sprayed despite treatment with progestins were effective in greatly reducing or stopping spraying in all six male cats but not in the three females. Although the stereotaxic surgery could be performed safely and the cats recovered well from the procedure, side effects not previously observed in laboratory cats render the procedure inefficacious as a clinical procedure. Many cats became hyperphagic and/or developed "startle reactions" during sleep.[20]

Olfactory tractotomy has been reported effective in reducing or eliminating spraying by 6 of 11 male and 1 female cat that had not responded to progestin therapy.[21] In this procedure the olfactory tracts and caudal parts of the olfactory bulbs are aspirated from a dorsal approach through the frontal sinuses. It is unclear whether spraying is reduced because of anosmia or effects of bulbectomy unrelated to anosmia. Side effects reported were cutaneous emphysema over the incision site, transient anorexia, increase in appetite, finicky appetite, and increase in affectionate behavior toward the owners.

Treatment for Urination and Defecation Problems

More than one of the causative factors listed in Figure 2 can be involved in a particular elimination behavior problem. One factor may have initiated the problem and yet another maintains it. The practitioner should be familiar with, and question the client about, all of the possible factors.

A treatment directed at the factor(s) presently maintaining the problem may be successful temporarily, but the problem may recur if the initiating factor(s) are not addressed. For instance, probably the most common technique (other than inappropriate punishment[22]), implemented by owners is to confine a cat to a small room or crate it with the litter box, food, and water for several days or more until it "relearns to use the litter box." Often, the cat will use the litter box under these conditions only to resume eliminating as before after it is released.

Aversions to either the litter box or the litter box location may interact with surface preferences. A cat may begin to eliminate on a carpet because the litter was dirty or a cover was placed on the box. Because of an acquired preference for the carpet, however, the cat may continue to use the carpet even after the box is cleaned more frequently or the cover removed.

The cat's current surface/location preferences can be used to advantage in designing a program that leads ultimately to the cat again using the litter box. Sometimes this involves such techniques as relocation of litter boxes or arrangement of carpet materials that the owner may find objectionable. It is important to counsel the owner that these specific treatment procedures are temporary and needed only until the hypothesized causative factors(s) can be verified and then the next appropriate procedure tried.

Follow-up information from the client (where and on what surface the cat next eliminates and digs) is critical in order to assess the cat's response to a particular procedure. Because the treatment program often involves the cat learning to associate particular surface/location characteristics with elimination behavior, each step in the program may have to be introduced gradually and/or be in effect for a period of a week or two before the next step is introduced. For instance, if placing a litter box on the currently used location is effective, then moving the litter box immediately back to the original location is not likely to work. Instead, the litter box may have to be moved several inches each day for about one week. If the cat continues to use the litter box, then gradually larger moves can be made until the box is back to the original location.

Most feline behavior problems can be treated successfully. Sometimes only a minimum amount of effort on the part of the owner is required, but other times a considerable amount of diligence and a logical systematic trial-and-error approach are necessary. As more information is gathered concerning the behavior of cats in general and elimination behaviors in particular, the success rate of treating and preventing feline elimination behaviors should increase.

REFERENCES

1. Voith VL: Attachment of people to companion animals. *Vet Clin North Am [Small Anim Pract]* 15(2):289-295, 1985.
2. Voith VL: Analysis of 2500 telephone calls about behavior problems of dogs and cats. Paper presented at Animal Behavior Society Meeting, Raleigh, NC, 1985.
3. Borchelt PL, Voith VL: Elimination behavior problems in cats. *Compend Contin Educ Pract Vet* 3(8):730-737, 1981.
4. Borchelt PL, Voith VL: Diagnosis and treatment of elimination behavior problems in cats. *Vet Clin North Am [Small Anim Pract]* 12(4):673-680, 1982.
5. Liberg O: Spacing patterns in a population of rural free roaming domestic cats. *Oikos* 35:336-349, 1980.
6. Leyhausen P: *Cat Behavior.* New York, STPM Press, 1979.
7. Hart BL, Barrett RE: Effects of castration on fighting, roaming, and urine spraying in adult male cats. *JAVMA* 163:290-292, 1973.
8. Hart BL, Cooper L: Factors related to urine spraying and fighting in prepubertally gonadectomized cats. *JAVMA* 184:1255-1258, 1984.
9. Pollock RVH: Anatomy of a diagnosis. *Compend Contin Educ Pract Vet* 7(8):621-628, 1985.
10. Pollock RVH: Diagnosis by calculation. *Compend Contin Educ Pract Vet* 7(12):1019-1034, 1985.
11. Gerber HA, Sulman FG: The effect of methyloestrenolone on oestrus, pseudopregnancy, vagrancy, satyriasis and squirting in dogs and cats. *Vet Rec* 76:1089-1092, 1964.
12. Pemberton PL: Feline and canine behavior control: Progestin therapy, in Kirk RW (ed): *Current Veterinary Therapy VII, Small Animal Practice.* Philadelphia, WB Saunders Co, 1980.
13. Hart BL: Objectionable urine spraying and urine marking in cats: Evaluation of progestin treatment in gonadectomized males and females. *JAVMA* 177:529-533, 1980.
14. Chastain CB, Graham CL, Nichols CE: Adrenocortical suppression in cats given megestrol acetate. *Am J Vet Res* 42(12):2029-2035, 1981.
15. Hernandez FJ, Fernandez BB, Chertack M, et al: Feline mammary gland carcinoma and progestagens. *Feline Pract* 5(5):45-48, 1975.
16. Eigenmann JE, Eigenmann RY: Influence of medroxyprogesterone acetate (Provera) on plasma growth hormone levels and on carbohydrate metabolism. Parts I–II. *Acta Endocrinol* 98:599-602, 602-608, 1981.
17. Eigenmann JE, Venker-van Haagen AJ: Progestagen-induced and spontaneous canine acromegaly due to reversible growth hormone overproduction: Clinical picture and pathogenesis. *JAAHA* 17:813-822, 1981.
18. Marder AR: Diazepam for the treatment of spraying in castrated male and spayed female cats. Paper presented at Animal Behavior Society Meeting, Raleigh, NC, 1985.
19. Hart BL: Medial preoptic-anterior hypothalamic area and sociosexual behavior of male dogs: A comparative neuropsychological analysis. *J Comp Physiol Psychol* 86:328-349, 1974.
20. Hart BL, Voith VL: Changes in urine spraying, feeding and sleep behavior of cats, following medial preoptic-anterior hypothalamic lesions. *Brain Res* 145:406-409, 1978.
21. Hart BL: Olfactory tractotomy for control of objectionable urine spraying and urine marking in cats. *JAVMA* 179:231-234, 1981.
22. Borchelt PL, Voith VL: Punishment. *Compend Contin Educ Pract Vet* 7(9):780-791, 1985.

UPDATE

An aversion to litter is a common cause of inappropriate elimination in cats. A logical environmental manipulation is to find a litter material that the cat prefers (e.g., one the cat digs in and uses to cover up its urine or feces). Coarse clay-based litters vary with respect to deodorants and additives, absorbency, and amount of dust; but all have about the same particle size (from about 5 to 6 mm to dust). Some cases of litter aversion are resolved by switching to another brand of commercial clay-based litter. Several other types of litter material, such as wood chips, sand, and topsoil, have been used with varying degrees of success in cases of litter aversion.

PREFERRED LITTER MATERIALS

One study found that cats strongly prefer a fine-grained, clumping litter.[1] For 10 days, the elimination behaviors of 48 cats in a shelter were observed individually; the number of urinations and defecations from these cats in various litter materials were counted three times daily. The overall results indicate at least a 3:1 or better preference for fine-grained litter over typical coarse-grained litter and a general disinclination for litter material with large particles. Switching to a fine-grained litter is an easy environmental manipulation and has successfully corrected aversion problems with litter and even some cases of urine marking.

This study supports the clinical observation that cats' preference for litter depends on its texture—granularity or coarseness. The cats preferred a fine-textured clay to clay with larger particle sizes. They did not like fine-textured sand, like that used in sandboxes, much more than coarse clay, perhaps because of the weight of the particles (i.e., the kinesthesia associated with this substance). The cats generally avoided litters with large wood or paper particles.

Several types of fine-textured nonclay litters are now commercially available. Litter made from wheat is readily used by cats, controls odor, and clumps well. We found that litters made from the absorbent portion of corncobs, paper, or one of several types of wood flakes usually do not control odor or clump well. All of these nonclay litters are reportedly flushable and do not contain silica dust.

A RECENT STUDY

Horwitz[2] compared 100 cats presented for elimination problems and 47 cats presented without elimination problems. Several factors were found to differ significantly between the two populations: cats with elimination problems were more likely to have scented than unscented litter, were more likely to have had urinary tract disease, and were less likely to cover their eliminations. There were no differences in likelihood of household changes, in litter box cleaning routine, or in number of cats in the household.

CASTRATION AND SPAYING

Early spaying or castration (i.e., at 2 to 3 months of age)[3] of kittens and puppies has been advocated for population control and behavioral effects. No deleterious physiologic side effects have been reported to occur in cases where appropriate anesthesia is used and the surgeon is skilled.[4] Several prospective longitudinal studies on the comparison of behaviors of kittens spayed or castrated early with those of cats spayed or castrated at or after puberty are currently underway; within a few years, the results of these investigations should be available.

A report of retrospective surveys of owners of cats spayed or castrated at between 6 and 10 months of age indicated that about 10% of the male cats and 5% of the female cats still sprayed frequently later in life.[5] This study found no indication that spayed female cats that sprayed in adulthood were more likely to be from litters with male siblings than from all-female litters.

DRUG THERAPY

Informed Consent

None of the drugs discussed below are marketed for use in cats, nor are we aware of any rigorously controlled efficacy or safety data regarding the use of these drugs in cats. Owners should be informed of this and about the potential side effects of these drugs. The use of any drug, particularly extralabel use, requires that the client be properly informed about potential adverse reactions and that the patient's response be monitored.[6] It is recommended that the owner sign a consent form when extralabel use of psychoactive drugs is prescribed.[7] For more information about these drugs, mechanisms of action, and potential side effects, see the Simpson and Simpson article on pharmacotherapy.

Diazepam

Two studies have examined the efficacy of diazepam (1 to 2 mg orally twice a day) in suppressing spraying or marking behavior. Whereas Marder[8] reported a 75% success rate in a study that used 23 cats (19 males and 4 females), Cooper and Hart[9] reported a 55% success rate in a study that used 20 cats (14 males and 6 females). The discrepancy in the results may be because Marder limited her study to cats that sprayed in a standing posture whereas Cooper and Hart also included cats that were believed to mark in a squatting posture.

Investigators in both of these studies reported that owners observed many of the diazepam-treated cats to be ataxic initially. Some cats were reported to exhibit increases in appetite, lethargy, sleep, and friendliness to people and to exhibit less aggression toward other cats. Both studies also indicated that cats unresponsive to previous progestin therapy responded to treatment with diazepam.

Diazepam can result in dependency; therefore, the dose must be tapered off gradually (e.g., halve the therapeutic dose for 1 week and then halve it again for another week before stopping).

Fatal hepatotoxicity has reportedly occurred in cats given 1 to 2 mg of diazepam orally once or twice a day for behavioral problems.[10,11] Despite intensive intervention and supportive therapy, animals have died within 2 weeks of the initial dose.

Many of the affected cats showed signs of anorexia within 96 hours of initiation of therapy. In addition to increases in hepatic transaminases and hyperbilirubinemia, some cats also had increases in creatine kinase, low serum cholesterol, low glucose concentrations, and abnormalities consistent with disseminated intravascular coagulation. It

Figure 1—A cat avoiding touching the litter in its box.

is recommended that baseline alanine aminotransferase and aspartate aminotransferase activities be evaluated before daily therapy is initiated and 5 days after treatment has been started.[11]

This idiosyncratic hepatic response is probably rare but apparently usually fatal. No predisposing factors have yet been identified. Owners should be warned of the possible side effects of diazepam, especially hepatotoxicity, and told to immediately stop medication and seek veterinary help if the cat exhibits anorexia following initiation of therapy.

Buspirone Hydrochloride

Buspirone hydrochloride is an antianxiety medication that affects serotonergic and dopaminergic activity. Buspirone hydrochloride was reported to stop or reduce urine marking in 55% of 62 neutered cats (47 males and 15 females).[12] Depending on the cats' responses to the drug, the doses ranged from 2.5 to 7.5 mg orally every 12 hours for 3 to 12 months. None of the owners reported that the cats were ataxic at the onset of treatment. Sedation reportedly occurred in 4 cats; 5 cats were "agitated" after administration of the medication; 17% of the cats were friendlier to people; and 17% were more aggressive to other cats in the household. The recidivism rate was considerably less after treatment with buspirone hydrochloride than after diazepam. In light of its lack of potential to cause physical dependency and its relative safety, buspirone hydrochloride (compared with diazepam or progestins) is generally the drug of choice for spraying problems in cats.

Amitriptyline Hydrochloride

Amitriptyline hydrochloride is a tricyclic antidepressant, anxiolytic, antihistaminic, and sedative drug that has been reported anecdotally to be successful in treating urination problems in squatting and standing postures in cats.[8] The dose generally administered is 5 to 10 mg orally once a day.[8] No systematic trials involving this drug have been reported. Amitriptyline hydrochloride has more potential for side effects than diazepam or buspirone hydrochloride.

URGE INCONTINENCE

In "urge incontinence," the phase of urine retention in the bladder is shortened because the animal is motivated to urinate frequently.[13] Voiding is basically normal, but the urge to urinate can come on so quickly that the animal may not engage in its typical elimination sequence (e.g., scratching before eliminating or traveling to its normal elimination site). Some of the conditions that can result in urge incontinence are infectious or noninfectious cystitis, a small bladder, a fibrotic bladder, or space-occupying lesions.[13]

It has recently been proposed that interstitial cystitis (IC) may be a cause of feline lower urinary tract disease (FLUTD),[14,15] which in turn could result in urge incontinence. Interstitial cystitis in humans has many similarities to idiopathic cystitis in cats. Interstitial cystitis is characterized by chronic irritative voiding symptoms, sterile and cytologically negative urine, and cystoscopic findings of submucosal petechiae after hydrodistention of the bladder.

Amitriptyline hydrochloride is one of the treatments for interstitial cystitis in humans. Its mechanisms of action may be related to inhibiting histamine release and decreasing transmission of nociceptive receptors in the bladder. Its anticholinergic effects might also reduce bladder spasms and increase bladder capacity. It also may be working at a higher central level as an antianxiety agent. Interstitial cystitis in humans and feline lower urinary tract disease have been correlated with environmental events (e.g., stress).

Clinical signs of hematuria and dysuria in cats with unobstructive idiopathic feline lower urinary tract disease often subside spontaneously without treatment and then reappear.[16] There may be multiple causes, including viral and interstitial cystitis, for idiopathic feline lower urinary tract disease. The cyclic nature of idiopathic feline lower urinary tract disease and the bacterially and cytologically negative urine that may be found in affected cats suggest extreme caution in assuming that cats with a negative urinalysis and that deposit urine in various horizontal locations in the home are marking or are doing so because they are "anxious." Such cats may have interstitial cystitis or may be in a remission phase of a cyclic idiopathic feline lower urinary tract disease. The nature of these diseases underscores the importance of frequent urinalyses and more extensive urologic evaluations of cats with "a negative urinalysis" that persist in voiding outside litter boxes despite conventional behavioral therapeutic techniques.

REFERENCES

1. Borchelt PL: Cat elimination behavior problems. *Vet Clin North Am [Small Anim Pract]* 21(2):257–264, 1991.
2. Horwitz D: Factors affecting elimination behavior problems in cats: A retrospective study (submitted for publication). *JAAHA*.
3. Lieberman LL: A case for neutering pups and kittens at two months of age. *JAVMA* 191:518–519, 1987.
4. Lane TJ, Lieberman LL: Prepubertal gonadectomy long term study of puppies and kittens: A preliminary report (unpublished data), University of Florida, Gainesville, Florida.
5. Hart BL, Cooper L: Factors relating to urine spraying and fighting in prepubertally gonadectomized cats. *JAVMA* 184:1255–1258, 1984.

6. Davis LE: Adverse drug reactions, in Ettinger JS (ed): *Textbook of Veterinary Internal Medicine: Diseases of the Dog and Cat*, ed 3. Philadelphia, WB Saunders Co, 1989, pp 499–510.
7. Wilson JF: *Law and Ethics of the Veterinary Profession.* Yardley, PA, Priority Press, 1988, pp 243–248.
8. Marder AR: Psychotropic drugs and behavioral therapy. *Vet Clin North Am [Small Anim Pract]* 21(2):329–352, 1991.
9. Cooper L, Hart BL: Comparison of diazepam with progestins for effectiveness in suppression of urine spraying behavior in cats. *JAVMA* 200(6):797–801, 1992.
10. Levy J, Cullen JM, Bunch SE, et al: Aversive reaction to diazepam in cats (letter). *JAVMA* 205:156–157, 1994.
11. Center SA, Elston TH, Rowland PH, et al: Hepatotoxicity associated with oral diazepam in 12 cats. *Proc ACVIM* 13:1009, 1995.
12. Hart BL, Eckstein RA, Powell KL, Dodman NH: Effectiveness of buspirone on urine spraying and inappropriate urination in cats. *JAVMA* 203:254–258, 1993.
13. Moreau PM, Lees GE: Incontinence, enuresis, nocturia, and dysuria, in Ettinger SJ, Feldman EC (eds): *Textbook of Veterinary Internal Medicine* ed 4. Philadelphia, WB Saunders Co, 1995, pp 164–169.
14. Buffington CAT, Chew DJ: Does interstitial cystitis occur in cats? in Bonagura JD (ed): *Kirk's Current Veterinary Therapy XII: Small Animal Practice.* Philadelphia, WB Saunders Co, 1994, pp 1009–1011.
15. Buffington CAT, Chew DJ: Idiopathic lower urinary tract disease in cats—Is it interstitial cystitis? *Proc ACVIM* 13:517–519, 1995.
16. Osborne CA, Kruger JM, Lulich JP, Polzin DJ: Feline lower urinary tract diseases, in Ettinger SJ, Feldman EC (eds): *Textbook of Veterinary Internal Medicine*, ed 4. Philadelphia, WB Saunders Co, 1994, pp 1805–1832.

Eliminating Urine Odors in the Home

Veterinary Behavior Consultants
San Diego, California
Patrick Melese-d'Hospital, DVM, MA

Urinating, defecating, and marking in unacceptable locations account for an enormous number of behavior problems in dogs and cats.[1–3] These soiling problems often lead to such frustration on the part of the owner that the pet can lose its home or be banished to living outside. Once a pet becomes an "outdoor pet," the bond between the pet and the family often diminishes and a host of new problems can arise. For dogs, these new problems include barking, destruction, digging, and escaping from the backyard, while outdoor cats are prone to an increased risk of injury and infectious disease.

Properly dealing with elimination problems in pets necessitates intervention in three fundamental areas:

- **Medical evaluation** to rule out or treat underlying physical disorders including infectious, inflammatory, neurologic, developmental, neoplastic, and hormonal problems. A thorough history, physical examination, and appropriate laboratory tests should be performed. Different problems will obviously suggest different levels of medical workup. For example, leg lifting with urine marking on furniture corners exhibited by a 1-year-old sexually intact male dog would warrant little or no medical workup. On the other hand, a middle-aged neutered male cat presenting with an acute onset of frequent, painful urinations in multiple, unusual locations in the home would suggest an underlying medical cause.
- **Neutralization of odor source** is necessary to reassure owners that their floor coverings and furniture can be saved and household odors controlled while the underlying problem is remedied. Odor control may be critical in keeping the pet with the family while the cause of the problem is being resolved. In addition, odor control is important in removing any olfactory or vomeronasal organ mediated cues that might attract the pet back to the inappropriate area and stimulate elimination.
- **Changing the inappropriate behavior** with a treatment plan that can include physical and social environmental changes; humane, effective behavior modification techniques; and surgical or pharmacological intervention (such as neutering or starting antianxiety medication).

Figure 1—Soiled carpet.

All three of these steps are important in successfully helping our clients resolve these serious problems.

BEHAVIORAL SIGNIFICANCE OF URINE DEPOSITION

Urine is used in both dogs and cats to mark territory. Cats often spray urine against a vertical surface whereas dogs use various forms of leg lifting to mark objects with urine.[4–6] Several studies have found that the composition of urine eliminated when voiding may be different than that eliminated when marking. This has been shown in lions and tigers, which spray a mixture of urine and anal sac secretions,[7] as well as in foxes, which spray urine containing small amounts of substances, absent in voided urine, that are used for marking purposes.[8] The greater interest that domestic cats show in sprayed versus voided urine has been well demonstrated.[8,9] This fact strongly implies that additional substances have been added to sprayed urine. Potential sources for these pheromones in cats include anal sacs, the prostate gland, the bulbourethral gland, and anal glands[7] in the anal skin below the tail. Tomcat urine has a particularly strong odor which is dependent on the presence of circulating levels of testosterone found in intact males. The odor (pheromone?) may be a sulfur compound, perhaps the amino acid felinine, which contains sulfur and enters the urine through the kidneys.[6] Information from these urine marks may be gleaned by the animal both via olfaction as well as performing flehmen (a gape/lip-curl-like motion occurring after investigation of a scent which is thought to facilitate transport of sampled material to be evaluated by the vomeronasal organ located in the nasal septum[10]). It has been shown that cats sniff sprayed urine more frequently than voided urine when each was deposited by a stranger tomcat.[8] The importance of urine odor for cats is shown by the tendency of tomcats to explore fresh urine marks more than the older ones, but to spray urine on the previous mark when the original deposit is more than two days old. This "overmarking" suggests a purpose for urine spraying: to freshen up the signal for other cats, which better allows several cats to share the same territory, but use it at different times ("time sharing").[11]

Figure 2—Cheetah urinating.

From the above discussion it is clear why odor elimination is critical to stopping ongoing soiling problems, especially for urine marking. The tendency to spray more on fainter urine deposits could stimulate a spraying cat to mark partially cleaned up urine deposits even more vigorously because, to the cat, they may appear to be old and fading marks. Using the more effective urine neutralizer products discussed below to completely eliminate the smell may be critical to stopping a spraying cat from "overmarking" inadequately treated spots. Even with problems that appear purely medical in origin, such as geriatric or other incontinence-related disorders,[12] soiled areas need to be detected and neutralized while the underlying cause is being treated. Any remaining odors can lead to a new soiling problem or perpetuate an existing one.[6,13,14]

The importance of olfaction in the persistence of elimination problems is strongly suggested by studies showing that olfactory tractotomy (causing the subject to become anosmic) stops many cats from continuing to spray urine in their homes.[15] In wolves, for exam-

Figure 3—Cat with rug turned upside down to show urine stains.

ple, it has been observed that areas marked with urine will trigger re-marking of the spot by the same wolf.[16]

Although olfaction is important, location and position preference also are very powerful, especially in cats. This is demonstrated by cases in which owners have rotated a soiled rug 180 degrees to find that the cat chooses to continue to soil the corner in the previous location, not the same part of the rug that still has the odor.[17] It must be remembered that using the owners' perception of urine odor as they stand some five to six feet above the soiled area is a poor indicator of the scent that pets sniff an inch or less from the surface. Cats and dogs may perceive minute quantities of contamination via olfactory[18] or vomeronasal mediated detection.[8,11]

Veterinary management consultants increasingly caution against relegating odor control or other "minor" behavior problems to nonpractitioners because such attitudes may send the message that veterinarians are not comprehensive pet care professionals.[19] Nonveterinarian behavior practitioners can face the same problem if they do not grasp the importance of odor control to their client. The remainder of this article deals with the critical subject of odor source detection and neutralization in an effort to enable practitioners to advise their clients more effectively. Sources of all odor control products mentioned in this article that are not commonly available are listed in the Appendix.

DETECTION OF THE SOILING PET IN A MULTIPET HOUSEHOLD

Because many households now have more than one pet, detection of the soiling culprit can be difficult. A detection method used with cats calls for using sodium fluorescein, a harmless, water-soluble dye available as a liquid (Fluorescite® Injection 10%) or ophthalmic strips used in small animal practice to stain the eye in search of corneal injury (Fluor-i-strip®). The dye is administered to the suspected cat either by subcutaneous (SQ) injection (0.3 ml SQ) or orally (as 0.5 ml solution or in two 9-mg-per-strip fluorescein strips in a No. 4 gelatin capsule). The client gives three capsules (six strips).[20] When given in this manner, the dye is readily excreted into the urine within two hours and remains **visible** for the next 20 to 24 hours. The client can then purchase (see Appendix for sources), rent, or borrow an ultraviolet (Wood's) lamp to scan suspected areas in a darkened house for the fluorescent urine, thus identifying the medicated pet as the culprit. Maximum fluorescence in urine-soiled areas is well retained for 24 hours. The fluorescence gradually deteriorates within a few days. When the dye is given in the late afternoon, urine deposited that night and the next morning will be labeled. Other possible suspects can be treated at two-day intervals to either confirm or eliminate them as problem animals. Although not yet attempted by the author, this technique could possibly also be adapted for use in dogs by slightly increasing fluorescein doses for larger body weights.

An alternative to use in dogs *only* (cats are very sensitive to aspirinlike products and can die from toxic exposure) is to administer a 5-grain (standard 325-mg tablet) aspirin tablet to the suspected dog.[21] When a fresh spot of urine is found, it is extracted using disposable paper towels, and ferric chloride (available from chemical supply companies or local school laboratories) is applied to the moist towel. If the aspirin-dosed dog is the source, the salicylate in the urine will turn the ferric chloride in the moist towel a burgundy color.

A third, intuitive option is to systematically confine various pets until the culprit is identified.

ODOR SOURCE DETECTION

It is important to learn to detect odor sources to help with the differential diagnosis of marking versus other house/litterbox training problems, and to assist the client in locating and removing the offending contamination. Veterinarians can learn from professional carpet cleaners, some of whom develop quite a knack for solving difficult urine soiling problems.[22] The choices include:

- **Simple Visual Inspection**. This method is substrate dependent as the eyes can miss urine in visible light.
- **Olfactory Inspection**. This method requires a sensitive sense of smell and an environment with little air circulation. It may be time consuming and it can seem less than professional if done in the most effective way, which is on one's hands and knees in order to be as close to the source of the odor as possible.

Figure 4—Devices used to detect urine: urine probe and ultraviolet light.

- **Inspection with Ultraviolet (Wood's) Lamp.** This is best for smooth surface contamination (i.e., hardwood, tile floors, or walls). This method can be extremely effective when used in conjunction with the sodium fluorescein dye technique described above. Its disadvantages, however, are that it is usually necessary to turn off room lights, close drapes, and obtain an extension cord. The results may be challenging to interpret on some surfaces and some false positives may be created. (See Appendix for sources of Wood's lamps.)
- **Inspection with a Moisture (Urine) Sensor.** This is a canelike mechanism with sharp metal prongs at the base, which penetrate the carpet and padding to indicate the presence of urine. Moisture and electrolytes in the fabric complete an electrical circuit to create an audible tone (or light an indicator bulb in some models). The signal is usually proportional to the amount of urine detected. This instrument is easy to use standing up, provides a quick answer if the user has a general idea of where the contaminated area is; and shows the extent of urine migration into the carpet pad even months after the area was soiled. It is, however, inefficient if large areas are to be surveyed with no idea as to suspected urine locations. Another disadvantage is that false positives may be detected from recent water spillage (rarely a problem in the author's hands). (See Appendix for sources.)

These last two instruments can be purchased and rented to clients who have pets with soiling problems. Our practice rents them for a nominal daily fee (one day is usually sufficient) and collects a refundable $100 deposit to assure that the unit is returned in good condition. With the use of the above techniques the real distribution and extent of urine contamination can usually be established quickly and accurately. Knowing the full distribution of urine soiling aids in obtaining a correct diagnosis of the problem (noting the pattern of contamination greatly helps to diagnose the type of elimination problem present), and assists the owner in planning effective environmental treatment to eliminate the odor.

- **Tactile Inspection.** The pet owner's bare feet are often inadvertent detection tools. Pet owners sometimes discover pet soiling by simply walking around their house, a technique used by a noted veterinary behavior practitioner.[a]

ODOR NEUTRALIZERS

When pets eliminate on household surfaces, they deposit carbon- and nitrogen-rich organic compounds which provide a ready food source for a diverse range of naturally occurring bacteria. The bacteria break down these compounds into ammonia, organic amines, sulfur, mercaptans, and a variety of other molecules.[23] Volatile forms of these by-products create the offensive odors we smell.

Although many products to neutralize urine are available commercially, little is known about them by those who recommend their use. For instance, most manufacturers neither list the ingredients, nor describe how the products work. Each of the dozens of products I have encountered claims to eliminate urine smell and assures us that theirs is the best. The following is a discussion of the types, advantages, and limitations of various products currently available. The only study of which I know that addresses individual products is by Dr. Bonnie Beaver et al.[24] In this report, Dr. Beaver tested 11 of these products as well as club soda, vinegar, Basic H® (a surface cleaner), Scope® Mouthwash, Ivory® Liquid, and Massengill® Douche Powder. My reading of her data supports conclusions that among the products tested, the Outright™ and Cat–Off™ products seemed to do the best overall. It must be noted that many of the pro-

[a]Hunthausen W: Personal communication, Animal Behavior Consultations, Westwood, Kansas, 1995.

Figure 5—Urine neutralization products.

prietary products currently available were not tested in this 1989 study. The preceding study was done on carpet in a "laboratory" environment maintaining controlled conditions under which product directions were carefully followed. Factors exist in home environments that can seriously affect the performance of various products. Problems encountered during treatment include underapplication of the product; an environment that is often too dry, warm, or cold for ideal performance of the product; the presence of previously used cleaners and residual chemicals (for example, factory stain-resistant carpet treatments and products applied during previous carpet cleanings); and overlooking "capillary" penetration into the adjacent carpet pad.

Effectiveness of home treatments can be improved in many cases by carefully evaluating contamination using urine detector systems and following manufacturer instructions carefully. If the underlying carpet pad is contaminated, the neutralizer must also penetrate below the carpet. The surface either can be flooded or the neutralization agent can be injected through the carpet into the pad with a large syringe (35 to 60 cc) and a large-bore long needle (14 to 18 gauge, 1.5-in. needle).[22]

Although specific ingredients are virtually never mentioned, a little investigation reveals that there are generally a few basic techniques that are used to eliminate, absorb, or mask odors. These are often used in combination.

- **Odor masking systems** cover odor with their perfume or prevent gas from forming from environmental bacterial action on urine and fecal substrate and are often carried over-the-counter in grocery stores as well as through veterinary distributors (e.g., Elimin-Odor™).
- **Odor absorbers and foaming detergent cleaners** absorb odors and are presumably vacuumed away. Examples include Woolite Spray® Foam Rug Cleaner, Arm and Hammer® Carpet and Room Deodorizer, Odorzout™ (a zeolite mineral odor absorbent), Petzorb (a polymer crystal absorbent system), and other products. These are designed to be applied and removed with the aid of a vacuum cleaner. Although odor is usually temporarily controlled, the urine often is not totally eliminated (especially if it has contaminated the carpet padding) and the scent may eventually return in that location.
- **Odor source chemical modifiers and disinfectants** such as Cat–Off™ and KOE™, X-O® and X-O® Plus, and Oxyfresh™ claim to use chemical reactions. Claimed reactions include counteraction using conjugates, absorption, "chemical binding," and various bacterial growth inhibitors. Many other products act principally as disinfectants to inhibit (often *temporarily*) the environmental bacteria which produce the odor. Products such as vinegar (5% acidity), club soda (carbonated water, sodium bicarbonate, sodium chloride, potassium sulfate), and others presumably work by acid-base changes on the urine substrate and make the environment inhospitable to bacterial growth. Although several of the specialized products in this group can be quite effective, they must be applied in a flood technique to contact all of the organic material or residual odor will remain. KOE™ has been reported to be effective using double the recommended concentration.[b]
- **Enzyme products** such as F.O.N.®, Nature's Miracle®, and others use direct enzymatic action to degrade the organic material. These can be effective and do remove the organic urine material but must be used in sufficient quantities to reach all parts of the contamination (including the carpet pad). Free-enzyme systems are very easily denatured by previously used chemical cleaners, disinfectants, and acid-base compounds, accounting for the variable success record reported by clients.

[b]Hunthausen W: Personal communication, Animal Behavior Consultations, Westwood, Kansas, 1993.

- **Bacteria/enzyme combinations** include Outright™ products and genetically selected/engineered products such as Anti–Icky–Poo® (AIP) and Enz-Odor®, in addition to several other products. These products use live biostrains of bacilli bacteria to produce large quantities of various protease, lipase, amylase, and cellulase enzymes to break down organic contamination that is, in effect, a food source for these beneficial microorganisms. In my experience, the bacteria/enzyme combinations seem to do the *best job overall.* Early generations of products within this group were limited by the need to produce an environment carefully suited to the beneficial bacteria's growth. This included proper temperature, humidity, lack of disinfectants and other chemicals, and transportation conditions (such as overheating), as well as adequate application. Like the free-enzyme systems mentioned above, these conditions inhibit the bacterial growth and enzymatic action. The genetically selected/engineered bacteria in some products (for example, AIP and probably also Enz-Odor®) are resistant to desiccation, disinfection, detergents, temperature extremes, and other chemical or environmental conditions that usually inactivate or inhibit the effectiveness of most other neutralizers. (Note: Common antibiotics used in cystitis cases can slow or inhibit the bacteria in these products during the course of treatment.)

Some of these products currently are used primarily by professional carpet cleaners and are not available through conventional retail outlets (e.g., AIP). This is a situation in which veterinarians and other behavior consultants can provide unique and highly effective products to help clients. This helps to keep the pet with the family and helps facilitate and maintain the human–animal bond which allows us all to practice our profession.

This discussion represents my experience and own limited investigation of the complex and somewhat secretive odor neutralization industry. It is not intended to be the final word on the topic but simply to stimulate additional discussion on the importance of this aspect of diagnosing and solving elimination problems in animals. More scientific investigations, such as those reported by Beaver et al,[24] need to be conducted on currently available products. This research would allow a less anecdotal reporting of the true effectiveness of these products. It must be emphasized that thorough odor detection and neutralization is an important but usually insufficient part of solving elimination behavior problems. Other factors that initiate and maintain the problem behavior also must be addressed to maximize success.

REFERENCES

1. Borchelt PL, Voith VL: Diagnosis and treatment of elimination behavior problems in cats. *Vet Clin North Am [Small Anim Pract]* 12(4):673–680, 1982.
2. Borchelt PL, Voith VL: Diagnosis and treatment of elimination behavior problems in dogs. *Vet Clin North Am [Small Anim Pract]* 12(4):637–644, 1982.
3. Borchelt PL: Cat elimination behavior problems. *Vet Clin North Am [Small Anim Pract]* 21(2):257–264, 1991.
4. Nott MR: Social behaviour of the dog, in Thorne C (ed): *The Waltham Book of Dog and Cat Behaviour.* Oxford, Pergamon Press, 1992.
5. Robinson IH: Social behaviour of the dog, in Thorne C (ed): *The Waltham Book of Dog and Cat Behaviour.* Oxford, Pergamon Press, 1992.
6. Beaver BV: *Feline Behavior: A Guide for Veterinarians.* Philadelphia, W.B. Saunders Co, 1992.
7. Macdonald DW: The Carnivores, Order Carnivora, in Brown RE, Macdonald DW (eds): *Mammalian Social Odours.* Oxford, Oxford University Press, 1985.
8. Natoli E: Behavioural responses of urban feral cats to different types of urine marks. *Behaviour* 94:234–243, 1985.
9. Passanisi WC, Macdonald DW: Group discrimination on the basis of urine in a farm car colony, in Macdonald DW, Muller-Schwarze D, Matynczuk S (eds): *Chemical Communication in Vertebrates (V).* Oxford, Oxford University Press,1990, pp 226–345.
10. Melese-d'Hospital PY, Hart BL: Vomeronasal organ cannulation in male goats: Evidence for transport of fluid from oral cavity to vomeronasal organ during flehmen. *Physiol Behav* 35:941–944, 1985.
11. De Boer JN: The age of olfactory cues functioning in chemocommunication among male domestic cats. *Behavioural Processes* 2:209–225, 1977.
12. Chapman BL, Voith VL: Behavioral problems in old dogs: 26 cases. *Compend Contin Educ Pract Vet* 7(3):209–221, 1985.
13. Voith VL, Borchelt PL: Fears and phobias in companion animals. *Compend Contin Educ Pract Vet* 7(3):209–221, 1985.
14. Borchelt PL: Cat elimination behavior problems. *Vet Clin North Am [Small Anim Pract]* 21(2): 257–264, 1991.
15. Hart BL: Olfactory tractotomy for control of objectionable urine spraying and urine marking in cats. *JAVMA* 179(3): 231–234, 1981.
16. Peters RP, Mech LD: Scent marking in wolves. *Am Sci* 63(2):628–637, 1975.
17. Borchelt PL, Voith VL: Elimination behavior problems in cats. *Compend Contin Educ Pract Vet* 3(8):730–737, 1981.
18. Moulton DG, Ashton EH, Eayrs JT: Studies in olfactory acuity. 4. Relative detectability of n-aliphatic acids by the dog. *Anim Behav* 8:117–128, 1960.
19. Becker M: No room for error: Preventing the mistakes that cost you clients. *Vet Econ* March:50–55, 1995.
20. Hart BL, Leedy M: Identification of source of urine stains in multi-cat households. *JAVMA* 180:77–78, 1982.
21. Karofsky PS: Identifying source of urine on rugs. *JAVMA* 191(8):917, 1987.
22. Newberry M: *A.I.P. Professional User's Guide,* 1992, 19 pp. MisterMax™ Products, Lakeside, CA 92040.
23. Bug-A-Boo Chemicals: Unpublished communication, Portland, OR, 1995.
24. Beaver BV, Terry ML, LaSagna CL: Effectiveness of products in eliminating cat urine odors from carpet. *JAVMA* 194(11):1589–1591, 1989.

APPENDIX
Sources for Odor Control Products

Urine Detection Products

Fluorescite® Injection 10%, equivalent to 100 mg/ml. Alcone Laboratories, Fort Worth, TX 76134.

Fluor-i-strip®, 9 mg of fluorescein per strip. Wyeth-Ayerst Laboratories, Philadelphia, PA 19101.

Ultraviolet (Wood's) Lamp:
Available from electrical suppliers and marketed by Nature's Miracle® (Joe Weiss Company, 5312 Ironwood Street, Rancho Palos Verdes, CA 90274) through various distributors (about $30).

Moisture Urine Sensors:
Probes run about $75 to $125 and are available through MisterMax™, Lakeside, CA 92040. 1-800-745-1671.

Odor Neutralizers

Basic H® by Shaklee Corporation, San Francisco, CA 94111.

Scope® Mouthwash by Procter & Gamble, Cincinnati, OH 45201.

Ivory® Liquid by Proctor & Gamble, Cincinnati, OH 45201.

Massengill® Douche Powder by SmithKline Beecham, Pittsburgh, PA 15205. 1-800-456-6870.

Outright™ Stain and Odor Remover, The Branton Co., P.O. Box 655450, Dallas, TX 75262-5450.

Cat–Off™ and KOE™, Thornell Corp., 160 Wheelock Road, Penfield, NY 14526.

Elimin-Odor™, SmithKline Beecham Animal Health, 812 Springdale Drive, Exton, PA 19341.

Woolite® Spray Foam Rug Cleaner, Rekitt and Coleman, Wayne, NJ 07474.

Arm and Hammer® Carpet and Room Deodorizer, Church & Dwight Co., Inc., Princeton, NJ 08543. 1-800-524-1328.

Odorzout™, No Stink, Inc., 6020 W. Bell Road, #E101, Glendale, AZ 85308. 1-800-887-8465.

Petzorb, ImmunoVet, 5910-G Breckenridge Parkway, Tampa, FL 33610-4253. 1-800-762-7648.

X-O® and X-O® Plus, X-O Corporation, 8330 Soberly Lane, Dallas, TX 75277. 1-800-442-9696; FAX 214-388-7140.

Oxyfresh™, Oxyfresh USA Inc., P.O. Box 3723, Spokane, WA 99220. 1-800-999-9551, extension 777.

F.O.N.®, Summit Hill Laboratories, P.O. Box 535, Navesink, NJ 07752. 1-800-922-0722.

Nature's Miracle®, Joe Weiss Company, 5312 Ironwood St., Rancho Palos Verdes, CA 90274.

Anti–Icky–Poo® (AIP), by Bug-A-Boo Chemicals, 1-800-326-3016; available through MisterMax™ distributor, Lakeside, CA 92040. 1-800-745-1671.

Enz-Odor®, by Alken Murray Corp., 417 Canal Street, New York, NY 10013; 1-800-431-4020; FAX 212-431-4944.

Ingestive Behavior: The Control of Feeding in Cats and Dogs

New York State College
Cornell University
Ithaca, New York
Katherine Albro Houpt, VMD, PhD

Clinical problems of ingestive behavior, such as obesity in dogs and anorexia in cats, are evident to the practitioner every day. Consideration of the physiologic and behavioral controls of food intake can aid the practitioner in alleviating these problems. Veterinarians encounter questions about feeding and appetite in companion animals far more often than they do actual clinical cases of failure to maintain normal body weight. Typical questions include:

- Will my dog get fat if it is spayed?
- How often should I feed my cat (or my dog)?
- Why doesn't my dog eat in hot weather?
- Why has my cat stopped eating a brand of cat food that it has been eating for months?
- Why did my older dog begin to eat more when we got a new puppy?
- Will putting sugar in my cat's water encourage it to drink more water?

The following discussion will provide the answers to some of these questions.

THE CAT

Central Nervous System

Cats have been used extensively in studies of brain function. Although there are many gaps in our knowledge of the controls of feline food intake, central neural involvement has been investigated. The classical means by which the function of a particular area of the brain is determined is to remove that area and observe the animal's postoperative behavior. Using this classical method, it has been observed that the hypothalamus is important in the regulation of body weight and in the control of food intake. Lesions in the ventromedial hypothalamus result in an obese animal[1]; lesions in the lateral hypothalamus sometimes result in fatal aphagia (failure to eat) or, in less severe cases, persistent anorexia (limited food intake) and a loss of body weight that is maintained even months after surgery.[2]

Based on the results of these studies and others in which various parts of the hypothalamus were electrically stimulated, it was concluded that the lateral hypothalamus is involved in the stimulation of feeding behavior or hunger and the ventromedial hypothalamus is involved in the inhibition of feeding or in the promotion of satiety; therefore, both areas are involved in the regulation of body weight.

Because the hypothalamus is not the only area of the brain involved in feeding behavior, the concept of discrete centers (each involved in control of a specific behavior) has been replaced by the concept of neural circuits and networks influencing behavior. Other areas of the brain—from the cortex to the midbrain—have also been found to influence feeding. The hypothalamus is important but not essential because cats with all but the most severe hypothalamic lesions survive and eat although the amounts eaten may be very abnormal. Recently, the paraventricular nucleus of the hypothalamus has been found to be important in feeding in rats, but this brain area has not been investigated in cats.[3] Presumably, the other areas of the brain (in particular the cortical areas) are able to take over the function of control of food intake. For these reasons, injuries to various parts of the brain may cause alterations in feeding behavior.

The brain areas integrate information on the fat reserves of the body, the reproductive state of the animal, the environmental temperature, and the characteristics of the food available. They also serve to stimulate or depress the motor actions of ingestive behavior.

Another outcome of neurophysiologic research is the finding that depressants of the central nervous system usually tend to increase food intake.[4] The observant practitioner may have noticed that an animal recovering from anesthesia often begins to eat voraciously even before it is able to stand. If the animal begins to eat before its pharyngeal reflexes have completely recovered from anesthesia, it may choke; therefore, food is usually removed from the cage until the animal is fully recovered. This effect of depressants indicates that feeding is actively inhibited by the brain such that unless the ventromedial hypothalamus and associated structures are suppressing intake the animal will eat.

The practitioner can take advantage of the effect of depressants to treat anorexic cats, which frequently stop eating when they are hospitalized or boarded. A minor illness can be exacerbated by the effects of semistarvation if the anorexia persists long enough. Many depressants of the central nervous system can be used to treat feline anorexia, but the drugs found to be most effective in stimulating intake are the benzodiazepine tranquilizers, such as diazepam. Recently, diazepam has been reported to cause hepatic toxicity in a few cats, so care must be taken with its use.

The effects of depressants are manifested in other clinical situations where increased intake is an undesirable side effect. The synthetic progestins have been found to have many therapeutic applications in canine and feline medicine. They are used for the inhibition of urine spraying, treatment of some dermatologic conditions, and suppression of estrus. Some progestins act as depressants of the central nervous system, and others have been developed as anesthetics. Probably for this reason an increase in food intake and consequent weight gain are sometimes seen as side effects of progestin therapy. Long-term use of progestin is associated with serious side effects, such as diabetes. The serotonin-antagonist cyproheptadine also can be used to stimulate appetite, but it is not as consistently effective as the other drugs.

Sex Hormones

The endocrinologic controls of feline food intake and body weight need to be investigated because of the many questions raised by veterinarians and cat owners concerning the effects of spaying and castration on the cat's appetite and weight. Although there is no direct evidence that neutering leads to an increase in body weight, a recent survey indicated that neutered cats are more likely to be obese.[5]

The effects of estrogen and testosterone on the ingestive behavior of rodents have been well investigated. In general, estrogen suppresses food intake and body weight. Progesterone in physiologic amounts (as opposed to the pharmacologic amounts used therapeutically in cats) has little effect on food intake. Testosterone is more complex in its action. As an anabolic steroid it stimulates lean-body growth and weight gain, but some testosterone is metabolized to estrogen and acts to suppress food intake. Consequently, male rats lose a little weight and eat less after castration. Female rats eat more and gain a large amount of weight after ovariectomy.

Cat owners should be told to observe an animal after neutering and, if necessary, restrict food intake. The benefit to the owner of having a cat that no longer fights, sprays, roams, or produces unwanted kittens far outweighs the risk of obesity. The incidence of obesity in cats may be increasing. It was estimated to be 9% in 1973[6] and 25% in 1994.[5] Scarlett et al[5] identified apartment living and special diets as risk factors. Apparently, cats would be active if

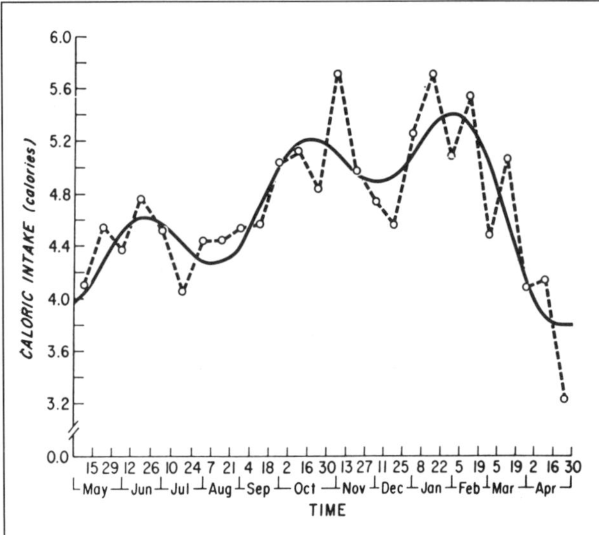

Figure 1—The cyclic variation in food intake in the cat. (From Randall WV, Lakso V, Liittschwager J: Lesion-induced dissociations between appetitive and consummatory behaviors and their relationship to body weight and food intake rhythms. *J Comp Physiol Psychol* 68:476–483, 1969. Copyright 1969 by American Psychological Association. Reprinted with permission.)

space were available but they tend to overeat highly palatable food.

Cycles of Body Weight and Feeding

A phenomenon that helps to explain a great deal of feline finickiness with regard to appetite is the cyclic nature of cats' body weight. By weighing cats frequently and measuring their ad libitum food intake daily, it becomes apparent that both weight and food intake fluctuate. Body weight peaks about three times a year as does food intake. These fluctuations represent a cycle of 3 to 4 months. The changes in body weight are not great, typically 200 g or less in a 4 kg cat. Food intake varies by 60 to 100 calories per day.[7] Upon entering a period of declining body weight, a cat would tend to eat less.

The anorexia might be of concern to the owner, who might decide to change cat food on the basis that the animal is *tired* of the original diet. A less concerned owner might ignore the variations in intake, and a more concerned owner might take the cat to a veterinarian. Unless there are other clinical signs or the anorexia persists and more than a few hundred grams of weight are lost, the cat should be considered normal. These fluctuations in body weight can occur even in a cat that is gaining weight over the long term. A cat that has gained 700 g from the beginning of one year to the next may still have had periods of weight loss during the year.

The rhythms of body weight have made it difficult to study long-term (day-to-day or week-to-week) controls of food intake in cats. One fundamental question has been whether cats eat for bulk, (i.e., a set amount or volume of food) or for calories (i.e., a set amount of energy). Being carnivores, cats are not exposed to food that varies much in caloric density as are omnivores; consequently, cats may not have developed mechanisms for increasing intake when food is low in calories (i.e., bulky).

To test cats' ability to eat for energy, three different experiments were performed. In the first two, cat chow was mixed with either kaolin or cellulose; in both studies the cats did not increase their intake and they lost weight.[8,9] In the third experiment, the diet consisted of canned cat foods with different percentages of water. The calories were diluted with water rather than with unpalatable solid diluents. In this experiment, volume of intake did increase while energy intake remained relatively constant.[10] Perhaps a way to increase the water intake of a male cat that is susceptible to urethral calculi is to offer it canned cat food that is high in water content. To maintain its weight, the cat will take in a greater volume and, thereby, consume more water.

In addition to the long cycles of weight fluctuations, there are shorter cycles that influence intake. These are ultradian rhythms (occurring more often than once a day but no more than every 30 minutes). These rhythms do not influence how much a cat eats but *when* it eats. When food is freely available, cats eat 9 to 12 meals a day, as do many species of animals.[10] Cats are often compared to lions. Because both species are predatory felids, domestic cats might be expected to eat large meals at infrequent intervals, as do free-ranging lions. The observed behavior of cats in the laboratory does not support that expectation. Cats are indeed predators, but they are just as likely to eat 9 to 12 mice a day as to eat one rabbit every other day, and it has been shown that the more effort a cat must expend to obtain food, the larger the meal it will consume.[11] Because cats appear to have a natural inclination for many small meals, food should probably be freely available to those that are not prone to obesity.

Palatability

The finicky cat is the delight of cat food manufacturers, who expect the owner to try several brands and flavors of cat food each year. Finickiness may be associated with the cycles of intake and body weight discussed above, but food intake preferences and aversions may also be involved. Several researchers have found that cats prefer a novel diet.[10,12] This preference is short lived if the new food is not more palatable than the old food, but it accounts for the fact that cats

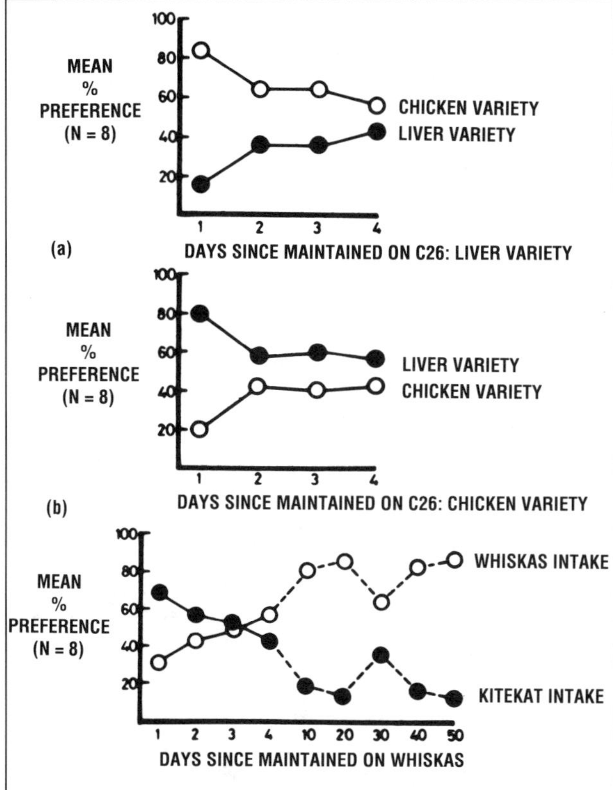

Figure 2—Preference of kittens for novel foods, in two choice preference tests. Note the initial strong preference for the novel food chicken *(top),* liver *(middle),* or Kitekat *(bottom).* After a few days the preference either disappears or is reversed so that the original food (Whiskas) is strongly preferred *(bottom).* (From Mugford RA: External influences upon the feeding of carnivores, in Kare M, Maller O (eds): *The Chemical Senses and Nutrition.* New York, Academic Press, 1977, pp 25–48. Reprinted with permission.)

may eagerly consume a new type of cat food, which may encourage owners to try many foods.

There is, no doubt, a great deal more known about the basic food preferences of cats than has been published, but the highly palatable ingredient of cat food is apparently a carefully guarded secret of the manufacturers. Nevertheless, some feline preferences are known. For example, cats generally prefer fish to meat. For this reason many cats that preferred to eat red tuna were afflicted with steatitis as a result of vitamin E deficiency before the cause was discovered.

The sucrose preference of cats has been investigated. Cats as well as lions and other large cats do not like sucrose solutions in water.[13] Adding sugar to its water supply will, in fact, inhibit a cat from drinking. Water intake can be increased by increasing the osmolality of the cat's body fluids, thereby, stimulating thirst. It may be wiser to feed a watery diet, as previously discussed, or add some tuna or chicken broth, rather than subject a cat with urinary problems to further stress.

Despite the fact that cats do not show a preference for sucrose in a water solution, many appear to prefer sweet foods. To be more scientifically accurate, they prefer some foods that taste sweet to *humans.* Many cats will eat an unattended pumpkin pie or a cherry cake left to cool. Many cats have also learned to wait patiently (or impatiently) for the opportunity to "clean" the ice cream dishes.

The reason for the conflict between the results of sucrose preference tests and observed solid food preferences is that cats can taste water. The water taste is not a preferred one in cats and is apparently stronger than or masks the taste of the sucrose.[14] Therefore, care must be taken not to extrapolate preferences for solutions to preferences for solid food. Certainly the combination of sweet and fatty or milky textures is highly palatable to most cats.

Cats are hunters. Although cats learn to hunt, some will not ordinarily kill a mouse or rat. Most will do so, however, if starved for three days. Nevertheless, predatory behavior is not the same as hunger. Brain stimulation experiments have revealed that stimulation of one area of the hypothalamus causes cats to eat even when satisfied whereas stimulation of another area causes hungry cats to ignore a dish of food and attack a rat. Therefore, it cannot be said that cats kill only to eat. The two behaviors are separate but related, and many well-fed pets still hunt and kill.

If cats are given a choice between freshly killed rats and canned cat food, they prefer the latter.[15] The separation of predatory and feeding behavior may have evolved so that a cat would hunt when hungry to feed itself, but would continue to hunt after it was satiated to bring food back to its kittens.

Food Aversion in the Cat

It is often stated, on the basis of very limited scientific information, that cats and other animals have innate *nutritional wisdom.* That is, they have specific appetites or cravings for nutrients that are deficient in their diets. Most examples involve sodium. Animals (cats have not been tested) have developed appetites for sodium; when they are sodium deficient, they ingest higher concentrations of NaCl or $NaHCO_3$ than normal.

Animals also eat for calories; consequently, they eat more of a diet that is dilute in calories. In contrast, if an animal is deficient in another type of nutrient (e.g., vitamin B_{12}), it will not be able to detect and choose a food rich in that vitamin. What the animal can learn is that it feels better after it eats certain foods and it will then eat more of that food.

The converse of learning what foods are safe is learning what foods cause illness. This phenomenon, called *taste aversion conditioning,* results when a cat

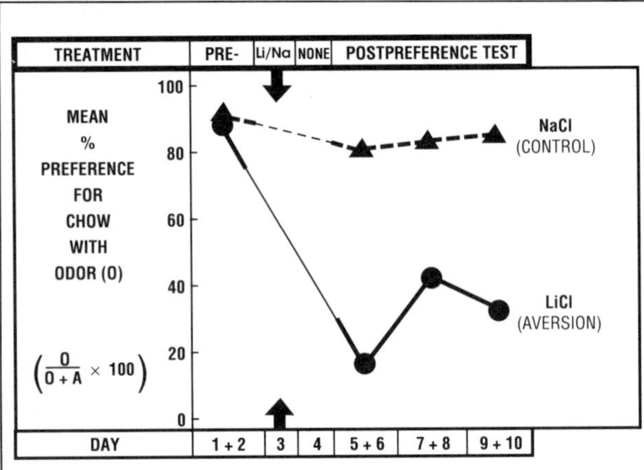

Figure 3—Dry food/odor preferences of cats demonstrating odor aversion learning. Initially, all cats showed a strong preference for chow with rabbit odor. Cats that were injected with LiCl immediately after eating chow with the odor of rabbit avoided food that smelled like rabbit on the next occasions it was presented. Control animals injected with NaCl continued to show a strong preference. (From Mugford RA: External influences upon the feeding of carnivores, in Kare M, Maller O (eds): *The Chemical Senses and Nutrition.* New York, Academic Press, 1977, pp 25–48. Reprinted with permission.)

tastes or smells a food that once made it sick. The benefit to the animal of this type of learning is obvious. If a cat ate a rat that had been dead for a long time and was decayed, the cat could vomit and have diarrhea after ingesting the rat. It would not eat rat meat that tasted or smelled rotten again. Laboratory studies of this type of learning have been performed to demonstrate that cats made ill when exposed to the odor of rabbit will not approach or eat food that smells like rabbit.[10] It has been suggested that cats and dogs eat grass to elicit vomiting. This may also be a form of learning, that is, learning to eat foods that lead to recovery from illness. Taste aversion conditioning also has practical applications. For example, it has been used to make wild coyotes averse to eating lamb.[16] A dead lamb is baited with a nonlethal emetic, such as LiCl or apomorphine. The coyote eats the lamb, becomes ill for a few hours, and learns not to eat lamb again or it will be sick. This method has been moderately successful in reducing coyote predation and is ecologically preferable to trapping or poisoning a predator that keeps rodent and rabbit populations in check.

THE DOG

Some of the controls of food intake in cats also have been investigated in dogs. For example, dogs with lesions in the ventromedial hypothalamus are hyperphagic (overeat) and become obese.[17] Thermostatic, gastrointestinal, and other controls have been studied in dogs but not in cats. Food preferences have been studied in both species, but dogs have been more thoroughly studied than cats.

Thermostatic Controls

Hunger and satiety result not only from changes within the dog but also from changes in the dog's environment. An environmental factor that strongly influences food intake is ambient temperature. Dogs eat more in the cold and less in the heat[18] and this is especially true for dogs kept outdoors. For example, a group of beagles in Alaska ate twice as much in cold weather but still lost weight in the harsh Arctic winter. The change in food intake with temperature is one type of behavioral thermoregulation. More energy must be expended to maintain body temperature and, therefore, more energy is taken in. In the heat, less energy is needed and the additional heat stress in the form of specific dynamic action, which accompanies digestion, is avoided.

The veterinarian may be consulted by an owner who is worried because his dog is not eating on a hot day or eats much less after a move to a warm climate; however, these are normal reactions because dogs need more food in cold weather. It is important that veterinarians advise owners of dogs kept in outside pens to provide more food in the winter. If the dogs have grown thick coats, the owners may not be aware of their weight loss until they shed in the spring.

Social Facilitation

Another external factor that influences food intake in the dog is the presence of other dogs. A dog—especially a puppy—eats more when another dog is present.[11] This phenomenon is called *social facilitation* of eating and can become a pathologic condition. Fox[19] has reported the case of a dog that overate and became obese when a second animal was brought into its owner's house.

Type of Food

A final external influence on canine food intake is the food itself. Given a choice between two foods, dogs make definite choices based on the sensory qualities of the food. Using the results of two choice preference tests, the relative palatability of various preparation methods and of various meats can be determined. Dogs appear to have a hierarchy of preferences for meats. Beef and pork are preferred to the leaner meats, and lamb and chicken are preferred to horsemeat.[20] They prefer cooked meat to raw meat; even canned meat is preferred to fresh.[21] They also prefer ground meat to chunks. Dogs have obviously adapted well to the dietary habits of humans. They

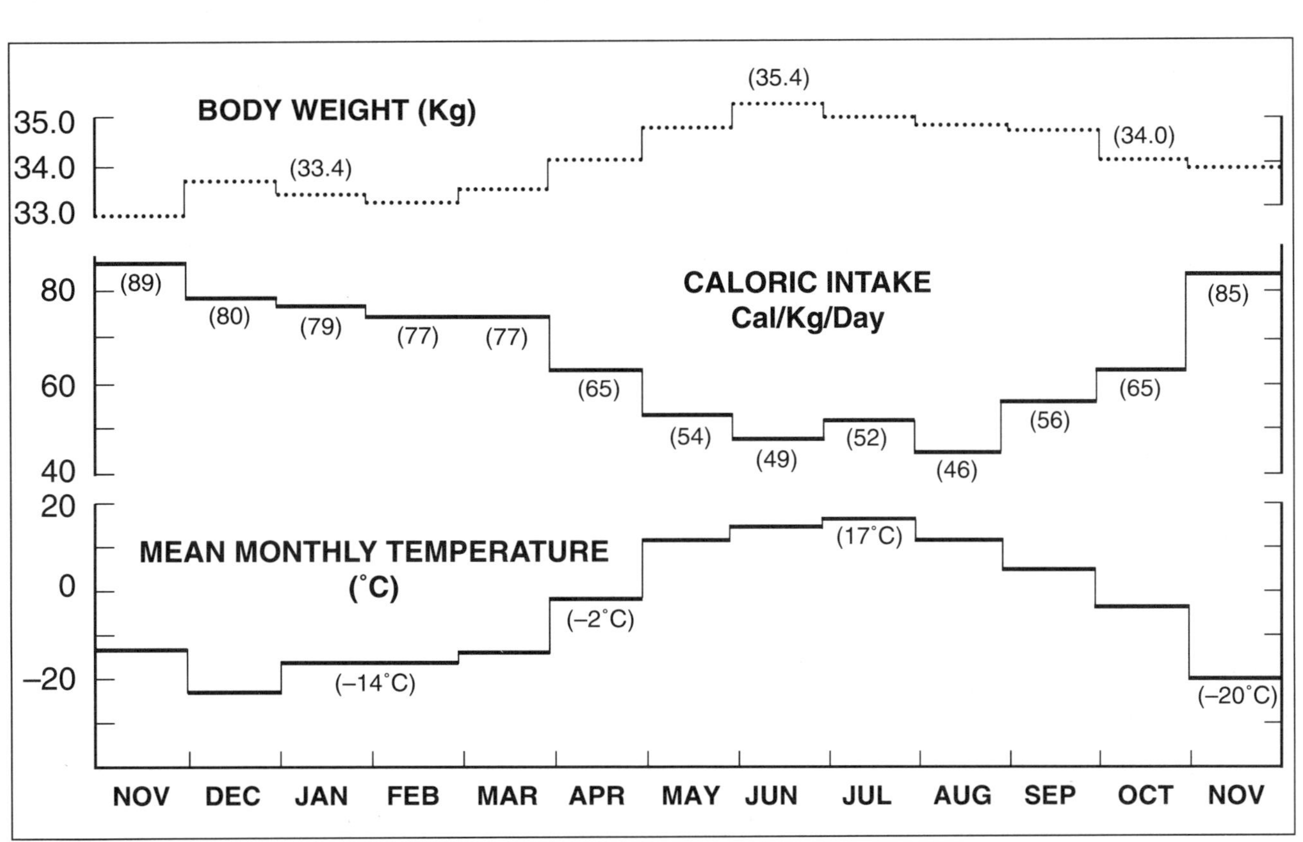

Figure 4—The inverse relationship of caloric intake and environmental temperature in a group of beagle dogs in Alaska. Note that the dogs gained weight in the summer when their food intake was lowest, but their energy requirements for thermoregulation were the smallest. (From Durrer JL, Hannon JP: Seasonal variations in caloric intake of dogs living in an Arctic environment. *Am J Physiol* 202:375–378, 1962. Reprinted with permission.)

prefer semimoist and canned preparations to dry chow, with canned and semimoist foods having approximately equal palatability.[22] Dogs are still carnivores because they prefer meat to cereal diets.

In contrast to cats, dogs show a liking for sucrose in water as well as for sucrose mixed with solid food.[23,24] They do not prefer saccharin, however; therefore, generalizations cannot be made about *sweet* taste responses from one species to another.

Dogs have a very well developed sense of smell and certainly use it to locate food, but the smell of food does not determine many canine preferences. If the sense of smell is temporarily eliminated, dogs still show preferences for sweet food and for meat rather than cereal; the only preference that is affected by lack of the sense of smell is the type of meat preference. For example, dogs that cannot smell do not discriminate between beef and lamb.[20] Taste and textural properties seem to be more important than odor in determining palatability.

Glucostatic Controls

One of the controls of food intake that has received a great deal of scientific and popular attention is the *glucostatic hypothesis.* According to this hypothesis, animals eat when the level of glucose utilization falls, particularly the level of glucose utilization in the brain. The importance of the rate of glucose utilization, rather than plasma level of glucose, explains the otherwise paradoxical hyperphagia of the diabetic. Plasma glucose is high but utilization is low so the animal is hungry.

Dogs show an increased food intake when hypoglycemic,[25] but it is generally assumed that glucostatic eating is an emergency mechanism that serves to increase energy when readily available energy stores have been acutely depleted. Therefore, a dog will be hungry after long and vigorous exercise, but the glucostatic mechanism probably does not control feeding in a relatively sedentary dog that has free access to food.

Meal Patterns

When dogs are eating ad libitum, they consume many small meals a day, as do cats. They differ from cats mainly in that they do not eat at night.[26] Frequent meals are the natural pattern, and owners should be advised to feed several small meals rather than a single

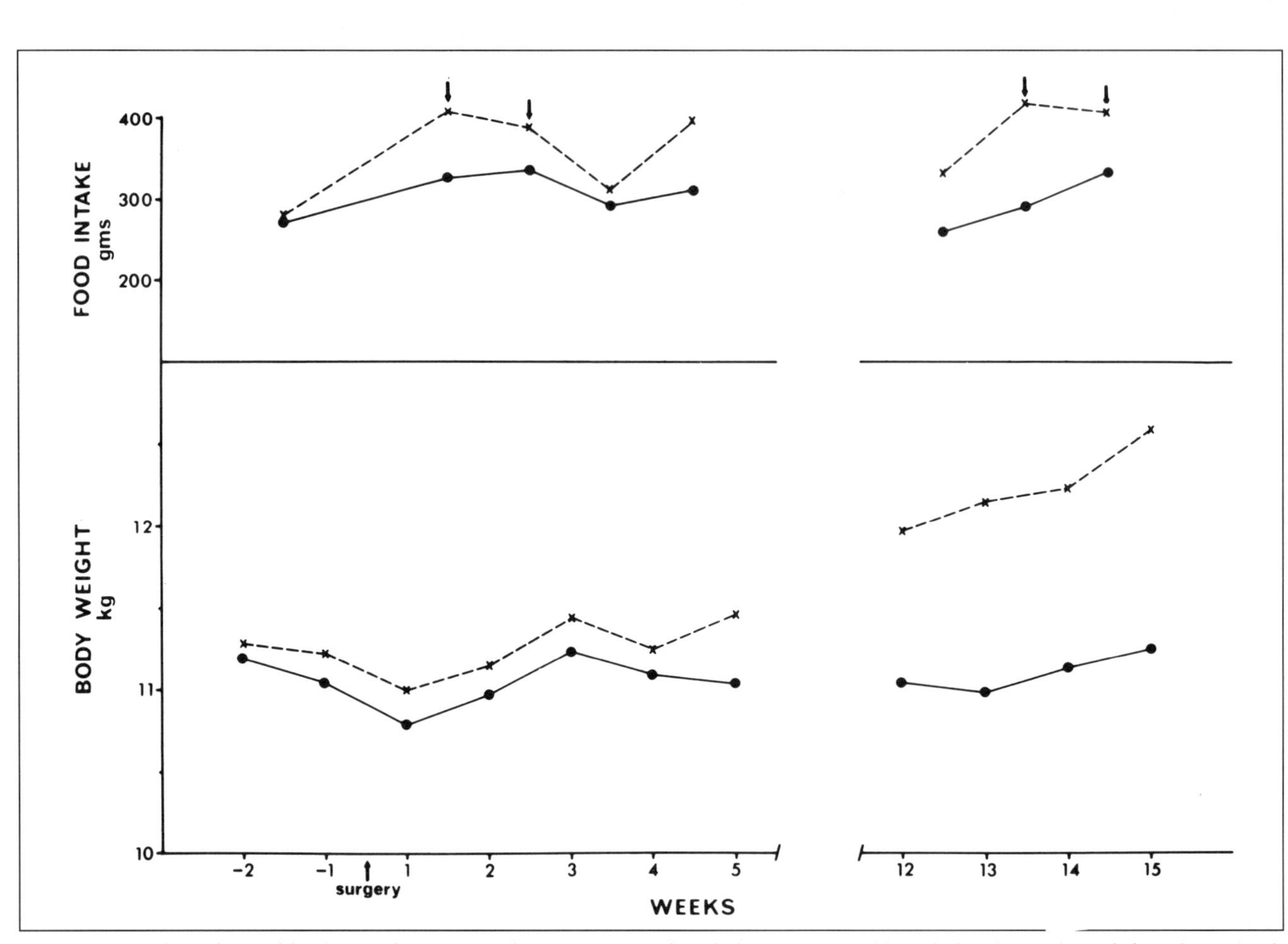

Figure 5—Food intake and body weight in ovariohysterectomized and sham-operated beagle bitches. The *solid circles* and *solid lines* represent food intake and body weight of the sham-operated bitches. The *asterisks* and *dashed lines* represent food intake and body weight of the ovariohysterectomized bitches. The *arrows* indicate weeks in which the mean daily food intake was significantly higher ($P < 0.05$) in the ovariohysterectomized animals than in the sham-operated controls. Food intake and body weight were not measured during the fifth to eleventh weeks after surgery. (From Houpt KA, Coren B, Hintz HF, Hilderbrant JE: Effect of sex and reproductive status on sucrose preference, food intake and body weight of dogs. *JAVMA* 174(10):1038–1088, 1979. Reprinted with permission.)

large one. This is especially important in large breeds of dogs, which are susceptible to gastric dilation.

Nevertheless, there are reasons for restricting intake to one or two meals. Dogs that sometimes defecate in the house should be fed only when the owner is able to take them outside shortly after eating to take advantage of the gastrocolic reflex. Most dogs defecate soon after eating and can be most easily housebroken by conditioning their elimination to mealtimes. Another reason for single meals is to overcome mild cases of anorexia. Dogs, unlike cats, often respond to any food after a 24 to 48 hour fast. A dog that has taught its owner to feed it omelettes and beef steak will eat dog food eagerly when its energy stores have fallen and the internal factors have overcome any inhibition of intake based on palatability.

Gastrointestinal Controls

Dogs with free access to food probably do not eat because they are hungry but because they are no longer satiated. The studies of central neural involvement in feeding previously outlined indicate that feeding may be actively inhibited. At least one of the inhibitions is from the gastrointestinal tract. A series of experiments on dogs with gastric fistulas revealed the role of the stomach and intestines in satiety.[27] Simple distention of the stomach did not inhibit feeding unless the organ was stretched to the point that the dog was in pain. If food was placed in the stomach just as the dog began eating, it would eat a normal-sized meal unless virtually an entire meal had been placed in the stomach. If food was placed in the stomach 20 minutes before feeding, subsequent intake was suppressed.

These studies indicate that intestinal and postabsorptive factors are important in canine satiety. One of the factors may be the hormone cholecystokinin, which is released when the products of protein and

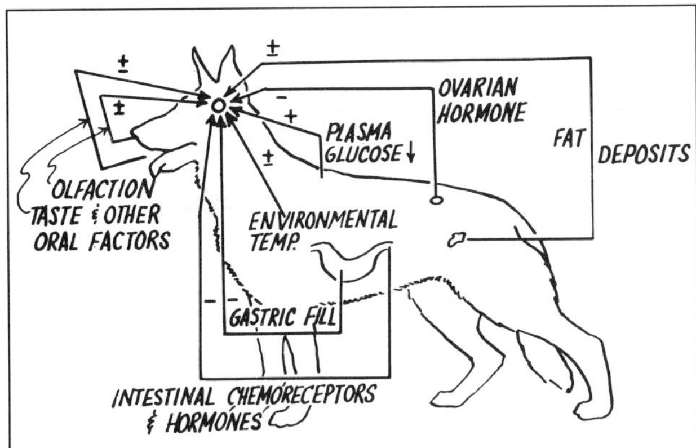

Figure 6—Factors that influence food intake in the dog. (From Houpt KA, Hintz HF: Obesity in dogs. *Canine Pract* 5:54–58, 1978. Reprinted with permission.)

fat digestion reach the intestine. Other factors may include glucoreceptors and osmoreceptors in the intestines or liver. When the signals from the intestinal tract no longer indicate the presence of food, the dog will probably eat again.

Hormonal Controls

Alterations in the endocrine balance in the dog may lead to alterations in food intake and body weight. For example, excess corticosteroid administration may lead to abnormal fat deposition and hyperphagia. It is the relation of gonadal hormones to feeding, however, that is of most interest to veterinarians. To help answer the frequently asked question "Will my dog become fat after it is spayed?," the following experiment was performed.[24] The food intake and body weight of two groups of beagles were measured before and after surgery. One group was ovariohysterectomized; in the other group, a sham operation (laparotomy alone) was performed. The body weight of the ovariohysterectomized group gradually increased over three months. Food intake also increased. The gain in weight of the ovariohysterectomized bitches is probably due to a decrease in physical activity as well as to an increase in food intake. Owners of ovariohysterectomized bitches should be warned that a tendency to gain weight is a side effect of the operation. It is their responsibility to limit the bitch's intake and to increase exercise so that a normal weight is maintained.

Obesity

The maintenance of normal body weight depends on all factors previously discussed. Glucostatic, thermostatic, gastrointestinal, hormonal, and gustatory factors are integrated in the central nervous system and determine whether the animal eats or does not eat. Most adult animals maintain a relatively constant body weight, even in the presence of unlimited food. A signal (recently identified as leptin) from the fat deposits is believed to inhibit intake when body weight increases.

It is obvious that some dogs do not maintain as lean a body as others. This does not imply that these obese dogs would become fatter and fatter if allowed to eat ad libitum. They would eventually maintain a constant weight, but this might be so great that the dogs would have an increased risk of developing circulatory, articular, and locomotor problems.[28] Although the tendency to gain weight is undesirable in pet dogs, it may be adaptive to dogs living in the wild. They could store energy in the form of fat to use when food was scarce. The incidence of canine obesity has been estimated to be 12% to 44% of the population.[6,26] A careful survey using objective criteria has not been made in the United States; however, there seems to be breed differences in susceptibility to obesity, with Labrador retrievers, spaniels, collies, terriers, and beagles at highest risk.[29]

The treatment of obesity is simple: reduce the dog's caloric intake to below its caloric output. The practitioner should determine how much the dog is eating. An owner may protest that the dog eats only half a cup of chow a day, but careful questioning reveals that it also eats two cups of ice cream a day. The veterinarian should be tactful and sympathetic because the dog may have developed highly effective strategies for obtaining food. The owner might prefer to have a fat dog than to have mealtimes constantly interrupted by barking and whining. Palatable, low calorie dog food should assist the owner to treat this type of obesity. Crowell-Davis et al[30] found that there was no increase in interdog aggression or coprophagia in dieting, pen-housed dogs in a laboratory setting.

Endocrine imbalance, brain injuries, or compulsive overeating in the presence of other dogs may lead to obesity. The most probable cause, however, is a highly palatable diet and too little exercise. The dog may eat only a few calories more than it expends each day, but the excess energy accumulates as fat. The dog cannot be condemned for gaining weight in an environment where there is little need to exercise and where there is much appetizing food.

REFERENCES

1. Anand BK, Shoenberg K: Hypothalamic control of food intake in cats and monkeys. *J Physiol* 127:143–152, 1955.
2. Wolgin DL, Teitelbaum P: Role of activation and sensory stimuli in recovery from lateral hypothalamic damage in the cat. *J Comp Physiol Psychol* 92:474–500, 1978.
3. Williams G, McKibbin PE, McCarthy HD: Hypothalamic regulatory peptides and the regulation of food intake and energy balance: Signals or noise. *Proc Nutr Soc* 50:527–544, 1991.

4. Feldberg W: A physiological approach to the problem of general anaesthesia and of loss of consciousness. *BMJ* 2:771–782, 1959.
5. Scarlett JM, Donoghue S, Saidla J, Wills JM: Overweight cats—Prevalence and risk factors. *Int J Obes* 18:S22–S28, 1994.
6. Anderson RS: Obesity in the dog and cat, in *The Veterinary Annual XIV.* Bristol, Wright, 1973, pp 182–186.
7. Randall WV, Lakso V, Liittschwager J: Lesion-induced dissociations between appetitive and consummatory behaviors and their relationship to body weight and food intake rhythms. *J Comp Physiol Psychol* 68:476–483, 1969.
8. Hirsch E, Dubose C, Jacobs HL: Dietary control of food intake in cats. *Physiol Behav* 20:287–295, 1978.
9. Kanarek RB: Availability and caloric density of the diet as determinants of meal patterns in cats. *Physiol Behav* 15:611–618, 1975.
10. Mugford RA: External influences upon the feeding of carnivores, in Kare M, Maller O (eds): *The Chemical Senses and Nutrition.* New York, Academic Press, 1977, pp 25–48.
11. James WT, Gilbert TF: The effect of social facilitation on food intake of puppies fed separately and together for the first 90 days of life. *Br J Anim Behav* 3:131, 1955.
12. Hegsted DM, Gershoff G, Lentini E: The development of palatability tests for cats. *Am J Vet Res* 17:733–737, 1956.
13. Beauchamp GR, Maller O, Rogers JG, Jr: Flavor preferences in cats *(Felis catus* and *Panthera* species). *J Comp Physiol Psychol* 91:1118–1127, 1977.
14. Bartoshuk LM, Harned MA, Parks LTL: Taste of water in the cat: Effect on sucrose preference. *Science* 171:699–701, 1971.
15. Adamec RE: The interaction of hunger and preying in the domestic cat *(Felis catus):* An adaptive hierachy. *Behav Biol* 18:263, 1976.
16. Garcia J, Hankins WG, Rusiniak KW: Behavioral regulation of the milieu interne in man and rat. *Science* 185:824–831, 1974.
17. Rozkowska E, Fonberg E: Salivary reactions after ventromedial hypothalamic lesions in dogs. *Acta Neurobiol Exp* 33:553, 1973.
18. Durrer JL, Hannon JP: Seasonal variations in caloric intake of dogs living in an Artic environment. *Am J Physiol* 202: 375–378, 1962.
19. Fox MW: Psychogenic polyphasia (compulsive eating) in a dog. *Vet Rec* 74:1023–1024, 1962.
20. Houpt KA, Hintz HF, Shepherd P: The role of olfaction in canine food preferences. *Chem Senses Flavor* 3:281–290, 1978.
21. Lohse C: Preferences of dogs for various meats. *JAAHA* 10:187–192, 1974.
22. Kitchell RI: Dogs know what they like. *Friskies Res Dig* 8:1, 1972.
23. Grace J, Russek M: The influence of previous experience on the behavior of dogs toward sucrose and saccharin. *Physiol Behav* 4:453–458, 1969.
24. Houpt KA, Coren B, Hintz HF, Hilderbrant JE: Effect of sex and reproductive status on sucrose preference, food intake and body weight of dogs. *JAVMA* 174(10):1083–1088, 1979.
25. Grossman MI, Cummings GM, Ivy AC: The effect of insulin on food intake after vagotomy and sympathectomy. *Am J Physiol* 149:100–102, 1947.
26. Mason E: Obesity in pet dogs. *Vet Rec* 86:612–616, 1970.
27. Janowitz HD, Grossman MI: Some factors affecting the food intake of normal dogs and dogs with esophagostomy and gastric fistula. *Am J Physiol* 159:143–148, 1949.
28. Buffington CAT: Management of obesity—The clinical nutritionist's experience. *Int J Obes* 18:S29–S35, 1994.
29. Legrand-Defretin V: Energy requirements of cats and dogs—What goes wrong? *Int J Obes* 18:S8–S13, 1994.
30. Crowell-Davis SL, Barry K, Ballam JM, Laflamme DP: The effect of caloric restriction on the behavior of pen-housed dogs: Transition from unrestricted to restricted diet. *Appl Anim Behav Sci* 43:27–41, 1995.

AGGRESSION

KEY FACTS

- The best treatment for play aggression is for the owner to initiate appropriate play regularly.
- The best way to deal with a defensively aggressive or frightened cat is to avoid the cat until it calms.
- When introducing cats to each other, they should initially be kept separated for several days.
- Information from surveys indicates that declawed cats are no more likely to bite than cats with claws.

Aggressive Behavior in Cats

Peter L. Borchelt, PhD
Animal Behavior Consultants, Inc.
Forest Hills, New York
The Animal Medical Center
New York, New York

Victoria L. Voith, DVM, PhD
Department of Clinical Studies
School of Veterinary Medicine
University of Pennsylvania
Philadelphia, Pennsylvania

The second most serious behavior problem reported by cat owners is aggression to people or other cats (Figure 1).[1] Although cat bites do not lead to human fatality, they can result in serious infection and hospitalization of the person bitten. Fights between cats in the wild rarely lead to immediate fatalities but can result in serious infection and even eventual death. Cat fights in the home often result in significant veterinary expenses for the owner.

We obtained data concerning the incidence of aggressive behavior in pet cats through a survey of cat owners, who filled out questionnaires while they were waiting for medical or surgical attention for their pets at one of four veterinary hospitals on the East Coast.[a] A total of 887 questionnaires was either completely or partially filled out; 41% of the questionnaires referred to cats brought to the clinic and 59% to cats at home.[2]

Table I shows that agonistic interactions among cats are common; about 80% of cats hiss and 85% swat at each other, and 70% fight with each other at least occasionally. Aggressive behavior by cats to people occurs less frequently (Table II). For instance, about 25% of cats were reported to growl or hiss periodically at either family members or strangers. The incidence of cats swatting at and biting or scratching is higher for family members than for strangers. Some cats that bite and break the skin do so repeatedly (Table III).

Aggression is a complex behavioral system that occurs in a variety of contexts and is affected by a multitude of factors. In previous articles,[1,3] we have discussed the definition and value of aggression, the facial and postural signals of aggression in cats, and a diagnostic classification scheme for feline

[a]The Veterinary Hospital of the University of Pennsylvania (VHUP), The Animal Medical Center in New York City, a feline specialty practice, and a suburban small animal general practice.

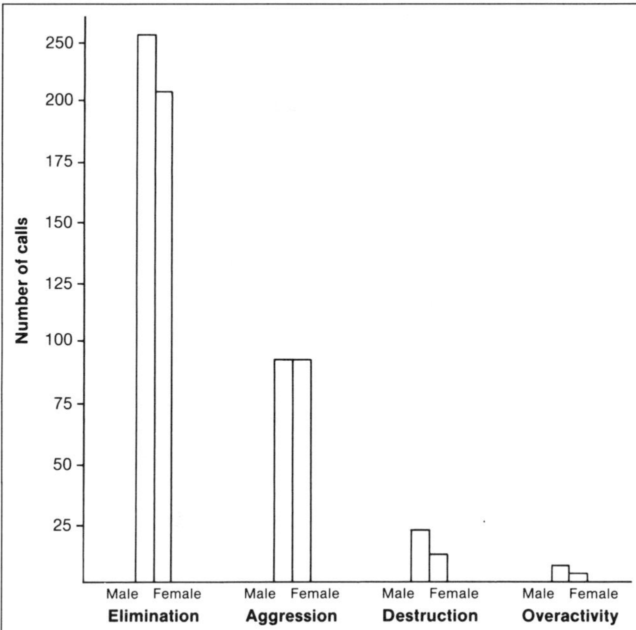

Figure 1—Distribution of the four common behavior problems in cats. Number of calls is based on the primary behavioral complaint made during a total of 721 telephone calls. (From Borchelt PL, Voith VL: Aggressive behavior in dogs and cats. *Compend Contin Educ Pract Vet* 7(11):950, 1985.)

aggression problems. In this scheme, feline aggression is divided into eight categories. In the following sections, we describe feline aggression problems and discuss treatment techniques as well as present some basic descriptive data from our cases.

Predation

Predation, or hunting, involves a sequence of searching or waiting for prey, stalking, pouncing, grabbing, and biting. Usually, predation is considered part of the feeding behavior system and not aggression. In domestic cats, predation is usually unnecessary to obtain food. Sometimes, however, cats that have access to the outdoors will hunt and often bring home prey, such as birds or mice. Bringing the prey home to the owner may be a socially motivated behavior or merely an attempt to cache food.

TABLE I

Aggressive Behavior Among Household Cats

Frequency of Interactions [a]	*Hiss (n=541)* [b]	*Swat (n=550)* [b]	*Fight (n=544)* [b]
Frequently	34.4%	43.1%	28.5%
Sometimes	19.2%	20.0%	16.0%
Infrequently	13.5%	13.1%	14.2%
Rarely	12.2%	9.5%	11.9%
Never	20.7%	14.4%	29.4%

[a]Frequently = once a week or more; sometimes = once a month or more; infrequently = less than once a month; rarely = once a year or more.
[b]Number of responses to the question.

It is difficult, if not impossible, to modify the behavior of a cat that is efficient at predation outdoors. Immediate punishment of the behavior is hard to implement, and associating an aversive stimulus with bringing the prey home is likely to have adverse consequences (e.g., inducing fear of the owner). The technique that is most likely to be successful is to deny the cat access to outdoors or "bell the cat" in such a manner that a prey has a better chance of escaping.

Predatory behavior of the indoor cat who stalks caged pet birds or rodents can sometimes be stopped by punishment. For example, wires conducting low-voltage current, as supplied by a battery (*not household current*), surrounding the cage of the small pet may deter the cat. Of course, the apparatus should be constructed so that the cage does not conduct electricity and so that the pet inside cannot be shocked.

A loud sound, such as that of a whistle or air horn, might be sufficiently aversive to frighten some cats and deter predatory behavior to small animals in the home. A plant mister or water pistol is not likely to be aversive enough. Regardless of method, delivery of the punishing stimulus by the owner could lead to cessation of predatory behavior only in the owner's presence.

TABLE II

Aggressive Behavior Between Cats and People

	Growls or Hisses		*Swats at But Does not Hit*		*Bites or Scratches Without Breaking Skin*		*Bites or Scratches and Breaks Skin*	
Frequency of Interaction[a]	*At Family (n=841*[b]*)*	*At Strangers (n=824*[b]*)*	*Family (n=842*[b]*)*	*Strangers (n=813*[b]*)*	*Family (n=841*[b]*)*	*Stranger (n=811*[b]*)*	*Family (n=830*[b]*)*	*Strangers (n=808*[b]*)*
Frequently	3.8%	4.2%	7.2%	2.5%	9.5%	3.1%	2.4%	0.7%
Sometimes	4.0%	3.8%	8.4%	4.1%	7.8%	2.6%	3.1%	1.2%
Infrequently	5.0%	5.1%	8.0%	4.1%	9.8%	4.3%	6.4%	1.7%
Rarely	12.4%	12.6%	9.6%	9.1%	10.1%	9.1%	10.7%	6.1%
Never	74.7%	74.3%	66.7%	80.3%	62.8%	80.9%	77.3%	90.2%

[a]Frequently = once a week or more; sometimes = once a month or more; infrequently = less than once a month; rarely = once a year or more.
[b]Number of responses to this question.

TABLE III
Frequency of Bites that Break the Skin of Family and Strangers

Number of Bites	*Family*		*Strangers*	
	n [a]	%	*n* [a]	%
None	595	77.5	687	87.5
1	47	6.1	36	4.6
2	32	4.2	26	3.3
3	19	2.5	9	1.1
4	7	0.9	1	0.1
5	22	2.9	14	1.8
6	5	0.7	1	0.1
7	1	0.1	2	0.3
8+	40	5.2	9	1.1

[a]Number of responses to this question.

Play Aggression

The most common type of aggressive behavior that cats exhibit toward their owners is a form of play (Table IV). Young mammals engage in considerable play behavior, and cats are no exception. Kittens play with each other, their mother, small moving objects, and objects that they can toss. They also play with their owners, particularly if no other young cats are available to play.

Play comprises components of other behavior systems, including predation, fighting, exploration, and investigation. Kittens explore dark corners and openings (e.g., boxes and bags) and then pounce or pat at seemingly nothing in the enclosure. These playful antics are part of *searching* and *catching* phases in a predatory sequence. The searching phase involves looking for game, and the catching phase relates to capture of the prey. Kittens also wrestle and attack each other as part of play-predatory behavior; these activities help them prepare for intraspecific (cat–cat) disputes and interactions later in life.

When animals play with each other, they give specific signals to indicate that the behavior is play and not to be confused with serious or real aggression. For example, although dogs growl and show their teeth during play, these behaviors are not misinterpreted because the dogs also exhibit "play bows" or paw at other dogs. During play, cats usually do not vocalize (hiss or growl) as they do during serious aggression. Although they *look* aggressive, the "attacks" are generally silent. Play bites are inhibited; that is, they are not done at full force and usually do not break through the skin. During play, the stronger animal modulates its behavior so that it does not injure the weaker animal. If play gets too rough, one animal quits or escalates aggression to serious biting; these actions terminate the play bout.

The typical play-aggressive cat is young—generally under two years of age—although play-aggressive behavior can be exhibited by cats of any age. Play aggression usually occurs in single-cat households or when other cats in the home are old and no longer playful. The cat may be selective as to whom it directs play aggression, often behaving that way only to a specific person. Play "attacks" also may occur in certain locations and only when the person is moving. Typical circumstances in which play attacks occur are as the person goes up or down the stairs, steps out of the shower, moves under the covers, or makes the bed. The cat often stalks or waits in ambush, then rapidly and silently runs toward a person. The cat may have a "nasty" look on its face, grab the person with its paws, and bite. Although play bites are typically inhibited, they may occasionally result in injury, particularly in the elderly or others with fragile skin.

Owners are often frightened by playfully aggressive cats because they "look dangerous." Owners may try to cope with an attack by running, putting a foot up, or trying to hit the cat. The cat usually interprets these antics as reciprocal play. If actually hit, the cat may immediately become defensive and seriously bite the owner.

Playfully aggressive cats are often very active, and some owners may try to cope with this problem by confining them. Sometimes owners also use confinement or "time out" as a punishment for being "bad." Restricting a cat's activity or isolating the cat is likely to worsen the problems because the cat will be even more motivated to play when released.

Playful animals must be allowed, and even encouraged, to play. If the owner tries to avoid or suppress a cat's play, the problem may only get worse or be manifested in other ways. Instead, the owner should initiate appropriate play several times a day using an object or toy that elicits a vigorous response. Not all commercial cat toys readily elicit play, and owners may need to experiment to find effective toys for their cats. Toys on strings that can be pulled and wiggled are usually effective, as are ping-pong balls or wads of paper that can be tossed and batted around. Owners can play-wrestle *indirectly* with their cats by using a stuffed sock. It is important that the owner *not* play-wrestle directly using hands or feet, because that type of play encourages the cat to grab and bite at human flesh. Providing small toys that the cat enjoys carrying around and playing with alone may occupy it during times when the owner is not present. Paper bags of different sizes can be provided for the cat to explore. Access to a window allows the cat to look out and entertain itself during the day.

If a cat's playful attacks can be predicted, the aggressive play could be directed to a moving toy before it becomes

TABLE IV
Incidence of Common Cat Aggression Problems Over a Two-Year Period [a]

Type of Aggression	*Total*	*To Cats*		*To People*	
		n	%	*n*	%
Play	22	3	13.6	19	86.4
Defensive	61	46	75.4	15	24.6
Territorial	9	8	88.9	1	11.1

[a]Data from Animal Behavior Consultants, Inc., New York, 1981 to 1983.

directed to the owner. For example, before going downstairs, the owner can throw a toy that the cat likes to chase; or the owner can keep a soft rope handy to wiggle when stepping out of the shower or when walking in "attack" areas in the home.

Punishment, properly timed and of just the right intensity, can be an appropriate technique in stopping aggressive play. It should not, however, be the only technique used; the cat also should actively be played with at other times. An effective punisher might be strong enough to disrupt the behavior and discourage the cat from trying it again but not so strong as to cause injury, anxiety, or defensive aggression. For punishment to be effective, it must be applied every time the cat begins to engage in the behavior. A startling noise, such as is produced by a small foghorn or canister of compressed air, meets these criteria quite well. For a few days, the owner may need to carry the noisemaking device constantly in order to use it each time *as* the cat attacks. Properly used, the device will probably only be required a few times. Whistles and water sprayed from a bottle may stop cats during an attack but usually neither is aversive enough to prevent future attacks.

Another possible solution to play aggression problems, particularly with a kitten, is to get another cat about the same age so they can play with each other. Two kittens are usually easier to raise than one.

Fear or Defensive Aggression

When a cat is frightened and defends itself from either another animal or a person, it may become aggressive. Defensive aggression may be exhibited, for instance, when a dog corners a cat or a person tries to discipline a cat by hitting it. Typical indications of defensive aggression are crouching, flattening the ears back against the head, hissing and spitting, piloerection (hair standing up), and scratching with the back feet. Although some of the postures are similar to those associated with submission in dogs, a cat in those postures is definitely *NOT* signaling submission. It is signaling that further approach by the other animal or person is likely to lead to a defensive attack.

The best way to deal with a defensively aggressive or extremely frightened cat is to avoid the cat until it becomes less afraid. An owner should not try to comfort the cat by approaching or picking it up or the owner may be attacked. The cat should be left alone until it relaxes enough to explore, play, or show affectionate behavior. In some instances, several hours or longer is required for the cat to calm.

Probably the most common problem involving defensive aggression occurs when two cats in the household who previously have gotten along well suddenly become aggressive toward each other. Neither of the cats seeks the other out; but, if they see each other, both will act startled and attack. Usually this problem begins "by mistake" or by accident. For example, Cat A may be sitting in a window as an outside cat walks by. Cat B, who is territorially aggressive toward outside cats, is in the middle of the room. Cat B sees the outside cat and rushes toward the window to attack it. Unfortunately, Cat A happens to be in the way and sees Cat B aggressively charging toward it. Cat A puffs up and hisses. Cat B redirects its attention to Cat A and puffs up and hisses. Then the cats attack each other. Each acts as if the other started the fight.

Occasionally, two friendly cats may be resting when a frightening incident occurs, such as a bookshelf falling over. Both cats become startled, puff up, and assume defensive postures. When they see each other in a defensive posture, they act as if the other is about to attack. Consequently, each reacts defensively, a fight ensues, and thereafter they are aggressive whenever they see each other.

Defensive aggressive behavior between two cats usually is treated successfully. Essentially, the cats must become accustomed to each other again without either cat becoming afraid or aggressive.[4] To implement this process, the cats should be kept separated so that they cannot see each other except during treatment procedures. One way to begin accustoming the cats to each other is while they are being fed. The cats can be brought out several times a day and put at opposite ends of the room where they are fed small amounts of food. If they are far enough apart and both are occupied with eating, each cat will be able to see the other in an unaggressive state. Gradually, over several days or weeks, the food dishes then can be placed closer together. Eventually, after eating, the cats can spend more time in the presence of each other, perhaps on leashes at a safe distance. The owners might pet or play with the cats to keep them relaxed while the cats are in proximity. It is important that this procedure be done gradually, without evoking fear or defense in either cat.

Another way to reaccustom the cats to each other is to expose them to each other for prolonged periods, but in such a way that one cannot get to the other. They might be kept in large cages placed at opposite ends of the room where they can see each other but cannot escape. Alternatively, the cats could be kept apart on leashes or in harnesses but within each other's view. After several hours, they might be brought closer together. After many such sessions, it should be possible for them to be next to each other, where they might groom or play together as they did before. When that status is achieved, they can be let loose.

Sometimes, however, a cat's defensive behavior can be so intense that even the sight of the other cat leads to an aggressive fear response. In such cases, it may be necessary to bring the cats gradually into each other's view; for instance, through a ½-inch gap of a partially open door or the space below the door. These limited visual presentations of one cat to the other may reduce the fear sufficiently that subsequent progress can be made. The door can gradually be opened wider, perhaps as the cats are fed, played with, or petted or kept at a distance on leashes or in wire cages.

If a cat is defensive and aggressive to a person, a treatment that is usually effective is for the person to minimize (ideally, avoid) all behaviors that frighten the cat and at the same time encourage nonfearful behaviors. One way to desensitize a cat's fear of a person is to associate the person with being fed. The person should be in the same room with the cat when it eats. Each time the cat is fed, the person should move a little closer until the cat will eat right next to the person. The person can then begin to engage in more activities around the cat, being careful not to frighten it. Alternately, the person could elicit play from the cat, for instance, with a rope that is wiggled along the floor or by tossing a toy that the cat likes to chase. Many desensitization sessions may be required before the cat no longer fears the person in other contexts.

Territorial Aggression

Territorial aggression between cats in a household often develops slowly and gradually. One cat is usually the aggressor and the other the victim. The aggressive behavior begins with hissing and growling; progresses to swatting, chasing, and relentless pursuit; and finally involves attacking and fighting. The victim may become progressively more afraid of the aggressor and defensive. It may hide on top of shelves or bookcases or in closets, coming out only when the territorially aggressive cat is not around. Occasionally, elimination problems occur because the fearful cat is too afraid to leave the vicinity of its hiding place.

We do not have a good understanding of the factors that determine which cats will become territorial and try to exclude other cats from the home. Territorial aggression between littermates, offspring and parent, or cats that have lived together for some time usually occurs when one or both of the cats are between one and three years of age. A territorial cat can be aggressive toward one cat in the household yet get along well with other cats in the home. The aggressor is not necessarily the first cat that was introduced into the household or the eldest.

Territorial aggression may be displayed immediately or soon after the introduction of a new cat to the home and may be displayed by either the newcomer or the resident cat. This type of problem may be difficult to distinguish from a typical defensive aggression problem, although chasing and pursuit of one cat by the other would suggest a territorial aggression problem.

In our experience, territorial aggression is rarely treated successfully. At present, the best solution for the problem often is to find one of the cats another home. A cat that is aggressive toward another cat is not necessarily aggressive toward people. Territorially aggressive cats can be good pets in a home without other cats.

Sometimes, providing a larger living area, for example, if the owner is moving from a small apartment to a large house or giving one of the cats access to outdoors, helps the problem. Another way to manage the problem is to restrict the cats to separate parts of the household or the yard so that they do not encounter each other.

People with cats that are aggressive toward other cats must be careful around them. If the owner interferes with a cat that is acting either aggressive or defensive, the owner may be bitten. A territorially aggressive cat may occasionally attack an owner who has recently handled another cat.

Occasionally, a cat that is friendly to family members is consistently aggressive to people who are visiting. Such cats do not exhibit fear or defensive behavior when a visitor enters but boldly approach and threaten (hiss, swat at, and chase) the stranger. This type of aggression may be restricted to the home; when the cat is in another environment, no aggression to strangers occurs. Aggressive behavior of this kind is uncommon, and we have had little experience treating it.

Intermale Aggression

As males reach adulthood, they may begin challenging one another. Roaming, intact tomcats frequently engage in ritualistic threats as well as actual fights. The cats sit or stand very stiffly and stare at each other. They rotate their heads slowly to a 45° angle and turn their ears so that the backs of the ears face forward. This posturing is accompanied by growls and very loud howling.[5] Cat owners and their neighbors often hear these vocalization duels occurring outside (or even inside) at night. Then one cat may leave very slowly, or one or both cats may attack. Sometimes an attack results in only one bite; but because cat bites frequently become infected, a single bite can be very serious.

Castration usually stops intermale fighting, particularly if *both* of the involved males are neutered. The effectiveness of castration *does not* appear to be related to the age of the cat or how long it has been fighting. A small percentage of castrated males will still engage in typically male fights.[6] Progestins can suppress the cat's motivation to engage in intermale aggression; but, when the drug is withdrawn, the cat probably will become aggressive again.

Redirected Aggression

When a cat is highly aroused and in an aggressive state (for instance, following a fight or having seen an outside cat), it may redirect its aggression to a person or another animal within reach. Generally, cats do not redirect aggression unless they are touched or closely approached by another animal or person. Redirected aggression can occur regardless of whether the aggression is intermale, territorial, fear-induced, or defensive in nature.

This type of behavior can occur in otherwise very affectionate and playful cats that the owner never would have expected to become aggressive. If a cat is in an aggressive state or mood, it is very dangerous for the owner to approach it, touch it, or try to pick it up. The owner should wait until the cat is out of its aggressive mood before interacting with it. Usually, it is unsafe to approach the cat until after it has engaged in another behavior, such as grooming, playing, or eating.

Aggression Related to Being Petted

Cats vary in how much they like being petted or held by people. Some cats enjoy being petted, held, carried, cradled, or even hugged lightly. Others merely tolerate these activities. Some do not mind being petted but do object to being held. Some like to be petted a little, but want the owner to stop after a few strokes. A few do not like to be

TABLE V
Number and Percentage of Declawed Cats and Cats with Claws Reported to Bite Family Members and Strangers[a]

Cat Bites	Family		Strangers		Population	
	n	%	*n*	%	*N*	%
One or more						
D[b]	45	(26.8)	27	(28.7)	252	(29.1)
C[b]	123	(73.2)	67	(71.3)	613	(70.9)
Eight or more						
D	9	(22.5)	2	(25.0)	252	(29.1)
C	31	(77.5)	6	(75.0)	613	(70.9)
Requiring treatment[c]						
D	6	(30.0)	2	(33.3)	252	(29.1)
C	14	(70.0)	4	(67.7)	613	(70.9)

[a]Numbers and percentages shown in comparison with the total questionnaire population.
[b]D = declawed; C = claws.
[c]Bites severe enough to break skin and require treatment by a doctor.

petted at all. Owners' feelings are often hurt when their cats resist physical contact, and they may think that the cats do not like them. Actually, cats that do not enjoy prolonged petting or being held often display other signs of attachment, such as playing with their owners, following them from room to room, sleeping on the bed with them, and even sitting in their laps.

When a cat nips or lightly bites while being petted, it is signaling the person to stop—much as it would signal another cat to stop grooming it. Hitting the cat or tapping it sharply on the nose is unlikely to teach it not to bite in this circumstance; rather, it may cause the cat to become defensive and perhaps retaliate with serious aggression. When a cat signals to stop petting, it is prudent to do so and petting should not be resumed at this time. An observant owner who knows a cat well should be able to predict when it is getting ready to bite. The cat may get restless, may flatten its ears, or its tail may begin to twitch. This is the time to stop petting, before the cat signals with an inhibited bite.

An owner who stops insisting that a cat accept petting or cuddling may actually find that the cat seeks the owner out more often and sits next to or on the owner's lap more frequently. Owner and cat may subsequently develop a better and more enjoyable relationship. Some cats are huggable, some cats are not; but both types of cats can be equally attached to their owners.

Pathophysiologic Disorders

It is important to differentiate normal, species-typical behavior and pathophysiologic aggressive disorders. In general, the description of a pathophysiologic aggression is not consistent with the descriptions of species-typical, normal aggressive behaviors. Furthermore, aggression related to pathophysiology would generally be accompanied by other abnormal neurologic and physiologic signs.

Neurologic disorders stemming from such conditions as infection, trauma, and infestation of parasites could all lead to abnormal behaviors. Rabies, a viral disease that can affect all warm-blooded animals, including cats and humans, usually results in behavioral changes in an infected animal.[7] Normally friendly animals become aggressive. The affected animal may bite at inanimate objects as well as other animals. Cats are reported more likely than dogs are to develop the "furious" form of rabies and can be very dangerous when handled. Clinical signs of the disease usually appear within 15 to 25 days of exposure; but, occasionally, extremely long incubation periods have been reported. Initially, the pupils may be dilated; but as the disease progresses, the pupils may become miotic. The affected animals may roam for long distances; exhibit both a change in voice and slobbering of saliva; and develop a dropped lower jaw, ataxia, progressive paralysis, and convulsions. Affected animals usually die within four days of the onset of characteristic clinical signs. Obviously, extreme care should be exercised in handling such animals to minimize risk of contracting infection. If the animal dies, its head should be sent to an appropriate laboratory for examination of nerve tissue. Although not required by law in most states or communities, cats allowed outdoors should should be vaccinated against rabies. Since 1981, the number of reported cases of rabies in cats has exceeded that reported in dogs in the United States.[8]

Idiopathic Aggressive Behavior

Occasionally, a cat is reported to suddenly and without warning become severely aggressive in the absence of any identifiable eliciting stimulus. Cats that do so may inflict multiple bites and scratches on owners during an attack episode. In these circumstances, the obvious first step is to avoid all interaction with the cat until a veterinary examination can be made to determine disease states. If the cat does not have an identifiable pathophysiologic disorder, a detailed behavioral evaluation should be made by a person knowledgeable about general animal behavior as well as feline behavior.

If the conditions (environmental or medical) responsible for the aggression cannot be identified, it is impossible to treat or manage the problem. Periodic attacks that cannot be treated or managed pose a great danger to the owner, and euthanasia of the cat is a reasonable option.

Clinicians should not, however, assume that cases initially presented by the owners as sudden unprovoked attacks actually are that. With careful questioning, such cases usually turn out to be redirected or defensive aggressive incidents that can be successfully managed or treated.

Preventing Aggression on Introducing New Cats or Reintroducing Cats to Each Other

To reduce the possibility of aggressive encounters when a new cat is brought into a household that has a resident cat, the cats should be kept separately for a few days. The cats should be alternated among rooms in the home to allow them to become familiar with each other's odor. This procedure also allows the new cat to acquaint itself with the environment, particularly with potential escape routes and hiding places. The cats also can be exposed to each other gradually using the techniques described previously in the section Fear or Defensive Aggression. Care should be taken not to introduce the cats too rapidly or they may attack each other. An introduction procedure usually takes only two or three days, particularly if the cats are young, but can occasionally take weeks.

If cats who are reintroduced (e.g., after one cat returns home from boarding or a visit to the veterinarian) act aggressively toward each other, the procedures used to introduce new cats to one another can be done.

Declawing and Aggressive Behavior

It is commonly believed by veterinarians and others that declawed cats are more likely to bite than cats not declawed. We know of no data that support this belief. Information from the questionnaire-based survey[2] (Table V) indicates that declawed cats are no more likely to bite, bite multiple times, or bite more severely than cats that are not declawed.

REFERENCES

1. Borchelt PL, Voith VL: Aggressive behavior in dogs and cats. *Compend Contin Educ Pract Vet* 7(11):949–957, 1985.
2. Voith VL, Borchelt PL: Social behavior of domestic cats. *Compend Contin Educ Pract Vet* 8(9):643–644.
3. Voith VL, Borchelt PL: *The Veterinary Clinics of North America: Small Animal Practice*, vol. 12, no. 4. Philadelphia, WB Saunders Co, 1982.
4. Voith VL, Borchelt PL: Fears and phobias in companion animals. *Compend Contin Educ Pract Vet* 7(3):209–218, 1985.
5. Leyhausen P: *Cat Behavior: The Predatory and Social Behavior of Domestic and Wild Cats*. New York, Garland STPM Press, 1979.
6. Hart BL, Barrett RE: Effects of castration on fighting, roaming and urine spraying in adult male cats. *JAVMA* 163(3):290–292, 1973.
7. Farrow BRH, Love DN: Bacterial, viral and other infectious problems, in Ettinger SJ (ed): *Textbook of Veterinary Internal Medicine*, ed 2. Philadelphia, WB Saunders Co, 285–281, 1983.
8. Burridge MJ: Wildlife rabies in the United States. *Avian/Exotic Pract* 1(1):17–22, 1984.

UPDATE

REDIRECTED AGGRESSION

Twenty-seven cats that had been presented to the Animal Behavior Clinic of the Veterinary Hospital of the University of Pennsylvania (ABC-VHUP) for aggressive behavior toward humans were classified according to the functional diagnoses in Table I in this update section.[1] Redirected aggression was the most common diagnosis, followed by play and defensive aggression. The most common stimulus for redirected aggression in the ABC-VHUP sample was another cat (see Figures 1 and 2 in this update). Other stimuli included high-pitched noises, dogs, visitors, and unusual odors. People were attacked when they approached or touched a cat that was aggressively aroused by one of these stimuli. The attacks were not inhibited, and the cats often inflicted severe injuries. The persons' responses to the attack often included screaming and grabbing the cat and quickly flinging it away. Subsequently, the cat was usually frightened of the people or locations involved in the redirected aggressive encounter and afterward continued to exhibit fear induced aggression in these situations. Follow-up of these redirected cases indicated that they could usually be successfully treated or managed.

Both the VHUP sample and the one depicted in Table IV of the original article indicate that play and defensive aggression are frequent reasons why cats manifest aggression toward people. The absence of redirected aggression in the New York data might stem from a population of cats housed in environments not conducive to situations that result in redirected aggression; that is, apartments in which outdoor cats are not seen. It also is possible that cases of

TABLE I

Functional Diagnoses of 27 Cats Presented for Aggression to Humans

Diagnosis[a]	*Number*	*%*
Redirected	9	33
Probable redirected	5	19
Play	7	26
Fear/defensive aggression	5	19
Biting when petted	4	15
Territorial-like aggression	2	7
Undiagnosed	2	7

[a]There was more than 1 diagnosis in some cases.

Figure 1—Nutmeg, the Abyssinian cat inside the door, is aggressively aroused by Frisbee, the Siamese cat outside.

Figure 2—Frisbee is not accessible to Nutmeg, but the good-natured, approaching Norwegian Forest Cat, Big Fuzzy, is. Guess what happens next?

redirected aggression were classified in the category of defensive aggression. Residual defensive aggression is a common sequela of a redirected aggressive attack and is often the reason the owner seeks help.

The most successful treatment is to identify the aggressive arousing stimuli and modify the cat's reaction to it (e.g., via castration, blocking the view of outdoor cats, or desensitization). It is also necessary to educate the owners about redirected aggression: what causes it, how to recognize it, how to prevent it, and what to do if it occurs. Owners should consider several factors when deciding whether to keep such a cat. These include the severity of the attacks, whether the stimuli that cause the aggression can be identified and controlled, the feasibility of changing the cat's response to arousing stimuli, and the ability of those in the household to recognize a potentially aggressive cat. Euthanasia is not the only option in such cases because favorable outcomes are possible.

The diagnosis of redirected aggression should not be made unless a link between an arousing stimulus and the attack can be directly identified or its likelihood be deduced. A redirected aggressive attack should be suspected if the cat has the following history: The cat is a totally indoor cat with a current and complete vaccination history and there is no indication of any neurologic abnormality. The cat is friendly and relaxed with its owners and visitors. The incident that prompted the owners to seek help was a violent attack on a visitor entering the house. The cat leapt at the visitor and hung onto her coat while growling, hissing, and biting the coat sleeves. Previously, the cat had always interacted amicably with this person. A detailed history revealed that a few times in the cat's past, the owners had tried to introduce a new cat or kitten. When the person carrying the new animal entered the house, the resident cat immediately attacked the incoming person and cat by leaping at them, growling, hissing, and biting. The visitor who was recently attacked had been playing with another friend's dog or cats shortly before entering the house. She was still wearing the same clothes. Although the aggression-arousing stimulus was not directly observable but was a likely possibility, this case could be considered probable redirected aggression.

It would be inappropriate and potentially dangerous to classify as redirected aggression those cases of sudden aggressive behavior for which no other diagnosis is apparent. The sudden onset of aggressive behaviors can be caused by neurologic disorders related to infectious agents, toxins, lesions, psychomotor epilepsy, or cerebral infarction[2,3]—as well as to physical maladies that cause pain.

INTERCAT AGGRESSION

A survey of owners of cats that had been spayed or castrated at or before 10 months of age indicated that there was less fighting in households with cats of the same sex than in households with cats of both sexes. This survey reported that 44% of 134 male cats and 30% of 152 female cats fought either occasionally or frequently with other cats.[4] These data are compatible with the high incidence of owner-reported intraspecific behaviors (hissing, swatting, and fighting), as discussed in our 1986 *Compendium* article. The question remains: What does this high frequency of apparently agonistic behavior mean? Are these serious threats and encounters or are they "not too subtle" means of communication? Perhaps these gross agonistic expressions reflect unpolished communication methods among cats, in contrast to the subtle and graded signals that have evolved in canids. This might be expected in a species (e.g., cats) with a flexible social system, as compared with a species that is almost obligatorily social. Perhaps the cat's behaviors reflect the initial stages of the evolution of communication signals in animals as they become progressively more social.

DECLAWING

Studies conducted on the effects of declawing since our original article was published have also failed to support the belief that declawing causes cats to become more aggressive.[5]

Pharmacology

See the Simpson and Simpson article on pharmacotherapy for an overview of case reports and review articles and possible mechanisms of pharmacologic influence on feline aggressive behaviors. Information on this topic is also in the update section of the article on cat elimination by Voith and Borchelt.

Genetics and Experience

See the Estep article on the ontogeny of behavior for a review of the influence of genetics and experience on feline aggression.

REFERENCES

1. Chapman BL, Voith VL: Cat aggression redirected to people: 14 cases (1981–1987). *JAVMA* 196(6):947–950, 1990.
2. Shores A, Cooper TG, Gartrell CL, et al: Clinical characteristics of cerebrovascular disease in small animals. *Proc ACVIM* 9:777, 1991.
3. Shepard DE, deLahunta A: Central nervous system disease in the cat. *Compend Contin Educ Pract Vet* 2(4):306–311, 1980.
4. Hart BL, Cooper L: Factors relating to urine spraying and fighting in prepubertally gonadectomized cats. *JAVMA* 184:1255–1258, 1984.
5. Landsberg GM: Feline scratching and destruction and the effects of declawing. *Vet Clin North Am Small Anim Pract* 21(2):265–279, 1991.

Aggressive Behavior in Dogs and Cats

Peter L. Borchelt, PhD
Animal Behavior Consultants, Inc.
Forest Hills, New York
The Animal Medical Center
New York, New York

Victoria L. Voith, DVM, PhD
Department of Clinical Studies
School of Veterinary Medicine
University of Pennsylvania
Philadelphia, Pennsylvania

Aggression is the most common and potentially the most serious behavior problem confronting the pet owner and small animal practitioner (Figure 1).[1-7] Dog bites occur in epidemic proportions; over one million people a year in the United States *report* being bitten by a dog,[8,9] and recent data indicate that dog bites are largely underreported.[10] Only 6.9% of bite incidents of dogs presented at the University of Pennsylvania School of Veterinary Medicine for behavior problems and 7.9% of bite incidents of dogs presented to Animal Behavior Consultants, Inc. are reported to local public health agencies.[11] Occasionally, dog bites even lead to human fatalities.[12,13] Aggressive behavior of cats (either toward people or other cats) is the second most common behavior problem reported by cat owners (Figure 2).[7]

What Is Aggression?

A veterinarian scratched by a growling cat that she is examining, an owner bitten when he tries to take a bone away from his dog, or a mail carrier that hears a dog barking and growling behind a door each can state correctly and with certainty that the animal in the situation is aggressive. A simple, precise, unitary definition of aggression has been impossible to formulate, however.

Despite thousands of research reports investigating aggression, no uniformly acceptable definition has ever been agreed on. By one definition, aggression[14] is behavior that leads to, or appears to an observer to lead to, the damage or destruction of some target. This is a fairly objective, observational definition of aggression and would include predation. Individuals in the field of human behavior are often reluctant to define aggression solely in terms of acts, such as killing or injury, and tend to consider the intentions of the attacker before labeling an act

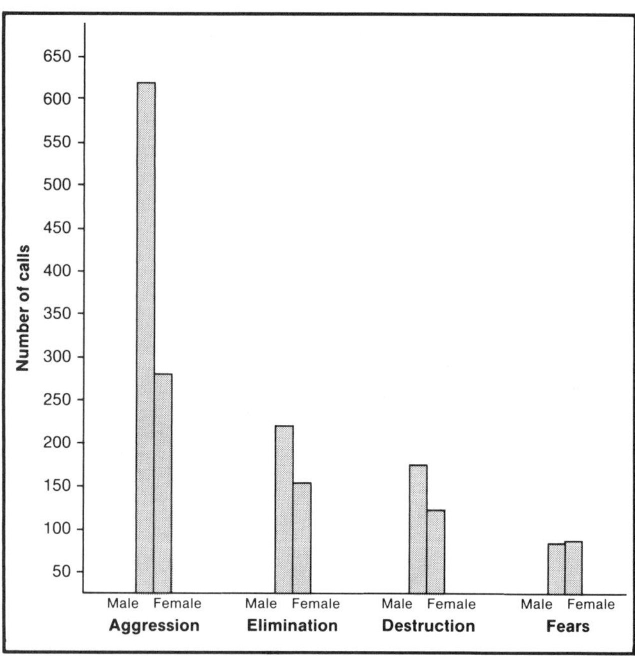

Figure 1—Distribution of the four most common behavior problems in dogs. Number of calls is based on the primary behavioral complaint made during a total of 1718 telephone calls.[7]

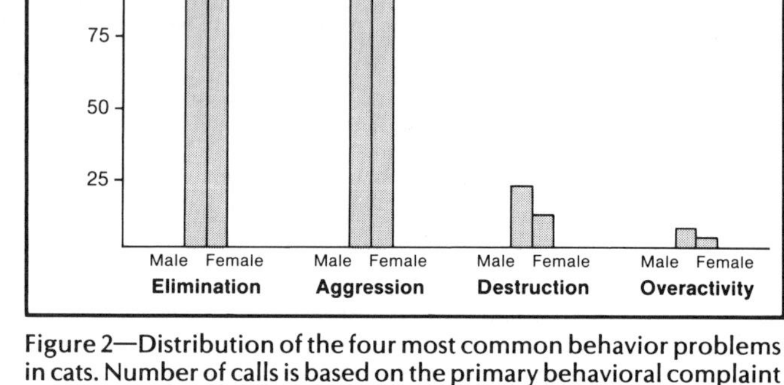

Figure 2—Distribution of the four most common behavior problems in cats. Number of calls is based on the primary behavioral complaint made during a total of 721 telephone calls.[7]

aggressive. There is, however, still considerable disagreement as to what actions constitute aggression. For example, not everyone would agree that each of the following actions is an example of aggression[15]:

- A boy swats at a hornet and gets stung.
- A cat kills a mouse, parades around with it, and then discards it.
- A wolf kills and devours a stray sheep.
- A farmer beheads a chicken and prepares it for Sunday dinner.
- A hunter kills an animal and mounts it as a trophy.
- A doctor gives a flu shot to a screaming child.
- A tennis player smashes his racket after missing a volley.
- A Boy Scout tries to assist an old lady but trips her by accident.
- A man commits suicide.
- An assassin misses his target.
- The President asks Congress for money for the Department of Defense.
- Two friends get into a heated quarrel after drinking too much.

If aggression in animals is defined solely as *inflicting injury* on another individual, then such threats as growling, baring the teeth, or snapping at an individual do not fit the definition. If aggression is defined solely in terms of certain *acts*, such as growling or biting, the definition is also unsatisfactory. For example, predation or food-getting behavior involves biting the prey, yet such behavior is not usually considered aggression. A dog or cat accidentally may bite its owner while playing; this, too, is not usually considered aggression. If aggression is defined solely as acts that occur when an individual is in an *emotional state*, e.g., "anger," the formidable task arises of defining and identifying the emotional state in order to assess whether or not an animal is aggressive. For example, consider the following sequence of behaviors. A young dog passes near the resting place of a dominant dog. The latter raises its lips and wrinkles its nose, and the younger dog averts its gaze and walks away. It is unclear if an emotional state is involved when such threats or social signals are exhibited.

The term *aggression* has many meanings and connotations. While avoidance of the term is unnecessary, it must be recognized that the word "aggression" has a commonsense, popular usage. Aggression is viewed by many people as a sinister act and is usually thought of as pathologic or maladaptive. Consequently, aggressive dogs are often erroneously labeled abnormal or as having a personality or temperament disorder.

Many animal behaviorists avoid the term *aggression*, believing it not very useful for a number of reasons. The word has been so colored by common usage and subjective interpretation that it immediately brings to mind a malicious intention or vindictive behavior that is inappropriately attributed to animals. Moreover, the term *aggression* does not always seem appropriate to use for subtle and noninjurious behaviors, such as threats, even though the ultimate results of these actions may be the same as the outcomes of overt attacks on an individual. Assertive actions by animals can be very subtle and may not involve direct or even indirect physical injury of another. For instance, a glance or a stare may be sufficient to assert dominance. Aggression is also a limited term in that it is often used to describe only one individual in a competitive encounter. Yet such encounters involve two or more individuals and a great amount of reciprocal communication.

Ethologists therefore developed their own term, *agonistic behavior*, to use in describing competitive encounters among animals. Agonistic is derived from a Greek word that means "to struggle." It is used in connection with any sort of behavior that involves a contest or confict between two animals, usually of the same species. The behavior could involve fighting, escape, or dominant and submissive gestures. The agonistic behavior system includes a variety of acts on a continuum of conflict. Agonistic behaviors include *threats* (sounds, postures, or even such a subtle facial expression as a stare), *offensive aggression* (such as chasing and biting) and *defensive behaviors* (including aggression, fleeing, submissive signals, threats, and biting).

Ethologists differ among themselves as to whether the term *agonistic behavior* should be used to describe conflict between species. Few behaviorists use the word as a synonym for predatory behavior; but many do use it to describe antipredator behavior, which involves defensive behaviors, such as threats, biting, and counterattacks as well as freezing and fleeing. Interspecific confict also can occur in the context of territorial defense. Most ethologists, however, confine their use of the term *agonistic* to intraspecific conflicts (conflict between members of the same species).

In all situations, though, animal behaviorists use terminology that describes the actual behavior of animals rather than inferred, subjective, internal states. For instance, the terms "vicious" and "angry" are not used in the animal behavior literature.

An individual animal can vary in its motivation to engage in aggressive behavior. Motivation can be influenced by physiologic states (e.g., hormones, hunger, state of health), environmental contexts, eliciting stimuli, and learning experiences.

An animal's postures and expressions correlate with its motivational state. Offensive dominant threats and submissive gestures are usually the antithesis of each other. An animal that is ambivalent or in conflict either assumes a composite expression of the conflicting expressions or will rapidly alternate between expressions (Figures 3 and 4). People are often surprised when bitten by a dog that is wagging its tail. Unfortunately, in such cases, people have ignored other signals the dog exhibits, such as piloerection, a wrinkled nose, or a low growl.

The motivation of an animal to attack, signal submission, or to flee can fluctuate depending on a number of variables, such as the size of the other animal, the distance from a specific location, previous social encounters with the other animal, presence of other animals in the vicinity, and an almost infinite number of other factors. For example, a bitch with her litter may unambiguously signal an approaching animal her intentions to attack and actually attack if the animal does not withdraw. Meeting the same individual several hundred yards from her nest, the bitch might instead give a

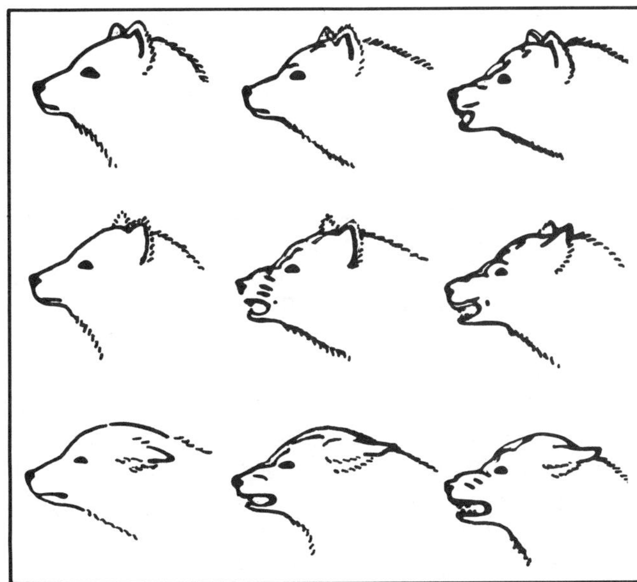

Figure 3—Facial expressions of dogs. A neutral expression is illustrated at top left. Increasing readiness to attack is shown across the top row. Increasing submission is shown down the left-hand column. The other expressions shown combine submissive and defensive signals.

Figure 4—Facial and body expressions of cats. A neutral expression is illustrated at top left. Increasing readiness to attack is shown across the top row. Increasing defensiveness is shown down the left-hand column. The other pictures show combinations of offensive and defensive signals. (Reprinted from Reference 20 with permission.)

submissive signal and avoid the other animal. At a distance intermediate from the puppies' location, the mother might give signals to the approaching intruder indicating both a willingness to attack and to inhibit attack.

Joggers, parents, and the general public frequently ask what they should do if confronted by an aggressive dog. The answer depends on the context, the circumstances, and the animal's degree of motivation to be aggressive. A three-year-old boy who threatens an aggressive dog might elicit an attack; an adult man who threatens the same dog may inhibit an attack. Major points to remember are that (1) whether or not an animal manifests aggression is dependent on a number of (intervening) variables, (2) an animal's aggressive tendencies can be graded depending on the variables, and (3) an animal can be in conflict about what to do because of simultaneously elicited conflicting motivations (i.e., escape or flight).

Aggression is a multidimensional concept and can occur in several contexts: between members of the same species in a social unit in order to establish a dominance/subordinance hierarchy; between members of the same species that are not in the same social group in order to acquire or defend a resource; and between animals of different species either to acquire or defend a resource or as a means of defense against predators. Moreover, aggression is influenced by numerous factors, such as sex, age, and size as well as an animal's hormonal status, its distance from another animal or from specific locations, whether the animal is in its territory, and the outcomes of previous encounters with other individuals. Often, an aggressive interaction involves a mix of behaviors; for instance, combinations of, or alternation between, offensive and defensive behaviors are common. The facial expressions and body postures assumed and specific aggressive behaviors exhibited usually vary in these different contexts.

The Value of Aggression

Competition between conspecifics (members of the same species), and competition between species is part of the process of survival. Animals that are able to obtain and maintain access to the necessities of life (food, mates, territory, etc.) tend to survive and hence pass their genes to future generations. Successful defensive behavior enables an animal to keep from being a meal for a predator and reduces its chances of injury in a fight with a conspecific.

In the natural environment, intraspecific competition generally does not lead to killing and does not always result in injury. There is survival value for both parties in settling conflicts with conspecifics using such social signals as threats, bluffing, and intimidation rather than risking injury in an actual fight. Social status, submission, and intent to defend resources can be communicated without involving actual physical encounters and permit social species to live in close proximity without the disruptive effects of actual fighting. It is, however, a mistake to believe that all conspecific fighting is ritualistic. Dogs, cats, and wolves living in the same social group can and do hurt each other. Telling owners to let their pets "work things out by themselves" could result in the death of one or more animals.

Animal behaviorists view most aggression as adaptive behavior that involves the assertion of an individual at the expense, or potential expense, of another. Aggressive encounters are a form of competition over a resource or an opportunity to enhance an animal's genetic representation (production or survival of offspring or relatives). Sometimes the resource is readily identifiable, such as food, shelter, or territory. Other times the resource or benefit is not immediate, but the aggression presently being manifested allows the animal to collect a payoff later. This is most evident in the aggressive encounters that periodically occur in the establishment and maintenance of a dominance hierarchy. Dominant animals usually have first access to resources, such as mates, food, water, and shelter.

Classification of Aggressive Behavior Problems

Successful treatment for any problem, whether medical or behavioral, requires an appropriate classification scheme for making a diagnosis. Behavior problems in general and aggression problems in particular can be classified in a number of ways.

Aggressive behavior problems in dogs can be classified by the target to which the aggression is directed, such as *owners, strangers,* or other *animals.*[4] An advantage to this type of classification scheme is that it is objective and easy to use. The major difficulty in classifying this scheme, however, is that dogs may exhibit aggressive behavior to these targets for more than one reason. For example, a dog may bite the owner in the context of either dominance or fear. Aggressive behavior problems in these two contexts are treated very differently. Contexts may be differentiated based on the sequence of actions that occur and the dog's facial and body expressions. Aggression in the context of dominance might involve the dog staring at or standing over the owner; whereas aggression in the context of fear might involve the dog withdrawing from the owner, gaze aversion, and tucking the tail between its legs.

Another classification scheme[16] loosely based on Pavlovian reflexology[17] generally dichotomized dogs into those that have nervous systems described either as excitable or inhibitable. In this scheme, aggressive problems occur as a result of frustration of so-called "active and passive defensive reflexes," the "freedom reflex," and the "chase reflex." "Stress" or conflict between various reflexes is implicated in the etiology of behavior problems. There are several difficulties with this classification scheme. First, Pavlovian reflexology, or nervous system typology, has not proved useful in either modern neurophysiology or behavior. Second, current terminology in animal behavior has moved away from the too simple notion of "reflexes" for complex sequences of behaviors. Third, the term *reflex* implies a fixed behavior pattern that is elicited in an all-or-none fashion and once initiated is impervious to environmental modulation. For example, a behavior sequence involving orienting, running toward, vocalization, threat, or biting directed to the movement of a small animal or a running person is too complex and variable to be described adequately as a "chase reflex." Furthermore, it is dubious whether any sequence of behavior can be accurately described as a "freedom reflex."

Some classification schemes dichotomize aggressive behavior into learned or "trained" aggression vs. "genetic" aggression. Attempting to classify aggression, or any other behavior, by the dichotomy of learned vs. genetic (or unlearned) is futile. All behaviors (aggressive or nonaggressive) are influenced by genes and the consequences of genetic/environmental interactions throughout an animal's life. Some of an animal's experience with the environment can be described as "learning," although even learned behaviors are based on species-typical responses that have a genetic base. Neither is behavior the result simply of genes; genes can only operate in an environment that in turn feeds back to influence the expression of the genes.

Classifying aggressive behavior problems by function seems most useful. A functional classification scheme is based on analysis of sequences of behavior patterns and the context or stimulus conditions in which they typically occur. Classification by function allows deduction of the motivation for the behavior and analysis of causative factors, which aids in formulation of treatment and prevention programs. A functional classification scheme can be objective and easy to use, assuming that the full range of an animal's behavior sequences and the prevailing stimulus conditions can be adequately observed or described. For aggressive (and other) behavior problems of companion animals, these criteria can be easily met, either through direct observation or careful history taking.

Facial and Postural Signals of Dogs and Cats

Familiarization with normal behavior patterns and the facial and body expressions of dogs[18,19] and cats[20] is necessary to use a functional classification scheme. In dogs, a neutral facial expression is one of erect but relaxed ears and the tail in a relaxed, downward position. As the dog becomes more aroused, the tail assumes a horizontal then erect position. As the dog becomes offensively aggressive, the ears become more erect, the hair may rise on the dog's neck and back, the dog stares directly at the target or threatening stimulus, the lips are retracted vertically exposing the canine teeth, and the dog barks threateningly or growls. A full threat involves a tail held still and erect. The dog also stiffens and appears to be walking on the tips of the toes.

When a dog assumes a fearful or submissive posture (the two are similar but not identical), the ears are drawn back against the head; the head and neck are lowered; the tail is kept low and may be tucked under the body; the dog avoids direct eye contact; and it retracts its lips in a horizontal manner. An extreme submissive posture involves the dog lowering itself to the ground, assuming a semilateral recumbent posture and raising a rear leg. A dog in conflict or ambivalent between motivational states might assume a composite expression. The dog might both vertically and horizontally retract the lips; growl but avert its gaze; flatten the ears but vertically retract the lips; approach with ears back, growling but wagging its tail; or bark aggressively, ears up, but tail tucked under the body. A display of ambivalence is not abnormal behavior (Figure 5).

Cats, in a feral state, usually live solitary life-styles and are not believed to have evolved dominant and subordinant social signals. Cats do, however, signal intentions to attack or defend themselves offensively or defensively. Variations in the size of the pupils, position of ears, carriage of tail, body posture, piloerection, and vocalization, all carry gradations of meaning; ambiv-

Figure 5A

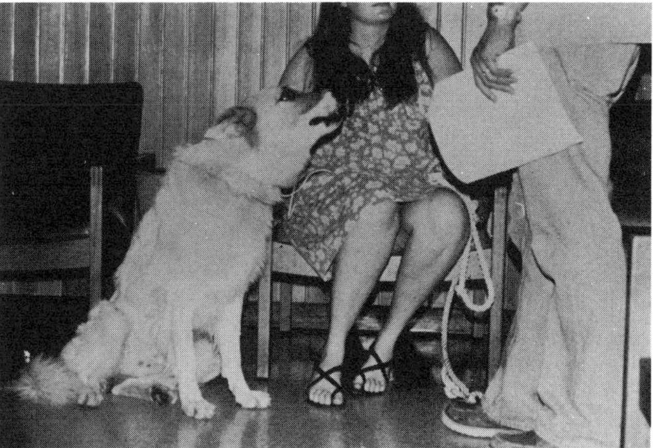

Figure 5B

Figure 5—The dog shown watched intently as the veterinary student approached (**A**). When the student reached a critical distance from the dog, it exhibited fearful/submissive signals (e.g., ears back, horizontal retraction of the lips). The dog was ambivalent, however. It was also growling, kept its head held high, and was leaning forward (**B**).

TABLE I

Functional Classification and Descriptions, Physiologic Variables, and Treatment Techniques of Normal Species-Typical Aggressive Behavior Problems of Dogs[a,b]

Classification	*Description*	*Circumstances*	*Sex*
Dominance aggression (can be very subtle and difficult to diagnose)	Bark, growl, bite directed to family members; often very friendly to strangers	Dog assumes dominance posture (stare, stand over, etc.); resists submissive postures (roll over or push, pet, groom, etc.); guards objects or areas in home, threatens at passageways; disturbed while sleeping or resting; may not be aggressive to all members of the family; punishment escalates aggression	Usually M
Possessive aggression (often associated with dominance aggression)	Bark, growl, bite directed to animals or humans	When person or animal approaches dog in possession of food, toys, objects	M or F
Protective aggression	Bark, growl, bite directed to animals or humans	When person or animal approaches area (home, room, or yard), owners, other animals	M or F
Predatory aggression	Chase or bite directed to animals or humans	Often preceded by stalking; elicited by quickly moving stimuli	M or F
Fear aggression	Bark, growl, bite directed to animals or humans (strangers or family); facial and body postures indicative of fear (e.g., ears back, tail down, crouch)	When dog is approached or reached for (especially when cornered), threatened, or punished	M or F
Intermale aggression	Bark, growl, bite directed to other male dogs; sometimes dominance and/or fear postures exhibited	When another male dog is observed at distance, approached; often in competitive or arousing situations	M
Interfemale aggression	Bark, growl, bite directed to other female dogs; sometimes dominance and/or fear postures exhibited	Usually directed to female dog in home; often in competitive or arousing situations	F
Pain-elicited aggression	Growl, bite directed to humans	When person attempts to groom, medicate, manipulate painful area	M or F
Punishment-elicited aggression	Bark, growl, bite directed to humans	When dog is exposed to aversive stimulus, conditioned aversive stimulus	M or F
Maternal aggression	Bark, growl, bite directed to animals or humans	When individuals approach puppies, puppy surrogates, nesting area	F
Redirected aggression	Growl or bite redirected to person or object other than that which evoked the initial aggression	Interference when dog is threatened or fighting	M or F

[a]This table originally appeared in Symposium on animal behavior, in Voith VL, Borchelt PL (eds): *The Veterinary Clinics of North America: Small Animal Practice*, vol. 12, no. 4. Philadelphia, WB Saunders Co, 1982. Modifications have been made.
[b]It is not uncommon for a dog to develop more than one type of aggressive behavior problem.
[c]Counterconditioning is teaching a behavior that is counter to the undesirable one. An example would be to teach a dog to lie down for a food reward when the dog's undesirable habit is to scratch at the door or leap up when food is offered.
[d]Habituation is getting the animal so used to the circumstances that it no longer pays attention or responds.

alent postures are common. Practitioners dealing with cats would be well-advised to study Leyhausen's feline behavioral descriptions[20] in detail, keeping in mind that no other studies have confirmed or questioned the conclusions.

Functional Classification of Aggressive Behavior Problems in Dogs and Cats

A functional classification scheme for aggressive behavior problems of dogs is presented in Table I and of cats in Table II. These tables list the types of behavior

Age	*Treatment*
Usually develops at about 1-3 yr; may occur earlier	Program involving: 1. Temporary avoidance of eliciting stimuli 2. Castration 3. Progestin therapy 4. Counterconditioning[c] or habituation[d] Physical punishment (hitting or even leash correction) is potentially dangerous and, if used, should be avoided until the problem has been reduced.
Not age related	Habituation; counter-conditioning; mild punishment for puppies
Usually develops at 1-3 yr	Habituation; counter-conditioning
Not age related	Habituation; counter-conditioning; punishment; confinement
Not age related	Gradual exposure techniques involving desensitization and counterconditioning
Usually develops at about 1-3 yr	Castration; habituation or counterconditioning; progestin therapy; separate dogs when not supervised
Usually develops at 1-3 yr	Spay if aggression is associated with phases of estrous cycle; habituation or counter-conditioning; separate dogs when not supervised
Not age related	Habituation or counter-conditioning; treat cause of pain; avoid eliciting pain
Not age related	Owner should avoid using punishment or conditioned punisher; habituation or counter-conditioning
When has litter or is in pseudocyesis	Avoid eliciting the aggression; habituation or counter-conditioning; spay
Not age related	Avoid directly interfering in aggressive situations; habituation or counterconditioning

problems with a brief description of the behavior patterns that typically occur, the typical circumstances in which they occur, information on whether sex and age are important factors, and the most effective types of treatment techniques. It has been the authors' experience that the vast majority of aggressive behavior problems of dogs and cats can be placed in this scheme. It should be noted that it is not unusual for an animal to display more than one aggressive behavior problem. In dogs, dominance and possessive aggression and dominance and punishment-elicited aggression occur together frequently. A dog may also exhibit combinations of unrelated problems, such as dominance, protective, and fear aggression, although the eliciting stimuli for these problems usually differ. It also is not unusual for an individual dog or cat to exhibit aggressive behavior as well as nonaggressive problems, such as phobias, separation anxiety, or overactivity.

It is important that the veterinarian be familiar with normal, species-typical aggression problems in order to differentiate them from pathophysiologic-based aggression. Laboratory research on a variety of species shows that stimulation or induced lesions of various regions of the brain can produce aggressive behaviors. It is thus reasonable to assume that some clinical cases of aggression might involve brain disorders. It is unlikely, however, that aggression resulting from neurologic or metabolic pathology would involve the same or similar sequences of behavior and eliciting conditions as normal aggressive behavior problems.

Few cases of pathophysiologic-based aggression have been documented. Sudden personality changes involving unprovoked serious aggressive attacks on people by dogs have been reported to be correlated with cardiac abnormalities and neural changes indicative of cerebral hypoxia.[21] It is possible that spontaneous electrical discharges in the brain could result in aggression similar to aggression elicited by electrical stimulation. The human literature regarding psychomotor seizures and aggression is inconclusive, however.

The following criteria would support a diagnosis of aggression caused by seizure activity:

- The aggression does not fit a description of a normal, species-typical behavior.
- Concurrent neurologic or pathophysiologic signs, such as an abnormal EEG or grand mal seizures, are seen.
- Epileptogenic drugs, such as phenothiazines, can elicit the aggressive behavior.
- Antiepileptic drugs can suppress the aggressive behavior.

A case that meets most of the above diagnostic criteria involved an intact, male Lhasa apso that barked, growled and seemingly attacked the air about him for a few minutes when presented with food or when a blanket was thrown over him (Figure 6).[22] An intravenous injection of chlorpromazine immediately elicited the aggressive behavior in the absence of food or a blanket, and oral administration of diazepam allowed the dog to eat without engaging in aggression. No EEG data were obtained. Other examples of pathophysiologic disorders and aggression involve possible neurochemical disturbances[23] and infectious diseases, such as rabies.

The term, "mental lapse syndrome"[1,24] has been applied to serious aggression cases in dogs, but no detailed

TABLE II

Functional Classification and Descriptions, Physiologic Variables, and Treatment Techniques of Normal Species-Typical Aggressive Behavior Problems of Cats[a]

Classification	*Description*	*Circumstances*	*Sex*
Predation	Stalking, chasing, killing, retrieving small game	Elicited by quick movements	M or F
Fear aggression[b]	Scratching, hissing, growling, biting directed to humans or other animals; defensive posture (ears back, body hunched) or combination of attack and defensive postures (ears back, back arched, piloerection)	When cat is approached or reached for (especially when cornered); when punished	M or F
Territorial aggression	Swatting, chasing, and attacking, progressing to relentless pursuit of another cat	When new cat is introduced to home; when resident cat reaches behavioral maturity	M or F
Intermale aggression	Scratching, hissing, growling, biting directed to another male cat; usually preceded by ritualized threat displays	Often related to territorial or competitive situation	Male, usually intact
Play aggression	Scratching and (usually inhibited) biting associated with stalking, pouncing, hopping sideways; directed to people	Moving stimuli, such as owner walking in home; may be directed to only one person in home; usually only one cat in home	M or F
Pain-elicited aggression	Scratching or biting	When person attempts to groom, medicate, manipulate painful area	M or F
Redirected aggression	Scratching, biting redirected to person or animal other than that which evoked the initial aggression	Interference when cat is threatening, fighting, or afraid	M or F
Undiagnosed serious aggressive attacks (rare)	Severe, repeated scratching, growling, biting people	Apparently unprovoked; with no indications of preceding threat, play, etc.	M or F

[a]This table originally appeared in Symposium on animal behavior, in Voith VL, Borchelt PL (eds): *The Veterinary Clinics of North America: Small Animal Practice*, vol. 12, no. 4. Philadelphia, WB Saunders Co, 1982. Modifications have been made.
[b]Often occurs when one cat is returned to the home after a short absence; may also arise from an event in the home that frightens one of the cats.

sequences of behaviors or contexts in which they occurred have been described and the EEG data presented as confirmation of an abnormality could also be interpreted as a muscle or movement artifact (e.g., 160 cycles/sec correlated with inspiration). Schizophrenia, purportedly described by Woodhouse[25] and in some commercially provided pamphlets for distribution to veterinary clients does not exist in animals.

Although most animals presented by owners because of aggressive behaviors are showing normative behavior

Figure 6A

Figure 6B

Figure 6C

Figure 6—This Lhasa apso would begin eating normally (**A**) then lift his head, growling (**B**), and begin whirling about growling and attacking the area around him. Diazepam suppressed this behavior, and chlorpromazine exacerbated it.

Age	Treatment
Not age related	Normal feline behavior; prevention of access to prey
Not age related	Desensitization and/or counter-conditioning; isolate fighting cats from each other except during treatment sessions
Usually develops after sexual maturity at 1-3 yr	May not be easily treated; progestins usually ineffective; counterconditioning and habituation techniques may be effective; in many cases, the only treatment is removal of one of the cats
Usually develops after sexual maturity at 1-3 yr	Castration; progestin therapy
Often in young cats, but may occur in adults	One or more: avoid situations that elicit the behavior; redirect play to an appropriate object; punishment, such as loud sound
Not age related	Habituation or counter-conditioning; treat cause of pain; avoid eliciting pain
Not age related	Avoid directly interfering in aggressive situations; habituation or counterconditioning of initial aggression
Not age related	Must be managed carefully or consider euthanasia

patterns for the species, this does not minimize the danger in keeping a pet that behaves aggressively. The aggression may be within normal limits for an animal, but still be unacceptable and undesirable from the owner's or society's perspective.

Before embarking on treatment of aggressive disorders, clients should be fully aware that they must assume responsibility for the animal's behavior during and after the therapy. They will have to take precautions to minimize risks during the course of therapy. The incidence and frequency of specific aggressive behaviors can often be reduced, and sometimes eliminated, if owners conscientiously apply the therapeutic regimens suggested and/or take appropriate precautionary measures. Clients must be instructed, however, that there is never any guarantee that an aggressive behavior can be completely eliminated. The most that may be accomplished is to reduce the probability that aggression will occur.

It is also important for a practitioner to realize that simply offering a client a suggestion or telling an owner to read a book as a solution to an aggressive behavioral problem may not suffice. Many behavioral techniques, even if slightly misapplied, can have the opposite effect of that intended (i.e., the behavior problem can get worse). A practitioner must be sufficiently knowledgeable about behavioral techniques to prevent placing the animal or client in a position of potential danger. Treatment of complicated behavior cases and aggressive behavior problems is usually complex and requires a commitment from the clinician to follow through and be available to correct and modify the therapy as indicated.

REFERENCES

1. Beaver BV: Clinical classification of canine aggression. *Appl Anim Ethol* 10:34-43, 1983.
2. Borchelt PO: Aggressive behavior of dogs kept as companion animals: Classification and influence of sex, reproductive status and breed. *Appl Anim Ethol* 10:45-61, 1983.
3. Polsky RH: Factors influencing aggressive behavior in dogs. *Calif Vet* 37:12-15, 1983.
4. Mugford RA: Behavior problems in the dog, in: Anderson RS (ed): *Nutrition and Behavior in Dogs and Cats*. Oxford, Pergamon Press, 1984, pp 207-215.
5. Houpt KA: Companion animal behavior: A review of dog and cat behavior in the field, the laboratory and the clinic. *Cornell Vet* 75:248-261, 1985.
6. Voith VL: Clinical animal behavior. *Calif Vet* 33:21-25, 1979.
7. Voith VL: Analysis of 2500 telephone calls about behavior problems of dogs and cats (presentation). Raleigh, NC, Animal Behavior Society Meeting, June 1985.
8. Harris D, Imperato PJ, Okern B: Dog bites—An unrecognized epidemic. *Bull NY Acad Med* 50:981-1000, 1974.
9. Beck AM, Loring H, Lockwood R: The ecology of dog bite injury in St. Louis, Missouri. *Public Health Rep* 90:262-267, 1975.
10. Beck AM, Jones BA: A study of unreported dog bites in children. *Public Health Rep* 100(3):315-321, 1985.
11. Voith VL, Richter C, Borchelt PL: Profile of dog bites presented for behavior problems, submitted to *Public Health Rep*.
12. Winkler WG: Human deaths induced by dog bites, United States, 1974-75. *Public Health Rep* 92:425-429, 1977.
13. Borchelt PL, Lockwood R, Beck AM, Voith VL: Attacks by packs of dogs involving predation on human beings. *Public Health Rep* 98:59-68, 1983.
14. Moyer KE: Kinds of aggression and their physiological basis. Part A. *Common Behavior Biol* 2:65-87, 1968.
15. Johnson RN: *Aggression in Man and Animals*. Philadelphia, WB Saunders Co, 1972.
16. Campbell WE: *Behavior Problems in Dogs*. Santa Barbara, CA, American Veterinary Publications, Inc, 1975.
17. Pavlov IP: *Lectures on Conditioned Reflexes*. New York, New York International Publishers, 1928.
18. Lorenz K: *On Aggression*. New York, Harcourt, Brace & World, 1953, p 96.
19. Fox MW: *Behavior of Wolves, Dogs and Related Canids*. New York, Harper & Row, 1971, p 190.
20. Leyhausen P: *Cat Behavior: The Predatory and Social Behavior of Domestic and Wild Cats*. New York, Garland STPM Press, 1979.
21. Meierhenry EF, Lui S: Atrioventricular bundle degeneration associated with sudden death in the dog. *JAVMA* 172:1418, 1979.
22. Voith VL: Behavioural problems, in Chandler EA, Sutton JB, Thompson DJ (eds) *Canine Medicine and Therapeutics*, Oxford, Blackwell Scientific Publications, pp 499-537.
23. Corson SA, Corson EO, Arnold LE, Knapp W: Animal models of violence and hyperkinesis. Interaction of psychopharmacologic and psychosocial therapy in behavior modification, in Serban G, Kling A (eds): *Animal Models in Human Psychology*, New York, Plenum Press, 1976.
24. Beaver BV: Mental lapse aggression syndrome. *JAAHA* 16(6):937-993, 1980.
25. Woodhouse B: *No Bad Dogs the Woodhouse Way*. New York, Summit Books, 1982.

UPDATE

Several other researchers/practitioners[1–6] have also presented data indicating that aggression is the most common behavior problem in dogs and that males are more likely to display aggressive behavior than are females. Table I below presents a more detailed analysis of the data depicted in Figures 1 and 2 in the original article. This table compares the percentage of consecutive telephone calls concerning categories of problem behaviors that were performed by male dogs and cats to the percentage of males presented for medical and surgical reasons. Several licensing agencies in the Philadelphia area also indicated that 50% to 52% of licensed dogs were male. The percentage of complaints about male dogs is disproportionate to their percentage in the canine population.

Table II presents the targets of canine and feline aggression, as depicted in Figures 1 and 2. Most of the calls about dogs were about aggression toward humans. Most of the calls about feline aggression problems were about aggression directed toward other cats; but a surprisingly high percentage of calls concerned aggression toward humans. Marder and Voith designed the data collection form included in this update. This form may help practitioners in history taking, evaluation, diagnosis, and treatment when a dog is presented because of aggressive behavior.

Many neurotransmitters are active in stimulation and suppression of aggressive behaviors. Attention has focused recently on the roles of dopamine receptor antagonists and serotoninergic mechanisms.[7,8] In general, studies show an inverse relationship between serotonin levels in the central nervous system and aggressive behavior in many species. Single case studies and a few small clinical trials suggest that selective serotonin-reuptake inhibitors, such as fluoxetine, may also reduce some types of aggressive behavior of companion animals (see the Simpson and Simpson article on pharmacotherapy elsewhere in this book).

TABLE I
Comparison of Sex Ratio Among Behavior Cases and Among Medical or Surgical Cases[a]

Behavior Problem	*Number of Calls*	*Percent About Males*	*z*	*P value*
Canine				
Aggression	885	(69.7)	8.75	<.0001
Elimination	381	(59.8)	2.93	.0017
Vocalization	103	(63.1)	2.25	.0122
Destruction	301	(56.8)	1.67	.0475
Fears	168	(49.4)	1.02	.1539
Overactivity	67	(56.7)	0.77	.2206
Feline				
Vocalization	13	(76.9)	1.59	.0559
Aggression	182	(50.0)	1.36	.0869
Destruction	36	(63.9)	1.07	.1423
Overactivity	16	(62.5)	0.60	.2741
Elimination	473	(54.8)	0.09	.4641

[a]Data from telephone calls to the Animal Behavior Clinic of the Veterinary Hospital of the University of Pennsylvania about behavior problems of dogs and cats in comparison with the percentage of males among canine and feline medical or surgical patients of the Veterinary Hospital of the University of Pennsylvania. (Z = The number of standard deviations away from the mean.)

REFERENCES

1. Adams GJ, Clark WT: The prevalence of behavioral problems in domestic dogs: A survey of 105 dog owners. *Aust Vet Pract* 19:135–137, 1989.
2. Beaver BV: Profiles of dogs presented for aggression. *JAAHA* 29: 564–569, 1993.
3. Beaver BV: Owner complaints about canine behavior. *JAVMA* 204: 1953–1955, 1994.
4. Knol BW: Behavioral problems in dogs: Problems, diagnoses, therapeutic measures and results in 133 patients. *Vet Q* 9:226–234, 1987.
5. Landsberg GM: Distribution of canine behavior cases at three behaviour referral practices. *Vet Med* 86:1011–1017, 1991.
6. Wright JC, Nesselrote MS: Classification of behaviors in dogs: Distribution by age, breed, sex, and reproductive status. *Appl Anim Behav Sci* 19:169–178, 1987.
7. Miczek KA, Weerts E, Haney M, Tidey J: Neurobiological mechanisms controlling aggression: Preclinical developments for pharmacotherapeutic interventions. *Neurosci Biobehav Rev* 18:97–110, 1994.
8. Ferris C: Using the golden hamster to understand the neurobiology of aggression. Paper presented at AVMA meeting, July 8, 1995.

TABLE II
Targets of Canine and Feline Aggression[a]

	Calls About Dogs	*Calls About Cats*
Target	*Number (%)*	*Number (%)*
Humans	409 (79.4)	47 (46.1)
Dogs	148 (28.7)	10 (9.8)
Cats	23 (4.5)	57 (55.9)
Other	2 (0.4)	1 (1.0)

[a]Targets identified during telephone calls made to the Animal Behavior Clinic of the Veterinary Hospital of the University of Pennsylvania about behavior problems of dogs (515 calls) and cats (102 calls). Some animals were reportedly aggressive to more than one target species.

Canine Aggression Evaluation

Amy R. Marder, Victoria L. Voith (1984)

Ask the client *all* of the following questions; use a checkmark to show that each has been asked. Clients should be asked specifically if they or anyone else has ever been involved in any of the listed circumstances. If the client says that no one has ever tried to interact with the dog in such a manner (e.g., no one has ever tried to clip the dog's nails or tried to take the dog's food away) then check the "never tried" column. If people have interacted with the dog in that way, then ask if the dog has ever exhibited any of the behaviors listed in the key box.

If the dog showed no signs of aggression to anyone under those circumstances, then check the *No* column. If the dog has shown any signs of aggression in the described circumstances, do the following:

1. Check the *Yes* column (if the dog is aggressive to any family members).
2. Indicate with the appropriate code (e.g., M or BT) which behavior the dog exhibited (optional).
3. Note to which family member the dog is aggressive.
4. If the owners volunteer that the dog is aggressive to strangers in the given situation, write "STR" in the column. It is unnecessary to ask about strangers in all the circumstances in part A unless the circumstances are unusual.

If any items are checked "yes" or marked "STR," then the more detailed questions on part B of the aggression form can be asked if it is appropriate to pursue the matter.

Key

M = mutter/grumble
BK = bark threateningly
G = growl
LL = lift lip
S = snap
BT = bite
F = family member
STR = strangers

PART A
Aggression to Family Members

Never Tried	*Asked*	*Circumstance*	*Reaction: Yes*	*Reaction: No*
		1. Take the dog's meal away while it is eating.		
		2. Add food to the bowl while the dog is eating.		
		3. Walk near the dog while it is eating.		
		4. Drop a piece of food near the dog and reach down to pick up.		
		5. Try to take away a real bone or a favorite (frequently stolen) food item.		
		6. Walk near the dog as it is chewing a real bone or favorite item.		
		7. Try to take away coveted (usually stolen) item; indicate item:_____________		
		8. Walk near the dog as it possesses coveted item.		
		9. Disturb (wake up or touch) dog while it is sleeping.		
		10. Physically disturb dog (touch, push off chair) while it is resting.		
		11. Walk by dog while it is resting.		
		12. Pet dog.		
		13. Reach toward dog.		
		14. Bother dog.		
		15. Bend down to hug or kiss dog.		
		16. Lift or try to lift.		
		17. Restrain dog (e.g., prevent from running through door or try to catch dog).		
		18. Groom dog's body (brushing, combing).		

Never Tried	Asked	Circumstance	Reaction: Yes	Reaction: No
		19. Bathe dog.		
		20. Towel-dry dog's body.		
		21. Wipe dog's face.		
		22. Wipe or pick up dog's feet.		
		23. Trim or try to trim nails.		
		24. Jerk or pull on leash while trying to make the dog sit or heel or while trying to perform leash correction.		
		25. Push or pull dog while trying to make it sit.		
		26. Push or pull dog while trying to make it lie down.		
		27. Grab dog by back of neck and pull.		
		28. Grab dog by the collar and pull.		
		29. Put on collar.		
		30. Take off collar.		
		31. Attach leash.		
		32. Remove leash.		
		33. Put on muzzle.		
		34. Take off muzzle.		
		35. Stare at dog.		
		36. Stare at dog and give it an order.		
		37. Scold verbally (with loud, threatening voice).		
		38. Threaten visually (shake finger, hold newspaper).		
		39. Hit or try to hit.		
		40. Try to go through door at same time as dog.		
		41. Bump into dog as it walks by.		
		42. Other circumstances that the owner might mention.		

Research Questions

43. In any circumstances in which the dog is aggressive to family members, does the hair rise on the dog's neck/back?

    If so, specify circumstances:

44. When the owner is sitting on a chair, does the dog ever "sit" on the owner's lap while keeping all four feet pretty much on the floor? (This is not the same as climbing on the owner's lap).

45. How would the owners rate dog's appetite?
    Excessive ____ normal ____ poor ____

46. Does the dog steal food from countertops? Does the dog steal dropped food?

47. Does dog frequently steal other items? Paper napkins? Slippers? Tissue?

48. Did owner spontaneously mention anything about the dog's eyes changing? Yes_____ no_____ Specify:

49. Did owner spontaneously mention anything about the dog having "moods"? Yes_____ no_____ Specify:

PART B
Aggression to Strangers or Family Members

Never Tried	Asked	Circumstance	Reaction: Yes	Reaction: No
		1. Person approaches, knocks at, or rings bell on outside door.		
		2. Person enters house when dog is already inside.		
		3. Person enters room when dog is already in it.		
		4. Person approaches dog in room.		
		5. Person enters yard when dog is already in yard.		
		6. Person gets up from sitting position or moves rapidly in room (not directed toward dog).		
		7. Person in room makes quick arm movement.		
		8. Person in house approaches family member.		
		9. Person touches or starts to touch family members.		
		10. Person threatens or pretends to threaten family member.		
		11. Dog reacts to startling noise outside.		
		12. Person leaves room dog is in.		
		13. Person leaves house dog is in.		
		14. Person approaches yard when dog is already in yard.		
		15. Person approaches dog loose in yard.		
		16. Person approaches dog tied up outside.		
		17. Person approaches dog on a leash on a walk (not near home).		
		18. Person approaches dog loose away from home.		

Part C
Dominance Exercises

These exercises are often helpful in assessing a dog's tendency to exhibit dominance aggression. They must, however, not be attempted except under the supervision of a person specifically trained to perform and interpret them. The exercises should be stopped at the slightest sign of aggression. The purpose of the exercise is to determine whether the dog has a tendency to respond—not to elicit a serious aggressive reaction.

For owner to perform (if safe to do so):

1. Stare at dog.
2. Hold dog's muzzle and neck.
3. Push dog over, hold dog's muzzle, and press on dog's shoulder.
4. Grasp dog's cheek.
5. Pull on dog's neck.
6. Pull on dog's collar.
7. Bump dog.

For clinician to perform (if safe to do so):

1. Stare at dog.
2. Hold dog's muzzle and neck.
3. Push dog over, hold its muzzle, and press on its shoulder.
4. Grasp dog's cheek.
5. Pull on dog's neck.
6. Pull on collar.
7. Bump dog.

Dominance Aggression in Dogs

Peter L. Borchelt, PhD
Animal Behavior Consultants, Inc.
Forest Hills, New York
The Animal Medical Center
New York, New York

Victoria L. Voith, DVM, PhD
Department of Clinical Studies
School of Veterinary Medicine
University of Pennsylvania
Philadelphia, Pennsylvania

The concept of dominance has been central to the study of social behavior of animals in general and has been particularly important in popular discussions of the social behavior of dogs living in the human family. A dominance order in a social group of animals is defined as a set of sustained relationships, based on threats or offensive aggression and submission or deference, among the individuals. The dominant animal is defined as the one that is the target of the least amount of threats and aggression, can initiate aggression with impunity, and receives the most amount and engages in the least amount of submission and deference.

A dominance order or hierarchy (Figure 1) can be simple, such as a *despotism* in which one individual rules over all other members of the group and in which there is no distinction in rank among the other subordinate members of the group. More commonly, dominance hierarchies involve a more or less linear sequence in which one animal (labeled *alpha*) dominates all the others, the second individual (labeled *beta*) dominates all animals but the alpha, and so on to the bottom of the hierarchy where the *omega* animal dominates no one and is dominated by everyone else.

In many social groups, however, it is common for nonlinear relationships to occur. For instance, a triangular relationship is present when animal A is dominant to animal B and B is dominant to C but C is dominant to A. Triangular or circular relationships can occur within an otherwise linear hierarchy. In some species, there is a hierarchy among males as well as among females. These social hierarchies are not mutually exclusive, and an individual female may dominate one or more males.

Some species organize themselves into social hierarchies that remain stable regardless of where the group goes or what its circumstances are. Such an organization is called an *absolute dominance hierarchy*. In

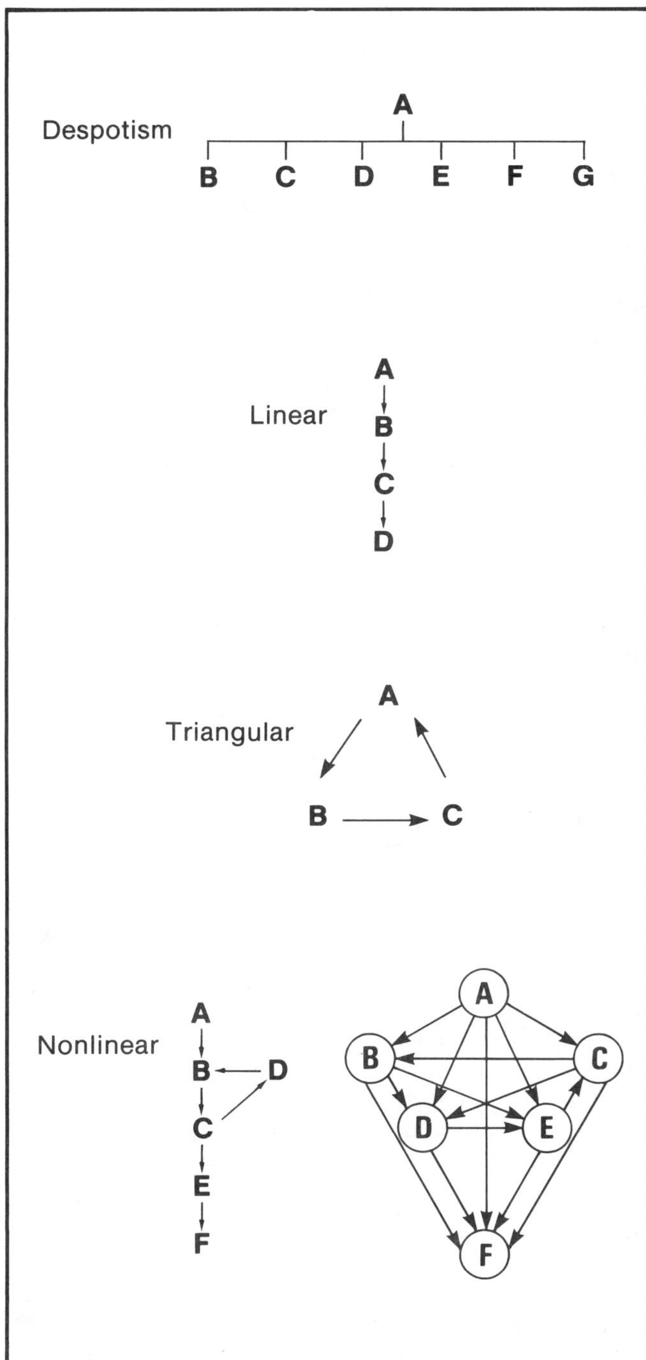

Figure 1—Types of dominance hierarchies.

contrast, other species (most notably groups of domestic cats) organize themselves into a *relative dominance hierarchy* in which a higher ranking individual defers to a lower ranking individual when the latter is in certain locations or areas, such as its sleeping or resting place. The hierarchy in these instances is relative to the location of the individuals that are interacting.

When a social hierarchy becomes stable and predictable, relatively little or no aggressive behavior is required to maintain social order. The dominant animal may simply signal with specific body postures or gestures, and a subordinate will respond appropriately. In some species, dominance behavior may be signaled by acoustic or chemical signals as well as by visual ones.

History of the Dominance Concept

The first systematic studies of social hierarchies involved domestic fowl,[1-3] and extensive research has been conducted subsequently. In brief, a new flock of chickens quickly forms into a *pecking order*. The chickens form and maintain the status or rank by pecking or threatening to peck an opponent. Higher ranking chickens gain priority of access to food, nest sites, and resting places. Dominant cocks mate much more frequently than subordinates do. If chickens are separated for longer than two to three weeks, then the social hierarchy has to be reformed; but a chicken can be separated from its group for shorter intervals and retain its rank when reintroduced. Flocks composed of approximately 10 or less individuals form linear hierarchies that are stable for long periods of time. In larger flocks, nonlinear relationships form and interfere with feeding efficiency, weight gain, and egg production.

The concept of a dominance hierarchy has been found to be more complex when applied to canids and primates. Wolves[4] and Rhesus monkeys[5] employ a variety of signals, other than overt aggression, that denote status. In fact, dominant individuals in the group may display the least aggression. Subordinate animals may maintain the social hierarchy by avoiding or deferring to the dominant animal.

The concept of dominance becomes even more complex when kinship or friendship relations between and among animals are involved in social interactions. In some species, for instance, it is sometimes common for two or more "subordinate" animals to join together to dominate another that the individuals could not dominate alone. Coalitions or alliances may be based on "friendship" or individual preferences or on kinship. In a number of primate species, the rank of an individual is influenced by the rank of its mother[6] or the rank of a mate.[7]

Advantages of Dominance

A dominant animal usually has priority in gaining access to food, mates, resting sites, grooming attention, help with care of young, and any other necessities of life or reproduction. A review of dominance in a variety of species[7] clearly showed that dominant animals had the advantage with respect to accessing food. In canids, possession of food is correlated with a number of other behaviors related to dominance[8]; but possession of food by a subordinate may be sufficient to override temporarily an already established dominant-subordinate relationship. For instance, a submissive wolf that has possession of a bone or a piece of meat may successfully defend it by growling or snapping at a "dominant" wolf that approaches.[9]

In some species, the dominant animal is the only one

to mate. In ungulates, such as elk, deer, and buffalo, the dominant male aggressively defends a harem of females and chases away approaching males. In other species, subordinates and juveniles mate; but access to females at times of high fertility is more or less restricted to dominant animals. Despite the vigilance of the dominant animal, fertile matings by subordinates do occur. These animals are called "kleptocopulators."

Dominant males that persistently and aggressively defend estrous females from other males may be under great stress and, in fact, may be able to remain dominant for only short periods of time. For instance, red deer do not eat when defending their harems, and after about two weeks they are easily defeated by younger and stronger stags.[7] In stable and enduring groups of animals, such as primates and wolves, the tenure of the dominant males may be only a few breeding seasons. The dominance hierarchies of females may be stable for a longer period of time.

Compensations of Being Subordinate

An animal that is not dominant still receives benefits from remaining in the group. Group living may result in an increase in feeding efficiency, particularly for species that must search or hunt for food. A group of animals searching for food has a higher probability than an individual does of finding food that is irregularly distributed in the environment. Moreover, for predatory species, cooperative hunting allows capture of bigger prey than an individual could capture. Group living also increases the degree of protection from predators. In a group, there are more individuals that are alert to the presence of, and can defend against, a predator. Groups also provide multiple individuals to care for the young.

Thus, it is often to a subordinate animal's advantage to remain in the group, where it can eat well, benefit from protection, and even occasionally mate and wait for a chance to achieve a higher rank as the dominant animals age and die. Social groups are usually composed of genetically related individuals. One adaptive strategy, when an individual is not dominant, is to contribute to the survival and reproductive success of related individuals. By remaining in the group, subordinate animals can increase their genetic representation by aiding related individuals.

Determinants of Dominance

A review of dominance in a wide range of species[7] provides a general summary of the qualities that influence the status of an animal in a social group. Generally, adult animals are dominant over juveniles and males are usually dominant over females. Size, weight, and strength may be important factors. In some species of primates, status is greatly influenced by the mother's rank and by membership in coalitions of friends or kin. Androgen and estrogen levels influence dominance status.[10] In complex social species, the multiplicity of determinants makes dominance difficult, if not impossible, to characterize completely.

The Social Structures of Wolves and Dogs

In the popular literature that is easily accessible by veterinarians, dog trainers, and the general public, great emphasis is placed on the necessity of owners establishing dominance over their domestic dogs. Failure to establish dominance is the purported cause of a wide range of behavior problems. This concept of dominance with respect to dogs and human interactions is often referred to as "pack theory"in the popular literature.

Popular discussions of any scientific concept are usually noted for oversimplification, and that is true of the concept of social dominance. For instance, a seemingly simple notion, such as a pecking order (which applies to domestic fowl), often is applied directly to more complex species, such as dogs, wolves, and humans. The concept of a linear hierarchy of social dominance is popularly applied to wolves and dogs; but no consideration is given to the problems discussed in the scientific literature with regard to defining what behaviors are the most appropriate measures of dominance, the frequent lack of correlation among various measures of dominance, and how circumstances or context may influence expression of dominance. In the following section, the authors review the social behavior of wolves—in light of the larger scientific views of social dominance—preliminary to discussion of the expression of dominance-related aggressive behavior in dogs in the human family.

The Social Structure of Wolves

Wolves, the presumed ancestors of domestic dogs, generally live in family groups. Although observations of wolves in the wild indicate group size can range from 1 to more than 30 individuals, most wolf packs appear to be composed of about 4 to 6 wolves of which usually only 2 are mature adults.[9] Wolves are seasonal breeders that reproduce once a year, generally between January and April (depending on latitude). They bear an average of five to six young per litter, but mortality is high during the first year of life and half the offspring may not survive to two years of age. Wolves do not breed during their first year of life. Females are reported to come into estrus for the first time at 22 months of age, and males do not show sexual interest until that time. Thus, a pack is composed of an adult breeding pair, perhaps a few mature subordinate animals, a few yearlings or juveniles, and pups. Most of the year, group activities center around hunting and care of the young.

The postures indicating dominance[4] include facial expressions and tail positions. A threat by a dominant wolf includes vertical retraction of the lips, bared teeth, ears erect and pointed forward, and the tail held high. Displays of dominance also include a "stare" and "standing over," in which the dominant animal stands stiffly over a recumbent individual or places its forepaws across the shoulders of a subordinate that is standing. Typically, subordinate wolves assume postures of either active or passive submission.[11] *Active submission* involves a crouched posture, with tail held low and often wagging; ears back and close to the head; and pushing with the nose, licking, or lightly nipping the muzzle of the dominant animal. Active submission is frequently initiated as a "greeting ceremony" when the dominant wolf rejoins the group or the group becomes active. *Passive submission* is a response by a submissive wolf that is threatened by a dominant one. The submissive wolf rolls over and lies on its side or back, exposing the back or abdomen. The ears are back and close to the head, and the tail may be tucked between the legs.

There are many gradations of these postures. In a social group in which individuals are familiar with each other over extended periods of time, the gestures and postures indicating social rank can be very subtle. If the social hierarchy is stable, little or no aggressive behavior is exhibited. In fact, the most frequent indications of relative social rank are submissive displays rather than dominant gestures. The concept of a submissive order or hierarchy rather than a dominance hierarchy has been proposed for primates[12] and would be equally appropriate for wolves in that the displays of submission are more widespread and frequent than dominant gestures.

Descriptions of wolf packs in either natural, seminatural, or enclosed or zoo environments suggest that the concept of dominance in wolves is as fraught with conceptual difficulties as it is for other complex species. For instance, consider the concept of a linear dominance hierarchy, wherein A is over B is over C, and so on (Figure 1). Presumably, then, the outcomes of encounters between any two animals are predictable (to the observer and to the animals).[a]

In a study of seven packs of wolves (total of 36 animals) in a seminatural environment in Alaska, Lockwood[8] found that rank orders based on aggressive behavior did not display a linear relationship; linearity values ranged from 0.10 to 0.69. Values based on submissive behaviors directed to individuals during interactions yielded values ranging from 0.09 to 1.00. One pack observed in 1973 yielded a value of 1.00, but the value for this pack in 1972 was 0.10 and in 1974 was 0.30. Thus, measures of dominance based on both percent of aggressive encounters in which the animal initiated aggression and on percent of encounters in which the animal received submissive displays from others failed to show anything approaching a linear hierarchy.

In another analysis of the social hierarchy of these wolves, Lockwood[8] subjected a correlation matrix of 27 variables (including age, weight, sex, various measures of aggressive behavior, marking, grooming, play, social motivation, activity, proximity, degree of synchronization of group behavior, and others) to factor analysis. A factor, subsequently interpreted as "dominance," was found to be heavily loaded with respect to initiation of aggressive behavior, winning contests in obtaining food, body weight, reception of submissive behaviors, and frequency of urine marking; but few dominant wolves scored high on all of the variables related to dominance. Thus, the statistical correlations derived from the overall population only imperfectly predict an individual wolf's behavior.

The association of dominance to winning contests over food is in contrast to frequent observations[9] that a subordinate wolf possessing meat or a bone will growl or snap at a dominant wolf that approaches. Thus, even though a dominant wolf may have the privilege of taking the choicest piece of the food, it may not be able to take away food from another subordinate animal that has possession of it.

A dominant animal usually is regarded as having priority in gaining access to mates. If a wolf pack comprises only one mature pair, obviously the top-ranking male and female will be the only ones to mate. Observations of packs of wolves containing several mature wolves indicate that the alpha male may not be the only one to mate and reproduce.[15-17] Kleptocopulations may occur, and a female may spurn the advances of a dominant but nonpreferred male. Individual mate preferences have been noted in wolves[9] as well as beagles.[18]

Dominance rivalry among females may increase greatly before and during the breeding season, and the alpha female may actually prevent other females from mating. There are reports, however, of more than one litter being born and successfully reared in a pack.[17]

Another presumed feature of dominant wolves is leadership or initiative as far as direction to move or when to hunt. Leadership, or the behavior of a wolf that controls, governs, or directs the behavior of several others, has been observed in packs of wolves in the wild. Mech[19] observed packs following a leader during movement as well as resting and then resuming hunting when the leader did so; however, these observations did not indicate whether the same individual led during each of these activities or what the status of the leader was in the pack's dominance hierarchy.

A study of two packs of wolves kept in large outdoor enclosures[20] indicated no relationship between the dominance order and the animals that hunted and captured rabbits. Large individual differences in time spent hunting were observed among the wolves, with some individuals hunting frequently and others very little. Large individual differences also were observed in specific behaviors, such as searching, chasing, and killing. No clear correlations were found between social rank, size, sex, and hunting efficiency. In fact, the animal that accounted for the most captures and kills was the lowest ranking individual in one of the packs.

[a]Statistical procedures[13,14] can be employed to derive a measure of predictability or degree of linearity. In a perfectly linear pecking order, a value of 1.00 indicates that there are no ties or reversals in the relationships. A value of 0 indicates that no predictions can be made about outcomes of interactions.

Understanding the social structure of wolves is limited by the extreme difficulty of observing specific behaviors of known individuals over long periods of time under natural conditions. Interpretation of data obtained under conditions of captivity, even in large seminatural enclosures, is constrained by not allowing sufficient room for the natural emigration of young, mature animals and the formation of new packs that is likely to occur in the wild. Interestingly, wild groups of primates have afforded a much better opportunity for the study of complex social groups. Even though the wolf's social structure may not reach the degree of complexity achieved by some primate species, it is useful to be aware of how recent research on social structure of primates has changed and elaborated the older views of dominance.

The Social Structure of Dogs in the Human Family

For the most part, dogs living as companion animals display intense affiliative, attachment behavior to people. They routinely display submissive, friendly, greeting behaviors, such as tail wagging and licking, on arrival of owners. They readily submit to being rolled over on their backs into submissive postures or having objects or food taken away from them. Most, if not all, household pets are "spoiled," that is, fed tidbits from the table, allowed to sleep on furniture or beds, and regularly held or petted "on request."[21] For most dogs, this type of human indulgence poses no particular problems either for the dog or for the owner. When behavior problems arise, they usually have no relationship to the social structure of the family, that is, the degree to which the dog or human is dominant. The concept of dominance is widely misconstrued to relate to all aspects of a dog's behavior. In fact, most of the common behavior problems involving elimination, fears and phobias, separation anxiety, and even many cases of aggressive behaviors, do not relate to the dominance structure of dogs.

Furthermore, far too much emphasis has been placed on the social structure of wolves as a model for the social relations that occur between the domestic dog and humans. Domestic dogs are not wolves (even though the two species will interbreed). For thousands of years the affiliated, submissive, and friendly aspects of wolf behavior have been strongly selected for among domestic dogs.

Despite selection for amicable behavior, some companion dogs direct aggressive behavior to humans (generally the owners) in the context of dominance. Dominance-related aggression[22] is directed to a person when the dog (1) guards a resource, such as food, bones, or a resting place; (2) assumes dominant postures, such as "standing over" or "staring"; and/or (3) resists behaviors of the person, such as petting, hugging, pushing, or pulling. The dog may interpret or perceive these behaviors as dominant gestures by the owner. A detailed history can differentiate dominance-related aggression from other classes of aggressive behaviors, such as fear-induced, pain-elicited, or predatory aggression.

In many cases, the presenting complaint of the owner of a dominant-aggressive dog is a "sudden, unprovoked" aggressive attack. Owners will also remark about the dog having a "funny, glazed" look in the eyes or that the "dog did not seem to know what it was doing." Usually, owners state that the dog became friendly and tried to "make up to the owner" immediately after the aggressive attack.

Detailed questioning of the owners typically reveals a distinct pattern regarding the situations and stimuli that elicit the aggressive behavior. Table I presents data listing the most common circumstances or situations in which dominance aggression was exhibited in samples of 19[23] and 84[24] dogs. It is immediately apparent that these circumstances naturally group into the three general categories of dominance-related aggression discussed previously. It is also apparent that there is considerable variation in the circumstances under which a dominant dog will display aggression. No particular circumstance is related to aggression in all dogs, and most dogs do not exhibit aggression in all or most of the circumstances listed in Table I. In each specific case of dominance aggression, owners were able to state confidently the circumstances in which the dog would become aggressive. Thus, it is clear that dominance aggression is context specific; that is, the behavior is elicited only in specific circumstances and these circumstances may differ among individual dogs.

A dominant-aggressive dog may not be aggressive to all family members. Furthermore, dominant-aggressive dogs are usually nonaggressive and very friendly to most people between confrontations. Dominant dogs typically show no aggression to a person over whom they are clearly dominant or to whom they are clearly submissive. Elderly persons or children in the home may never assert themselves with the dog or even interact with it. If these individuals never present a "challenge" to the dog, the dog rarely demonstrates aggression toward them. Only persons who confront or challenge the dog in some way (Table I) present potential eliciting stimuli for aggression.

It is important to point out that, contrary to popular belief, owners of dominant-aggressive dogs typically are not first-time dog owners nor are they "wimps."[23-25] Often, owners have had dogs previously, even of the same sex and breed, without encountering the problem. Many dog presented for dominance-aggression problems have been taken to obedience school by the owners and have done well. In fact, some dogs presented for dominant-aggressive behavior would be problems for any pet owner or even for professional dog handlers.

Table II presents data from two samples of canine behavior problem cases,[23,26] in which dominance aggression appears to represent about 20% of the cases of

TABLE I

Situations in Which Dominance Aggression Was Elicited Based on Two Samples of Cases Presented to Animal Behaviorists

Cases Presented to Line and Voith[23] **(N = 19)**			***Cases Presented to Coppola and Borchelt***[24] **(N = 84)**	
N	*Percent*	***Situations***	*N*	*Percent*
		Dominant postures, gestures shown		
3	13	Stand over	19	23
—	—	Stare	35	42
		Possessive elicitors		
12	50	Food taken away by owner	19	23
12	50	Animal guarded/stole objects	27	32
—	—	Objects, toys taken away by owner	51	61
7	29	Person left, went through doorway	10	12
		Preceding owner behaviors to animal		
18	75	Discipline (verbal, physical)	55	65
15	63	Disturbed when resting	52	62
14	58	Petted or touched	44	53
5	21	Pulled at	23	27
5	21	Hugged	9	11
14	58	Reached for, stood, or bent over	31	37
10	42	Groomed	39	46
2	8	Cleaned, wiped	9	11
4	17	Put on, took off leash	16	19
4	17	Picked up, held	23	27

behavior problems in dogs. Beaver[27] reported dominance-related aggression made up 30% of canine behavior problems. The majority of dogs exhibiting dominance aggression are males, and purebred dogs are presented for dominance aggression more often than mixed-breed dogs. One author (Borchelt)[26] found dominance aggression to be the only aggressive behavior problem for which the difference between purebred and mixed-breed categories was statistically significant. Line and Voith[23] found that the proportion of purebred and mixed-breed dogs in their sample of dominance-aggression cases was significantly higher than the proportions presented for general medical and surgical reasons at the same veterinary hospital.

Dominance aggression to people can begin to develop in puppies (under six months of age) or can first occur in dogs four to six years of age or even older. Most cases, however, begin to occur only after the dog has become sexually mature (about one year of age) or behaviorially mature (about two years of age). Because this behavior does not usually manifest until the dog is mature, it is not possible with a high degree of reliability to assess the likelihood that a puppy will exhibit dominance-aggression problems at a later age.[27] It is true that puppies 6 to 12 weeks of age in a litter begin to establish a dominance hierarchy related to possession of food or perhaps expression of dominance gestures, such as "standing over."[28] There is, however, no reason to believe that the relationship that exists among these puppies will relate in any way to the dominance-subordinance structure of these dogs in the human family. In fact, Scott and Fuller[28] remarked that the dominance order of puppies in a litter bore little or no relationship to the animals' puppy or adult behavior to people in the laboratory.

Treatment of Dominance Aggression

The goals of treating a dominance-aggression problem are to (1) ensure the owner's safety, that is, prevent a person from being bitten; (2) develop procedures so that the problem dog can eventually be placed, without displaying serious signs of aggression, into the context or stimulus situation that previously elicited aggression; and (3) have the dog assume a subordinate role in all circumstances in which a dominance-subordinance interaction is likely to occur.

Physiologic Intervention

Approximately 90% of dogs presented for dominance aggression are males.[23,26] Thus, it is reasonable to assume that castration may affect dominance aggression, and it is the authors' clinical impression that castration is indeed helpful. As with other behaviors influenced by hormones,[10,29] however, not all dominant-aggressive dogs respond equally to castration. Some castrated dogs continue to display dominance aggression as well as marking behaviors, roaming, urine marking, and intermale aggression.

Most females presented for dominance aggression are spayed. Perhaps more spayed than intact females are

TABLE II
Statistics on Dominance-Aggression Problems from Two Samples Presented to Animal Behaviorists

	Line and Voith[23]	*Borchelt*[24]
Total number of canine cases in sample	103	372
Number of cases of dominance aggression	24	73
Percent of cases involving dominance aggression	23.3	19.6
Percent of males	87.5	90.0
Percent of purebred dogs	87.5	82.2

kept as pets or neutering the animal was the first approach to treating the problem. It is also possible that spaying allows the expression of masculine behavior traits in some female dogs, an effect that is supported by some of the authors' cases in which a rapid onset of dominance aggression occurred within a few weeks after spaying. Female dogs that exhibit dominance aggression might have been androgenized in utero and thus might be further predisposed to masculine behavior after spaying.

Behavioral Intervention

The first step in a behavioral treatment plan is obtaining a detailed history to identify and specify all the stimuli and circumstances that elicit aggressive behaviors (fighting, growling, and staring or subtler signs, such as stiffening of the body or a sideways look). To prevent the owner from being bitten, he or she should be counseled to avoid temporarily the situations in which these signs are present. The more frequently dominance aggression is elicited by a particular circumstance, the more likely the dog is to learn to anticipate the occurrence of the eliciting stimuli and the more likely the dominance-aggression problem is to either escalate or generalize.

Use of physical force as a preventive or treatment procedure is not as simple as it might seem. It is true that physical force (hitting) may permanently suppress, for instance, a puppy from growling at the owner in a specific circumstance, and may further prevent the development of subsequent dominance-related behavior. It is not true, however, that doing so will *necessarily* prevent the same animal from developing a dominance-aggression problem after it matures. Moreover, the use of physical force as a treatment for a fully developed, or even a developing, dominance-aggression problem in a mature dog can be dangerous. Pain has long been recognized as an elicitor of aggression.[30] Threatening gestures associated with pain (for instance, raising an arm, reaching for a newspaper, yelling, and approaching) also can elicit aggression. In fact, as is evident in Table I, discipline or attempts to discipline a dominant-aggressive dog are the most likely stimuli to elicit aggression.

It is worthwhile to note that even though pain or the threat of pain may be used by wolves to assert and maintain dominance in a wolf pack, it is a wise wolf that knows when to back off and avoid injury. Because of the potential for serious injury to the owners, the authors firmly counsel them not to get into a dogfight with their dog but instead to approach the problem rationally and safely so that systematic behavior modification and/or physiologically based treatment can be implemented. Hanging dogs by the leash or the use of supposedly "natural" discipline techniques, such as rolling the dog over on its back or grasping the dog by the scruff of the neck and shaking it, are potentially dangerous—particularly if the dog is highly motivated to be dominant.

It is also worth noting that strangers to the dog may easily engage in dominance-related behaviors in the same stimulus situation(s) that lead to the owners being bitten. Owners rather than strangers are the more likely targets of aggression because of the dog's relationship with an owner as part of the normal social group. A stranger is usually perceived by most dominant-aggressive dogs as a stimulus for play and attention. If, however, the stranger remains in the social group for any period of time or confronts the dog directly, then dominance-related aggression is likely to be exhibited. Owners should be cautioned that because a person outside the social group may be able to safely confront a dog (several times) does not ensure that the owner can do so continually. Professional dog handlers and others experienced in handling dogs are likely to have the physical skills to avoid physical injury or to interpret correctly subtle signals that might predict aggression.

Of the literally infinite number of variations on behavioral techniques available for treating dominance aggression, a common feature is approach by indirect or gradual exposure. The general procedure is to devise ways for the owner to interact with the dog so that dominance on the part of the owner and submission on the part of the dog is gradually achieved. Stimuli that initially elicit aggression are gradually introduced. The precise details of the exposure technique depend on the specific eliciting stimuli involved in the particular case, the magnitude of aggression exhibited by the dog, and the capabilities and preferences of the owner. No specific techniques are likely to work in all cases. To achieve a high rate of success, individual programs are usually required and constant refinement of procedures should be expected. Brief examples of specific techniques that have been used in some cases are described in the literature.[21,22]

Progestin Therapy

Because progestins tend to suppress typically masculine behaviors and because dominance-aggression problems are clearly typical masculine behaviors, it is logical to assume that administration of progestins may be an effective therapy. In the authors' experience, use of megestrol acetate can lead to a reduction or even a temporary cessation of dominance aggression. Sometimes, as the dosage is decreased, the intensity or frequency of aggression increases and some form of concurrent behavioral therapy is then required to ensure that dominance aggression does not recur after the drug is withdrawn.

A typical starting dosage is 2.2 mg/kg/day orally for one to two weeks, followed by 1.1 mg/kg/day for another two weeks. The dose is then progressively reduced and withdrawn. The dose and duration of progestin treatment vary among cases and depend on a dog's response to the drug and concurrent behavioral therapy. Most dogs are treated for one to two months. The general principle is to use the minimum dose that will reduce or suppress the behavior and to withdraw the drug as quickly as possible. Veterinarians prescribing the drug should be thoroughly familiar with the side effects and contraindications of its use.

Euthanasia

For a number of reasons, it is sometimes beneficial for the clinician to bring up the topic of euthanasia. It may make the owner aware of the potential seriousness of the problem, allows the clinician to determine whether euthanasia is an option, and provides an opportunity to discuss euthanasia procedures. Some owners have already reached the conclusion that they would like to euthanatize the animal but feel guilty and believe the euthanasia procedure is inhumane. Some people do not want to euthanatize an animal because they believe the procedure is traumatic and the dog will experience pain. The clinician can discuss with owners the validity of their beliefs and offer other perspectives regarding the situations.

Sometimes it is necessary to advise an owner not to keep a dominant-aggressive dog, particularly if the dog is extremely aggressive and likely to injure a person severely. If a dog is aggressive to children or infirm individuals, or if the household includes people who simply cannot understand or are unable to implement therapeutic procedures, then even a mildly aggressive dog may prove too dangerous.

If the owner expresses a desire to euthanatize an untreatable aggressive animal, the request should generally be supported. Euthanasia ensures that the animal will not injure anyone and removes intact animals from the breeding pool. To encourage owners to work with a dog that they do not want is counterproductive and probably dangerous.

Sometimes a dog that is dominant in one household can be a submissive and acceptable pet in another. Placement of a dog with a dominance-aggression problem must, of course, be done carefully. A person obtaining the dog should be aware of the dog's complete history. The authors believe that the majority of dogs presented because of dominance aggression would be difficult for most people to manage. Sometimes, however, there are exceptions.

Owners who decide to treat a dominant-aggressive dog should be advised that they are responsible for the dog during and after the treatment procedures. Behavioral therapy can sometimes, but not always, reduce the intensity and/or frequency of dominance aggression. A guarantee that the dog will never bite or threaten again cannot be given.

REFERENCES

1. Schjelderup-Ebbe: Beitrage zur Sozialphyschologie des Haushuhns. *Zeit Fur Psychol* 88(3-5):225-252, 1922.
2. Schjelderup-Ebbe: Weitere Beitrage zur Sozial-und Individual psychologie des Haushuhns. *Zeit Fur Psychol* 92(1,2):60-87, 1923.
3. Schjelderup-Ebbe: Social behavior of birds, in Murchinson CA (ed): *A Handbook of Social Psychology.* Worcester, MA, Clark Univeristy Press, 1935, pp 947-972.
4. Schenkel R: Ausdrucks-Studien an Wolfen. Gefangenschafts-Beobachtungen. *Behaviour* 1(2):81-129, 1947.
5. Altmann SA: A field study of the sociobiology of Rhesus monkeys, *Macaca mulatta. Ann NY Acad Sci* 102(2):338-435, 1962.
6. Marsden HM: Agonistic behavior of young Rhesus monkeys after changes induced in social rank of their mothers. *Anim Behav* 16(1):38-44, 1968.
7. Wilson EO: *Sociobiology: The New Synthesis.* Cambridge, MA, Beltknap Press of Harvard University Press, 1975.
8. Lockwood R: Dominance in wolves: Useful construct or bad habit, in Klinghammer E (ed): *The Behavior and Ecology of Wolves.* New York, Garland STPM Press, 1979, pp 225-244.
9. Mech LD: *The Wolf: The Ecology and Behavior of an Endangered Species.* New York, The Natural History Press, 1970, pp 71-73.
10. Hart BL: Gonadal androgen and sociosexual behavior of male mammals: A comparative analysis. *Psych Bull* 81(7):383-400, 1974.
11. Schenkel R: Submission: Its features and functions in the wolf and dog. *Am Zool* 7:319-330, 1967.
12. Rowell RE: The concept of social dominance. *Behav Biol* 11:131-154, 1974.
13. Landau HG: On dominance relations and the structure of animal societies: I. Effect of inherent characteristics. *Bull Math Biophys* 13:1-19, 1951.
14. Chase ID: Models of hierarchy formation in animal societies. *Behav Sci* 19:374-382, 1974.
15. Murie A: *The Wolves of Mount McKinley.* U.S. National Park Service Fauna Service No. 5, 1944.
16. Ginsburg BE: Coaction of genetical and nongenetical factors influencing sexual behavior, in Beach F (ed): *Sex and Behavior.* New York, John Wiley & Sons, 1965, pp 53-75.
17. Rabb GB, Woolpy JH, Ginsburg BE: Social relationships in a group of captive wolves. *Am Zool* 7:305-311, 1967.
18. Beach FA, LeBoeuf BJ: Coital behaviour in dogs. I. Preferential mating in the bitch. *Anim Behav* 15(4):546-558, 1967.
19. Mech LD: *The Wolves of Isle Royale.* U.S. National Park Service Fauna Service No. 7, 1966.
20. Sullivan JO: Individual variability in hunting behavior of wolves, in Klinghammer E (ed): *The Behavior and Ecology of Wolves.* New York, Garland STPM Press, 1979, pp 284-306.

21. Voith VL: Human/animal relationships, in Anderson RS (ed): *Nutrition and Behaviour in Dogs and Cats.* Oxford, Pergamon Press, 1984, pp 147-156.
22. Voith VL, Borchelt PL: Diagnosis and treatment of dominance aggression in dogs, in Voith VL, Borchelt PL (eds): *Vet Clin North Am Small Anim Pract* 12(4):655-663, 1982.
23. Line S, Voith VL: Dominance aggression of dogs toward people: Behavior profile and response to treatment. *Appl Anim Behav Sci,* in press, 1986.
24. Polsky RH: Factors influencing aggressive behavior in dogs. *Calif Vet* 37:12-15, 1983.
25. Mugford RA: Behaviour problems in the dog, in Anderson RS (ed): *Nutrition and Behaviour in Dogs and Cats.* Oxford, Pergamon Press, 1984, pp 207-215.
26. Borchelt PL: Aggressive behavior of dogs kept as companion animals: Classification and influence of sex, reproduction status and breed. *Appl Anim Ethol* 10:45-61, 1983.
27. Beaver BV: Clinical classification of canine aggression. *Appl Anim Ethol* 10:35-43, 1983.
28. Scott JP, Fuller JL: *Dog Behavior: The Genetic Basis.* Chicago, The University of Chicago Press, 1965.
29. Hopkins SG, Schubert TA, Hart BL: Castration of adult male dogs: Effects on roaming, aggression, urine marking, and mounting. *JAVMA* 168:1108-1110, 1976.
30. Azrin NH, Holz WC: Punishment, in Honig WK (ed): *Operant Behavior: Areas of Research and Application.* New York, Appleton-Century-Crofts, 1966, pp 380-447.

UPDATE

Does spaying increase aggressive behavior in female dogs? One study[1] analyzed questionnaire data and found that the incidence of dominance aggression increased in spayed females under one year of age that were already showing some aggression.

Another study[2,3] of a series of 84 cases of dominance aggression (82% males, 18% females) suggests that the onset or escalation of dominance aggression and typically masculine behavior in females after spaying may be related to androgenization in utero.[4] The litter sex ratio for 14 of the 15 females in this sample was obtained from American Kennel Club or breeders' records. Table I in this update shows the sex ratio of the litter, whether dominance aggression increased after spaying, and whether typically masculine behaviors appeared or escalated after spaying.

Females from litters that were predominantly male were more likely to show increased dominance aggression and typically masculine behaviors than were females from litters that were predominantly female or equal in sex ratio. None of the females displaying dominance aggression came from all-female litters.

BEHAVIORAL TREATMENT TECHNIQUES

Several approaches have been applied in the treatment of dominance aggression. Few reports have assessed treatment outcomes.[5,6] There has been no uniform method of comparing the efficacy of various treatment techniques; such an endeavor would be valuable but extremely difficult given the number of variables involved (e.g., characteristics of the aggressive behavior, temperament and abilities of the owner and the dog, family and environmental circumstances, practitioner preference and skill with various treatment techniques, and types of cases referred to a specific practice).

Probably the most frequent approach (for dominance aggression as well as for other behavior problems) is obedience training. Most dog–owner dyads that graduate from a basic obedience course maintain a dominance–subordinance relationship that is acceptable to dog and owner. The dog becomes reasonably obedient (responds to some commands) and compliant (gives up possessions and accepts physical manipulation and reprimands from the owner). Such success may have one or several causes: the compliance to obedience commands is generalized to other situations, the owner develops greater ability to "countercommand" the dog when it begins to show dominance aggression, and most likely the majority of dogs may simply never be motivated to exhibit dominance aggression.

Nevertheless, some dogs that have done well in obedience training, including advanced degrees, still develop dominance aggression. In such cases, harsh punishment techniques sometimes used in obedience training (e.g., strong leash corrections, "hanging," or "helicoptering") may be counterproductive and dangerous for dog and owner. Some frequently recommended techniques that could elicit an attack by a dominant aggressive dog are shaking the dog by the scruff, holding the dog's head and staring into its eyes, and rolling the dog over and holding it down.

Sometimes head halters are used to obtain obedience and compliance in the contexts in which dominance aggression occurs. This approach is safer (although not totally without risk) than use of a choke (or prong/spike) collar. Head halters provide control over the dog's head (so obtaining compliance is easier) and some control over the mouth and jaws (thus reducing the likelihood of a bite).

Enforced compliance has been used in selected cases. Initially, a dog (muzzled or on a halter, if necessary) is moved, physically manipulated, rolled over, and so on, *in situations in which it is unlikely to resist* (with even low levels of aggression) until compliance readily occurs. Then, contexts in which dominance aggression is typically displayed can be gradually introduced. Remote, low-level electrical stimulation has been used in some cases for noncompliance, but never at the start of treatment. Such a technique should be used only in selected cases where there is strong evidence that electrical stimulation will not escalate the level of aggression and *only* by individuals well educated about escape/avoidance learning principles and experienced with aversive procedures.[7]

A more subtle approach is gradual restructuring of the social relationship with the dog.[8] The dog first is required to exhibit elements of submissive behavior (such as sitting or lying down) to obtain any and all things that it wants (e.g., to go out, to come in, to be petted, to be fed, or to play). Gradually, the dog is counterconditioned in the contexts that elicit dominance aggression.

PHARMACOLOGIC APPROACHES

Recent reports indicate an inverse relationship between serotonin levels in the central nervous system and aggressive behaviors in many species.[9] Lower concentrations of serotonin and dopamine metabolites have been found[10] in

TABLE I
Data Suggesting Prenatal Androgenization in 14 Female Dogs Exhibiting Dominance Aggression After Spaying

Litter sex ratio (male to female)	*Litter Size*		*Increased aggression after spaying*	*Masculine behavior (mount, leg-lift urine mark)*
	Males	*Females*		
>1	8	4	No	No
	2	1	Yes	Yes
	3	1	Yes	No
	5	3	Yes	No
	4	1	Yes	Yes
	5	2	No	Yes
	8	4	No	No
1	2	2	No	No
	6	6	No	Yes
<1	2	4	No	No
	1	4	No	No
	1	4	Yes	No
	2	6	No	No
	1	3	No	No

the cerebrospinal fluid of dominant-aggressive dogs than in laboratory control dogs. Aggressive dogs with a history of biting without warning had significantly lower levels of these metabolites than did dogs that gave warning. Fluoxetine, a drug that increases the availability of serotonin, reportedly reduces dominance-related aggression in dogs[11] (also see the Simpson and Simpson article on pharmacotherapy).

CAUTION

The degree of danger and the limitations of the owner and family must be carefully considered in developing a treatment plan, and the owners must be advised of the risks involved. Not all dominance aggression cases can be treated successfully. The likelihood of euthanasia for dominance aggression was reported to be higher in dogs over 18 kg (40 lb) and in dogs with unpredictable aggression or severe aggression in response to benign dominance challenges.[12]

Reasonably successful treatment of dominance aggression requires knowledge, skill, and a great deal of commitment from owner and behaviorist. Treatment techniques are not mutually exclusive, and more than one technique may be necessary for a specific case. Each dominance aggression case (and each dog–owner dyad) is unique; and specific techniques must be carefully tailored to individual situations. The preferences, knowledge, and skills of the behaviorist also determine which techniques are recommended.

REFERENCES

1. O'Farrell V, Peachey E: Behavioral effects of ovariohysterectomy on bitches. *J Small Anim Pract* 31:595–598, 1990.
2. Borchelt PL, Coppola MC: Characteristics of dominance aggression in dogs. Paper presented at Annual Meeting of the Animal Behavior Society, North Carolina, June 1985.
3. Coppola MC: Dominance aggression in dogs. Master's thesis, Department of Psychology, Hunter College, New York, NY, 1986.
4. Vom Saal E, Bronson E: Sexual characteristics of adult female mice are correlated with their blood testosterone levels during prenatal development. *Science* 208:597–599, 1980.
5. Line S, Voith VL: Dominance aggression of dogs towards people: Behavior profile and response to treatment. *Appl Anim Behav Sci* 16:77–83, 1986.
6. Knol BW: Behavioral problems in dogs: Problems, diagnoses, therapeutic measures and results in 133 patients. *Vet Q* 9(3):226–234, 1987.
7. Tortora DF: Safety Training: the elimination of avoidance motivated aggression in dogs. *J Exp Psychol [Gen]* 112(2):176–214, 1983.
8. Voith VL: Aggressive behavior and dominance. *Canine Pract* April:8–15, 1977.
9. Ferris C: Using the golden hamster to understand the neurobiology of aggression. Paper presented at the annual meeting of the AVMA, Pittsburgh, 1995.
10. Reisner IR, Mann JJ, Stanley M, et al: Comparison of cerebrospinal fluid monoamine metabolite levels in dominant-aggressive and non-aggressive dogs. Paper presented at annual meeting of American Veterinary Society of Animal Behavior, Pittsburgh, July, 1995.
11. Dodman NH, Mertens PA: Fluoxetine (Prozac) for the treatment of dominance-related aggression in dogs. Paper presented at annual meeting of American Veterinary Society of Animal Behavior, Pittsburgh, July, 1995.
12. Reisner IR, Erb HN, Houpt KA: Risk factors for behavior-related euthanasia among dominant-aggressive dogs: 110 cases (1989–1992). *JAVMA* 6:855–863, 1994.

Canine Aggression: Dog Bites to People

Mercer University
John C. Wright, PhD

Canine aggression toward people has been cited as an underrecognized public health problem since the 1950s.[1–3] Recent estimates of the annual number of dog bites delivered to people range from one to three million, and of these, an estimated 585,000 bites result in injuries serious enough to require medical attention.[4,5] Children are bitten more often than adults, and younger children are bitten more often than older children.[4,6] Children are particularly susceptible to serious and fatal injuries resulting from dog bites.[7–9] More than 60% of severe injuries and 85% of fatal injuries are delivered to children 12 years of age and younger.[6] Serious bites to younger children commonly involve severe lacerations to the face, wounds may become infected, and children may experience disability as a result of the bite.[1,10–12] Thus, substantial emotional and financial costs can be associated with dog bites for the bite victim and the victim's family.

In addition to the potential loss of use and loss of life to the victim, the consequence for the biting dog, even in nonsevere bite cases, is a reduction in its quality of life or loss of life (i.e., euthanasia). Dangerous or vicious dog laws may require, rightly or wrongly, the lengthy impoundment, confinement, banishment to another county, or destruction of a biting dog. As a result, significant personal costs can be incurred by the biting dog and dog owner's family.

An important goal of this chapter is to present a brief description of the factors that characterize commonly occurring dog-bite events and to identify some important causes. A second goal is to identify and discuss several important methodological issues surrounding research on dog bites and their causes. Measures for reducing the rate of dog bites to people have been proposed elsewhere.[5,6,9,13]

AGGRESSION AS A CONSTRUCT

To classify dogs as if they were either "aggressive" or "nonaggressive" is for the most part inaccurate because aggressive behavior is not a *unitary phenomenon* or *unitary construct.*[14] Seven different "kinds" of aggression in animals initially described by Moyer[15] were revised by Borchelt and Voith,[16] Borchelt,[14] and others,[17–19] resulting in eight functionally classified kinds of canine aggressive behavior. The eight major types of canine aggression observed by Borchelt include aggression related to fear, dominance, possessiveness, protectiveness (of people and territory), predation, punishment, pain, and intraspecific aggression.[14] Several of the categories reflect different neurobiologic mechanisms of aggression in dogs or other domesticated species. For example, in rats and mice different genetic and physiologic mechanisms have been shown to affect intraspecific territorial aggression, irritable or pain-induced aggression, and predatory aggression (although the latter category is often not included as "aggression" because of its association with feeding).[20] (Note, however, that Scott[21] reports a positive relationship between canine predatory and intermale attacking.) In dogs, different mechanisms affect the probability of eliciting biting when a dog is mildly restrained,[22] and eliciting reactivity to fearful stimuli that might result in defensive biting when an animal cannot escape.[22,23] It is also likely that different neurobehavioral mechanisms influence both the frequency and intensity of biting in competition over resources as in *competitive aggression*[19] or *possessive aggression.*[14]

If one is to accurately describe "an aggressive dog," one should refer to its *behavior*, not to the dog itself. Further, the behavioral event (the dog-bite event) should be classified within the functional category or categories that best represent(s) the aggressive behavior(s) exhibited. A dog whose behavior reaches the threshold for exhibiting fear-induced aggression might exhibit only that kind of aggression even though the dog is exposed to higher intensities of stimulation in other settings. Another dog might be equally likely to exhibit both fear-induced aggression and possessive aggression, and still another dog might exhibit three or more types of aggression. The more different kinds of aggression a dog exhibits, the more likely one is to suspect a single underlying or moderating cause. The developmental onset of each kind of aggression can be gradual (i.e., increasing in frequency and/or intensity with age), or acute, resulting from a single traumatic event, even in dogs that have received the best "socialization" and training.

The categories of aggression proposed by Borchelt and others are open to revision as additional research on canine aggression becomes available.[24] Applied animal behaviorists who have used these categories to design effective treatment programs for the reduction of dog bites to people have been successful in reducing the frequency and intensity of canine aggression.[24,25] Treatment programs designed to reduce dominance aggression could backfire, however, if the source of the dog's biting is actually fear, and its response tendency is to avoid an individual rather than to confront and control. A good behavioral history leading to an accurate diagnosis of the cause(s) of aggression should be obtained before *any* recommendation for treatment.

FACTORS IN DOG BITES TO PEOPLE

Within each category of aggression, different aspects of a dog-bite event can influence the probability of biting. Among the many factors that contribute to a dog-bite event are those classified as characteristics of the dog, the victim, the dog-person relationship, and the bite-event setting.[6,9]

Dog Characteristics

Dog characteristics include the dog's genetic preparedness to exhibit different kinds of aggressive behavior[13,20,26] and the dog's medical health,[27] age,[3,28,29] sex,[3,28,30,31] reproductive status,[5,14,25,32,33] and size.[1,4,6,31,33,34]

Breed

Several studies have indicated that German shepherd dogs (or dogs that phenotypically resemble German shepherd dogs) are the breed most likely to bite.[2,4,14,28,35,36] Attempts to describe various dog *breeds* as more or less genetically prepared to bite, however, have failed to take into account the extent to which dogs might have been misidentified as representing a specific breed. Misidentification may involve dogs that are genotypically but not phenotypically close to German shepherd dogs, resulting in an underestimate of dog bite risk attributable to those genotypes. (For example, F2-generation cocker spaniel × basenji crosses do not much resemble either cocker spaniels or basenjis.[22]) More importantly, misidentification of dogs that are phenotypically but not genotypically close to German shepherd dogs may have the effect of overestimating the number of bites, and thus bite risk, attributed to German shepherd dogs. For example, any medium or large-sized, black and tan dog may be inappropriately identified as a German shepherd dog.[3,30,37,38]

Other attempts to describe the "most aggressive" breeds have failed to take into account the possibility of independent genetic mechanisms controlling different kinds of aggression,[18,20] the possibility of a dog's genotype interacting with other factors (e.g., early experience; see Elliot and King[39] for a good example), and the possibility of different breeds consisting of different diversities of genotype.

Determining which breeds are most aggressive is a difficult endeavor for a number of practical reasons as well. Computing breed-specific bite rates (i.e., the annual number of bites delivered by German shepherd dogs/the number of German shepherd dogs) may be a problem because the number of reported bites might underrepresent the actual number of bites delivered[34] or registration figures used to compute breed frequencies might not be accurate.[34] Furthermore, breed-specific bite rates might change over time because of changes in breed popularity (affecting the number of dogs "available" to deliver bites); breed distributions might differ among different states, cities, and counties within a state; and line-bred dogs within a geographic location for a particular breed might skew the breed bite rate for that location. For example, if a pair of golden retrievers in Atlanta whelp generations of biting dogs, and those offspring are bred by other Atlantans, how long would it take for those dogs to affect the golden retriever bite rate in Atlanta?

In another example, in Palm Beach County, Florida, German shepherd dogs, Labrador retrievers, and chow chows delivered the highest percentage of bites in 1992 (13.5%, 7.6%, and 7.2%, respectively); no information was provided on numbers of dogs registered per breed.[40] The highest percentage of severe bites per breed-specific bites, however, was delivered by cocker spaniels: 23.6% of the 59 cocker spaniel bites (2.6% of all reported dog bites) were classified as "severe"; only 13.8% of the 301 bites from German shepherd dogs were classified "severe."[a] The same statistics taken in 1986–1989 indicated the highest percentage of severe bites/delivered bites in three of those four years were by golden retrievers (mean = 25% of 110 bites)! It is difficult to provide a clear answer to the question "Are German shepherd dogs, cocker spaniels, or golden retrievers the 'most aggressive' breed?" even if one restricts the geographic location to Palm Beach County.

As a third example, a case-control study of risk factors associated with first bites to nonfamily members resulting in medical treatment showed that German shepherd dogs were 16 times more likely to have delivered a bite than were any other breeds investigated, followed by chow chow dogs.[4] The report did not include information on the kind of aggression that was exhibited and the results were based on a relatively small sample (18% of the reported bites in Denver in 1991). Clearly, an answer to the question "Which dogs (breeds) are most genetically prepared to bite people?" awaits further study.

Medical Factors

Medical factors associated with the probability of aggression, such as liver, thyroid, or visual dysfunction, may directly affect aggressive behavior. Other medical factors, such as any condition resulting in heightened pain or irritability, for example, skin allergies or hip dysplasia, may indirectly influence biting. (See Reisner[27] for a review of the pathophysiologic basis of aggression.)

[a]DiPace G: Personal communication, Field Operations Supervisor, Animal Care and Control, Board of County Commissioners, Palm Beach County, FL, 1993.

Age

Statistics on age reveal the clear majority of bites involve dogs younger than 5 years of age. Approximately 50% to 70% of reported dog bites implicate dogs in this age group.[3,4,31,41] Severe bites, however, are delivered by dogs 3 to 4 years of age (mean = 3.5 years; range: 8 months to 10.5 years).[9,42,43]

Age-specific bite percentages have been reported to be representative of the licensed population,[3,31] although there is some indication that pets 6 to 11 months of age may exhibit the highest bite rate (number of bites/year).[3] In Pittsburgh, dogs from that age group represented 4.8% of all licensed dogs, yet inflicted 13.1% of all reported bites. Other studies indicate that all age groups are equally likely to bite.[31]

Inspection of reports from applied animal behaviorists indicates the mean age of dogs referred for all kinds of aggression ranges from 2 years[18] to 3 years,[44] and that young dogs (6 months to 2 years) may be especially prone to exhibit dominance aggression toward a family member.[25,29,45] Dominance aggression and protective-territorial aggression commonly appear between 1 and 3 years of age or earlier, but other kinds of aggression may not be related to age.[29]

Sex

Male dogs inflict approximately 70% to 76% of all reported dog bites,[3,9,28,30,31,46] and deliver at least 80% of severe bites.[9,43,b] Sex-related bite rate estimates (i.e., number of male dog bites/number of male dogs) also show people are at a higher risk for bites from male than from female dogs,[4,28,38] although one estimate indicates bite rates to be in direct proportion to the number of male and female dogs in the population.[31]

Investigations of dog-bite fatalities show unneutered male dogs to be the most frequent perpetrators, especially in the breeds most frequently cited for fatal bites (i.e., German shepherd dogs, "pit bulls," chow chows, and rottweilers).[5]

Dogs presented to applied animal behaviorists for reduction of aggression are predominantly male.[14,18,41,44] Males seem to be especially prone to dominance and protective/territorial aggression,[14,18,24,25,29] but there appears to be no clear relationship between sex and fear-induced aggression.[14,18]

Reproductive Status

Few statistics exist on the effects of gonadectomy on dog bites reported to health departments nationwide. A recent Centers for Disease Control and Prevention (CDCP) study that determined the dog-spe-

[b]Texas Zoonosis Control Division: Personal communication, 1990.

cific factors independently associated with a dog biting a nonhousehold person, however, revealed that compared with controls, biting dogs were 6.2 times more likely to be male and 2.6 times more likely to be reproductively intact males and females.[4]

Inspection of the canine behavioral literature also indicates that more intact males are aggressive than are neutered males[14,29,41] and that neutering seems to be effective in reducing male dogs' *dominance aggression* or *dominance-related aggression* toward people.[25,29] Dominance-related aggression includes both possessive aggression (biting exhibited within the context of possessing a toy or food) and dominance aggression (biting elicited by, e.g., a member of the dog-owner's family standing over, staring at, and reaching toward the dog), although the two kinds of aggression are frequently exhibited in the same dog.[14] The sex distribution of dogs exhibiting only dominance aggression, however, is different from that of dogs displaying dominance-related aggression, resulting in statistics that are often difficult to interpret clearly. For example, of the 174 male dog aggression cases seen by Borchelt,[14] dominance aggression was diagnosed in 62 intact males and only 4 neutered males; possessive aggression occurred in 34 intact males (approximately 54% of the number of dogs diagnosed as dominant-aggressives) and 9 neutered males (more than twice the number of dominant-aggressive neutered males). It is not clear whether neutering affects agonistic behavior exhibited within the context of guarding/obtaining items (possessive aggression), within the context of a family member's "threatening" communicative behaviors (dominance aggression), or whether neutering affects biobehavioral mechanisms (including biting) associated with both kinds of aggression (as in "dominance-related" aggression).

Further, sampling error can affect the sex distributions reported in a specific study and might indirectly lead to spurious information regarding the influence of neutering on different kinds of aggression. For example, of the 110 dogs diagnosed for dominance-related aggression in a recent study,[24] only 28 dogs were intact males and 48 dogs were neutered males. Taken alone, it would appear that neutered males are more likely to exhibit dominance-related aggression than are intact males. (Note: there is no indication that the authors intended their sample to be used for this convenient example.)

The relationship between gonadectomy and aggression becomes even less clear for spayed and intact females. Recall that unneutered *males and females* were reported as an independent risk factor for a dog biting a nonhousehold person.[4] Others have reported that with respect to dominance-related aggression, *spayed* females may be *more* likely to bite than sexually intact females,[14,24,29] especially if female dogs show aggressive tendencies before spaying.[32] Overall[33] presents a discussion of these results.

At least for males, gonadectomy is likely to result in a reduction of the overly assertive, controlling behaviors leading to biting, analogous to the reduction reported for canine intraspecific aggression.[47] Neutering is not likely to reduce the behavioral components related to defensive aggression (i.e., biting in the context of fear or pain; see Borchelt[14] for a further discussion).

Size

The biting dog's *size* has been cited as a factor associated with reported bites to people,[1,34] especially in recent cases involving fatal attacks.[5] Both a dog's weight and an observer's impression have been used to determine dogs' size as "large" (>50 pounds), "medium" (15–50 pounds), and "small" (<15 pounds).[1,31,34] Large dogs commonly deliver a higher percentage of reported bites than do medium or small dogs, although the "actual" bite rates (as opposed to those based on reports to health departments) may reflect the availability of different-sized dogs in the population.[31,48] For example, of the 1480 free-ranging dogs observed on both public and private property in a California study, 44% were classified as large, 30% as medium, and 26% as small. The rank orderings were the same for dogs who bit people and for dogs who were nonaggressive to people.[48] It may also be that bite reports overestimate the number of bites from large dogs because of reporting bias,[6,34] especially if the dogs are owned (as opposed to stray).[34] Serious bites to children, however, are commonly inflicted by large dogs because large dogs cause more serious wounds that require medical attention.[4,5,42,49,50]

Non-Dog Characteristics

Bite likelihood is also a function of the characteristics of the victim and other factors, including the dog-victim relationship, the attack setting, and the rearing environment.

The Victim

Risk factors associated with the bite victim include the person's age, sex, and behavior in the dog's presence.

Age

Children aged 12 years and younger are victimized by dog bites more than any other age group. Reported bites to younger children, especially those 5 to 9 years of age, are higher than bites delivered to older children; 5- to 9-year-olds constitute 25% to 30% of all dog bite victims.[1,3,4,11,30,51] Not only are children bitten more frequently than adults, they are also bitten disproportionately to their representation in the population.[30,31,38] For example, in St. Louis, 5- to 9-year-olds received 27.4% of all reported bites but represented only 8% of the population.[30]

Fatal injuries are delivered to children 10 years and younger at an alarming rate. Children 5 days to 10 years of age were involved in approximately 72% of the 60 fatalities resulting from dog attacks in the four-year period from 1990 to 1993.[c]

Sex

Both bite frequencies and bite rates are higher for male than for female victims. Bite rates allow one to compare risk by equating the number of males and females available to bite (i.e., the number of bites of males/100,000 males in the population). Boys and men account for approximately 65% of all reported bites,[4,30,31,34,51,52] and are 1.5 to 2.2 times more likely to be bitten than are females.[3,36,38] It may be that compared with females, males make themselves more available to bite by coming into contact with dogs more often.[48]

The "frequency of contact" hypothesis provides a convenient explanation for the differential bite rate, but has yet to be verified empirically. Indirect evidence in support of the hypothesis comes from studies reporting that boys and men prefer dogs as pets,[2,46] even after being bitten by a dog,[46] and as a result are more likely to come into contact with potential biters (see Wright[6] for a discussion of other mediating factors).

Victims of fatal bites are also more likely to be male.[8,43,53,c] When age and sex are combined, however, a more meaningful picture emerges. Victims of fatal bites are those least able to defend themselves from dog attacks: the very young of both genders. Approximately 37% of the 60 fatal bites from 1990 to 1993 were delivered to boys 5 years of age and younger, compared with 22% for same-aged girls[c]; taken together, boys and girls less than 6 years of age account for almost 6 of every 10 fatal bites.

Victim Behavior

A person's movement in the presence of a dog or dogs increases the likelihood of a dog bite. Case studies of serious and fatal attacks consistently show that when people move or try to defend themselves before and during an attack, the attack escalates; when movement ceases, the dogs release their grip and discontinue the attack.[9,54] There may be exceptions to this "rule" for certain dogs, (e.g., "pit bulls").

Almost any kind of stimulation provided in the presence of some dogs, such as yelling or sudden movements, increases bite likelihood. Results from behavioral assessment procedures designed to evaluate "nonfearful dogs" that have attacked people (used by this author and Borchelt; also see Wright & Lockwood[55]) show that increased stimulation exacerbates aggression in dogs that have attacked people, but increases submissive behavior in "subordinate dogs." Even the mildest visual stimulation, such as staring at a dog, is sufficient to elicit growling and snapping in some biting dogs, whereas the same stimulation elicits submissive tail wags, look-aways, and "submissive grins" in comparison dogs.

Other Bite-Event Factors

Because dogs deliver a majority of bites to people's extremities, it is tempting to speculate about which *victim behaviors* precede common bite events. Approximately 76% (median value; range = 69%–79%) of the reported bites cited in large-sample studies were delivered to peoples' extremities[1,4,28,34,51,52]; bites to the lower extremities (median = 42.8%) commonly exceeded bites to the upper extremities (median = 33.5%), followed by facial bites (median = 15%). According to one survey, however, the bites inflicted by *owned and unowned* dogs differ with respect to bite location and the bite event scenario: Owned dogs delivered almost three times the number of facial bites than did stray, unowned dogs (pets, **16%**, strays, 5.9%), and owned dogs delivered fewer bites to the fingers and hands (pets, 20.5%, strays, 36.3%).[34] Do people behave differently in the presence of owned and unowned dogs? Are the bite scenarios (i.e., the situations or contexts within which a bite is delivered) different for owned and stray dogs?

The Dog-Victim Relationship

Recall that children are the recipients of a majority of dog bites. The behaviors exhibited by children in the presence of a potential biter may help to explain the relatively high percentage of facial bites from pets and finger and hand bites from strays. It has been reported that people believe strays are more likely to bite, and strays tend to be more fearful of people.[30,46] Fearful dogs may exhibit defensive aggression in the presence of a child and bite the child's hand, the most salient moving object, when the child reaches above its nose to pet it.

An owned dog may also bite to defend itself from pain or a perceived threat, but companion dogs are also likely to bite in scenarios that are less common in strays. Unlike strays, owned dogs are likely to exhibit

[c]Lockwood R: Personal communication, Humane Society of the United States, 1995.

dominance and possessive aggression to people, and people known to the dog are its most frequent victims.[14,29,56] An increased risk of biting a nonhousehold member has been reported for owned dogs *residing in households with at least one child*; perhaps these dogs have a greater opportunity to express protective/territorial aggression and possessive aggression as well as fear-induced aggression toward playmates visiting a household child.[4] Further, an unusually high number of children are bitten by dogs chained for long periods of time (defending resources or dominance related, or fearful?), and these bites frequently involve a child's face.[4,9]

Because young children compete with a pet for common resources, such as a toy or comfortable location in the home, and because those resources are usually located on the floor, as is the dog, children are likely candidates for facial bites. Children less than 5 years of age seen at a Chicago emergency room were reported to be at risk for head, face, and neck bites delivered by their own dog in their own home.[57]

Other circumstances associated with the onset and escalation of dominance-related aggression may lead to a likelihood of facial bites, especially in children. Bending over or leaning on a dog, hugging or pushing it, reaching for or taking away an object, and other activities that place the child's face at the level of a dog's mouth increase the risk of facial bites. Bites from strays, on the other hand, are unlikely to occur in the context of these eliciting victim behaviors.[34] Unfortunately, bites from owned dogs account for 85% to 90% of reported bites.[34]

Dog ownership, another dog-victim relationship factor, is associated with an increased risk of dog bites. People who own dogs are more frequently bitten than nonowners, but not necessarily by their own dog.[58,59] Although bite rates have been reported not to differ for victims who like and dislike dogs,[58] compared with nonowners, people who own dogs are less fearful of dogs, are more likely to approach and interact with dogs, and are thus more likely to be bitten.[58,59]

People's perception of a dog's "friendliness" may also be a contributing factor. People may misperceive the communicative behaviors that are associated with an increased likelihood of "dominance biting"; they are more likely to approach dogs displaying dominant signals than dogs displaying subordinate signals, and wrongly believe that dogs displaying submissive communicative signals are more likely to bite them than dogs displaying dominant signals.[59] Additional factors that contribute to the likelihood of a dog bite include geographic location, weather, time of year, time of day (more bites are delivered in the spring and summer months, and in the late afternoon), and quality of care (including early rearing and experience with people, and recent traumatic events occurring in the presence of people).[6,34]

SUMMARY

Canine aggression toward people is a complex, multivariate phenomenon. Informed explanations of bite likelihood necessitate more than simplistic statements regarding breed ("it's a rottweiler!"), teasing ("the dog was provoked"), or other univariate causes. Rather, explanations leading to an understanding of and a reduction in the dog-bite epidemic will require a clear description of the risk factors associated with dog-bite events.

Treatment procedures designed to resolve or reduce instances of canine aggression in clients' pets will require practitioners to obtain a clear description of each dog-bite event (i.e., a behavioral history), including an identification of the kind of aggressive behavior exhibited and an understanding of the underlying neurobehavioral mechanisms, as well as a clear description of the characteristics of the dog, the victim, the dog-victim relationship, and the attack setting. Prevention of future aggression should be aimed at eliminating or controlling those factors that contribute to the acquisition and maintenance of each dog-bite event.

REFERENCES

1. Harris DH, Imperato PJ, Oken B: Dog bites: An unrecognized epidemic. *Bull NY Acad Med* 50(9):981–1000, 1974.
2. Lauer EA, White WC, Lauer BA: Dog bites: A neglected problem in accident prevention. *Am J Dis Child* 136:202–204, 1982.
3. Parrish HM, Clack FB, Brobst D, Mock JF: Epidemiology of dog bites. *Publ Health Rep* 74:891–903, 1959.
4. Gershman KA, Sacks JJ, Wright JC: Which dogs bite? A case-control study of risk factors. *Pediatrics* 93(6):913–917, 1994.
5. Lockwood R: The ethology and epidemiology of canine aggression, in Serpel J (ed): *The Domestic Dog: Its Evolution, Behavior & Interactions with People.* Cambridge University Press, 1994.
6. Wright JC: Canine aggression toward people: Bite scenarios and prevention, in Marder AR, Voith VL (eds): *Vet Clin North Am [Small Anim Pract]* 21(2):299–314, 1991.
7. Lockwood R, Beck AM, Hornberger K: Dog bites among letter carriers in St. Louis. *Publ Health Rep* 90(3):267–270.
8. Sacks JJ, Sattin RW, Bonzo SE: Dog bite: Related fatalities from 1979 through 1988. *JAMA* 262:1489–1492, 1989.
9. Wright JC: Severe attacks by dogs: Characteristics of the dogs, the victims, and the attack settings. *Publ Health Rep* 100:55–61, 1985.
10. Karlson TA: The incidence of facial injuries from dog bites. *JAMA* 251:3265–3267, 1984.
11. Kizer KW, Town M: Epidemiologic and clinical aspects of animal bite injuries. *JACEP* 8:134–141, 1979.
12. Berzon DR, DeHoff JB: Medical costs and other aspects of dog bites in Baltimore. *Publ Health Rep* 89:377–381, 1974.
13. Lockwood R, Rindy K: Are "pit bulls" different? An analysis of the pit bull terrier controversy. *Anthrozoos* 1:2–8, 1987.

14. Borchelt PL: Aggressive behavior of dogs kept as companion animals: Classification and influence of sex, reproductive status and breed. *Appl Anim Ethol* 10:45–61, 1983.
15. Moyer KE: Kinds of aggression and their physiological basis. Part A. *Commun Behav Biol* 2:65–87, 1968.
16. Borchelt PL, Voith VL: Classification of animal behavior problems. *Vet Clin North Am [Small Anim Pract]* 12:571–585, 1982.
17. Hart BL, Hart LA: *Canine and Feline Behavioral Therapy.* Philadelphia, Lea & Febiger, 1985, p 275.
18. Landsberg GM: The distribution of canine behavior cases at three behavior referral practices. *Vet Med* October:1011–1018, 1991.
19. Wright JC: The development of social structure during the primary socialization period in German Shepherds. *Dev Psychobiol* 13(1):17–24, 1980.
20. Popova NK, Nikulina EM, Kulikov AV: Genetic analysis of different kinds of aggressive behavior. *Behav Genet* 23(5):491–497, 1993.
21. Scott JP: Aggression in canine females, in Bjorkqvist K, Niemela P (eds): *Of Mice and Women. Aspects of Female Aggression.* San Diego, Academic Press, 1992, pp 307–316.
22. Scott JP, Fuller JL: *Genetics and the Social Behavior of the Dog.* Chicago, University of Chicago Press, 1965, p 468.
23. Blanchard DC, Blanchard RJ: Inadequacy of pain-aggression hypothesis revealed in naturalistic settings. *Aggress Behav* 10:33–46, 1984.
24. Reisner IR, Erb HN, Houpt KA: Risk factors for behavior-related euthanasia among dominant-aggressive dogs: 110 cases (1989–1992). *JAVMA* 205(6):855–863, 1994.
25. Line S, Voith VL: Dominance aggression of dogs towards people: Behavioral profile and response to treatment. *Appl Anim Behav Sci* 16:77–83, 1986.
26. Podberscek AL, Blackshaw JK: Dog bites: Why, when and where. *Aust Vet Pract* 20:182–187, 1990.
27. Reisner IR: The pathophysiologic basis of behavior problems, in Marder AR, Voith VL (eds): *Vet Clin North Am [Small Anim Pract]* 21(2):207–224, 1991.
28. Hanna TL, Selby LA: Characteristics of the human and pet populations in animal bite incidents recorded at two Air Force bases. *Publ Health Rep* 96:580–584, 1981.
29. Voith VL, Borchelt PL: Diagnosis and treatment of dominance aggression in dogs. *Vet Clin North Am [Small Anim Pract]* 12:655–663, 1982.
30. Beck AM, Loring H, Lockwood R: The ecology of dog bite injury in St. Louis, Missouri. *Publ Health Rep* 90:262–267, 1975.
31. Daniels TJ: A study of dog bites on the Navaho reservation. *Publ Health Rep* 101:50–59, 1986.
32. O'Farrell V, Peachey E: Behavioral effects of spaying in bitches. *5th International Conference on the Relationships Between Humans and Animals.* Monaco, November, 1989.
33. Overall KL: Sex and aggression. *Canine Pract* 20(2):16–18, 1995.
34. Wright JC: Reported dog bites: Are owned and stray dogs different? *Anthrozoos* 4(2):111–119, 1990,
35. Spence G: A review of animal bites in Delaware - 1989 to 1990. *Del Med J* 62(12):1425–1433, 1990.
36. Maetz HM: Animal bites, a public health problem in Jefferson County, Alabama. *Publ Health Rep* 94:528–534, 1979.
37. Marcy SM: Infections due to dog and cat bites: Special series. Management of pediatric infectious diseases in office practice, in Klein JO, Marcy MS (eds): *Pediatric Infectious Disease.* Baltimore, Williams & Wilkins, 1982, pp 351–356.
38. Morton C: Dog bites in Norfolk, VA. *Health Serv Rep* 88:59–64, 1973.
39. Elliot O, King JA: Effects of early food deprivation upon later behavior in puppies. *Psycholog Rep* 6:391–400, 1960.
40. Dog Bites. *Anthrozoos* 7(1):71–72, 1993.
41. Wright JC, Nesselrote MS: Classification of behavior problems in dogs: Distribution of age, breed, sex and reproductive status. *Appl Anim Behav Sci* 19:169–178, 1987.
42. Wilberger JE, Pang D: Craniocerebral injuries from dog bites. *JAMA* 249:2685–2688.
43. Winkler WG: Human deaths induced by dog bites, United States, 1974-75. *Publ Health Rep* 92:425–429, 1977.
44. Beaver BV: Clinical classification of canine aggression. *Appl Anim Ethol* 10:35–43, 1983.
45. Voith VL: Diagnosing dominance aggression. *Mod Vet Pract* 62:717–718, 1981.
46. Beck AM, Jones BA: Unreported dog bites in children. *Publ Health Rep* 100:315–321, 1985.
47. Hopkins SG, Schubert TA, Hart BA: Castration of adult male dogs: Effects on aggression, roaming, urine marking and mounting. *JAVMA* 168:1108–1110, 1976.
48. Westbrook WH, Allen RD: Animal field research, in Allen RD, Westbrook WH (eds): *Handbook of Animal Welfare.* New York, Garland Press, 1979, pp 148–167.
49. Jacobs R: Summer summary of bite wound care: 1. The first crucial steps. *Mod Med* May:127–135, 1982.
50. Schultz RC, McMaster WC: The treatment of dog bite injuries, especially those of the face. *Plast Reconstr Surg* 49: 494–500, 1972.
51. Brobst D, Parrish HM, Clark FB: The animal bite problem. *Vet Med* 54:251–256, 1959.
52. DeHoff JB, Ross L: Animal bites. *Maryland State Med J* 30:35–44, 1981.
53. Pinckney LE, Kennedy LA: Traumatic deaths from dog attacks in the United States. *Pediatrics* 69:193–196, 1982.
54. Borchelt PL, Lockwood R, Beck AM: Attacks by packs of dogs involving predation on human beings. *Publ Health Rep* 98:57–66, 1983.
55. Wright JC, Lockwood R: *Behavioral testing of dogs implicated in a fatal attack on a young child.* Paper presented at the Animal Behavior Society meeting, Williamstown, MA, June 1987.
56. Wright JC: Canine Behavior, in Rhoades JD (ed): *National Animal Control Training Guide.* Baton Rouge, LA, National Animal Control Assoc, 1989, pp 61–76.
57. Chun Y, Berkelhamer JE, Herold TE: Dog bites in children less than four years old. *Pediatrics* 69:119–120, 1982.
58. Lockwood R, Beck AM: Dog bites among letter carriers in St. Louis. *Publ Health Rep* 90:267–269, 1975.
59. Moss SP, Wright JC: The effects of dog ownership on judgments of dog bite likelihood. *Anthrozoos* 1:95–99, 1987.

MISCELLANEOUS READINGS

Social Behavior of Domestic Cats

Victoria L. Voith, DVM, PhD
Department of Clinical Studies
School of Veterinary Medicine
University of Pennsylvania
Philadelphia, Pennsylvania

Peter L. Borchelt, PhD
Animal Behavior Consultants, Inc.
Forest Hills, New York
The Animal Medical Center
New York, New York

Domestic cats are often described as relatively asocial animals, and thus the concept of social behavior of domestic cats may appear to be oxymoronic. Domestic cats are, however, variable with respect to social behaviors; they have the capacity of living as stereotypic "solitary hunters" or of interacting within social units of other cats or humans. In this article, we briefly review animal social behavior in general, emphasizing the flexibility of social behavior as a function of ecologic variables. We then discuss the social systems of cats and present new data on cat–human interactions.

Social Systems

A *social system*, or *society*, is generally defined as a cooperating or interacting group of individuals of the same species that live in close proximity to each other. An *aggregation* of animals, such as a swarm of insects attracted to a light or birds following a plow, is not considered a social group. Sociality requires that animals engage in reciprocal interactions and communications that are of a cooperative nature and that transcend sexual activity.[1]

There are many benefits to sociality. It facilitates encounters with members of the opposite sex and, therefore, sexual reproduction. Sociality also allows the potential for learning and exchange of information. In a group, animals can share in care of the young, protection from predators, food getting, and defense of resources against other animals. A group of animals may accomplish laborious tasks, such as modifying and/or monitoring the environment.[2] For instance, social insects, such as bees, maintain and repair the structure of the hive as well as regulate the humidity inside it. Prairie dogs and marmots have extensive tunneling systems that offer protection against inclement weather as well as escape routes from predators. These animals also spend a considerable amount of time keeping the brush low around their burrows, perhaps in order to better view the approach of predators.

A group of animals may also distribute tasks among specific members. Social insects, such as ants and termites, consist of separate castes composed of individuals specialized both morphologically and behaviorally for specific tasks, e.g., nest defense or foraging and care of the young. Different age groups and sexes of monkeys are likely to assume different roles in a troop. For example, in vervet monkeys, the adult males and juveniles engage in 99% of the territorial displays and chasing of intruders.[3] Dominant adult males in some primate groups suppress aggressive interactions within the group before the dispute becomes disruptive.[4] In primate troops, the adult males are often the individuals that engage in defense against predators. Female giraffes past calf-bearing age may tend nurseries of calves while the mothers browse away from their young.[5]

Group living also can result in social facilitation of many behaviors (e.g., eating and drinking) and synchronization of reproduction. Synchronized mating, parturition, and rearing can offer advantages in the successful rearing of offspring.

Being solitary has advantages as well. If food resources are scarce and occur in small concentrated units, it is advantageous to be alone rather than have to fight over or share a limited resource. In addition, a single animal may be able to avoid predators by hiding instead of belonging to a larger, more visible group.

Essentially, animals have evolved specific social organizations that maximize their ability to survive and propagate. Whether an animal is social or solitary is a reflection of a number of factors, such as availability of food and shelter, predator pressure, finding a mate, and requirements of rearing offspring. The social system of an animal is directly related to the environmental structure or setting in which it finds itself.

Describing Social Systems

A description of social systems is complex, because such systems can be described in many ways and the descriptive categories overlap. Social systems can be described according to group size and composition, type of mating system, patterns of parental care, use of space and territory, and social hierarchies.

Group Size

The simplest social system is the asocial individual or solitary animal. Except for periods of reproduction, the single individual is neutral toward, avoids, or even attacks conspecifics. Even though animals live apart, however, that does not mean they are not in communication with each other. Frequent communication may occur by means of auditory, visual, and/or olfactory signals. Feral or free-ranging domestic cats often live in a solitary state as adults but periodically vocalize and frequently deposit odors on conspicuous objects with secretions from sebaceous glands along the tail, forehead, lips, chin, and pedal areas[6] and by spraying urine. Scratching with claws is another means of leaving conspicuous marks on objects. Olfactory and visual marks allow reception of a signal long after the sender has left the area. Thus, a solitary animal may not really be *asocial* but may actually engage in frequent communication with conspecifics, either over distances or over periods of time, by leaving odors or visual signals.

Group Composition

Most species group into typical organizations that are popularly recognized and labeled,[7] for example, a pack of dogs, a herd of deer, a pride of lions, a pace of asses, or an exaltation of larks. Groups may consist of a mated pair, a mother and her offspring, all females, all males, or different ratios of males and females. The typical group composition may be organized year-round, seasonally, or temporarily.

Small, functioning groups may congregate into larger units. For example, bands of horses or groups of gazelles, which consist of one-male harems including several females, may group into larger herds. A herd of gazelles or zebras can number in the thousands. Albatross pair every year with the same individuals to raise young. Each pair has its own nest and territory within a temporary colony of hundreds or thousands of birds.

Some species form *casual groups* or *subgroups*, which come together for specific reasons (e.g., feeding, roosting, grooming, or breeding) and then break up and reform, sometimes with different individuals. Leyhausen described an interesting phenomenon involving evening social gatherings of urban cats in Paris.[8] Male and female cats came to meeting areas near the peripheries of their territories and stayed together for varying periods of time, generally dispersing by midnight. These cats sat near, looked at, and sometimes groomed each other. The tomcats sometimes "paraded" before the group. There was little hostility other than an occasional growl. Whether or not these cats were related or raised together is unknown. The function of these gatherings can only be surmised.

Mating and Parental Systems

In monogamous systems, one male and a single female may remain together for a single breeding season or for life. *Polygamy* refers to any form of multiple mating. *Polygyny* involves a single male mating with multiple females; *polyandry* involves a single female mating with multiple males. Animals may group temporarily for mating purposes only or for extended periods, sometimes as long as the life spans of the individuals. Some species form groups that live together until the offspring can survive on their own, and then the group disperses. In other species, adults may remain together after the young leave. Some groups form extended families in which the offspring stay for several generations.

Harems are composed of a single breeding male and several females, who may or may not be accompanied by offspring from the previous year. Harems can be seasonal (as in camels) or permanent (as in vicuñas). Other groups of animals, such as baboons, may include several adult males and females and their offspring.

Matriarchal families occur in many species. These societies are composed of a female and her offspring of the last one or two reproductive seasons and often several generations of her female offspring and their offspring. One or more males may temporarily join the system and then leave.

An atypical mating system involves leks, which are communal display and breeding areas used exclusively for reproduction.[1] The males vie for optimal territories in the leks in which they display to the females and which they vigorously defend from other males. The females are attracted to the displays or the area itself. The majority of matings takes place in a small percentage of the territories within a lek, and competition among the males for these optimal areas is intense. The females leave the area shortly after being bred.

Most wild felids, as well as the feral domestic cat, live solitary adult lives. The female raises her young without help from the male. Kittens stay with their mother until six months to one year of age, at which time the group disperses.[9] The kittens may stay together for several more months before establishing independent home ranges. If food and shelter resources are plentiful, the female offspring may stay with the mother indefinitely, establishing a stable matriarchal group. Estrous cycles of the adults tend to be synchronized, and the mothers often nurse each other's kittens or pool the kittens in a communal nest.[10] The mothers also assist each other in repelling from the area strange tomcats that may engage in infanticide. Owners of household cats with kittens have observed not only nursing queens care for each other's kittens but also report observations that other cats in the household, including spayed females, castrated males, and even the sire of a litter, curl up with the kittens in the nest area.

Lions form a very different social system from that of other felids.[11] Prides often consist of two adult males of breeding age, usually siblings, and several females and offspring. Lions engage in cooperative hunting strategies and bring down large game. Even if the male lions do not engage in most of the hunting, they provide an important function in the pride by protecting the young from nomadic male lions, which may try to establish themselves in a pride and kill the cubs. After death of the cubs and subsequent cessation of nursing, the females go into estrus. The new males can then sire their own offspring and invest their guarding energies in protecting their own cubs.

In some species, members of the society other than the parents help care for the young. These helpers are commonly referred to as "aunts" or "uncles" and are usually related to the young, although a genetic relationship is not a prerequisite for this type of behavior. Such alloparenting behavior occurs in the higher social insects, birds, and mammals and is especially common among primates. Some of the functions of alloparenting can include allowing young animals to gain experience with parental care before they reproduce as well as to establish social alliances with other animals. Obviously, with alloparenting, a mother and/or offspring may benefit from increased protection from predators, increased time for foraging, added food resources, and perhaps protection from the environment.

Use of Space or Territory

Societies can vary with respect to use of space. Many animals have a *home range*, that is, an area over which they routinely travel and patrol during hunting and exploration. Home ranges can overlap. Among male domestic cats, home ranges overlap each other as well as encompass the territories of several females. A *territory* is generally a smaller area, which is defended against encroachment from conspecifics. The territories of female cats usually do not overlap, although they may be shared (usually by female siblings, littermates, or mothers and offspring) on a temporal basis. The cats may use the areas consecutively, either several hours or a day apart, and thus rarely meet. It is possible that the regular schedules followed by the cats are monitored by the cats assessing each other's urine marks.[9] Large groups of migrating animals usually do not defend a territory, although they may repel conspecifics that approach too closely to their family or social units.

Social Hierarchy Systems

Some form of social hierarchy usually develops in all social systems. Social hierarchies range in complexity from despotisms, in which one animal is dominant over all others, to linear hierarchies or pecking orders to complex social networks influenced by kinship and social alliances. Some societies are organized into absolute dominance hierarchies in which the social hierarchy is the same regardless of where the group goes or what the circumstances. Other societies are organized into a relative dominance hierarchy in which the social rank depends on particular locations or circumstances. For example, in feral domestic cats, a high-ranking individual defers to a subordinate when near the subordinate's sleeping places. Circumstances, such as time of day, may determine social status, in that a cat may enjoy high rank at one time of day in a particular location but defer to another cat in that location at a different time of day.[8] Liberg observed that rural tomcats with overlapping home ranges would routinely defer to each other in specific areas.[12] A particular cat obviously had the right of way or was dominant in one location, while the reverse applied in another location. In laboratory colonies where cats are kept in higher concentrations than occur in household or feral environments, a despotic system may develop involving both despots and social pariahs who are the recipients of most of the aggressive displays.[8]

Usually, dominance/subordinance is indicated by specific signals or postures. For instance, a dominant wolf or dog may engage in direct stares and "stand-over" postures with tail and ears erect. In contrast, the subordinate averts its gaze, rolls over on its side or back into a submissive posture, and displays facial postures of submission (such as horizontal retraction of the lips and flattening of the ears). Cats, on the other hand, do not have specific dominance or subordinance displays. Relative social status is indicated by combinations of offensive and defensive aggressive behaviors, avoidance, immobility, and deference.[8]

Flexibility of Social Systems

It was a common assumption that the social system of a species was a fixed product of evolution and natural selection and, thus, inevitable and unmodifiable. Recent evidence, however, particularly from socioecologic studies, indicates that such assumptions are untenable.[13] Depending on the environment, usually in terms of food distribution and presence of predators, animals of a given species may actually alter the social system in which they operate.

There is evidence that all of the basic features of social systems (e.g., composition, defense of territory, establishment of dominance hierarchy, and types of mating and parenting systems) are flexible, at least to some degree, and that some species are more flexible socially than others.[13] For instance, a specific species of fish may be solitary, live in loose groups, or live in more tightly condensed groups depending on the likelihood of predator pressure. Guppies of a particular species in Trinidad do not group together if they live above waterfalls, which act as natural barriers to certain predators; however, in other areas of the same stream, the guppies group together as a means of predator defense.[14] A species of surgeonfish that used to feed alone now forms foraging groups because of increased pressure from predators.[15] Sociality as a response to predator pressure also occurs in white-tailed deer. They tend to be solitary in wooded areas but form groups when living in open habitats.[16]

Coyotes provide an interesting example of how sociality depends on availability of food resources.[17,18] Social organization of coyotes is influenced by the size of available prey, the spatial distribution of the prey, and its seasonal or temporal distribution. During the winter, coyotes that live in areas where numerous large dead prey (hunter-killed elk) are available tend to live in groups and to have larger group sizes than coyotes living in areas of less food availability (e.g., occasional dead deer). The social structure of a single population may change seasonally. During the summer, when coyotes sustain themselves mostly by catching rodents, group size is significantly smaller and many animals become solitary.

Coyotes living in packs, which form when there are numerous large dead prey available, are more likely to defend a specific territory. Territoriality, thus, is related to the presence of a large clump of defendable food resources.

A dramatic example of variation within a social system involves golden-winged sunbirds,[19] a nectarivoris species that winters in East Africa. These birds feed on flowers that vary in nectar production throughout the day. When the nectar production is high, individual sunbirds guard specific areas and remain solitary. When the flowers produce less nectar, the birds stop defending specific areas and begin interacting with each other in a different, organized way involving a dominance hierarchy system. The type of social system displayed is a direct function of the amount of food available; the birds can shift from one social system to another in a matter of minutes, depending on the balance of energy benefits accrued by one system over another.

TABLE I

Household Cat Interactions with Each Other

Frequency of Interactions[a]	*Play (n=554*[b]*)*	*Sleep (n=547*[b]*)*	*Groom (n=543*[b]*)*	*Eating (n=548*[b]*)*
Frequently	70.8%	65.6%	57.8%	80.8%
Sometimes	9.7%	9.3%	11.8%	4.4%
Infrequently	4.0%	4.2%	4.4%	3.3%
Rarely	4.7%	4.8%	3.5%	2.2%
Never	10.8%	16.1%	22.5%	9.3%

[a]Frequently = once a week or more; sometimes = once a month or more; infrequently = less than once a month; rarely=once a year or more.
[b]Number of responses to this question.

TABLE II

Household Cat Interactions with Owners

Frequency of Interactions	*Sleep With*[a] *(n=875*[c]*)*	*Table Tidbits*[b] *(n=875*[c]*)*	*Share Snacks*[b] *(n=877*[c]*)*	*Greet Owners*[b] *(n=870*[c]*)*
Always	38.5%	8.5%	6.5%	52.1%
Usually	20.3%	7.9%	5.2%	28.6%
Frequently	14.4%	13.0%	13.2%	9.1%
Sometimes	15.4%	37.8%	43.0%	7.1%
Never	11.3%	32.8%	32.0%	3.1%

[a]Usually = 5 to 6 nights/week; frequently = 2 to 5 nights/week; sometimes = less than 2 nights/week.
[b]Usually = 80% to 90% of the time; frequently = 30% to 70% of the time; sometimes = less than 30% of the time.
[c]Number of responses to this question.

TABLE III

Time Owners Spend Daily Interacting with Their Cats by Talking to, Playing with, Exercising, Training and/or Care Giving[a]

Time (hr)	*Percent of Respondents*
Less than 1/2	19.3
1/2–1	34.6
1–2	23.2
2–3	14.2
4 or more	8.8

[a]n = 862.

Brown hyenas switch their rearing strategies depending on environmental conditions. In low-quality territories they care only for their own young, occasionally rear their young communally in higher-developed territories, and always rear young communally in the central Kalahari (Africa).[20,21] It is not clear, however, whether these alternate rearing strategies are a result of fluctuations in resources or of increases in pressure from predators.

The Social Systems of Cats

Most field studies of feral domestic cats have indicated that cats are solitary hunters of small prey. As adults, their friendly social behavior is generally restricted to the interactions of courtship and mating. Domestic kittens may stay with the mother until they are 6 to 12 months of age before dispersing and establishing their own areas of living. Free-ranging domestic cats that are provided with food and/or shelter may congregate in groups, usually consisting of large numbers of females as well as their female offspring.[8-10,12,22,23] It is interesting to note, however, that even when there are abundant resources, adult intact males do not live together.

Kittens between 8 and 16 weeks of age may accompany their mother on excursions in her home range.[9] During these expeditions she may lead the kittens in scavenging or may bring them prey, although she apparently does not let them hunt with her. The trips appear to be to familiarize the kittens with the area. Kittens then begin hunting alone but continue to rest together and groom each other throughout their first year of life. Thereafter, they establish their own foraging ranges. If resources are plentiful, the female kittens may remain in their maternal group until they die. Even in a well-provisioned area, some females emigrate. When they do so, they establish a residence in a location free of other cats and may start a new female group.[12] Unfamiliar intact female cats will not join an established female group. Between 10 and 14 months of age, the male kittens begin staying away from the core area for progressively longer periods of time and eventually disperse.[12] It is thought that permanent dispersal is related to harassment by the adult breeding tomcat that patrols the female's core area. If the young male is protected by periodically being allowed in a house, the age at which he leaves the area is delayed.[12]

When the young males emigrate they settle in areas away from the focal areas of social activities. As they mature, however, they begin to visit the female units. Between two and four years of age, they gradually begin challenging the adult tomcats that patrol the female group. If the central breeding male does not tend the estrous female constantly, a peripheral male may breed her.

Between 3 and 5 years of age and after reaching a critical weight, a challenger may establish a home range that encompasses one or more female groups where he is the central breeding male. Home ranges of breeding males often overlap. An individual male may be the central breeding male in some of the female groups and a peripheral male in others. The tenure of a central breeding male is rather short—a few years at most. In his study area, Liberg was not aware of any breeding males over 6 years of age.[12] He did observe that breeding males shifted their home ranges over the years. This trait would decrease incest, because after a year in the same home range a breeding

TABLE IV
Percentage of Cats That Owners Considered As Knowing Tricks[a]

	Fetch	*Interesting Behavior*	*Learned Trick*	*Meow on Command*	*Play Games*	*Understand Everything*	*Come When Called*
Percent of Responses	15.8	12.3	7.9	6.0	4.2	1.7	4.0

[a]n = 887.

TABLE V
Percentage of Cats Indicated by Breed as Engaging in Categories of Behaviors That Owners Considered Tricks[a,b]

		Percent							
Breed	*n*	*Fetch*	*Meow on Command*	*Play Games*	*Show Interesting Behavior*	*Learned Trick*	*Come on Call*	*Understand Everything*	*Other Tricks*
DSH/DLH[c]	485	15.3	7.0	4.3	11.1	9.3	4.9	1.4	3.5
Mixed Breed[d]	114	11.4	3.5	5.3	15.8	9.6	3.5	0.9	2.6
Siamese	113	29.2	8.0	4.4	14.2	4.4	3.5	3.5	3.5
Persian	33	9.1	0.0	9.1	15.2	9.1	3.0	3.0	3.0
Abyssinian	11	45.5	18.2	0.0	9.1	27.3	18.2	0.0	9.1
Himalayan	11	27.3	18.2	0.0	27.3	0.0	0.0	0.0	0.0
8 other purebreds	52	9.6	1.9	3.8	15.4	1.9	3.8	0.0	7.7

[a]n = 819.
[b]Tricks are not mutually exclusive.
[c]DSH = Domestic Shorthair; DLH = Domestic Longhair.
[d]Half purebred or a cross of two purebreds.

TABLE VI
Percentage of Cats Taken with Owners on Overnight Trips or on Errands

Frequency of Cat Accompanying Owner[a]	*Overnight (n=866)*	*Errands (n=873)*
Always	4.6%	0.1%
Usually	5.0%	0.1%
Frequently	3.8%	0.9%
Sometimes	16.1%	6.5%
Never	70.6%	92.3%

[a]Frequently = once a week or more; sometimes = once a month or more; infrequently = less than once a month; rarely = once a year.

male's daughters would be of reproductive age. Wolski noticed that in free-ranging situations females often did not bear their first litter until 18 to 24 months of age.[9] Tomcats may experience higher mortality than females in that their extensive home ranges take them across more roads and they engage in fights with other males. Ethologists often mention that tomcat fights are rarely serious, but a more accurate description would be that immediate mortality is rare. Certainly few cats seriously mutilate or kill each other during a fight, but a single bite can develop into an abscess and accompanying septicemia over the course of several days.

When sufficient food and shelter are available, domestic cats display friendly social behaviors to each other on a more extended basis. Leyhausen described the friendly approach behavior displayed by cats toward other cats, based on observations of both free-living groups and laboratory colonies. The social behaviors observed included (1) sleeping together, (2) grooming each other, (3) rubbing against each other, (4) friendly greeting after a prolonged absence, (5) running beside each other and purring and rubbing against each other with tail raised, and (6) playing together. Owners have always insisted that cats engage in amicable social behaviors with one another. A recent survey we conducted among owners of household cats indicated that the majority of cats frequently play, sleep, groom, and eat with other cats (Table I).

Moelk[24] described many years of observations on the development of vocal and friendly approach behaviors in house cats. The author of the paper described four patterns of friendly approach behavior:

1. *Murmuring*—sounds like an extended single-stroke purr; usually uttered in one to four distinctive rolls and represented as "mhrn," "mhrnhrn," or "mhrnhrn mhrnhrn"
2. *Purring*—a continuous vibration of both the inhaled and exhaled breath, represented as "hrn-rhn-hrn-rhn"
3. *Rubbing*—rubbing head, shoulder, and body against a vertical object, such as a doorpost, piece of furniture, or human leg
4. *Rolling*—cat lying on the side and rolling the body from side to side with or without clutching or batting movements of the paws.

These four types of friendly behavior develop in the kitten by the age of five weeks. Each behavior can be used as a response to the touch, the voice, or the sight of the mother cat or a human.

Purring, rubbing, and rolling do not vary much other than in length of time and intensity; whereas "mhrn" murmurs exhibit wide variation and are sensitive to the response of a social partner (cat or human). Murmuring can provide the cat a broad range of vocal expression.[25] "Mhrn" vocalizations appear with high frequency between mother and young. They serve as a greeting when the

TABLE VII
Percent of Cats That Alert Owners to Visitors[a]

	Frequency of Alerting Behavior				
	Always	*Usually*[b]	*Frequently*[b]	*Sometimes*[b]	*Never*
Percent of Responses	26.3	14.3	8.4	17.3	33.8

[a]n = 868.
[b]Usually = 80% to 90% of the time; frequently = 30% to 70% of the time; sometimes = less than 30% of the time.

TABLE VIII
Types of Feline Behaviors That Owners Considered Protective[a]

	Type of Protective Behavior					
	Alert to Noises	*Threaten People*	*Threaten Animals*	*Follow Owners*	*Sleep with Owner*	*Other*
Percent of Responses	65.2	12.2	9.4	2.6	2.2	13.8

[a]n = 283.

TABLE IX

Percentage of Owners That Believed They Are Aware of Their Cat's Moods and Percentage of Owners That Believed the Cats Were Aware of the Owner's Moods

Frequency[a]	*Owner Aware of Cat's Moods*[b]	*Cat Aware of Owner's Moods*[c]
Always	29.7%	21.4%
Usually	44.7%	33.7%
Frequently	13.2%	16.1%
Sometimes	10.9%	19.5%
Never	1.4%	9.3%

[a]Usually = 80% to 90% of the time; frequently = 30% to 70% of the time; sometimes = less than 30% of the time.
[b]n = 868.
[c]n = 851.

mother returns to the nest and attract the kittens to her. "Mhrn" vocalizations greatly increase in frequency during courtship and mating. The "mrhn" is not simply a mating call but also appears to function as a friendly approach communication that may coordinate and synchronize courtship and mating behaviors.

"Mhrn" greetings between kittens decrease around the time of weaning and when the mother starts bringing prey back to the nest. Small prey items, such as mice, can only be possessed by one kitten; hence, they elicit extreme competition. The kitten that manages to seize the prey item runs off with it and will growl, hiss, and swat at other kittens who approach. Kittens raised in a household where food is plentiful usually do not display such possessiveness and intense aggression toward each other.

Occasionally, however, regardless of food supply, a very palatable morsel, such as a piece of chicken, may cause a normally docile kitten to hiss and spit as it defends its find. This aggression may be directed toward another cat or to a person who tries to take away the food. At this time, kittens also become increasingly interested in small moving objects and object play. Solitary play and intense aggression over scarce, palatable prey probably begins to prepare the kitten for its life as a solitary hunter. In situations where food and shelter are abundant, intense competition does not arise among the kittens over these resources and friendly behaviors continue to develop.

New Data on Cat–Owner Interactions

Owners have always insisted that cats engage in amicable social behaviors with one another and with people. Evidence supporting the social behavior of household cats was obtained from a survey of cat owners who filled out a questionnaire while they were waiting for medical or surgical attention for their pet at four veterinary hospitals on the East Coast.[a] A total of 887 questionnaires was either completely or partially filled out; 41% of the questionnaires referred to cats brought to the clinic and 59% to cats at home. Thirteen percent of the respondents filled out two questionnaires concerning their cats, 3% filled out three questionnaires, and 1% filled out more than three questionnaires.

According to the survey, the majority of cats frequently engage in social behavior with other cats in the household (Table I) as well as with the owners. The majority of owners indicated that their cats slept with a person in the family, shared food with people, and greeted the owner on return to the home (Table II). Several respondents wrote in comments indicating that their cat also greets friends, rubs against their legs, or plops itself down on the rug in the middle of the activity area when visitors are present.

According to this questionnaire, 46% of cats and owners spend more than one hour daily interacting with each other (Table III). Although talking to, playing with and exercising, training, and/or caring for the cat each day may be primarily owner initiated, the cat has to be cooperative and interact reciprocally with the owner for the behaviors to occur. Forty percent of the cats were reported as knowing tricks. Types of behaviors that the owners listed as tricks were fetching, retrieving, meowing on command, playing games (such as chasing strings or batting pencils), coming when called, learned actions (such as sit or roll over on command or jump over a fly swatter) as well as any interesting behavior that the cat did, such as wake the owner, watch birds, or roll over to be petted. Some owners simply answered the question with the statement that the cat understands or obeys everything (Table IV). The responses to the survey confirmed the belief that Siamese cats retrieve or fetch more than most other breeds. Abyssinians and Himalayans also appear to be good retrievers (Table V).

Most owners interacted with their cats in ways that per-

[a]The Veterinary Hospital of the University of Pennsylvania (VHUP), The Animal Medical Center in New York City, a feline specialty practice, and a suburban small animal general practice.

TABLE X

Frequency of Owners Talking to Cats About Matters Important to the Owners[a]

	Frequency of Confiding				
	Very Often[b]	*Frequently*[c]	*Sometimes*[d]	*Occasionally*[e]	*Never*
Percent of Responses	11.6	13.1	3.2	19.2	43.0

[a]n = 865.
[b]Once a day or more.
[c]Once a week or more.
[d]Once a month or more.
[e]Less than once a month.

TABLE XI

Number and Percentage of Cats Addressed with Pet Talk, as a Child, as an Adult, or Other[a]

Type of Talk	*Number*	*Percent*
Child only	320	36.8
Adult only	178	20.5
Pet only	113	13.0
Other (one answer)	22	2.5
Pet talk and child	52	6.0
Pet talk and adult	28	3.2
Pet talk and other	6	0.7
Child and adult	70	8.1
Child and other	8	0.9
Adult and other	6	0.7
Pet, child, and adult	54	6.2
Pet, child, adult, and other	12	1.4

[a]n = 869.

TABLE XII

Type of Family Member That the Cat was Considered[a]

Type of Member	*Number*	*Percent*
Animal member only	376	49.6
Child only	224	29.6
Other single answer[b]	53	7.0
Animal and other answer(s)	83	10.9
Child and other answer(s)	90	11.9
Animal and child only	56	7.4

[a]n = 758.
[b]Brother/sister or other.
[c]Child, parent, brother/sister, and/or other.
[d]Animal, parent, brother/sister, and/or other.

mitted or encouraged the cats to remain close to them throughout the day. Not only did owners allow the cats to sleep on the bed, feed them tidbits and snacks (Table II), and spend time playing with and caring for them (Table III); but also 95% of the cats were allowed on furniture, and some accompanied the owners on errands or on overnight trips (Table VI). Most cats (96%) were talked to at least once a day, and less than 1% were talked to less than once a week. No respondents answered that they never talked to their cats.

Many owners celebrated their cat's birthdays, although only 23% knew their cat's exact birth date. Twenty-five percent of the cats were assigned a birthday if the exact date was unknown. People usually celebrated the cat's birthday by giving the cat a present or a special treat (70%), and a few did so with parties (6%).

Owners' Perceptions of Their Cat's Behavior

Thirty-three percent of the cats were considered to be protective of the owners or a good "watchcat" for the home. About 26% of the cats were described as always alerting their owners (Table VII). A total of 283 people responded to the question "*what does your cat do that makes you consider the cat protective?*" The majority considered cats that were alerting the owner to noises as having protective value. Some felt that their cat's aggressive behavior to people or other animals was motivated by protection. A few people believed that following and sleeping with the owner was protective, and a few considered killing bugs or mice protective behaviors (Table VIII).

Most owners believed that they and their cats were able to communicate to each other their respective moods. For instance, more than 70% of the owners believed that they were usually or always aware of their cat's moods, and more than 50% of the owners believed that the cats were aware of their owner's moods (Table IX).

The majority of the people surveyed answered affirmatively to the question "*do you confide in your cat (talk to him/her about problems and events important to you)?*" (Table X). When asked how they talked to their cat, the majority indicated they conversed as if the cat were a person, usually a child, rather than a pet (Table XI). Some cats, however, were addressed in more than one manner, usually both as a pet and as a child.

A total of 99% of the cats in the survey was considered members of the family. Although most cats were talked to as people, the majority were thought of as animal members of the family (Table XII). In response to the question "*what kind of family member do you consider him? (check all that apply),*" about 50% of the cats were considered strictly animal members and 11% as human members as well as animal members. About 30% of the cats were considered as children.

Our survey suggests that most owners of household cats do not consider their pet unfriendly and aloof. In fact, cats may be interacted with and perceived as social companions in much the same way dogs are, at least by a population of owners that seek sophisticated veterinary care. A number of surveys indicate increasing cat ownership in the United States.[26] Veterinarians should be aware that most pet cats interact with each other and with their owners in a far more social manner than has been generally appreciated.

REFERENCES

1. Wilson EO: *Sociobiology, The New Synthesis*. Cambridge, MA, The Belknap Press of Harvard University Press, 1975.
2. Barash DP: *Sociobiology and Behavior*, ed 2. New York, Elsevier, 1982, p 200.
3. Gartlan JS: Structure and function in primate society. *Folia Primatologica* 8(2):89–120, 1968.
4. Bernstein IS, Sharpe LG: Social roles in a rhesus monkey group. *Behaviour* 26(1,2):91–104, 1966.
5. Mochi U, MacClintock D: *A Natural History of Giraffes*. New York, Charles Scribner's Sons, 1973.
6. Fox MW: The behaviour of cats, in Hafez ESE (ed): *The Behaviour of Domestic Animals*, ed 3. Baltimore, The William & Wilkins Co, 1975, p 413.
7. Lipton J: *An Exaltation of Larks*, ed 2. New York, Viking, 1977.
8. Leyhausen P: *Cat Behavior*. New York, Garland STPM Press, 1979.
9. Wolski TR: Social behavior of the cat, in Voith VL, Borchelt PL (eds): *Vet Clin North Am [Small Animal Pract]* 12(4):693–706, 1982.

10. MacDonald DW, Apps PJ: The social behaviour of a group of semi-dependent farm cats, *Felis catus*: A progressive report. *Carnivore Genetics Newsletter* 3:256-267, 1978.
11. Schaller GB: The Serengeti lion: A study of predator-prey relations. Chicago, University of Chicago Press, 1972.
12. Liberg O: Predation and social behaviour in a population of domestic cats. An evolutionary perspective (dissertation). Sweden, University of Lund, 1981, pp 81-82.
13. Lott DF: Intraspecific variation in the social systems of wild vertebrates. *Behaviour* 88(3,4):265-325, 1984.
14. Seghers BH: Schooling behavior in the guppy (*Poecilia reticulata*). An evolutionary response to predation. *Evolution* 28:486-489, 1974.
15. Barlow GW: Extraspecific imposition of social grouping among surgeonfishes (*Pisces*: *Acanthuridae*). *J Zool* 174:333-340, 1958.
16. Hirth DL: Social behavior of white-tailed deer in relation to habitat. *Wild Mono* 53, 1977.
17. Bekoff M, Wells MC: The social ecology of coyotes. *Sci Am* 242:130-148, 1980.
18. Bekoff M, Wells MC: Social ecology and behavior of coyotes, in *Advances in the Study of Behavior*, vol 16. New York, Academic Press, 1986, pp 252-338.
19. Gill FB, Wolf LL: Economics of territoriality in the golden-winged sunbird. *Ecology* 56:333-345, 1975.
20. Mills MG: The mating system of brown hyaena, *Hyaena Brunnea*, in the southern Kalahari. *Behav Ecol Sociobiol* 10:131-136, 1982.
21. Owens DC, Owens NJ: Communal denning and clan association in brown hyenas (*Hyaena Brunnea*, Thunberg) of the central Kalahari desert. *Afr J Ecol* 17:35-44, 1979.
22. Laundre J: The daytime behavior of domestic cats in a free-ranging population. *Anim Behav* 25:990-998, 1977.
23. Natoli E: Spacing pattern in a colony of urban stray cats, *Felis catus L.* in the historic centre of Rome. *Appl Anim Behav Sci* 14:289-304, 1985.
24. Moelk M: The development of friendly approach behavior in the cat: A study of kitten-mother relations and the cognitive development of the kitten from birth to eight weeks, in *Advances in the Study of Behavior*, vol 10. New York, Academic Press, 1979, pp 164-224.
25. Moelk M: Vocalization in the house cat: A phonetic and functional study. *Am J Psychol* 57:184-205, 1944.
26. *Pet Food Institute Fact Sheet*. Washington, DC, Pet Food Institute, 1985.

UPDATE

Penny L. Bernstein, PhD
Biology Department
Kent State University

Within the past few years, four valuable texts on cat behavior have been published.[1-4] Turner and Bateson's book[1] is the most basic and research oriented of the recent volumes. It has excellent reviews of several subject areas by the major researchers currently studying cat behavior, many of whose previous works were cited in the original Voith and Borchelt article.[5] Articles include updates by Leyhausen on cat social life in general, Kerby and Macdonald and Liberg and Sandell on social organization in feral outdoor populations, and Natoli and De Vito on feral cat mating systems; an article on domestication of the cat by Serpell; and an overview of the human–cat relationship by Karsh and Turner to name a few.

John Bradshaw's *The Behaviour of the Domestic Cat*[2] is a slim volume summarizing current research in several major areas, including history and domestication of the cat, behavioral development, sensory capabilities, feeding and hunting behavior, social behavior, cat–human interaction, cat personality "types," and cat welfare. In the final chapter, Peter Neville, a London-based behavior consultant, discusses diagnosis and treatment of behavior problems.

The book is an easy read and an articulate overview of the behavioral biology of domestic cats and is based on published studies rather than anecdotes. The last chapter is based primarily on his personal philosophy and experiences from his own caseload and gives fewer references to or discussions of other published articles.

Overall, this volume provides a quick way to review the classic literature, catch up on recent developments, and gain a better understanding of the domestic cat—where it comes from, how it acts, and why. There is also an excellent bibliography.

Beaver's book[3] has a more veterinary slant. She uses some of the same information sources as Bradshaw, so there are some areas of overlap; but she provides a more extensive bibliography of cat behavior topics, many of which are related to veterinary medicine and are not covered in the other volumes.

The O'Farrell, Neville, and Ross text[4] is primarily devoted to the diagnosis and treatment of feline behavior problems. It also includes chapters on factors that influence a cat's behavior and may underlie problem development, such as owner attitudes, early development, senses, instinct, cognition, anxiety, and stress. The final chapter is devoted to helping owners prevent problems by providing practical information, based on O'Farrell and Neville's experience and the few recent studies, about such topics as how to choose a breed of cat for a pet, whether to allow cats outdoors, socializing cats to one another, acclimating cats and babies to one another, and helping cats cope with a move.

IN-HOME STUDIES

Although an estimated 60 million cats in the United States currently reside as pets in people's homes, little research has focused on how cats actually live and interact in this setting. At the University of Georgia,[6] a study of the congeniality of pairs of household cats by gender is discovering, for example, that male dyads engage in many amicable behaviors. A colleague and I recently reported a study[7,8] of seven castrated male and seven spayed female housecats living in a one-story ranch house; this study showed several new or revealing aspects of how cats actually live within a home.

We found, for example, that cats indoors formed individually distinct but overlapping home ranges (areas of habitual use), with adult males having slightly larger ranges than adult females. Although these findings resembled those for adult feral cats outdoors[9]; they also suggested that sex, age, individual personalities, and relationships played a role in how cats formed and maintained their home ranges with respect to one another.

We could also identify particular spots within home ranges where individuals were likely to be found at particular times, using the areas repeatedly for sleeping, resting, and grooming. Such "favored spots" are familiar to all cat owners but are rarely discussed in the literature. Individuals in our population were either the only cat to use a spot or they shared the spots with others. In this population, cats shared the spots by using them in temporal sequence, each cat using the same spot at different times. This "time-sharing" aspect of housecat behavior has not, to our knowledge, been formally studied previously. Because we could identify groups, usually two to three individuals, who predictably used the same spot at different times, our findings suggested that individuals "knew" with which other cats they shared a spot. Gender played a role in these groupings, with most spots being shared either by all-female or all-male groups.

There is little agreement in the literature about the concept of dominance in domestic cats. Some studies find obvious evidence for dominant individuals and even hierarchies, at least in outdoor populations or in laboratory settings.[10–13] Others find mixed evidence or no evidence at all to support this concept of dominance within the species.[a,2,3,5] I am not aware of any published formal studies that focus on dominance among cats in the home.

Despite the presence of 14 cats in the small home (124 m^2) in our study, overt aggression was rare; it was therefore difficult to identify individuals as dominant. One or two cats indicated a dominance, at least in some situations, by controlling resources or supplanting other individuals—two commonly cited general indicators of dominance in animals.[14,15] One individual seemed subordinate to all others, withdrawing from the resource at the approach of any other cat. The death of a dominant individual, however, led to changes in the behavior of the other cats. These changes, coupled with the lack of overt aggression, suggested that individual relationships, situation, temperament, and even breed played important roles in shaping dominance–subordination interactions in this population and that subtle cues, rather than overt aggression, may have governed these relationships.

Our preliminary findings[7,8] also indicated that tail signals in this population may have helped "tag" individuals at a distance as more or less likely to interact and in what ways (e.g., aggressive or not), such that recipients might have tailored their responses in ways that reduce aggressive encounters. My colleagues and I have shown or speculated that tail signals play such a role in other social groups.[16–18]

Density calculations for the population in our study indicated that they lived at densities some 50 times greater than those found in most studies of feral cats outdoors,[9] yet they formed a stable grouping with little overt aggression. Although the possible interactions between density and social organization have been addressed for populations of feral cats outdoors[9,19] and explored informally for various animals indoors (including nondomestic felids in zoo or circus settings),[20–23] we have not found any studies of housecats that investigated the effects that living at different densities might have on social interaction. As part of a follow-up project, I am currently pursuing such a study as part of a follow-up project. In this study, cat owners are being surveyed to investigate the general applicability of our findings to other in-home populations of various compositions and densities.

[a]Borchelt PL: Personal communication, Animal Behavior Consultants, Inc., Brooklyn, New York, 1995.

REFERENCES

1. Turner DC, Bateson P (eds): *The Domestic Cat: The Biology of its Behaviour.* New York, Cambridge University Press, 1988.
2. Bradshaw J: *The Behaviour of the Domestic Cat.* Tucson, CAB International, 1992.
3. Beaver BV: *Feline Behavior: A Guide for Veterinarians.* Philadelphia, WB Saunders Co, 1992.
4. O'Farrell V, Neville P, Ross CStC (ed): *BSAVA Manual of Feline Behaviour.* Cheltenham, England, British Small Animal Veterinary Association, 1994.
5. Voith VL, Borchelt PL: Social behavior of domestic cats. *Compend Contin Educ Pract Vet* 8(9):637–646, 1986.
6. Berry K: The social behavior of the indoor domestic cat (abstract), in *AVSAB Annual Scientific Session.* Chicago, American Veterinary Medical Association, 1995.
7. Bernstein PL, Strack M: Home ranges, favored spots, time-sharing patterns, and tail usage by 14 cats in the home. *Anim Behav Consultant Newsletter* 10(3):1–3, 1993.
8. Bernstein PL, Strack M: A game of cat and house: Spatial patterns and behavior of 14 cats (*Felis catus*) in the home (in press). *Anthrozoos.*
9. Liberg O, Sandell M: Spatial organisation and reproductive tactics in the domestic cat and other felids, in Turner DC, Bateson P (eds): *The Domestic Cat: The Biology of its Behaviour.* New York, Cambridge University Press, 1988, pp 83–98.
10. Natoli E, De Vito E: Agonistic behaviour, dominance rank and copulatory success in a large multi-male feral cat, *Felis catus* L., colony in central Rome. *Anim Behav* 42:227–241, 1991.
11. Winslow CN: Observations of dominance and subordination in cats. *J Genetic Psychol* 52:425–428, 1938.
12. Cole DD, Shafer JN: A study of social dominance in cats. *Behaviour* 27:39–53, 1966.
13. De Boer JN: Dominance relations in pairs of domestic cats. *Behavioural Processes* 2:227–242, 1977.
14. Alcock J: *Animal Behavior*, ed 4. Sunderland, MA, Sinauer Associates, 1989.
15. Grier JW: *Biology of Animal Behavior.* St. Louis, Times Mirror/Mosby College Publishing, 1984.
16. Bernstein PL: Abundantly performed displays: The tail as a source of information in black-tailed prairie dogs *Cynomys ludovicianus* (doctoral thesis). Madison, WI, Microfilms, 1978.
17. Bernstein PL: Nearly continuous displays and animal communication (abstract), in *XVIth International Ethological Conference.* Vancouver, University of British Columbia, 1979.
18. Bernstein PL, Smith WJ, Krensky A, Rosene K: Tail positions of *Cercopithecus aethiops. Z Tierpsychol* 46:268–278, 1978.
19. Kerby G, Macdonald DW: Cat society and the consequences of colony size, in Turner DC, Bateson P (eds): *The Domestic Cat: The Biology of its Behaviour.* New York, Cambridge University Press, 1988, pp 67–81.
20. Hediger H; Sircom G (trans): *Studies of the Psychology and Behaviour of Captive Animals in Zoos and Circuses.* London, Butterworths Scientific Publications, 1955.
21. Hediger H; Sircom G (trans): *Wild Animals in Captivity.* New York, Dover Publications, 1964.
22. Hediger H, Vevers G, Reade W (trans): *Man and Animal in the Zoo.* New York, Delacorte Press, 1969.
23. Price EO: Behavioral aspects of animal domestication. *Q Rev Biol* 59(1):1–32, 1984.

Ecological Aspects of Urban Stray Dogs

Alan M. Beck, ScD
Director, Center on the Interactions
of Animals and Society
The School of Veterinary Medicine
University of Pennsylvania
Philadelphia, Pennsylvania

The dog population is composed of three interacting, and at times interchangeable, subpopulations: (1) pets that never roam without human supervision; (2) straying pets that roam continuously or sporadically; and (3) ownerless animals usually referred to as *strays*. Social policy towards strays is ambivalent. On one hand, they are protected by the fact that society is unwilling to either socially or financially support the dog catcher and, on the other hand, they are unfairly blamed by popular literature as the major cause of dog bite injuries.

It is important to distinguish straying pets from ownerless strays, because they cause different problems for society and are managed or controlled by different means.[1] Straying pets are best managed by encouraging and enforcing responsible ownership, while strays are controlled by capture and alterations of the environment, e.g., the boarding of vacant buildings, leasing of dumps and clearing of urban lots.[2]

As a general rule, straying pets are more common in high human density, low- to middle-income areas, especially where people have direct access to the streets (e.g., areas of low- to middle-income private housing or row houses). Ownerless strays are more common in low-density areas where there is shelter and fewer requests for animal control, such as around parks, dumps or abandoned parts of the inner city.

Origin

There is little evidence that strays are an actively self-perpetuating population. The author hypothesizes that the stray population is maintained by being seeded with straying pets or permanently released or escaped pets.[1] Litter survivorship in dogs living under stress is low: It is likely that people protect and feed pregnant and lactating bitches until the pups are weaned, and then the young either escape or are released, abandoned or given away and then subsequently released.[3,4] In wild areas, truly feral dogs may exist, as observed on a wildlife refuge.[5,6]

The author has observed strays mate, *tie* and even whelp; however, no young ever survived past a few weeks of age.[3,4] It appears, from general observation and the increased number of calls about packs received by dog control agencies, that there is a slight increase in mating activity in March and October. This is somewhat consistent with wild canine breeding activity, although there is also evidence that dogs have no discernible breeding season.[7]

Fig 1—Straying pets on a college campus. This is a typical example of free-running pets.

Fig 2—A stray dog utilizing an abandoned car in St. Louis.

There are no statistics concerning the extent of pet escape or abandonment. Considering that only 3 to 6% of stray animals captured by animal wardens are ever claimed by owners, these occurrences may be more common than is presently believed. It is possible that owners are ignorant of the fact that their animals may have been captured, or perhaps some owners harbor an unconscious desire that their animals not return. Encouraging an animal to escape or run free may be an alternative to surrendering an animal for euthanasia.

Population

There is a general impression, both in the cities and particularly in rural areas, that strays and feral dogs are more common than they really are. In truth, most urban and rural uncontrolled dogs are straying pets. Surveys taken from 1952 to 1965 showed a wide range in the ratio of unowned to owned dogs, e.g., the ratios were 1:2.6 in Atlanta and 1:40.5 in Denver City and County.[8]

There are approximately 41 million owned dogs in the United States, with ownership being greater among the higher income groups. The number of strays is unknown. However, through the use of modified wildlife estimating techniques, the author was able to calculate the total population in one study of Baltimore City dogs: Much of Baltimore had 450 dogs per square mile.[3]

It is also important to know the age distribution of a *wild* animal population. Estimating the age of a dog by tooth eruption and wear pattern can be inexact for an individual dog, but has value when pooling the data to assess the general population. After euthanasia, it is possible to judge the age of dogs collected from the streets. The data can then be handled statistically, as with life tables such as those used by life insurance companies.

The mean age of the uncontrolled dog population appears to be younger than that of pets: In both Baltimore and St. Louis, the mean age was only about two and one-half years old, with longevity under six years (even though all younger animals may not have been counted, due to the difficulty in finding them).[3,4]

Activity

Free-running dogs are active in urban areas sporadically through the night and mostly in the early morning. It was determined by direct observation of individual strays in Baltimore and St. Louis that home ranges are less than 0.1 square mile about the dog's homesite.[3,4] Pets that are always permitted to run free have ranges almost as large.

Stray dogs occupy a variety of places. Vacant garages and lots, shrubbery, abandoned cars and narrow spaces between buildings are used for shelter. The author observed one stray who lived in the hallway of an occupied brownstone, in which case each tenant probably assumed it was a neighbor's pet.[9] Other strays used an abandoned building for shelter, and even after the accessible entries to the building were boarded up to discourage the dogs, they jumped through a closed window and resumed occupancy.[4]

Fig 3—An aggregate of stray dogs utilizing an abandoned house in a low-density, low-income area in St. Louis.

Fig 4—A stray dog resting on a discarded mattress in a vacant lot in Baltimore. Human trash serves as additional cover for the dog.

Free-ranging dogs primarily use alleys for most activities, including resting, exercising, feeding and mating. For example, 40% of all dogs observed in Baltimore were in alleys. Streets and sidewalks were utilized only half as often.[3]

Social Organization

Strays form small, permanent groups which are periodically joined by other strays and straying pets. In general, about half the free-roaming dogs on the streets are in the company of other dogs, usually in twos or threes. About 26% of the stray dogs are in groups of two, and 16% in groups of three and 7% in groups of four or five, with larger groups occurring very infrequently. The author has never observed more than one adult female in a stable group.[3] On occasion, loose pets will join groups, but not usually as permanent members. A bitch in estrus often attracts 10 or more followers.[3,10]

Food

Pets are fed in and around the home, and strays appear to find ample nutrition on the streets. Serious malnutrition is rarely observed, though lean strays are the norm.

Dogs spend only 11% of their time engaged in food procurement, mostly in alleys. More dogs were observed in alleys before trash collection than on the following days; presumably, they alter their *foraging* pattern on those days. Strays routinely rummage through trash, which is one of the motivations for dog catching. (Overturned garbage cans are esthetically displeasing, encourage the breeding of flies and rats and add greatly to the expense of trash collection.) Strays also chase cats, squirrels and birds, although this author has never observed a capture. Sometimes people put food out for strays or feed particular favorites.[3,9,10] On one low-income block in Baltimore, 20% of interviewees stated that they observed people feeding strays or did so themselves.

Loose dogs find water in gutters and puddles filled by rain, car washing, and leaking fire hydrants and air conditioners. The ponds and lakes in city parks are used by dogs whose ranges include the park.

Fig 5—A stray dog rummaging through trash—its main source of food.

Mortality

Death is primarily the result of being hit by cars, disease, worm infection and being caught and subsequently killed by animal control officers. In the United Kingdom, over 6% of all automobile accidents involve free-running dogs.[11] In the United States, 12% of the pet dog population is killed in animal shelters yearly. In Baltimore and St. Louis, more than a quarter of the estimated dog population is killed in shelters.[3,4] The actual mortality for strays must be even greater, but is unknown. The average age of strays caught by the shelters in Baltimore and St. Louis was only two years, with 60% of the population under the average. This young average age of the stray population, which is generated by the high mortality rate, has serious implications. Young animals are more susceptible to disease (e.g., rabies) and worm infection. It was found that 70% of all strays and 41% of pets brought to veterinarians are parasitized by nematodes—19% with roundworms, 19% with hookworms and 26% with whipworms. In addition, younger animals more often bite people than do older animals. True strays account for only about 15% of the reported bites, with straying pets being the major problem.[12,13] Because it is often difficult to distinguish between the two, dog owners are quick to blame the strays and community control programs are encouraged to capture strays, attenuating their efforts to enforce leashing laws.

Management

The most widespread social response to the need to manage dogs has been the leash law, i.e., dogs must be supervised on public property. If strictly enforced, this would be the single most beneficial policy for both man and dog. Dogs on leashes bite less, do not disrupt trash, and do not defecate in unsuspected areas that might be used by children; pregnancy and impregnation are also controlled when dogs are leashed. In addition, leashed dogs would be spared from being hit by cars, being caught by control officers, and from coming in contact with diseased dogs and other animals. Even though this simple management tool would solve many problems, it receives far less attention than new ideas, such as nonprofit sterilizing clinics.[14,15] These clinics infer the assumption that dogs can run loose if they are sterile. Finally, people must be discouraged from abandoning unwanted dogs to the streets.

The major impetus for concern, at this time, is the free-running dog. When running free, urban dogs are at best pitiful and at worst destructive. Socially aware and responsible ownership would make it possible for the dog to continue its role as the foremost ambassador of animal life in the city.

REFERENCES

1. Beck AM: Ecology of unwanted and uncontrolled pets, in *Proceedings*. National conference on the ecology of the surplus dog and cat problem, AVMA, Chicago, 1974.
2. Beck AM: Guidelines for planning for pets in urban areas, in Fogel B: *The Human-Companion Animal Bond. A BSAVA Symposium*. London, Bailliere Tindall, 1980.
3. Beck AM: *The Ecology of Stray Dogs: A Study of Free-ranging Urban Animals*. Baltimore, York Press, 1973.
4. Fox MW, Beck AM, Blackman E: Behavior and ecology of a small group of urban dogs (*Canis familaris*). *Appl Anim Ethol* 1:119-137, 1975.
5. Nesbitt WH: Ecology of a feral dog pack on a wildlife refuge, in Fox MW (ed): *The Wild Canids*. New York, Van Nostrand Reinhold Co, 1975, pp 381-396.
6. Scott MD, Causey K: Ecology of feral dogs in Alabama. *J Wildlife Management* 37:252-265, 1973.
7. Engle ET: No seasonal breeding cycle in the dog. *J Mammal* 27:79-81, 1946.
8. Marx MB, Furcolow ML: What is the dog population? *Arch Environ Health* 19:217-219, 1969.
9. Beck AM: The life and times of Slag, a feral dog in Baltimore. *Natural History* 80(8):58-65, 1971.
10. Beck AM: The ecology of "feral" and free-roving dogs in Baltimore, Ch 26, in Fox MW (ed): *The Wild Canids*. New York, Van Nostrand Reinhold Co, 1975.
11. Carding AH: The significance and dynamics of stray dog populations with special reference to UK and Japan. *J Small Anim Pract* 10(7):419-446, 1969.
12. Beck AM, Loring H, Lockwood R: The ecology of dog bite injury in St. Louis, Missouri. *Public Health Rep* 90(3):262-267, 1975.
13. Harris D, Imperato PJ, Oken B: Dog bites—an unrecognized epidemic. *Bull NY Acad Med* 50:981-1000, 1974.
14. Spay clinics: The other side of the story. *Mod Vet Pract* (April): 29-34, 1973.
15. Spay clinics: Boon or Boondoggle? *Mod Vet Pract* (March):23-29, 1973.
16. Purvis MJ, Otto DM: *Household demand for pet food and the ownership of cats and dogs: An analysis of a neglected component of US food use*. (Staff paper P 76-33) Dept. of Agriculture and Applied Economics, University of Minnesota, St. Paul, 1976, p 47.

UPDATE

Alan M. Beck, ScD
Center for Applied Ethology and
Human-Animal Interaction
School of Veterinary Medicine
Purdue University

Since this article was originally published, there have been relatively few additional studies of urban stray or feral dogs; specific populations have been studied in New York City,[1] Berkeley, California,[2] Italy,[3] Newark, New Jersey,[4,5] and some areas of Mexico and the American Southwest.[6]

SOCIAL ORGANIZATION

Personal observations since my original studies have found that one adaptation of unowned stray dogs in an urban environment is to behave like socialized pet dogs. In that way they are indistinguishable from owned stray dogs and are tolerated as loose pets and not wild dogs—a form of "cultural camouflage." There are differences between owned and unowned stray dogs that are not easily observed without extensive study (Table I).

POPULATION

There is no official census of the pet or feral dog population or the numbers that are killed in animal shelters. Rowan, however, estimates that between 2 million and 6 million dogs and cats are euthanized in animal shelters each year.[7] The popular press often quotes in excess of 12 million. Estimates of the total dog population come from local, regional, or national surveys, and there is considerable variation in their methods and results. All indicate, however, that the owned dog population is decreasing.[8] Most methods rely on surveys of consumer panels and estimate about 52 million owned dogs.[8] Data from the only true statistical probability sample indicate that about 28% of U.S. households have at least one dog per household—with the average being 1.5 dogs—for a total estimated population of 41 million dogs.[9] There are still no data on how many pets are abandoned to the streets and added to the stray population, although methods exist for such determination for those interested in animal control or public health.[10,11] Because the urban stray dog population comes mostly from the pet population, we assume that the stray population is also decreasing.

MORTALITY

Stray dogs continue to have little impact on human public health, be it as a result of animal bites[12] or rabies.[13,14] The major threat facing them remains humans; these dogs are killed by cars or collected by animal control. For some people, ownerless dogs that live out their lives as wild canids are part of the urban scene; for others, they are animals that must be captured as pests or removed from the streets. After capture or removal, they are usually killed, although some are adopted and become owned pets.

TABLE I
The Urban "Stray" Dog

Characteristic	*Owned Dog*	*Unowned Dog*
Origin	Offspring of owned dogs or from dog breeders	Released or escaped pets
Abundance	~⅓ of owned population	~1/40 of stray population
Distribution	High human density, low or high socioeconomic urban area	Low human density, low socioeconomic urban and rural areas
Home range	0.02–0.1 sq mile	0.2–11.0 sq mile
Activity	Before and after human activity	Nocturnal and crepuscular
Food	From owner, garbage	Garbage, little hunting
Social behavior	Solitary or human companionship	In groups of 2–3 isolated packs
Behavior with people	Friendly → aggressive	Wary, secretive
Mortality	Owner intolerance, old age, injury, disease, dogcatcher	Injury, disease, dogcatcher
Dog bite problem	Very serious	Less serious
Dog waste problem	Increasing social and health problem	Less serious, unknown
Noise problem	Serious	Less serious
Control	Owner responsibility	Self-limiting

REFERENCES

1. Rubin HD, Beck AM: Ecological behavior of free-ranging urban dogs. *Appl Anim Ethol* 8:161–168, 1982.
2. Berman M, Dunbar I: The social behaviour of free-ranging suburban dogs. *Appl Anim Ethol* 10:5–17, 1983.
3. Hansen J: The wild dogs of Italy. *New Scientist* 3:590–591, March 1983.
4. Daniels TJ: The social organization of free-ranging urban dogs. I. Non-estrous social behavior. *Appl Anim Ethology* 10:341–363, 1983.
5. Daniels TJ: The social organization of free-ranging urban dogs. II. Estrous groups and the mating system. *Appl Anim Ethol* 10:365–373, 1983.
6. Daniels TJ, Bekoff M: Populations and social biology of free-ranging dogs, *Canis familiaris*. *J Mammal* 70(4):754–762, 1989.
7. Rowan AN: Shelters and pet overpopulation: A statistical black hole. *Anthrozoos* 5(3):140–143, 1992.
8. Patronek GJ, Glickman LT: Development of a model for estimating the size and dynamics of the pet dog population. *Anthrozoos* 7(1):25–41, 1994.
9. Crispell D: Pet projections. *Am Demographics*, p 59, Sept 1994.
10. Beck AM: Free-ranging dogs, in Davis DE (ed): *CRC Handbook of Census Methods for Terrestrial Vertebrates.* Boca Raton, FL, CRC Press, pp 232–234, 1982.
11. Bogel K: *Guidelines for Dog Population Management.* Geneva, Switzerland, World Health Organization (WHO) & World Society for the Protection of Animals (WSPA), 1990.
12. Beck AM: The epidemiology and prevention of animal bites. *Semin Vet Med Surg Small Anim* 6(3):186, 191, 1991.
13. Beck AM, Felser SR, Glickman LT: An epizootic of rabies in Maryland, 1982–1984. *Am J Public Health* 77:42–44, 1987.
14. Torrence ME, Beck AM, Glickman LT, et al: Raccoon rabies in the mid-Atlantic (epidemic) and southeastern states (endemic) 1970–1986: An evaluation of reporting methods. *Prev Vet Med* 22:197–211, 1995.

Providing Environmental Enrichment to Captive Primates

KEY FACTS

- **Behaviors of a captive nonhuman primate are qualitatively abnormal when they differ in pattern in a maladaptive way from those of a free-ranging animal; behaviors are quantitatively abnormal when they are qualitatively normal but occur at an abnormal frequency.**
- **The captive nonhuman primate spends much less time in feeding than does a free-ranging animal and may spend considerable time in repetitive stereotypic self- or cage-directed behavior.**
- **The environment of captive primates can be enriched by introducing other primates or by providing furniture or toys to provide a broader range of locomotion or manipulation behaviors.**
- **Foraging boards covered with artificial turf or shearling lengthen the time that the captive nonhuman primate spends in feeding and may reduce excessive self-grooming.**

National Institutes of Health
Bethesda, Maryland
Kathryn Bayne, MS, PhD, DVM

THE COMPLEX nature of nonhuman primates is well illustrated by the broad range of behaviors exhibited by this diverse group of animals. Some species are renowned for the following attributes: their locomotive abilities, especially the speed at which they travel (patas monkeys have been clocked at 50 kph[1]) or their precision in leaping from tree to tree; their manipulative tendencies (e.g., the capuchin monkey [organ grinder's monkey], which participates in programs to assist quadriplegic persons); or their cognitive abilities (e.g., language studies with chimpanzees and gorillas).[2,3]

Each species has specific abilities that it has developed to survive in its particular ecologic niche. Removal of an animal from its dynamic, stimulating natural environment to a less complex captive environment can result in that animal exhibiting a number of behaviors that are considered abnormal for the species.[4] This abnormality may be the manifestation of behaviors of a different and maladaptive pattern (or quality) than is normal for the free-ranging animal. Also, an animal may demonstrate a behavior that is qualitatively normal but occurs at an abnormal frequency (quantitatively aberrant). Such behaviors can be considered signs of some degree of psychological distress.[5]

Although aberrancies in an animal's immunophysiologic profile (e.g., heart rate, cortisol level, or T-cell function) are often present, the diagnosis of psychological disorders is usually based on qualitative and quantitative behavioral abnormalities. Qualitatively abnormal behaviors that are considered markers for psychological distress include depression,[6] self-orality,[7] self-clasping, stereotypic behaviors (e.g., palate stroking and eye poking), self-abusive behavior (e.g., self-slapping, head banging, and self-biting), and some locomotor patterns (e.g., somersaulting and repetitive rocking).

Quantitatively abnormal behaviors are motor patterns that are often exaggerated. Examples are circling or pacing that is so focused that the animal is not distracted from the behavior when novel stimuli occur in the environment (e.g., someone entering the room). This lack of attention to surroundings or inability to respond to changes in the environment is considered pathologic because of the maladaptive consequences. Excesses in manipulation and self-grooming behavior (often producing bald spots on the body) reflect a similar exaggeration of otherwise normal behaviors resulting in narrow attentiveness to the environment. Polydipsia of purely behavioral origin is another quantitative abnormality evident in captive primates.[5]

The definitive explanation for the expression of qualitatively and quantitatively abnormal behaviors is elusive. Theories that these behaviors serve as coping strategies or provide a degree of control over the kind and amount of stimulation in the captive environment appear to be the most valid.[8] In either case, the animal seems to be striving for homeostasis in an aberrant physical environment. The ubiquitousness of abnormal behavior in numerous species in captive environments[4,9] suggests that the need to achieve homeostasis in these environments is critical.

The most efficient approach to managing psychological disturbances is similar to that used for physical illnesses. A preventive, diagnostic, and therapeutic approach commonly used for managing animals with physical disorders can be successfully applied to animals with psychological disorders. Consideration of psychological disorders in a clinical framework encourages a more scientific approach to their elimination. For example, environmental conditions that elicit behavioral pathology can be identified in the medical record. Environmental enrichment techniques that were successful in treating the pathology should also be noted in the permanent record.

A PREVENTIVE APPROACH to providing for the psychological well-being[10] of captive primates may include the following: (1) the provision of compatible social partners if the species is social in nature and (2) increasing the complexity of the cage environment by nonsocial methods, thereby providing the opportunity to express species-appropriate activity.[11] These methods can also be applied therapeutically.

SOCIAL ENRICHMENT

Pair Housing

Creating compatible social groups of primates has had variable success. The age, sex, rearing history, and species of primate play significant roles in the ease with which compatible animal groups are formed. These social units often consist of a pair of animals.[12] Pair formation can be accomplished by moving two conscious primates into a new primary enclosure. In this way, neither animal enters the established territory of the other and much subsequent fighting is avoided.

An alternative method involves maintaining the proposed pair in a cage with a removable central divider that allows visual but not physical interaction. The panel can be made of solid acrylic plastic, polycarbonate, or wire mesh to prohibit digits and tongues from protruding into the neighboring cage (preventing them from being bitten off by the neighbor). The animals are paired by removing the panel when it has been determined that the animals are not continually exhibiting aggression through the divider, exhibiting excessive fear responses, or avoiding visual contact.

One primate should be dominant over the other. The respective social roles should be clear. Often, abnormal rearing conditions prohibit communication or understanding of species-normal signals. This situation can result in unabated fighting. Food sharing has been suggested as a means of assessing compatibility.[13]

Other social behaviors that are considered positive signs of compatibility include social grooming and sleeping in physical contact. Although vigilant monitoring of primate pairs must continue for as long as the animals are together, a fair judgment of the initial success of the pairing can be made as early as 15 minutes after the animals are introduced. I have observed, however, that relationships are not necessarily stable for the long term. Two pairs of female rhesus monkeys (*Macaca mulatta*) were compatible for 13 months and 15 months, respectively, before aggression necessitated permanent separation.

Group Housing

Much larger groups of unfamiliar primates also have been successfully introduced.[14] The safest manner in which to introduce macaque species is to place conscious, unfamiliar animals together in a new enclosure. In this way, no previously formed cohorts act together against a single primate. On the contrary, the disorganization that occurs tends to inhibit outright, immediate conflict. Disputes occur later as the dominance hierarchy is established over the course of weeks. The degree of ensuing hostility between animals is unpredictable. Therefore (as with the formation of pairs of monkeys), close monitoring of body weight, appetite, general physical condition, and behavior is necessary for newly established groups of primates.

CLEARLY, a social partner is a dynamic stimulus that can evoke the full range of species-typical behavior in the captive primate. The issue of compatibility, however, is essential. Mature primates that are initially compatible may not remain so. The animals' behaviors toward each other must be closely monitored.

NONSOCIAL ENRICHMENT

Nonsocial enrichment may also be provided by allowing the animal to express its manipulative proclivities, locomotor patterns, or foraging behavior in the more restricted captive environment. Nonsocial types of enrichment are often easier to implement than social types because of the aggressive tendencies of many species of primates during introduction to unfamiliar, nonspecific animals.

Several nonsocial enrichment techniques to reduce the

occurrence of abnormal behavior have been developed and tested in the laboratory setting, in which many species of nonhuman primates are housed in various social and non-social caging conditions. Because the nonhuman primates that the private practitioner may encounter spend much or all of their time housed in cages (because of often aggressive and destructive natures), it is reasonable to extrapolate methods that were successful in enhancing the laboratory environment to other holding conditions. These cage complexities are generally designed to encourage a broader range of species-typical behaviors.

In general, the devices included in the caged primate's environment target manipulative foraging and locomotor behaviors. These devices are useful for individually housed and socially housed primates if enough devices are given to group-housed animals to avoid conflict over the device.

Toys

Various toys that were typically manufactured for non-primate animals, such as dogs, have been used with laboratory primates. Some commonly used nonsocial enrichment items for primates are listed on this page. These toys encourage the natural manipulative behaviors of many species and have demonstrated a beneficial effect on the behavioral profile of the laboratory primate.[15,16] Many of the toys are bright-colored and inexpensive and have a varied topography.

The toys that represent the best monetary value can withstand chewing and repetitive cleaning. The toy should be too large to be swallowed by the animal, thereby avoiding intestinal obstruction by a foreign body. If the toy is gradually destroyed by the primate, the remaining pieces should be removed before they too cause a problem because of sharp edges or small size. Any enrichment project—whether tailored for a laboratory, a zoo, or the private owner—should include a process whereby the enrichment devices are monitored for safety and efficacy.

Visual and Auditory Enrichment

The provision of enrichment by varying sight and sound stimuli has anthropomorphic appeal. Notwithstanding the potential cost of this type of enrichment, studies examining the effects of light quality, television, and radio on behavior of captive primates have been conducted. The results have been mixed, possibly because of the differences in species, sex, and housing condition.

In a draft of regulations regarding psychological well-being of primates, the United States Department of Agriculture (USDA) recommended that full-spectrum lighting be incorporated into animal holding rooms. The hypothesis was that the ultraviolet portion of the spectrum would enhance well-being.[17] When this hypothesis was tested,[18] it was determined that socially housed rhesus monkeys exhibited increased aggression and elevated urine cortisol levels (indicating increased stress) under such lighting.

The effect of television on the range of captive primates typically maintained in laboratories has not been fully evaluated. A preliminary examination of the effect of color television on the behavior of chimpanzees indicated only a trend ($P < 0.10$) toward a reduction in behavioral pathology.[19] The exhibition of high levels of nonrepetitive stereotypies (e.g., self-directed aggression, clapping, covering the eyes, flipping, poking the eye or face, spinning, or self-directed sucking) was correlated with high levels of television watching. Thus, chimpanzees that manifested more self-oriented stereotypies also watched television more than other animals.

GREATER SUCCESS in modifying primate behavior has been associated with a musical feeder box.[20] A radio was set to a popular music station and connected to a

Enrichment Items for Primates

Swings
- Flexible polyvinylchloride tubing (varying diameters)
- Polyvinylchloride pipe (varying diameters)
- Prima-hedrons® (Double LL)

Perches
- Custom-designed perches for existing caging (provided by commercial cage companies)

Treats
- Prima-treats® (BioServe)
- Monkey Yums® (Ralston Purina Company)
- Banana pellets

Toys
- Kong® Toys (The Kong Company)
- UltraRubber™ Toys (rings, tugs, bones, balls, flying disks) (Nylabone Products)
- Clutch balls
- Odd Bounders® (Fundex)
- Boomer Balls® (Boomer Ball Company)
- Stall-mate® apples
- Bird music boxes
- Stuffed animals
- Preschool-age children's toys (only for infant and juvenile monkeys; toys geared for birth to three years of age are most appropriate)

Foraging Devices
- Artificial turf boards
- Foraging and grooming boards
- Puzzle feeders

pellet dispenser attached to the cage of some rhesus monkeys. The feeder resulted in a significant reduction in urine cortisol levels, abnormal behavior, cage manipulation, sitting, and self-grooming; standing and nonstereotypic movement increased. The radio/dispenser was available to the primates for 12 weeks, during which time the radio was played an average of 76 minutes per day.

Locomotion

Enhancing typical locomotor patterns requires an understanding of the normal postures and gaits of the species. This usually necessitates sufficient vertical space for the animal to stand quadrupedally and sit upright. Some species (e.g., such brachiating species as gibbons and spider monkeys) require vertical space to suspend from a horizontal support with full body extension. Brachiating species should be provided with ample space to express this unusual type of locomotion.

In addition to providing adequate cage space to encourage species-typical locomotor patterns and postures, furniture can be included in the living quarters. Swings made of flexible polyvinylchloride tubing[21] or metal or polyvinylchloride pipes are frequently used by younger primates. Prima-hedrons® are sturdy, hollow, plastic geodesic globes that have been used successfully for laboratory primates.[22] The primates swing from, sit in, and climb on them. Prima-hedrons® can be suspended from a chain, thereby providing a swinging action; or they can be mounted rigidly to a ceiling or wall. Young rhesus monkeys readily use either configuration; adult macaques apparently spend more time in the stable units.[a] These devices can be bolted together to form complex units of various shapes and sizes.

Numerous styles of cage furnishings have been used to accommodate primate body postures. I have found that laboratory and zoo primates given the opportunity to perch readily do so.[23] Perching and climbing behaviors can be encouraged by including artificial trees with many levels of horizontal projections, single or multiple bars, or rest boards (i.e., shelves) in the home enclosure.[24,25] The perches should be made of a material that is readily cleaned and durable (e.g., polyvinylchloride or metal) or readily disposable (e.g., wood). Placement of perches at varying levels encourages the primate to use three-dimensional space in the enclosure with jumping and leaping forms of locomotion.

Cage furnishings can be designed to provide a protective visual barrier. Some species (e.g., owl monkeys) are more retiring by nature and use a hiding place as a visual break to increase their sense of security. Other species, such as macaques, use visual breaks to arrest interanimal conflicts. Large and open-ended barrels, solid panels behind which the primate can retreat from view,[26] treated-wood nest boxes, polyvinylchloride elbow joints, cat carriers, and even concrete culverts have been used as barriers in primate enclosures.

Foraging

Most species of free-ranging primates spend a large portion of their waking hours in search of, processing, and consuming food.[27,28] In the captive environment, feeding occurs once or twice daily because ad libitum feeding can result in some species succumbing to gastric bloating. The captive primate thus spends a very small portion of its daily activity budget in feeding-related behaviors. It has been suggested that the behavioral void created by this type of feeding regimen can be partially filled by providing the captive primate with artificial foraging opportunities.[29] Various foraging exercises can be designed. Many apparatuses are commercially available.

Essentially, foraging devices function in one of two ways. Either the food item is placed in a maze (also known as a food puzzle) that the primate must solve to obtain the food, or the food item is in particulate form in a container or spread out in a substrate (in which case the animal must spend more time retrieving small pieces of food). Foraging boards made of artificial turf[30] or shearling substrate increase the amount of time macaques spend in normal feeding-related activities and decrease the amount of abnormal behavior.[29,31]

IN THESE STUDIES, the foraging material was provided as a supplement to the twice-daily feeding regimen. The subjects still foraged for a mean of 40% of the observation session. In addition, there were significant reductions in repetitive locomotor patterns, excessive self-directed activities, and stereotypic cage-directed behaviors. Although all subjects groomed the artificial fleece board, one animal repeatedly assumed a body posture identical to that exhibited to another primate when grooming is being solicited. In other words, the primate was requesting that the board exhibit reciprocal grooming.

Similar findings of reduced behavioral pathologies have been reported for secreting food items in wood shavings on the enclosure floor,[32] for termite feeders used with captive chimpanzees,[33] and for various food puzzles.[29,34,b] An additional benefit attributed to the artificial shearling foraging device is the induction of social grooming behavior; primates have been observed grooming the shearling long after the food items have been removed.[29]

REDIRECTING ABNORMAL BEHAVIORS

The variety and scope of environmental enrichment strategies available for preventing and treating psychologi-

[a]McCully C: Personal communication, National Cancer Institute, Bethesda, MD, 1991.

[b]Brent L: Personal communication, Southwest Foundation for Biomedical Research, San Antonio, TX, 1989.

cal disorders allow targeting of the enrichment technique to a specific disorder. For example, I have found that primates that engage in excessive grooming may benefit from an artificial shearling board to deflect some of the grooming activity. Others have observed that primates that manifest self-biting behaviors redirect this aberrant pattern to a cage object, such as a Kong® toy.[c] Caged primates that exhibit excessive repetitive locomotor behaviors often disrupt their activity to sit and feed from a foraging device. By targeting the behavioral therapy to the specific behavioral abnormality, resolution of the pathology can be accomplished in a more timely and successful manner.

ACKNOWLEDGMENTS

The author thanks Ms. Sandy Dexter and Ms. Amy DeMouy for their assistance in the preparation of this manuscript.

[c]Brown V: Personal communication, Veterinary Resources Program, National Institutes of Health, Bethesda, MD, 1991.

About the Authors

Dr. Bayne is affiliated with the Veterinary Resources Program of the National Institutes of Health, Bethesda, Maryland.

REFERENCES

1. Tylinek E, Berger G: *Monkeys and Apes*. New York, Arco Publishing Co, 1985.
2. Gardner RA, Gardner BT: The role of cross-fostering in sign language studies of chimpanzees. *Hum Evol* 3(1-2):65-79, 1988.
3. Hayes KJ, Hayes C: The cultural capacity of chimpanzees. *Hum Biol* 26(3):288-303, 1954.
4. Mason GJ: Stereotypies: A critical review. *Anim Behav* 41:1015-1037, 1991.
5. Erwin J, Deni R: Strangers in a strange land: Abnormal behaviors or abnormal environments? in Erwin J, Maple T, Mitchell G (eds): *Captivity and Behavior*. New York, Van Nostrand Reinhold Co, 1979, pp 1-28.
6. Harlow HF, Novak MN: Psychopathological perspectives. *Perspect Biol Med* 16:461-478, 1973.
7. Harlow HF, Harlow MK: Psychopathology in monkeys, in Kimmel HD (ed): *Experimental Psychopathology: Recent Research and Theory*. New York, Academic Press, 1971, pp 203-229.
8. Novak M, Suomi S: Psychological well-being of primates in captivity. *Am Psychol* 43(10):765-773, 1988.
9. Meyer-Holzapfel M: Abnormal behavior in zoo animals, in Fox MW (ed): *Abnormal Behavior in Animals*. Philadelphia, WB Saunders Co, 1968, pp 476-503.
10. Animal welfare: Standards—final rule (9 CFR, part 3). *Fed Register* 56(32):6369-6505, 1991.
11. Line S: Environmental enrichment for laboratory primates. *JAVMA* 190(7):854-859, 1987.
12. Reinhardt V, Houser D, Eisele S, et al: Behavioral responses of unrelated rhesus monkey females paired for the purpose of environmental enrichment. *Am J Primatol* 14:135-140, 1988.
13. Reinhardt V, Cowley D, Eisele S, et al: Pairing compatible female rhesus monkeys for cage enrichment has no negative impact on body weight. *Lab Primate News* 27(1):13-15, 1988.
14. Bernstein IS, Gordan TP, Rose RM: Factors influencing the expression of aggression during introduction to rhesus monkey groups, in Holloway RL (ed): *Primate Aggression, Territoriality and Xenophobia*. New York, Academic Press, 1974, pp 211-240.
15. Line SW, Clarke AS, Markowitz H: Adult female rhesus macaque responses to novel objects. *Lab Anim* 18(4):33-40, 1989.
16. Weld K, Metz B, Erwin J: Environmental enrichment for *Macaca fascicularis*: Effects of shape and substance of manipulable objects. *Proc 14th Annu Meet Am Soc Primatol*, 1991.
17. Animal welfare: Proposed rules (9 CFR, parts 1, 2, and 3). *Fed Register* 54(49):10,822-10,954, 1989.
18. Novak MA, Drewsen KH: Enriching the lives of captive primates: Issues and problems, in Segal EF (ed): *Housing, Care and Psychological Well-Being of Captive and Laboratory Primates*. Park Ridge, NJ, Noyes Publications, 1989, pp 161-182.
19. Brent L, Lee DR, Eichberg JW: Evaluation of two environmental enrichment devices for singly caged chimpanzees (*Pan troglodytes*). *Am J Primatol* 1(Suppl):65-70, 1989.
20. Line SW, Clarke AS, Markowitz H, Ellman G: Responses of female rhesus macaques to an environmental enrichment apparatus. *Lab Anim* 24:213-220, 1990.
21. Bayne K, Suomi S, Brown B: A new monkey swing. *Lab Primate News* 28(4):16-17, 1989.
22. O'Neill P, Novicky P, George E: Preliminary evaluation of Primahedron play structures for non-human primates. *Lab Anim* 19(5):40-41, 1991.
23. Schmidt EM, Dold GM, McIntosh JS: A perch for primate squeeze cages. *Lab Anim Sci* 39(2):166-167, 1989.
24. Reinhardt V, Smith MD: PVC pipes effectively enrich the environment of caged rhesus monkeys. *Lab Primate News* 21(3):4-5, 1988.
25. Wolff A: Polyvinylchloride piping as perch material for squirrel monkeys. *Lab Primate News* 28(1):7, 1989.
26. Reinhardt V: A privacy panel for iso-sexual pairs of caged rhesus monkeys. *Am J Primatol* 20:225-226, 1990.
27. Herbers JM: Time resources and laziness in animals. *Oecologia* 49:252-262, 1981.
28. Malik I, Southwick CH: Feeding behavior and activity patterns of rhesus monkeys (*Macaca mulatta*) at Tughlaqabad, India, in Fa JE, Southwick CH (eds): *Ecology and Behavior of Food-Enhanced Primate Groups*. New York, Alan R Liss Co, 1988, pp 95-111.
29. Bayne K, Mainzer H, Dexter S, et al: The reduction of abnormal behaviors in individually housed rhesus monkeys (*Macaca mulatta*) with a foraging/grooming board. *Am J Primatol* 23:23-35, 1991.
30. Bayne K, Dexter S, Mainzer H, et al: The use of artificial turf as a foraging substrate for individually housed rhesus monkeys (*Macaca mulatta*). *Anim Welf*, submitted for publication.
31. Lam K, Rupniak NMJ, Iverson SD: Use of a grooming and foraging substrate to reduce cage stereotypies in macaques. *J Med Primatol* 20(3):104-109, 1991.
32. Anderson JR, Chamove AS: Allowing captive primates to forage, in *Standards in Laboratory Animal Management*, part 2. Potters Bar, England, Universities Federation for Animal Welfare, 1984, pp 253-256.
33. Maki S, Alford PL, Bloomsmith MA, Franklin J: Food puzzle device simulating termite fishing for captive chimpanzees (*Pan troglodytes*). *Am J Primatol* 1(Suppl):71-78, 1989.
34. Bloom KR, Cook M: Environmental enrichment: Behavioral responses of rhesus to puzzle feeders. *Lab Anim* 18(5):25-31, 1989.

Feline Purring

Lea Stogdale, BVSc
Diplomate, ACVIM
Winnipeg, Manitoba, Canada

John B. Delack, PhD, DVM
Metropolitan Veterinary Services
Saskatoon, Saskatchewan, Canada

The nature of purring has fascinated everyone who has had any experience with *Felis catus.* Although purring is very common feline behavior, few articles about it have been published in the veterinary literature. Charles Darwin recorded the occurrence of purring in the domestic house cat, cheetah, puma, and ocelot[1]; and tigers also appear to purr.[2] Domestic cats and cheetahs purr—the latter very loudly—when they, anthropomorphically, appear to be pleased and contented.[2-6] Cats commonly purr in the presence of humans, although the sound is not always audible.[7] Female cats purr during mating,[2] and both a queen and her kittens purr during suckling.[2,4,5] Cats have been observed to purr while they are sick[3,8] and purr even during the terminal stages of an illness.[3-5] Purring does not occur while a cat is asleep.[7]

Two theories about the mechanism of purring have been prominent. The earlier hypothesis suggested that purring was hemodynamic in origin, with the vibration that causes the blood turbulence arising from the wall of the thoracic caudal vena cava.[3,4] Bending of the aorta when a cat arches its back has also been suggested as a mechanism.[4] The authors found no scientific evidence to support either of these theories.

Experimental Studies

In 1972, Remmers and Gautier[7] studied the physical mechanisms and the neurologic control involved in the production of feline purring, using standard physiologic procedures to record tracheal air pressures as well as electromyography. Those authors showed purring, as recorded by a microphone designed for the human voice, to be a sound burst with a frequency of 150 to 200 Hertz (cycles/sec). The sound burst occurred regularly every 30 to 40 milliseconds throughout the respiratory cycle, except during a brief pause between the phases (inspiration and expiration) of the cycle (Figure 1).[7]

Only the laryngeal and diaphragmatic electromyograms showed a phasic pattern similar to the pattern of purring. In the larynx, the cricoarytenoid muscle or the thyroarytenoid muscle or both were involved. With the onset of purring, a regular pattern of activity began: 10 to 15 milliseconds of muscle activity every 30 to 40 milliseconds (Figure 2). The pattern of the cranial laryngeal muscular discharges was constant irrespective of the loudness (intensity) of the purr. The diaphragmatic electromyograms indicated that, besides the electrical discharge of respiration, purring initiated an electrical discharge, which lasted 10 to 20 milliseconds and was repeated every 40 milliseconds. The diaphragmatic activity alternated with that of the cranial laryngeal musculature (Figure 2).[7]

Purring caused more rapid and larger pressure changes in the trachea than respiration in an animal that was not purring. Cyclical pressure fluctuations were superimposed on the respiratory pressure changes. These pressure fluctuations corresponded with the electromyographic findings and the sound bursts.

Bilateral sectioning of each of the nerves involved in laryngeal or diaphragmatic, or respiratory muscle activity failed to abolish purring. Laryngeal discharge was eliminated by section of the vagus or recurrent laryngeal nerves. Diaphragmatic activity ceased after section of both phrenic nerves. Electromyographically, however, purring activity continued in the nondenervated recording sites.[7]

The heart and respiratory rates increased consistently during purring. The minute volume of air almost doubled as a result of the increased respiratory rate (decreased duration of both inspiration and expiration) with no change in the tidal volume. This hyperventilation associated with purring caused a 20% decrease in end-tidal carbon dioxide partial pressure (pCO_2).[7]

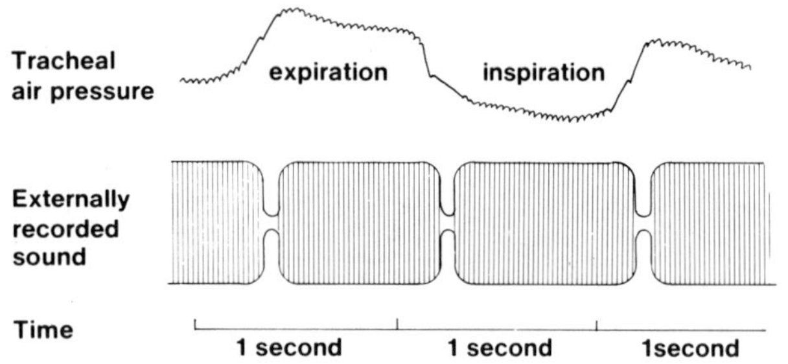

Figure 1—Diagrammatic representation of the purring sound and the concurrent respiratory phases.[7]

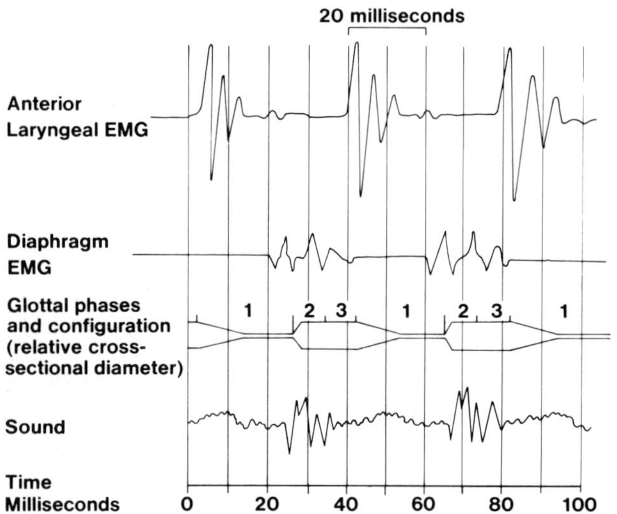

Figure 2—Diagrammatic representation of the events that occur during a purring sound burst.[7]

Discussion

By correlating the activities shown on the laryngeal and diaphragmatic electromyograms, the tracheal pressures recorded, and the phases of the respiratory cycle that were determined, the authors postulate that purring is produced by changes in the configuration of the glottis.[7] Three phases are evident (Figure 2). During phase 1 (20 to 30 milliseconds in length), the glottal opening is actively narrowed, causing increased transglottal pressure and resistance. This phase is followed by phase 2 (5 to 10 milliseconds in length), a rapid opening of the vocal cords with a corresponding rapid dissipation of the transglottal pressure. The change in pressure causes air resonance in the respiratory system, and thus the production of sound. In phase 3 (5 to 10 milliseconds in length), the glottis remains open, permitting a high airflow either into or out of the trachea. This 30- to 40-millisecond cycle of events is virtually identical during inspiration and expiration (except for the direction of the airflow) and produces the continuous purring sound. Thus, cats can ventilate while they purr.

Purring results from an extremely regular and rapid (25 to 30 times per second) alternating activation of the intrinsic laryngeal muscles and the diaphragm; this activation creates air turbulence, and the air turbulence produces the sound. Both the air turbulence and the muscle vibrations can be felt by a person holding a cat.

The persistence of purring despite sectioning of afferent nerves strongly suggests that the neural oscillation originates within the central nervous system.[7] Stimulation of the infundibular region within the midbrain causes purring.[9] While this is not a startling finding, it is interesting because of the high frequency and extreme regularity of the neural oscillations. No other similar tremors or repetitive neurologic activities are known to involve such rapid and regular neurologic coordination of various muscle groups.[10] Thus, while muscular mechanisms of purring have been elucidated, the central control and the neural interactions still remain a mystery.

Summary

The work of Remmers and Gautier[7] has not been disputed. It has occasionally been discussed and approved as explaining the purring phenomenon.[2,4-6] The fact that purring can continue when the larynx has been bypassed by either an endotracheal tube or by a tracheostomy[3,4] can be explained by the diaphragmatic vibrations.[7] Similarly, purring can occur following partial laryngectomy.[11] The dual origin of the purr explains why the vibrations can be felt at both the larynx and the cranial abdomen of a purring cat.

In summary, Remmers and Gautier have studied a commonly observed phenomenon in cats. Purring is controlled by an oscillatory mechanism within the central nervous system. This mechanism was most frequently activated in domestic cats when humans approached, spoke to, or petted the animals. The purr is caused by a very regular, rapid, alternating activation of the intrinsic laryngeal muscles and the diaphragm.

REFERENCES

1. Darwin C: The expression of emotions in man and animals. London, J. Murray, 1904, pp 130-131.
2. Prescott CW: Purring, in *Refresher Course on Cats*, vol 53. Sydney, The Post-Graduate Committee in Veterinary Science, The University of Sydney, 1980, pp 459-461.
3. McCuistion WR: Feline purring and its dynamics. *VM SAC* 61:562-566, 1966.
4. Beaver BV: Purr-fect communication. *VM SAC* 78:41, 1983.
5. Beaver BV: *Veterinary Aspects of Feline Behavior*. St. Louis, The C.V. Mosby Co, 1980, pp 54-55, 141.
6. Stogdale L: Feline purring. *Control and Therapy*:998, 1980.
7. Remmers JE, Gautier H: Neural and mechanical mechanisms of feline purring. *Respir Physiol* 16:351-361, 1972.
8. Cook TF: The relief of dyspnoea in cats by purring. *New Zealand Vet J* 21:53-54, 1973.
9. Gibbs EL, Gibbs FA: A purring center in the cat's brain. *J Comp Neurol* 64:209-211, 1936.
10. Stuart DG, Eldred E, Wild WO: Comparisons between the physiological, shivering and Parkinsonian tremors. *J Appl Physiol* 21:1918-1924, 1966.
11. Hardie EM, Kolata RJ, Stone EA, Steiss JE: Laryngeal paralysis in three cats. *JAVMA* 179:879-882, 1981.

Facilitating Euthanasia Decisions Regarding Animals with Behavior Problems*†

Animal Behavior Associates, Inc.
Littleton, Colorado
Suzanne Hetts, PhD

Behavior problems in companion animals are a common occurrence. So common, in fact, that it has been said that it is uncommon to find a companion animal without one.[1] Owners vary dramatically in their willingness to tolerate problem behavior and to invest resources to resolve it. At one end of the spectrum are those who insist on doing everything to resolve a problem and keep the animal, even if it presents a potential danger to itself or others, and for whom having the animal leave the home or be euthanatized are not options. At the other extreme are those who quickly "get rid of" the pet for relatively minor or easily resolved problems, such as destructive or elimination problems in young animals.

Both pet owners and veterinarians may struggle with whether, or when, to euthanatize otherwise healthy pets due to behavior problems. This struggle often centers around an inherent difference between euthanasia for medical reasons as compared with euthanasia for behavioral reasons. Pets die due to terminal illnesses or injuries or when they reach the end of their normal life spans. Discussing euthanasia when an animal is ill, injured, or aging may seem more natural and acceptable because death can be a naturally occurring outcome, perhaps even an inevitable one in the near future. Behavior problems are not inherently life threatening—pets do not naturally die because of them. An otherwise healthy animal may have years of normal life ahead. A discussion of euthanasia in light of a behavior problem is usually either because the animal presents a danger to others (aggression) or to itself (escaping, self-mutilation, phobias that result in escape attempts) or represents an intolerable inconvenience or management problem to its owners. The animal is usually not suffering, in pain, near death, or afflicted with a terminal disease. When the decision is made to euthanatize elderly or ill animals, owners and veterinarians may feel they are ending or preventing pain

*This presentation is based on information from Lagoni L, Butler C, Hetts S: *The Human-Animal Bond and Grief.* Philadelphia, WB Saunders Co, 1994.

†Readers are referred to a related article by Butler C, Lagoni L: Facilitating euthanasia decisions. *Compend Contin Educ Pract Vet* 16(11):1469–1475, 1489, 1994.

and suffering and helping animals to die in a peaceful, painless manner. Destructive cats and barking dogs are not ill nor are they suffering, and they are not near death. Consequently, it may be difficult to rationalize or justify euthanatizing animals for behavior problems. When they are considering euthanasia for behavior reasons, owners may have even stronger feelings that they are responsible for taking the lives of their pets than they would in the face of medical problems. Veterinarians may feel the same way.

The first step in helping owners to come to terms with these feelings and decide how to proceed is to provide them with accurate information. Good decisions cannot be made without good information. Just as owners need to have an understanding of the causes, treatments, and prognosis for any disease, illness, or injury that is threatening the life of their pet, they need similar information about behavior problems. Second, owners need to know what options are available to them before they can decide whether euthanasia is indeed the choice they need to make.

PROVIDING INFORMATION

Causes of Behavior Problems

Other articles in this book have clearly discussed the causes of common behavior problems. Most owners, however, anthropomorphize the causes of their pets' problem behaviors. They may often view problem behaviors as analogous to human mental illnesses and state that their pets are psychotic, neurotic, or schizophrenic. The implication is that the pet is abnormal or that nothing can be done for the problem. Another common misconception is that the problem behavior is occurring due to spite or vengeful motives. The animal seems to be "getting back at them" for leaving it alone, not providing it with the tastiest food, or some other perceived slight. These views negatively affect owners' attitudes toward their pets and color their judgment about the pet's future. When owners understand that the animal is displaying normal animal behavior (this is most likely the case) and that the pet is not mad at them, they are better able to make an informed decision rather than one based on misconceptions.

Finding Help

Because pet owners have a variety of types of people from whom to choose when they seek help with behavior problems, the veterinary professional's role as facilitator is a crucial one. Veterinarians should provide pet owners with the opportunity to be referred to behavior specialists in a timely fashion. While not all behavior problems are resolvable, owners should be given the choice to seek specialized help. Good veterinary practitioners have long made it a practice to set their own limits with specialized or difficult cases. Each practitioner establishes a set of guidelines regarding when to refer an orthopedic, oncologic, or other case. These guidelines are usually based on the practitioner's knowledge, experience, and interest in the particular aspect of veterinary medicine. Similar guidelines should be set for behavior referrals. To ensure that clients are making decisions about euthanasia based on accurate and appropriate information, veterinarians should either refer cases to or work closely with board-certified behaviorists or certified applied animal behaviorists if they feel they do not have the specialized training, experience, or desire to provide the latest scientific information about a particular behavior problem.

DISCUSSING OPTIONS

Most owners do not want to consider euthanasia as the first or only option when behavior problems arise. Those that do may not be aware of what other choices are available to them. Veterinarians can initially be most helpful to clients by offering a variety of choices, rather than focusing on any one option that they themselves would prefer. Recommending euthanasia without discussing other alternatives, even in aggression cases, may leave owners feeling as though their pet's life is not important. On the other hand, if veterinarians do not discuss the possibility of euthanasia, owners may not be confident enough to initiate discussions about it themselves. Veterinarians can ask "What options have you considered?," "Are you considering euthanasia if the problem doesn't resolve?," or "Would you like to discuss all the options that are available to you?" Before making a decision, owners need time to consider all their options and the possible consequences, advantages, and disadvantages associated with each choice. Thus, the first step in facilitating a euthanasia decision is to help owners sort through this information.

Resolving Behavior Problems

Myths and misconceptions about animal behavior and behavior problems are widespread. For example, owners may believe or may have been told that problem behaviors cannot be changed. Common examples of such beliefs are that after a dog has "tasted blood," it will always be an aggressive, dangerous animal and that once a cat begins to eliminate outside the box, nothing can be done to stop the behavior. In addition, sometimes "popular cures" for behavior problems are worse than the problems themselves. Because they are often not knowledgeable about other techniques, many people rely on some sort of aversive approach to attempt to stop an animal from misbehaving. Examples of popular aversive solutions include taping the dog's mouth shut around an object it has chewed (dogs have suffocated from this procedure), administering severe leash and collar corrections to an aggressive dog to the point of "stringing the dog up" until it passes out (resulting in a damaged trachea), or confining cats for weeks or months at a time in a crate with nothing more than food, water, and a litterbox (even laboratory cats often have more space). These approaches rarely resolve the problem, often exacerbate it and, in my opinion, border on cruelty. Owners may invest significant time and money on methods that have very little chance of success if they rely on nonscientific approaches.

Veterinarians should discuss scientific approaches to behavior problems with clients. If the veterinarian's knowledge about or experience with a particular type of problem is limited, referral to a behavior specialist should be an option. Owners need to have a reasonable idea of what resolving the problem would entail before deciding how to proceed.

Managing the Problem

Owners may choose to live with or manage the problem. For example, they may crate or in other ways confine destructive or house-soiling dogs or allow cats to destroy old furniture with their claws. Psychoactive drugs may also provide a means of managing a problem, at least on a short-term basis.

Managing the problem may work well in some situations; however, some ways of managing problems may have a negative impact on the pet's well-being. Pets may be confined for long periods in barren rooms or small crates or tied to stationary objects. If the animal cannot avoid problems when left inside, it may become a backyard dog or a strictly outdoor cat. Typically, when dogs are left alone outside, their social needs are not met and they often develop more problems. Outdoor cats are subject to disease and injury and likely do not live as long as indoor cats. Owners who attempt to manage an aggressive animal by not allowing the animal to come in contact with other animals or people for fear that it may bite them may be assuming huge liability risks. In any case, management techniques that do not allow the owner to enjoy the pet as a true companion animal will most likely negatively affect the animal as well as the human-companion animal bond. The "manage it or live with it" approach may last indefinitely, or the situation may ultimately escalate to a crisis level, at which point the owners decide they must either take steps to resolve the problem or have the animal leave the home.

Having the Pet Leave the Home

When owners want a guarantee that a particular behavior will never occur again in their household, the only option that provides this security is for them to not keep the pet. Owners may elect to give the animal away, surrender it to an animal shelter for possible adoption or euthanasia, or have it euthanized by a private veterinarian.

If the pet leaves the home as the result of a behavior problem, the human-companion animal bond is permanently broken, and the veterinarian loses a client (if another pet is not obtained). If the animal goes to another home directly or through a humane society, there will be a period of adjustment as the animal adapts to its new environment. This may be relatively easy for many animals and more difficult for others. Some problem behaviors may not persist in a different environment, but more often than not the new owners "inherit" the pet's existing problems. If the new owners adopted the pet in spite of existing problems, they may either be more willing to tolerate the problem or to work with it. Alternatively, they may not have realistic expectations about what these efforts may entail. Some owners are fearful about finding new homes for pets with behavior problems either because of ethical concerns about burdening someone else with a problem pet or because they worry that the pet will be mistreated because of the problem. The pet also could be passed from home to home if successive owners cannot manage or resolve the problem. When a pet is surrendered to an animal shelter, owners usually do not ever know whether the animal was adopted or euthanatized. Some owners may not want to live with this uncertainty. When owners are concerned about the possible negative aspects of finding the pet a new home, they may elect to euthanatize the pet instead.

ACCEPTING EUTHANASIA AS AN OPTION

Preventing Pain or Suffering

In some situations it could be argued that the consequences of the animal's behavior problem may result in pain or death, which could be prevented by euthanasia. Possible mistreatment by future owners is one example. Other examples include dogs that escape their yards and cats that have been made outdoor-only pets due to elimination problems; these animals could be hit by cars or be injured in fights with other animals. Dogs that chase livestock could be shot. Dogs that suffer from fear of thunderstorms (brontophobia) or separation anxiety sometimes injure themselves while attempting to find a "safe place" or reunite with their owners. Owners may also question the quality of life of pets that are constantly fearful.

Preventing Injury to Others

Aggressive behavior problems are one of the most common types of cases seen by behavior consultants.

The only choice that guarantees a pet with an aggression problem will not ever harm anyone is euthanasia. If owners choose to work with aggressive behavior, there is always the risk that a person or other animal may be injured. The degree of risk varies depending on many factors.[2] Owners also vary tremendously in their willingness to assume these risks. A reasonable assessment of the potential danger involved cannot be made without obtaining a behavioral history and observing the animal. Many owners want to pursue treatment before considering euthanasia. Except in rare cases, owners should be offered options in addition to euthanasia, as long as the potential risks and liabilities are also explained. Due to the risks and liabilities involved, veterinarians may want to refer these cases to behavior specialists experienced in working with aggression rather than attempting to work with the case themselves. If owners choose to euthanatize a pet with an aggression problem, they can be reassured that they have prevented a possible tragedy.

Inability to Manage or Treat the Problem

Despite the best efforts of owners, veterinarians, and behavior specialists, some behavior problems may not be readily resolvable or easily manageable. When problems do not respond acceptably to behavior modification, drug therapy, or environmental management, owners may have no other choice than to find the pet another home or have it euthanatized. In these cases it may be helpful for owners to view the problem as analogous to an untreatable disease or illness. Owners can be reassured that they have indeed done everything possible for their pet. Chronic animal behavior problems can affect the family's quality of life if the pet is the source of continual friction between family members, repeatedly causes destruction or damage to the house and possessions, or necessitates family members constantly scheduling their every activity to accommodate the problem behavior.

MAKING THE DECISION

Both pet owners and veterinarians struggle with knowing when the time has come to euthanatize a pet. Lack of appetite, lethargy, and other signs of deteriorating health that are often helpful when pets are ill or aging are not useful for healthy pets with behavior problems. Veterinarians can suggest, however, other signs or checkpoints to help owners decide when the time for euthanasia has come.

Frequency of Occurrence

One such guideline would be to monitor the frequency of the problem behavior. Having owners keep a written tally of how often the behavior actually occurs gives them a more objective view of the situation. If owners are very frustrated with the pet and the problem, they may not see progress when it is actually occurring. Alternatively, they may have falsely convinced themselves the problem is improving because they do not want to face a euthanasia decision.

Intensity of Problem

Owners may also be encouraged to evaluate the intensity of the behavior. For example, perhaps the cat is now only hissing in situations in which it has previously bitten, or maybe the dog is no longer tearing up large sections of carpet when left alone, but only leaving a few scratch marks by the door. Monitoring the frequency and/or intensity of the problem allows owners to set their own limits or "bottom lines" as to how much improvement they need to see within a set amount of time to continue working with or managing the problem.

Family's Quality of Life

Veterinarians can provide a different perspective by gently asking owners how many changes they are willing to make in their family's life-style to accommodate the pet's problem. Devoted owners will sometimes go to great lengths to avoid circumstances that elicit the problem behavior. Because these changes in routine occur over time, owners may not realize how much their life-style has been altered until it is gently pointed out to them. This questioning should be done in a sensitive, caring communication style not a judgmental one.

EUTHANASIA PLANNING

There is no denying that making the decision to euthanatize a pet is painful. In an attempt to avoid further pain, owners may try to distance themselves from the euthanasia process. They may want to avoid discussing how they are going to say good-bye to their pet, their options about being present during the euthanasia, viewing the body afterward, or choices for body care. Not attending to these issues now is likely to result in more pain later. Owners may subsequently regret that they did not make choices that would have eased the emotional pain related to their pet's death. Veterinarians have been through the euthanasia process many times but this may be the first time for many owners. In addition, guilt, anger, sadness, and other emotions are probably compromising their ability to make decisions. Veterinarians can help owners by offering them choices rather than avoiding such discussions.

Helping Owners Say Good-Bye

Most owners may take longer than veterinarians to

Many other guidelines can be suggested based on the specific details of the situation. The following are examples of how veterinarians can respond to common questions and concerns when owners are struggling with euthanasia decisions.

"What would you do, if it was your pet?

Owners who ask this question are not actually asking you to tell them what they should do. They are asking to be reassured of their own ability to make a decision that will be best for them. Responding to this question by describing your own personal preference tells owners that you believe this is what they should do. They may then be uncertain about pursuing a different course of action for fear of displeasing you. Later, they may feel as though they were not allowed to make their own decision and be resentful toward you or your staff.

Examples of more helpful responses are:

- "If Sam were my pet I'd be just as confused and torn as you are. I'd probably be angry that I would have to make such a difficult decision."
- "It would be unfair of me to answer that question directly for you. What I would do might be different from what you want to do. My circumstances are different than yours. You know Sam better than anyone, and you also know what's best for him and for your family. Whatever you decide, it will be the best decision and I'll support you through whatever comes next."
- "If it were me, I wouldn't want someone to tell me what to do. I'd want the support and confidence of the people around me. And that's what I'm offering you. I can't make the decision for you, but I'll support whatever decision you make."

"How can I go through with this! How can I choose a time to end my pet's life?"

Owners making these and similar remarks are feeling panic and anxiety. Imagining themselves acting on their decision seems overwhelming. Helping owners think through the consequences of not acting while providing both concrete plans and the freedom to change them is often helpful. For example:

- "I know how hard it is to actually make an appointment to euthanatize Button. It seems so final. Let's go ahead and schedule the appointment now. If you decide you don't want to go through with it, you can call and cancel and we'll talk more then."
- "I know how painful it is to schedule an appointment for euthanasia. It's easy to put it off for another day or another week. But putting it off doesn't make it any less painful for you. No matter how long you wait, you'll never be completely ready to say goodbye to Button. Waiting provides more opportunities for something else to happen *(you can provide examples specific to the problem—"someone else may be bitten," "she may get out again and this time she could be hit by a car," etc.)* and I know you don't want that. Let's schedule an appointment and see if we can't get through this together."

"But she's not sick! She could live many more years if it wasn't for this! Isn't there something else we can do?"

Owners are feeling responsible for ending the life of their pet. It is common for owners to feel guilty after making a euthanasia decision under any circumstances, but this may be even more likely if the pet is relatively young and otherwise healthy. It may be helpful to remind owners of everything they have already done to try and resolve the problem:

- "In some ways, it may be helpful to think about Pepper's problem as a kind of incurable illness. We haven't been able to successfully treat it, and your lives together are no longer enjoyable because of it. Since there's nothing more we can do for Pepper's problem, let's talk about how you can say goodbye to Pepper and plan for her death."
- "I know how frustrating it is that we haven't been able to resolve Pepper's problem. But I want to tell you sincerely that you have done everything you could for her. Not every problem or disease can be cured. Pepper has been lucky to have an owner who tried so hard for her. It might be time to think about how you can make your last days with Pepper special for you both."

make the transition from trying to help resolve the behavior problem and keep a pet alive to planning its death. Veterinarians can help by acknowledging the emotional pain and difficult circumstances associated with making a euthanasia decision. As owners are focusing more on euthanasia as their choice, veterinarians can help the process by suggesting that they begin to think about how to say good-bye. This can be facilitated by encouraging owners to explore their feelings and concerns about letting go of their pet. Examples of questions that may help owners are:

- "What do you need to do to say goodbye to Sam? What do you want to say or do that you haven't done yet?"
- "When you look back on Button's death, what kinds of things will be important to you about how she died?"
- "What do you want to remember about your last days with Pepper? What kinds of activities can you do with her now that will make those memories pleasant?"

- "What is your biggest fear or worry about Sam's death? What will help to lessen your fear?"

Choosing to Be Present during Euthanasia

Many owners wish to be present when their pets are euthanatized but never have the confidence to ask their veterinarians if this is possible; therefore, it is up to the veterinarian to offer this choice to owners. Common reasons why owners want to be present are because the pet has always been there for them and they want to be there when their pet needs them most, they don't want to feel as though they abandoned their pet, and they want to know for themselves that their pet's death was peaceful and painless. Because owners may be guilt ridden about choosing euthanasia due to a behavior problem, it may be tempting for veterinarians and owners alike to avoid or minimize discussions about the details of the euthanasia. This may inadvertently send the message to clients that the veterinarian is also feeling guilty and actually contribute to their own normal doubts. Being present may comfort many owners, as it may help to lessen their feeling that they have "given up" on their pet. A good way to initiate a discussion about client presence is to clearly explain what the euthanasia process is like.

Owners will make different choices about these issues. Some will want to be highly involved and plan a meaningful process. Others will feel comfortable with a good-bye pat on the head as they leave the examination room. Whatever choices owners make should be respected, but they must have the opportunity to choose.

Veterinarians' Choices

In some cases, even after detailed discussion with owners, veterinarians may not feel that euthanasia for a behavior problem is appropriate. Some owners request euthanasia for reasons that to others seem trivial or unjustified. The traditional view in veterinary medicine has been that the veterinarian's first obligation is to the client and that what the client wants done is what the veterinarian should do. Not all veterinarians agree with this position. Whether the veterinarian is more likely to lose a client for refusing rather than agreeing to perform a euthanasia has never been studied. Veterinarians must decide what their own "bottom lines" will be as far as when they will or will not agree to euthanatize an animal. It may be helpful to think about how your choices will affect your feelings about performing euthanasia and the practice of veterinary medicine in general as well as the impact your choices will have on your staff.

ADDITIONAL RESOURCES

Veterinarians wishing to learn more about euthanasia planning, grief education, the human-animal bond, and their roles as helpers when companion animals die can find a variety of additional resources. *The Human-Animal Bond and Grief*[3] published by WB Saunders is perhaps the most comprehensive text on these subjects. The American Veterinary Medical Association (AVMA), American Animal Hospital Association, and ALPO PetFoods all have educational brochures for clients about euthanasia.

Pet loss support hotlines for clients are available at veterinary medical schools at the University of California–Davis and the University of Florida. The Delta Society in Renton, Washington, and the AVMA's Human-Animal Bond Committee and Association are additional sources of information on pet loss, euthanasia, and grief. The Behavior College of the AVMA, the American Veterinary Society of Animal Behavior, and the Animal Behavior Society can provide lists of board-certified veterinarians and certified applied animal behaviorists. Addresses for these and other resources can be found in Lagoni et al[3] or by contacting the organizations directly.

CONCLUSION

The "success rate" for resolving behavior problems varies considerably based on a variety of factors including the type of problem, its etiology, the competency of the behavior consultant, and the ability of owners to work with the problem as directed. More often than not, however, behavior problems can be resolved with competent assistance and a committed pet owner. With the recent growth in research and knowledge and the increased availability of specialists in applied animal behavior and behavior medicine, the prognosis for resolving animal behavior problems is better than ever before.

However, if the numbers of animals that die in animal shelters are taken into account, it is likely that more pets die due to behavior problems than due to medical causes. Consequently, veterinarians will inevitably be faced with euthanasia requests because of behavior problems. By providing owners with expert behavioral help, accurate information, and other choices, veterinarians can sometimes prevent deaths attributable to behavior problems. When euthanasia must be considered, facilitating decisions about euthanasia and supporting owners through the process is a helpful and healthy approach to grief and loss.

REFERENCES

1. Hart BL, Hart LA: *Canine and Feline Behavioral Therapy.* Philadelphia, Lea & Febiger, 1985.
2. Reisner IR, Erb HN, Houpt KA: Risk factors for behavior-related euthanasia among dominant-aggressive dogs: 110 cases (1989–1992). *JAVMA* 205(6):855–863, 1994.
3. Lagoni L, Butler C, Hetts S: *The Human-Animal Bond and Grief.* Philadelphia, WB Saunders Co, 1994.